MICROBES IN THEIR NATURAL ENVIRONMENTS

SYMPOSIA OF THE SOCIETY FOR GENERAL MICROBIOLOGY*

1 THE NATURE OF THE BACTERIAL SURFACE
2 THE NATURE OF VIRUS MULTIPLICATION
3 ADAPTATION IN MICRO-ORGANISMS
4 AUTOTROPHIC MICRO-ORGANISMS
5 MECHANISMS OF MICROBIAL PATHOGENICITY
6 BACTERIAL ANATOMY
7 MICROBIAL ECOLOGY
8 THE STRATEGY OF CHEMOTHERAPY
9 VIRUS GROWTH AND VARIATION
10 MICROBIAL GENETICS
11 MICROBIAL REACTION TO ENVIRONMENT
12 MICROBIAL CLASSIFICATION
13 SYMBIOTIC ASSOCIATIONS
14 MICROBIAL BEHAVIOUR, 'IN VIVO' AND 'IN VITRO'
15 FUNCTION AND STRUCTURE IN MICRO-ORGANISMS
16 BIOCHEMICAL STUDIES OF ANTIMICROBIAL DRUGS
17 AIRBORNE MICROBES
18 THE MOLECULAR BIOLOGY OF VIRUSES
19 MICROBIAL GROWTH
20 ORGANIZATION AND CONTROL IN PROKARYOTIC AND EUKARYOTIC CELLS
21 MICROBES AND BIOLOGICAL PRODUCTIVITY
22 MICROBIAL PATHOGENICITY IN MAN AND ANIMALS
23 MICROBIAL DIFFERENTIATION
24 EVOLUTION IN THE MICROBIAL WORLD
25 CONTROL PROCESSES IN VIRUS MULTIPLICATION
26 THE SURVIVAL OF VEGETATIVE MICROBES
27 MICROBIAL ENERGETICS
28 RELATIONS BETWEEN STRUCTURE AND FUNCTION IN THE PROKARYOTIC CELL
29 MICROBIAL TECHNOLOGY: CURRENT STATE, FUTURE PROSPECTS
30 THE EUKARYOTIC MICROBIAL CELL
31 GENETICS AS A TOOL IN MICROBIOLOGY
32 MOLECULAR AND CELLULAR ASPECTS OF MICROBIAL EVOLUTION
33 VIRUS PERSISTENCE

* Published by the Cambridge University Press, except for the first Symposium, which was published by Blackwell's Scientific Publications Limited.

MICROBES IN THEIR NATURAL ENVIRONMENTS

EDITED BY

J. H. SLATER, R. WHITTENBURY

AND J. W. T. WIMPENNY

THIRTY-FOURTH SYMPOSIUM OF
THE SOCIETY FOR GENERAL MICROBIOLOGY
HELD AT
THE UNIVERSITY OF WARWICK
APRIL 1983

Published for the Society for General Microbiology

CAMBRIDGE UNIVERSITY PRESS

CAMBRIDGE

LONDON NEW YORK NEW ROCHELLE

MELBOURNE SYDNEY

Published by the Press Syndicate of the University of Cambridge
The Pitt Building, Trumpington Street, Cambridge CB2 1RP
32 East 57th Street, New York, NY 10022, USA
296 Beaconsfield Parade, Middle Park, Melbourne 3206, Australia

First published 1983

Printed in Great Britain at the Pitman Press, Bath

Library of Congress catalogue card number: 82-19859

British Library Cataloguing in Publication Data

Microbes in their natural environments.–
(Symposia of the Society for General Microbiology; 34)

1. Micro-organisms – Congresses
I. Slater, J. H. II. Whittenbury, R.
III. Wimpenny, J. W. T. IV. Series
576 QR41.2

ISBN 0 521 25063 3

CONTRIBUTORS

ARMITAGE, J. P., Department of Botany and Microbiology, University College London, London WC1E 6BT, UK

BROWN, C. M., Department of Brewing and Biological Sciences, Heriot-Watt University, Edinburgh EH1 1HX, UK

BURNS, R. G., Biological Laboratory, University of Kent at Canterbury, Canterbury CT2 7NJ, UK

CARR, N. G., Department of Biochemistry, University of Liverpool, Liverpool L69 3BX, UK

COOMBS, J. P., Department of Microbiology, University College Cardiff, Cardiff CF2 1TA, UK

DOW, C. S., Department of Biological Sciences, University of Warwick, Coventry CV4 7AL, UK

FLANNIGAN, B., Department of Brewing and Biological Sciences, Heriot-Watt University, Edinburgh EH1 1HX, UK

GOWLAND, P. C., Department of Environmental Sciences, University of Warwick, Coventry CV4 7AL, UK

KING, C., Department of Botany and Microbiology, University College London, London WC1E 6BT, UK

KONINGS, W. N., Laboratorium voor Microbiologie, Universiteit van Groningen, Haren (GR), The Netherlands

LOVITT, R. W., Department of Microbiology, University College Cardiff, Cardiff CF2 1TA, UK

NEIJSSEL, O. M., Laboratorium voor Microbiologie, Universiteit van Amsterdam, Amsterdam 1018WS, The Netherlands

ORMEROD, J. G., Botanical Laboratory, University of Oslo, Oslo 3, Norway

REANNEY, D. C., Department of Microbiology, La Trobe University, Bundora, Australia 3083

ROWBURY, R. J., Department of Botany and Microbiology, University College London, London WC1E 6BT, UK

SLATER, J. H., Department of Environmental Sciences, University of Warwick, Coventry CV4 7AL, UK

STEWART, W. D. P., Department of Biological Sciences, University of Dundee, Dundee DD1 4HN, UK

TEMPEST, D. W., Laboratorium voor Microbiologie, Universiteit van Amsterdam, Amsterdam 1018 WS, The Netherlands

VELDKAMP, H., Laboratorium voor Microbiologie, Universiteit van Groningen, Haren (GR), The Netherlands

WARDELL, J. N., Department of Brewing and Biological Sciences, Heriot-Watt University, Edinburgh EH1 1HX, UK

WHITE, D. C., Department of Biological Sciences, Florida State University, Tallahassee, Florida 32306, USA

WHITTENBURY, R., Department of Biological Sciences, University of Warwick, Coventry CV4 7AL, UK

WIMPENNY, J. W. T., Department of Microbiology, University College Cardiff, Cardiff CF2 1TA, UK

ZEIKUS, J. G., Department of Bacteriology, University of Wisconsin 53706, USA

ZEVENBOOM W., Laboratorium voor Microbiologie, Universiteit van Groningen, Haren (GR), The Netherlands

CONTENTS

EDITORS' PREFACE

Twenty-two years ago the eleventh symposium of the Society for General Microbiology concentrated on microbial reaction to the environment. Eight years later the symposium was devoted to various aspects of microbial growth, whilst more recently, in 1976, the questions of the survival of and death of microorganisms were the topics of debate. These symposia sought to examine in great detail the multitude of interactions which exist between microorganisms and the environment in which they find themselves: elsewhere the importance of these interactions has never been far from the surface even though other specific aspects of microbiology have been under discussion. The effects of the relationships between a microbial population and its environment are often considered to be axiomatic, although often the magnitude and significance of environmental effects are not fully appreciated. Microbes have a remarkable capacity to come to terms with a wide range of environmental conditions: indeed, to such an extent that, as Tempest, Neijssel and Zevenboom have observed in their present article, 'it is quite impossible to specify any microbe in precise terms of structure and function without, at the same time, specifying the environmental conditions prevailing during its growth'.

The thirty-fourth symposium of the Society for General Microbiology returns to the general theme of the interaction between microorganisms and their environment, in particular to the growth of microorganisms where they are most at home in their natural environments. It is a complex area of study because of the immense number of related variables, which in the past has led to the formation of two broad approaches. On the one hand there is the experimental approach which concentrates on measurements and extrapolations made directly from the natural environment; alternatively physiological and biochemical experimentation under highly defined conditions, usually in laboratory-based systems, is related to the process occurring in nature. In recent years there has been substantial progress in attempting to fuse these two strategies into a more unified approach. The symposium was planned to concentrate on the reality of microbial life in a number of different ecosystems and to eschew on the one hand the contribution that was pure physiology with only passing reference to ecology and, on the other hand, the ecologically orientated presentations which emphasize microbial behaviour in highly specialized environments or concentrate on general properties such as energy flow or biogeochemical cycles. We hope that through the contributions in this volume, progress will continue to be made in unravelling the properties of microbes growing in their natural habitats.

We are grateful to the contributors for agreeing to take part in this venture and providing the material within the general framework of the symposium's theme. Our sincere thanks are due to the staff at Cambridge University Press for constructing order from chaos; and to Roger Berkeley, the Society's Meetings Secretary, for applying a guiding hand (usually benevolently!).

J. Howard Slater, Coventry
Roger Whittenbury, Coventry
Julian W. T. Wimpenny, Cardiff

NATURAL ENVIRONMENTS – CHALLENGES TO MICROBIAL SUCCESS AND SURVIVAL

WILLIAM D. P. STEWART

Department of Biological Sciences and ARC Research Group on Cyanobacteria, University of Dundee, Dundee, Scotland, UK

INTRODUCTION

About 70% of the Earth's surface is covered by the sea; freshwaters, apart from the polar icecaps, cover less than 1% of the land surface. On the land and within the oceans and freshwaters is a multitude of living organisms. Bergey (1974) lists 247 genera of bacteria, there are over 4000 genera of fungi (Ainsworth, 1968; Ainsworth *et al.*, 1973) and there are 42 genera and 3 sub-groups of photosynthetic prokaryotes (see later). To these lists must be added eukaryotic microalgae, protozoa, etc. New additions are regularly being found: for example, *Prochloron*, a photosynthetic prokaryote with higher plant pigments, has been described by Lewin (1977). This 'missing link' between photosynthetic prokaryotes and photosynthetic eukaryotes has created taxonomic controversy, phylogenetic speculation (Lewin, 1981) and irritation by the fact that it cannot yet be grown in axenic culture. *Legionella pneumophila*, the etiological agent of Legionnaires' disease (McDade *et al.*, 1977), and other *Legionella* species, *L. bozemanii*, *L. microdadei* and *L. dumoffii* are now well characterised (Balows & Brenner, 1981). *L. pneumophila*, which has been isolated from the water of cooling towers, humidifiers and ponds has been found associated with mats of the cyanobacterium *Fischerella* and may feed on the cyanobacterial extracellular products (Tison *et al.*, 1980). However, recent evidence has shown that *L. pneumophila* is pathogenic to a variety of free-living freshwater and soil amoebae (*Acanthamoeba* and *Naegleria* spp.) and it may be that transmission to humans is via amoebae rather than as free-living legionellae (Rowbotham, 1980). There has been the discovery of hydrothermal vent microorganisms of the Galapagos rift and elsewhere, some fixing carbon dioxide via ribulose 1,5-bisphosphate carboxylase, as in other autotrophs, and using reduced sulphur compounds as energy and reductant sources (Jannasch & Wirsen, 1979; Karl *et al.*, 1980; Cavanaugh *et al.*, 1981; Felbeck, 1981; Jones, 1981). Some of these prokaryotes contain heterocyst-like cells, as in nitrogen-fixing cyanobacteria (Stewart, 1980). Rau (1981) has evidence from $^{15}N/^{14}N$ fractionation studies that biological nitrogen fixation may be occurring in such vents. It will be of interest to determine whether these prokaryotes fix nitrogen and, if they do, whether the heterocyst-like cells are involved.

Microorganisms can be found in virtually any natural environment on Earth. The problem for the microbial ecologist is how best to achieve an understanding of the interactions of microorganisms with their environment. The tendency within recent years, has rightly been to consider the microbial niche and Schlegel & Jannasch (1981) consider that, as a result of the microbial niche concept, the sharp distinction between aquatic, terrestrial and even medical microbiology can be 'de-emphasised'. In this chapter I shall broaden the debate, since elsewhere others have emphasised the microenvironment. I shall consider the macrohabitats of the soil, freshwaters and the sea, the major groups of microorganisms and then attempt to inter-relate the two. Particular attention will be paid to the ways in which the availability of carbon, nitrogen and other major ions affects the ecological distribution of microorganisms.

MICROBIAL MACROHABITATS

Freshwaters

Freshwaters are aquatic ecosystems which, with a few exceptions such as the African crater lakes (Melack & Kilham, 1972) and ion-rich hot spring waters (White *et al.*, 1963), have a low ionic content. The ionic content is influenced by the land over which the water accumulates or through which it passes. Agricultural drainage and urbanisation, for example, may have considerable effects on water quality (Stewart *et al.*, 1977, 1982). Table 1 lists some physico-chemical features of mesotrophic, eutrophic and polyeutrophic waters in Scotland. The dominant microorganisms of the water column are primary producers (planktonic eukaryotic algae and cyanobacteria). The over-riding physico-chemical features of importance are thus light quantity and quality, turbulence, temperature, inorganic carbon availability and oxygen content. Numbers of chemoheterotrophic microorganisms are relatively low, increase with eutrophication and are generally associated with the primary

Table 1. *Some major inorganic ions and physico-chemical features of three freshwater lochs in Eastern Scotland*

Parameter	Lowes Loch	Balgavies Loch	Forfar Loch
Sodium (mg l^{-1})	4.6–5.5	12.8–13.9	25.8–28.2
Potassium (mg l^{-1})	0.6–0.7	1.7–1.9	7.1–8.5
Calcium (mg l^{-1})	4.0–4.5	24.0–27.5	35.6–48.9
Silica (mg Si l^{-1})	<0.10–2.08	<0.010–6.08	0.34–5.23
Phosphate (mg P l^{-1})	<0.01–0.20	<0.01–0.43	0.44–4.70
Ammonium (mg N l^{-1})	0.0–0.24	0.0–1.02	0.04–3.50
Nitrate (mg N l^{-1})	<0.01–0.50	<0.01–7.20	0.32–6.67
Nitrite (mg N l^{-1})	<0.01–0.044	<0.01–0.064	0.03–0.50
Alkalinity (mg $CaCO_3$ l^{-1})	16–23	75–125	116–165
Conductivity (μS cm^{-1})	60–100	250–330	420–600
Oxygen (mg l^{-1})	8.5–15.0	8.6–20.0	7.0–18.6
pH	7.00–9.31	7.50–9.95	7.20–9.67

All three lochs are small shallow lochs, further details of which are available (Stewart *et al.*, 1982). Lowes is mesotrophic, Balgavies is eutrophic and Forfar Loch is polyeutrophic. The data which are mean values during 1974–76 were obtained by Dr F. Sinada.

producers and aquatic animals. Planktonic microorganisms on decay or settling-out serve as major routes of nutrient transport from the water column to the sediment. (Preston *et al.*, 1980).

The microorganisms of freshwater sediments contrast sharply with those of the water column. They tend to be prokaryotic and chemoheterotrophic, although on the surface of sediments or shallow lakes photosynthetic bacteria, benthic algae and cyanobacteria may be found. Apart from the sediment-water interface, conditions tend to be anaerobic and physico-chemical features of importance include redox potential, oxygen content and the availability of fixed carbon. Organic degradation is the major microbial process of importance with carbon dioxide, methane, hydrogen sulphide and hydrogen gas being major end products.

Terrestrial environments

The soil which covers much of the land surface is a complex of water, air, mineral matter and organic matter. The dominant microorganisms are chemoheterotrophs. Alexander (1977) gives microbial numbers in the top 3 to 8 cm of a temperate soil as (numbers per gram of soil): aerobic eubacteria, 7 800; anaerobic

Table 2. *Major inorganic ions in 'normal' seawater*

Constituent	Concentration (g l^{-1})	Constituent	Concentration (g l^{-1})
Cl^-	19.630	H_3BO_3	0.022
SO_4^{2-}	2.701	$H_2BO_3^-$	0.005
Br^-	0.066	Na^+	10.770
F^-	0.001	Mg^{2+}	1.298
CO_2 } H_2CO_3 }	0.001	Ca^{2+}	0.408
		K^+	0.387
HCO_3^-	0.116	Sr^{2+}	0.014
CO_3^{2-}	0.012	Total salinity	35.161

After Rubey (1951).

eubacteria, 1 950; actinomycetes, 2 080; and algae, 25. Overall the abundance of the microorganisms relates directly to the amount of available fixed carbon provided by primary and secondary producers. The dominant microbial events are those associated with organic degradation. Physico-chemical features of importance include the soil particulate matter and the soil pore space. Attachment to the particulate material involves cation bridges, van der Waals forces, extracellular polysaccharides, fimbriae, etc. The soil pore space affects oxygenation, redox potential, water-holding capacity, etc. Excess moisture may limit oxygen diffusion, produce an anaerobic environment and set in train a series of anaerobic events including denitrification, sulphate reduction, methanogenesis and the release of adsorbed nutrients, just as in freshwater sediments.

Marine ecosystems

Marine ecosystems are aquatic habitats of high ionic content (Table 2). Seawater in general contains 35 g l^{-1} of total salt, with 99.5% of the total being due to nine ions which are present in fairly constant proportions; the most abundant of these are Na^+ and Cl^-. The marine ecosystem is dominated by primary producers, particularly planktonic microalgae (Round, 1981). Collectively they contribute about 35% of the world's primary production (Bassham, 1976). The physico-chemical features of importance in open seawaters, other than salinity, thus include light quantity and quality, inorganic carbon availability, turbulence and temperature (over 90% of the sea by volume has a temperature of less than 5 °C). Highest numbers of chemoheterotrophic bacteria occur in intertidal and estuarine habitats and in marine sediments. Marine sediments are dominated by detrital material, biogenic material and anthigenic minerals (Malcolm & Stanley, 1982). They provide dynamic ecosystems with interactions and fluxes associated primarily with organic degradation, nutrient cycling and nutrient assimilation. As in soils the microorganisms present are usually adsorbed to particulate material (Billen, 1982) with redox potential being a major ecological factor of importance. Nitrate dissimilation in marine and estuarine sediments is discussed in detail by Herbert (1982) and sulphur cycling by Nedwell (1982).

In summation, freshwaters and soil water have low total ionic contents while marine habitats have high ionic contents; fixed carbon availability and inorganic nitrogen availability on the other hand are highest in soil and freshwaters and are lowest in oceanic waters.

MICROBIAL DIVERSITY AND ENERGY GENERATION

Distinctions between microorganisms can be based on many criteria. Cohn (1872) was probably the first to emphasise morphology, Orla-Jensen (1909) advocated the use of physiology and Stanier & van Niel (1962) distinguished prokaryotes and eukaryotes depending on whether or not the cell nucleus had a limiting membrane. It is virtually certain that prokaryotes and eukaryotes will remain as the major microbial taxonomic division of importance. Prokaryotic/eukaryotic features are also emphasised as major determinants of ecological niche (Schlegel & Jannasch, 1981). Working, as I do, with cyanobacteria which show both prokaryotic features of cell structure and eukaryotic features of photosynthetic oxygen evolution, I am inclined to the view that a powerful determinant of ecological distribution is the division between those organisms which can harvest light energy, and which in most cases obtain their fixed carbon requirement by fixing carbon dioxide, and non-photosynthetic organisms. The major groups of photosynthetic microorganisms are listed in Tables 3 and 4, together with some of their major features (see also below on the Halobacteriaceae).

There are four families of anoxygenic photobacteria (Anoxyphotobacteria – Gibbons & Murray, 1978) (Table 3). The Rhodospirillaceae (non-sulphur purple bacteria; Athiorhodaceae) are photosynthetic, unicellular Gram-negative, straight, curved or helical, usually non-gas-vacuolated rods, with well-defined guanine plus cytosine (G + C) ratios (61 to 70%), which when motile are flagellated (usually polar except in the swarmers of *Rhodomicrobium* species which have peritrichous flagella) (Biebl & Pfennig, 1981; Trüper & Pfennig, 1981). The Chromatiaceae (sulphur purple bacteria, Thiorhodaceae) are Gram-negative, frequently gas-vacuolated rods, although some show cellular aggregation, for example, *Thiocapsa*. They have heterogeneous G + C ratios, deposit elemental sulphur intracellularly and have only a limited capacity to use organic compounds (Pfennig & Trüper, 1981; Trüper & Pfennig, 1981). The Chlorobiaceae (Copeland, 1956) are anaerobic, non-motile, frequently gas-vacuolated, photolithotrophic unicells which divide by binary fission (except for *Pelodictyon clathratiforme* which shows ternary cell division). Their photosynthetic apparatus is located in discrete vesicles (chlorosomes) which are contiguous with the cell membrane. They deposit elemental sulphur extracellularly. They are incapable of assimilatory sulphate reduction. Simple organic compounds, for example, acetate, can be assimilated and metabolised but only in association with sulphide reduction and carbon dioxide assimilation (Pfennig & Trüper, 1981; Trüper & Pfennig, 1981). The Chloroflexaceae (Trüper, 1976) (photosynthetic flexibacteria – Pierson & Castenholz, 1974) are Gram-negative, filamentous, gliding, anoxygenic phototrophs with flexible walls. Only *Chloroflexus aurantiacus* has been studied in axenic culture (Pierson & Castenholz, 1974); other genera are *Chloronema* (Dubinina & Gorlenko, 1975) and *Oscillochloris* (Gorlenko & Pivovarova, 1977). They show anoxygenic photosynthesis using reduced sulphur compounds as electron donors but the best growth occurs in the light on fixed carbon compounds (Castenholz & Pierson, 1981). Light carbon dioxide fixation in the Thiorhodaceae, Athiorhodaceae and probably in the Chloroflexaceae is by a typical Calvin cycle. In the Chlorobiaceae a mechanism of carbon dioxide fixation involving a reverse tricarboxylic acid cycle has been proposed (Evans *et al.*, 1966) but is not universally accepted as the major route of fixed carbon production (Fuller, 1978).

There are three major groups of oxygenic phototrophic microorganisms: the Cyanobacteria, the Prochlorophyta and the eukaryotic Algae. Table 4 lists their general features. The cyanobacteria are unicellular colonial or filamentous photosynthetic prokaryotes which use water as electron donor in photosynthesis and evolve oxygen. They contain chlorophyll *a* as in eukaryotic microalgae, with the phycobiliproteins c-phycocyanin, c-phycoerythrin and

Table 3. *Features of anoxygenic phototrophic bacteria*

Family	Genera	Bacteriochlorophylls	Photosynthetic electron donors	Deposition of elemental sulphur	Carbon sources	Useful references
Rhodospirillaceae	*Rhodocyclus*, *Rhodomicrobium*, *Rhodopseudomonas*, *Rhodospirillum*	*a* or *b*	H_2, H_2S, S	None	CO_2 and organic	Pfennig (1977) Trüper & Pfennig (1981) Biebl & Pfennig (1981)
Chromatiaceae	*Amoebobacter*, *Chromatium*, *Lamprocystis*, *Thiocapsa*, *Thiocystis*, *Thiodictyon*, *Thiopedia*, *Thiospirillum*, *Ectothiorhodospira*	*a* or *b*	H_2, H_2S, S	Intracellular (except *Ectothiorhodospira* which deposits S extracellularly)	CO_2	Pfennig (1969, 1977) Trüper & Pfennig (1981) Pfennig & Trüper (1981) Trüper & Imhoff (1981)
Chlorobiaceae	*Ancalochloris*, *Chlorobium*, *Pelodictyon*, *Prosthecochloris*	*c*, *d* or *e*	H_2, H_2S, S	Extracellular	CO_2	Pfennig (1977) Trüper & Pfennig (1981) Pfennig & Trüper (1981)
Chloroflexaceae	*Chloroflexus*, *Chloronema*, *Oscillochloris*	*c* or *d*	H_2, H_2S, S	None	CO_2 and organic	Madigan & Brock (1977) Pierson & Castenholz (1974); Castenholz & Pierson (1981); Gorlenko (1975); Gorlenko & Pivovarova (1977).

Table 4. *Features of oxygenic phototrophic microorganisms*

Group[a]	Major sub-groups	Typical genera	Chlorophylls	Accessory pigments	Reductant supply	Carbon reserves
Cyanobacteria	I Unicells showing budding or binary fission	*Gloeocapsa*, *Gloeothece*	*a* + *b*	C-phycocyanin, C-phycoerythrin, allophycocyanin, carotenes, xanthophylls	H_2O (H_2S in some)	glycogen
	II Unicells showing multiple fission	*Dermocarpa*, *Xenococcus*				
	III Non-hetero-cystous filamen-tous forms	*Spirulina*, *Oscillatoria*				
	IV Heterocystous filaments dividing in one plane	*Anabaena*, *Nostoc*				
	V Heterocystous filaments dividing in more than one plane	*Chlorogloeopsis*, *Fischerella*				

				xanthophylls		
Algae	Chlorophycophyta (green algae)	*Chlamydomonas*, *Chlorella*	$a + b$	carotenes, xanthophylls	H_2O	starch, oil in some
	Euglenophycophyta (euglenoids)	*Euglena*, *Trachelomonas*	$a + b$	carotenes, xanthophylls	H_2O	paramylon
	Chrysophycophyta	*Ochromonas*, *Cyclotella*	a, $a + c$	carotenes, xanthophylls	H_2O	chrysolaminarin, oil
	Pyrrhophycophyta (dinoflagellates)	*Ceratium*, *Gymnodinium*	$a + c$	carotenes, xanthophylls	H_2O	starch, oil in some
	Cryptophycophyta (cryptomonads)	*Cryptomonas*, *Chroomonas*	$a + c$	carotenes, xanthophylls, phycobilins	H_2O	starch
	Rhodophycophyta (red algae)	*Porphyridium*	a	R- and C-phycocyanin, R- and C-phycoerythrin, allophycocyanin, carotenes, xanthophylls	H_2O	floridean starch

[a] For further information see Rippka *et al.* (1979) for the Cyanobacteria; Lewin (1981) for the Prochlorophyta and Bold & Wynne (1978) for the Algae.

allophycocyanin, which are aggregated in discrete phycobilisomes attached to the outer surface of the thylakoid membranes, serving as accessory pigments. Certain strains can perform anoxygenic photosynthesis using reduced sulphur compounds as reductant sources (Cohen *et al.*, 1975; Garlick *et al.*, 1977) as in photosynthetic bacteria. There are five main groups of cyanobacteria (Rippka *et al.*, 1979) and certain strains from all five groups fix nitrogen (Stewart *et al.*, 1980; Stewart, 1982) (see also below). The Prochlorophyta (Lewin, 1981) are spherical unicells, containing chlorophylls *a* and *b* as in higher plants; they lack phycobiliproteins, use water as electron donor in photosynthesis and evolve oxygen photosynthetically. In cellular structure they are akin to cyanobacteria with an outer chromatoplasm, inclusion bodies such as polyhedral bodies and a central nuclear-containing region. They are found as extracellular symbionts on certain tropical and subtropical marine invertebrates, usually in locations of low light intensity. They were previously considered as zoochlorellae or cyanobacteria. *Prochloron didemni* is the only recognised species although other forms have been described. *Prochloron* has not yet been grown axenically. Eukaryotic microalgae belong to six main groups (Table 4). Their photosynthetic system, which is located in distinct chloroplasts, is rather similar to that of cyanobacteria with water serving as electron donor and with carbon dioxide fixation via the Calvin cycle. All contain chlorophyll *a* and another chlorophyll species, and accessory pigments, which in the case of the Rhodophycophyta and Cryptophycophyta include phycobilins as in cyanobacteria.

The non-photosynthetic microorganisms include eubacteria, flexibacteria and actinomycetes, as well as fungi, protozoa and colourless algae. The only forms other than the phototrophic microorganisms listed in Tables 3 and 4 which are known to harvest light energy are archaebacteria (Woese *et al.*, 1978) belonging to the halophilic genera *Halobacterium* and *Halococcus* (Halobacteriaceae) (Larsen, 1981). These are chemoheterotrophic microorganisms which under oxygen-deficient conditions can convert light energy to ATP via the membrane bound pigment bacteriorhodopsin (Oesterhelt, 1976). When oxygen is available the pigment is not synthesized and ATP-generation is by oxidative phosphorylation. The chemolithoautotrophic bacteria, on the other hand, while unable to harvest light energy, obtain their energy by the oxidation of inorganic compounds and can obtain fixed carbon from carbon dioxide via ribulose 1,5-bisphosphate carboxylase. These organisms have been reviewed by Smith & Hoare (1977), Whittenbury & Kelly, (1977) and Kelly (1981) and considered elsewhere in this volume.

In summation phototrophs can harvest light energy and can often generate fixed carbon via the Calvin cycle; the Halobacteriaceae can use light energy at low oxygen tensions but require organic carbon; and the chemolithoautotrophic bacteria while unable to use light energy to generate ATP, generate energy by chemical oxidation of inorganic compounds and can obtain fixed carbon via ribulose 1,5-bisphosphate carboxylase.

FIXED CARBON DISSIMILATION IN PHOTOTROPHS AND HETEROTROPHS

The success of microorganisms in their natural environment depends not only on the availability of fixed carbon but also on how the carbon is used to generate ATP and reductant and to provide carbon skeletons for growth. There is a need to store fixed carbon when it is available in excess, for example, as starch, glycogen, poly β-hydroxybutyrate, etc., for catabolism during periods of carbohydrate deficiency (darkness in the case of photosynthetic forms). The theoretical efficiencies with which ATP, NAD(P)H and FADH may be generated from hexose, for example, by different major groups depend on the route(s) of carbon dissimilation, in particular the extent of involvement of glycolysis, the oxidative pentose phosphate pathway, fermentative metabolism and whether or not the tricarboxylic acid cycle is complete. Autotrophs in general have an incomplete tricarboxylic acid cycle (Smith & Hoare, 1977) which is less efficient in ATP plus reductant provision than is the complete cycle. While this is of little disadvantage to a light-harvesting phototroph, it may be disadvantageous to most chemolithoautotrophs which have to use a large part of their energy budget in fixing carbon dioxide (Kelly, 1981).

MAJOR ELEMENTS, MACROHABITATS AND ECOLOGICAL SUCCESS

The major geochemical cycles are those of carbon, nitrogen, phosphorus and sulphur (Dugdale, 1976; Svennson & Söderlund,

1976; Jørgensen, 1981; Paul & Voroney, 1981). I do not intend to consider these cycles yet again but to consider ways in which the distribution of microorganisms, described above, may relate to the abundance of carbon, nitrogen and other major elements in the environment.

Carbon

In habitats where there is light, for example, terrestrial habitats and the photic zones of freshwaters and seas, the dominant organisms are primary producers (plants and microorganisms). Chemoheterotrophic bacteria predominate over phototrophs within the soil and in lake sediments because on land primary and secondary producers provide the necessary fixed carbon to sustain the chemoheterotrophic organisms and because phototrophs generally do not compete well with obligate chemoheterotrophs in the dark.

The availability of fixed carbon is a major reason why chemoheterotrophs accumulate particularly in and near the rhizosphere of plant roots in soils and salt marsh regions (Bowen, 1981; Paul & Voroney, 1981; Waughman *et al.*, 1981) and why marine chemoheterotrophs associate particularly with macroalgae in intertidal habitats, with marine animals (Baumann & Baumann, 1981) and with coastal upwellings where primary production is high (Floodgate *et al.*, 1981). It helps explain why in freshwater lakes intimate associations between epiphytic chemoheterotrophs and phytoplankton occur and why bacterial numbers increase as primary production increases with increase in eutrophication. It is a reason why associative symbioses are better developed with tropical C_4 plants than with temperate C_3 plants (Dobereiner & DePolli, 1980).

The relative abundance of chemoheterotrophs in soil and in the dark is aided by the fact that many photosynthetic microorganisms, for example, cyanobacteria, are obligate photoautotrophs (Fogg *et al.*, 1973; Smith, 1973). Cyanobacteria may lack the necessary permeases or transport systems (Stewart, 1980) and there is little competitive advantage in assimilating, for example, organic acids if the Krebs cycle is incomplete (Smith & Hoare, 1977) and they cannot be used efficiently. In cyanobacteria, organic compounds, when assimilated, tend to be hexoses (Smith, 1973; Bottomley & Van Baalen, 1978*a*,*b*) which can be effectively catabolised by the oxidative pentose phosphate pathway. Also, on many soil surfaces photosynthetic eukaryotes, particularly chlorophytes, tend to be shaded out by higher plants. This is partly why soil algae are often most abundant where there is little competition from higher plants, for example, in desert ecosystems (Metting, 1981). Photosynthetic bacteria are restricted in ecological distribution by their need for anaerobic conditions in the light; in the light, in the presence of oxygen, bacteriochlorophyll is not synthesized (Cohen-Bazire *et al.*, 1957).

The marine environment: ion-rich habitats dominated by phototrophs

The seas and oceans contribute over a third of the primary productivity of the Earth's surface (see above). Yet genera of truly marine chemoheterotrophic microorganisms are rare. Baumann & Baumann (1981) list only four indigenous genera of marine chemoheterotrophic bacteria (*Vibrio* including *Beneckea*; *Alteromonas*, *Pseudomonas* and *Alcaligenes*) and these are most abundant in association with marine animals from which they obtain fixed carbon; numbers of chemoheterotrophic microorganisms in the open ocean are lower per unit weight of chlorophyll than are numbers in freshwater habitats and halophilic microorganisms tend to be photosynthetic eukaryotes such as *Dunaliella*, cyanobacteria such as *Aphanothece halophytica* or *Synechocystis* DUN52, or Halobacteriaceae such as *Halobacterium* and *Halococcus*. Under anaerobic conditions in the light it is photosynthetic bacteria rather than anaerobic chemoheterotrophs which tend to develop; in tropical sediments where extremes of salinity occur stromatolites dominated by cyanobacteria are found (Kendall & Skipwith, 1968; Golubic, 1976; Krumbein *et al.*, 1977; Potts & Whitton, 1980).

Why are marine habitats dominated by light-harvesting, frequently photoautotrophic microorganisms? I believe that at least part of the reason is because seawater has a high ionic composition (Table 2) and that to sustain an osmotic balance which permits metabolism and growth in such an environment there is an energy and fixed carbon demand which can best be met by photosynthetic microorganisms.

Osmoregulatory mechanisms in marine microorganisms may involve any, or several, of five main groups of compounds: inorganic ions, amino acids, disaccharides, polyols and betaines (Hellebust, 1976*b*, 1980; Kauss, 1977). Inorganic ions have been widely implicated in osmoregulation. In chemoheterotrophic microorganisms they may serve as the major osmotica. For example, Larsen (1967) obtained evidence that in some marine heterotrophs the external Na^+ concentration could be internally balanced by K^+; others have evidence of chemoheterotrophs requiring up to 4 M salt for optimum growth (Lanyi, 1974; Brown, 1976). Marine fungi, such as *Asteromyces cruciatus* and *Dendryphiella salina*, accumulate K^+ and exclude Na^+ (Joshi *et al.*, 1962; Jennings, 1974; Flowers *et al.*, 1977), the K^+ usually being exchanged for H^+ (Slayman & Slayman, 1968; Shere & Jacobsen, 1970).

In chemoheterotrophs, organic compounds may also be involved. Tempest *et al.* (1970*a*), Brown & Stanley (1972), Stanley & Brown (1974) and Measures (1975) have all implicated amino acids in ionic regulation. Stanley & Brown (1974) reported that an increase in salinity resulted in increases in the glutamine, glutamate and proline pools and postulated that the glutamate anion may be synthesized in proportion to intracellular K^+ accumulation; however the quantitative or relative importance of either the amino acids or the inorganic ions as intracellular osmotica was not established. In chemoheterotrophic yeasts and fungi, glycerol serves as internal osmoticum in sugar-rich medium (Anand & Brown, 1968; Brown & Simpson, 1972; Brown, 1974). Mutants of *Salmonella typhimurium* which accumulate abnormally high proline concentrations show an increased osmotolerance (Csonka, 1980). The overall conclusion, particularly from early work, is that inorganic ions, especially K^+, are important intracellular osmotica in many chemoheterotrophs but that there is little evidence, in general, that they serve as sole osmotica. More recently, the possible importance of organic osmotica has been realised, but their quantitative significance, as yet, has not been properly evaluated.

In photosynthetic microorganisms both inorganic and organic osmotica are again involved. Kirst (1975*a*,*b*), using *Platymonas subcordiformis*, noticed that on increasing the external salinity from 0.1 to 0.6 M-NaCl, intracellular K^+ increased from 40 to 50 mM, Cl^- increased from 5 to 40 mM and Na^+ increased from 10 to 100 mM. However, according to Borowitzka & Brown (1974) Na^+ is effectively excluded from *Dunaliella tertiolecta* and *Dunaliella viridis*; in *Chlamydomonas* K^+ and Na^+ are poorly accumulated (Okamoto & Suzuki, 1964). Using various marine cyanobacteria and the halophile *Synechocystis* DUN52, Batterton & van Baalen (1971) and Mohammad *et al.* (1982), respectively, obtained no evidence of an increase in inorganic ions, or change in the internal K^+ to Na^+

ratio in response to salinity increase. Miller *et al.* (1976) and Tindall *et al.* (1978), however, found an increased intracellular K^+ to Na^+ ratio in the halophilic *Aphanothece halophytica* but concluded, as did workers on *Chlamydomonas* (Okamoto & Suzuki, 1964; Greenway & Setter, 1979*a*), *Dunaliella* (Borowitzka & Brown, 1974; Ben-Amotz & Avron, 1980) and *Platymonas subcordiformis* (Kirst, 1977*a,b*) that intracellular inorganic ions could not serve as the sole osmotica.

The major internal organic compounds found in photosynthetic microorganisms are listed in Table 5. In *Ectothiorhodospira halochloris*, betaine accumulates (Galinski & Trüper, 1982); in the marine *Synechococcus* RRIMP–N–100, 2-O-α-D glucopyranosylglycerol (glucosylglycerol) is the major organic osmoticum (Borowitzka *et al.*, 1980; Norton *et al.*, 1982); other marine cyanobacteria also accumulate glucosylglycerol (Richardson *et al.*, 1982). Data for the halophilic *Synechocystis* DUN52 (Mohammad *et al.*, 1982) are presented in Table 6. This organism has been isolated from the calcareous stromatolites of the upper intertidal flats of the Al-Khiran area of Kuwait (F.A.A. Mohammad & W.D.P. Stewart, unpublished observations) (an area subjected to very high salinities – supersaturated salt concentrations, and high temperatures – up to 55 °C). It does not grow in freshwater and has a broad salinity range for optimal growth of between 15 to 120‰ with maximum growth occuring near 60‰, and with some growth at 300‰; it is thus a moderate to extreme halophile (Kushner, 1978). As the external salinity increases up to 200‰ (Table 6) there are osmotically insignificant changes in the internal concentrations of Na^+ or K^+ or in the Na^+ to K^+ ratio. K^+ or Na^+ are clearly not major internal osmotica. On examining the intracellular pools of amino acids (Table 6), no marked changes were found, although the glycine and serine pools increased. On examining for intracellular glucosylglycerol concentrations there was a 10-fold increase as external salinity increased from 60 to 200‰; nevertheless the internal concentrations were less than 0.3% of those required if glucosylglycerol was to serve as sole osmoticum. However, the internal concentrations of quaternary ammonium compounds trebled on increasing the external salinity from 60 to 200‰ (Table 6) and were 10 to 200 times higher than those of any other intracellular solute measured. Nuclear magnetic resonance spectroscopy showed that glycine-betaine was the major quaternary ammonium compound present. The increased pools of glycine and serine noted were probably related to their role in glycine-betaine biosynthesis. *Synechocystis* DUN52 thus resembles *Ectothiorhodospira halochloris* (Galiniski & Trüper, 1982), and indeed halophilic angiosperms (Ladyman *et al.*, 1980; Wyn Jones, 1980; Wyn Jones & Storey, 1981; Coughlan & Wyn Jones, 1982) in having glycine-betaine as an organic osmoregulator. Possible pathways of glycine-betaine biosynthesis and their relationship to photorespiration and lipid metabolism in higher plants are considered by Coughlan & Wyn Jones (1982). It will be of interest to determine the extent to which such pathways operate in glycine-betaine accumulation in photosynthetic prokaryotes. Cyanobacteria thus have organic compounds as the major osmotic regulators, although these differ in different organisms. Whether these differences are species-specific or habitat-specific requires further study (Mohammad *et al.*, 1982).

In eukaryotic microalgae organic osmoregulators are also important. Sucrose, simple sugars and polyols such as glycerol, mannitol, glucosylglycerol etc. may all be important (Table 5). Proline is the sole amino acid of possible major importance (Hellebust, 1976*a*; Kirst, 1977*a*; Brown & Hellebust, 1978; Greenway & Setter, 1979*b*) although in *Cyclotella cryptica*, a marine diatom, glutamic acid and two unidentified compounds accumulate (Hellebust, 1980).

It appears that while inorganic osmotica may be of major importance in large-celled, large vacuolated macroalgae where Na^+, K^+ or Cl^- accumulations in the vacuoles serve as major osmotica (MacRobbie, 1974; Raven, 1975; Cram, 1976; Bisson & Gutknecht, 1977; Bisson & Kirst, 1979; Kirst & Bisson, 1979) their relative importance is reduced in microalgae which tend to have a smaller vacuole/cytoplasm ratio and where the organic osmotica are probably located in the cytoplasm, as in halophilic angiosperms (Hall *et al.*, 1978; Wyn Jones, 1980).

Most studies on how organic osmotica respond to salinity change have been carried out with *Dunaliella* where the osmoregulator is glycerol (Ben-Amotz & Avron, 1980; Hellebust, 1980, Norton *et al.*, 1982). Equilibration using glycerol is rapid with new intracellular equilibria being established within minutes of changing the external salinity. Such changes occur in the light and dark, the glycerol can be produced from stored starch, and starch and glycerol are alternatives into which fixed carbon can be channelled (Kaplan *et al.*, 1980; Gimmler & Möller, 1981). Einhuber & Gimmler (1980) consider that there is a significant loss of glycerol from *Dunaliella parva* by extracellular diffusion but Kessly & Brown (1981) conclude that in *D. viridis* and *D. tertiolecta* glycerol disappearance is due mainly to glycerol dissimilation. Norton *et al.* (1982) using ^{13}C-NMR to characterise the *in vivo* states of the osmoregulatory solutes in *Dunaliella* (glycerol) and *Synechococcus* (glucosylglycerol) concluded from elegant studies that the intracellular mobilities of these molecules were only 2 to 2.4 fold slower than their mobilities in free solution.

Inorganic ions may serve as triggers or activators of osmotic change and osmotic response. Such a trigger probably exerts its effect by activating certain enzymes and/or deactivating others, for example, those involved in amino acid or polyol biosynthesis and degradation. Salt stress stimulates the production of glutamate, a precursor of proline, it stimulates proline synthesis and it inhibits proline breakdown (Liu & Hellebust, 1976); isofloridoside phosphate synthetase is activated at low water potential in *Ochromonas malhamensis* (Kauss *et al.*, 1975; Kauss, 1977). In *Aphanothece halophytica* K^+ accumulation may be the primary mechanism which results in deplasmolysis after salinity increase and which triggers organic metabolite accumulation (Miller *et al.*, 1976). However, Kessly & Brown (1981) obtained no evidence that energy-induced proton fluxes across the plasmalemma acted as triggers or sensors in either *Dunaliella tertiolecta* or *Dunaliella viridis* and the trigger is not an external inorganic ion because glycerol accumulation occurs whether the external osmoticum is sugar or salt (Borowitzka *et al.*, 1977). The nature of the biochemical sensors which activate the production of organic osmotica is quite unresolved although specific metabolites and membrane bound enzymes which may respond to changes in membrane tension as a result of osmotic change have all been suggested as being important (Kauss, 1977; Ben-Amotz & Avron, 1980; Hellebust, 1980).

In addition to serving as osmotic regulators there is increasing evidence that internally synthesized organic solutes such as glycerol and glycine-betaine may act as compatible solutes, that is, compounds which have no inhibitory effect on enzymic reactions even at high concentrations and which play a role by protecting enzyme systems from inhibition by, for example, high concentrations of inorganic ions (Brown, 1976, 1978). Enzymes including glucose-6-phosphate dehydrogenase (Borowitzka & Brown, 1974; Pavlicek & Yopp, 1982) and isocitrate dehydrogenase (Brown, 1976) may be protected from the adverse effects of inorganic ions by glycine-betaine and glycerol which thus serve both as organic osmoregula-

Table 5. *Microorganisms and organic compounds which they accumulate in response to increasing salinity*

Group/organism	Compounds	Reference
Heterotrophic bacteria	GABA (γ-aminobutyric acid), glutamate, proline	Tempest *et al*. (1970*a*); Brown & Stanley (1972); Stanley & Brown (1974); Measures (1975)
Chromatiaceae	betaine	Galinski & Trüper (1982)
Cyanobacteria		
Aphanothece halophytica	betaine	Pavlicek & Yopp (1982)
Synechococcus	glucosylglycerol	Borowitzka *et al*. (1980)
		Norton *et al*. (1982)
Synechocystis DUN52	betaine	Mohammad *et al*. (1982)
Algae		
Asteromonas	glycerol	Ben-Amotz & Avron (1980)
Chlorella pyrenoidosa	sucrose	Hiller & Greenway (1968)
Chlorella sp.	proline	Hellebust (1976*b*)
Cyclotella spp.	proline	Mirg & Hellebust (1976); Schobert (1974)
Chlamydomonas	glycerol	Hellebust (1980)

		Borowitzka & Brown (1974)
Monochrysis lutheri	1,4/2,5-cyclohexanetetrol	Craigie (1969); Kauss (1969, 1973)
Ochromonas malhamensis	isofloridoside (o-α-D-galactopyranosyl-(1,1)-glycerol)	Schobert *et al*. (1972); Kauss *et al*. (1975); Quader & Kauss (1975)
Phaeodactylum tricornutum	proline	Besnier *et al*. (1969)
Platymonas	mannitol	Kirst (1975*a*,*b*); Hellebust (1976*a*); Kirst (1977*b*)
Pyramimonas	mannitol	Hellebust (1976*b*)
Scenedesmus obliquus	sucrose, and glucose	von Willert (1974)
	fructose, raffinose	Wetherell (1963)
Fungi		
Asteromyces cruciatus	arabitol, glycerol, mannitol	Flowers *et al*. (1977)
Chaetomium globosum	arabitol, glycerol, mannitol	Flowers *et al*. (1977)
Dendryphiella salina	arabitol, mannitol	Jennings (1974)
Saccharomyces rouxii	arabitol, glycerol	Brown (1974, 1978)
Saccharomyces cerevisiae	glycerol	Brown (1978)

Table 6. *Changes in intracellular concentrations of Na^+, K^+ and organic solutes in relation to salinity change in* Synechocystis *DUN52*

Intracellular compound (mmol dm^{-3})	Salt concentration (‰)				
	15	30	60	100	200
Na^+	—	188	—	218	230
K^+	—	170	—	274	218
Na^+/K^+	—	1.1	—	0.8	1.1
Aspartate	0.3	—	—	0.2	u
Serine	0.3	—	—	0.5	1.1
Threonine	0.4	—	—	u	0.3
Glutamate	6.4	—	—	1.7	0.4
Glycine	0.7	—	—	2.5	4.8
Alanine	0.4	—	—	0.4	0.6
Citrulline	u	—	—	u	0.2
Glucopyranosylglycerol	—	—	1.9	4.9	21.3
Quaternary ammoniun compounds	—	—	1180	2430	2980

For further details see Mohammad *et al.* (1982). – not tested; u undetectable.

tors and as compatible solutes. It is likely that such compounds also play roles in protecting the organisms from other environmental stresses. For example, glycine-betaine can protect higher plant membranes against heat destabilization (Jolivet *et al.*, 1982).

In sum, the energy cost necessary to maintain an osmotic balance is likely to be an important factor mitigating against the success of chemoheterotrophic bacteria in marine environments. Chemoheterotrophic microorganisms short of fixed carbon are unlikely to compete well if superimposed on any energy shortage for growth, they have, additionally, to divert ATP and/or fixed carbon to sustain ion pumps, or high internal concentrations of organic solutes in marine habitats. Phototrophs being able to harvest light energy may be more able to do so.

Such a hypothesis accords with various ecological observations, for example the facts that: (1) there is a paucity of truly marine chemoheterotrophic bacteria (Baumann & Baumann, 1981) although there are many halotolerant forms; (2) marine fungi are halotolerant rather than halophilic; (3) halophilic microorganisms are mainly those which harvest light energy; usually they also fix carbon dioxide photosynthetically, for example, photosynthetic bacteria, cyanobacteria and eukaryotic algae; (4) cyanobacteria-rich stromatolitic regions, where the extremes of salinity and desiccation are perhaps the greatest on Earth, are located in tropical regions of high light intensities; (5) marine microorganisms all store carbohydrate when it is available as: poly-β-hydroxybutyrate in marine heterotrophs (Baumann & Baumann, 1981), glycogen in cyanobacteria and starch in *Dunaliella* (Kaplan *et al.*, 1980). This fixed carbon can be catabolized to provide an osmoticum under carbon-depleted conditions, for example, in darkness in the case of photosynthetic microorganisms; (6) phototrophs are the dominant microorganisms of other salt-rich habitats such as Ethiopian soda lakes where *Spirulina* is abundant (Talling *et al.*, 1973), thermal hot spring regions such as Yellowstone National Park where cyanobacteria and flexibacteria predominate (Castenholz, 1981) and in alkaline salt-rich Usar lands of India where cyanobacteria are abundant (Singh, 1961).

NITROGEN

For sustained nitrogen cycling on Earth, there has to be a major and sustained input of newly fixed nitrogen from the atmosphere. This comes mainly from biological nitrogen-fixation which contributes annually around 139 million tonnes of nitrogen, and which on a global scale is substantially greater than the inputs from chemical nitrogen fertilizer, rainfall and other sources (Söderlund & Svennson, 1976).

The organisms which fix nitrogen are all prokaryotes; a recent report of nitrogen-fixation by a thermophilic eukaryotic microalga (Yamada & Sakaguchi, 1980) has not been confirmed. All nitrogen-fixing prokaryotes possess the enzyme nitrogenase. This enzyme is composed of two redox proteins, neither of which fixes nitrogen alone but which do so in combination. One is an Fe-Mo protein of two dissimilar pairs of sub-units, with an overall molecular weight of 220 000 to 240 000. The other is an Fe protein of two identical sub-units, with an overall molecular weight of 60 000 to 65 000 (Mortenson & Thorneley, 1979). The enzyme requires ATP for activity, and nitrogen-fixing organisms require carbon skeletons to remove the fixed nitrogen, because among other things NH_4^+ inhibits nitrogenase synthesis and activity (Eady, 1981); the primary route of NH_4^+ assimilation, the glutamine synthetase – glutamate synthase pathway (Tempest *et al.*, 1970*b*) requires ATP. Nitrogenase is extremely oxygen sensitive and functions only under strictly anaerobic conditions (Robson & Postgate, 1980).

Table 7 lists those prokaryotic genera which have nitrogen-fixing representatives. Relative to the total number of prokaryotic genera (almost 300), 57 genera have nitrogenase. Nitrogenase is widely distributed among different taxonomic groups (Postgate, 1981*b*). Schlegel & Jannasch (1981) comment that there is 'almost no correlation with taxonomic units'. However, a close inspection of the information in Table 7 reveals some interesting features. First, it is seen that of the 12 known genera of photosynthetic bacteria, 11 of these have nitrogenase; the twelfth has not yet been tested – it may be able to fix nitrogen. Second, of 25 genera including the three sub-groups of cyanobacteria, 19 of the genera tested have nitrogenase – the presence of nitrogenase tends to be the rule rather than the exception in phototrophs. It will be of interest to determine whether members of the Halobacteriaceae fix nitrogen under anaerobic conditions. Third, among the nitrogen-fixing non-photosynthetic forms 10 out of the 27 genera are chemoautotrophs: *Alcaligenes, Desulfovibrio, Desulfotomaculum, Xanthobacter, Mycobacterium, Methylococcus, Methylosinus, Methylobacter, Thiobacillus* and a CO-utilising organism (Dalton, 1980; Stewart, 1982). In addition strains of three of the well-established nitrogen-fixing chemoheterotrophs *Azospirillum* (Sampaio *et al.*, 1981), *Derxia* (Pedrosa *et al.*, 1980) and *Rhizobium* (Hanus *et al.*, 1979; Evans *et al.*, 1980, 1981) contain ribulose 1,5-bisphosphate carboxylase. The point is not that nitrogenase is rare in prokaryotes; it is that nitrogenase is rarely found in chemoheterotrophic prokaryotes.

Why should this be? It may be related to the fact that, compared with NH_4^+, the use of nitrogen, as nitrogen source, demands an additional energy input in converting it to NH_4^+ prior to assimilation. According to Postgate (1981*a*) this may not be much greater than that required to reduce NO_3^-. However, in the case of nitrogenase there is an additional energy and/or fixed carbon demand in aerobic organisms because of the need for an oxygen-free environment in which nitrogenase can function. Such a mechanism involves extremely high rates of respiration in *Azotobacter* which effectively scavenges oxygen, or O_2-excluding mechanisms, such as the gummy cell-envelope of *Derxia* (Robson & Postgate, 1980). In general, aerobes fix nitrogen best at oxygen concentrations below ambient and even then the rates are usually low per unit of carbohydrate consumed (5 to 40 mg N (g of carbohydrate consumed)$^{-1}$). On the other hand chemoheterotrophs which fix nitrogen under anaerobic conditions only, for example

Table 7. *Genera of N_2-fixing prokaryotes and how they relate to broad nutritional types*

Group	Total number of genera	Genera with nitrogen-fixing[a] representatives	% of total genera with nitrogen-fixing representatives
Phototrophic bacteria	12	*Amoebobacter*, *Chlorobium*, *Chromatium*, *Ectothiorhodospira*, *Pelodictyon*, *Prosthecochloris*, *Rhodomicrobium*, *Rhodopseudomonas*, *Rhodospirillum*, *Thiocapsa*, *Thiocystis*	92%
Cyanobacteria	22 + 3 groups	*Gloeothece, Synechococcus, Dermocarpa, Xenococcus, Myxosarcina, Chroococcidiopsis,* Pleurocapsa group, *Oscillatoria, Pseudanabaena* LPP group A, LPP group B, *Anabaena, Nodularia, Cylindrospermum, Nostoc, Scytonema, Calothrix, Chlorogloeopsis, Fischerella*	76%
Chemotrophic bacteria	~260	*Alcaligenes,*[b] *Aquaspirillum, Arthrobacter, Azomonas, Azospirillum, Azotobacter, Bacillus*, *Beijerinckia*, *Campylobacter*, *Citrobacter*, *Clostridium*, CO-utilising organism, *Derxia*, *Desulfotomaculum*, *Desulfovibrio*, *Enterobacter*, *Erwinia*, *Frankia*, *Klebsiella*, *Methylobacter*, *Methylococcus*, *Methylosinus*, *Mycobacterium*, *Propionibacterium*, *Rhizobium*, *Thiobacillus*, *Xanthobacter*	~10%

[a]For further details see Postgate (1981b); Stewart (1982); Stewart *et al.* (1980).
[b]See Malik & Schlegel (1981).

Bacillus and *Clostridium* are disadvantaged in that anaerobic energy-generating mechanisms are less efficient than aerobic ones. It is scarcely surprising, therefore, that nitrogenase is rarely present in genera of aerobic chemoheterotrophs and that when it is present, such organisms tend to be associated with oxygen-deficient carbohydrate-rich environments, for example, legume root nodules (Evans *et al.*, 1980, 1981) and the roots of tropical C_4 grasses (Dobereiner & De Polli, 1980).

When the macro-ecological distribution of nitrogen-fixing agents is examined, it is evident that despite the paucity of fixed nitrogen in the sea, nitrogen-fixing forms are uncommon there. An excellent review on aquatic nitrogen-fixation has recently been published (Paerl *et al.*, 1981) and I have no need to detail the literature here. Chemoheterotrophs when they occur seem to be mainly terrestrial forms which have been washed into the sea (Herbert *et al.*, 1977). Nitrogen-fixing cyanobacteria of supra-littoral and intertidal habitats (Stewart, 1962, 1967, 1975; Potts, 1980) tend to be halotolerant rather than halophilic species (Stewart, 1964, 1975); reports of nitrogen-fixing *Nodularia* blooms come mainly from Baltic regions of low salinity (Ostrom, 1976); *Calothrix* in coastal Pacific waters occurs near shore (Allen, 1963). However *Anabaena* CA one of the fastest growing nitrogen-fixing cyanobacteria known, grows optimally in full strength seawater (Stacey *et al.*, 1977). The only truly oceanic, free-living, nitrogen-fixing cyanobacterium is *Trichodesmium* (*Oscillatoria*) but little is known of its origin or physiological state (see however Goering *et al.*, 1966; Mague *et al.*, 1974; McCarthy & Carpenter, 1979; Carpenter & Price, 1977; Mague *et al.*, 1977; Li *et al.*, 1980). Its photosynthetic activity is inhibited at surface irradiances in the Caribbean Sea (Li *et al.*, 1980) and it may be that the excessive phycoerythrin which it contains enables it to harvest light energy efficiently in sub-surface waters. The extent to which fixed carbon is channelled into organic osmotica is not known. In addition to *Trichodesmium* there are reports of nitrogen-fixation by cyanobacterial endophytes of certain diatoms, for example, *Rhizosolenia* (*Richelia intracellularis*) (Mague *et al.*, 1974, 1977). In these symbioses photosynthesis by the eukaryote may protect the cyanobacteria from adverse effects of salt and possibly contribute additional fixed carbon to sustain nitrogenase activity; in the case of a nitrogen-fixing *Dichothrix* which is epiphytic on *Sargassum* (Carpenter, 1972), the *Sargassum* may provide additional fixed carbon which can be assimilated by the cyanobacterium. Overall despite the paucity of fixed nitrogen in the marine environment there appears to be little competitive advantage in fixing nitrogen. Possibly this is because nitrogen-fixation is a drain on the ATP and fixed carbon pools necessary to sustain an internal osmoticum. It will be of interest to compare the growth rates of nitrogen-fixing forms grown on nitrogen and NH_4^+ at different salinities to check out this possibility.

In this paper, I have tried to show that although the microniche concept is clearly of great fundamental importance it is also important not to lose sight of the wood for the trees (if one pardons a botanical expression!) and to consider whether the differences in microbiological forms found in the sea compared with freshwater and terrestrial environments, are not related, in part at least, to the need to expend energy and carbon in providing an osmoticum in the latter habitat.

Acknowledgements

Work from my own group, reported here, was supported, in part, by the Agricultural Research Council, the Science Research Council and the Natural Environment Research Council. I am grateful to Dr R. H. Reed and Dr F. A. Mohammad for useful discussions.

REFERENCES

Ainsworth, G. C. (1968). The numbers of fungi, In *The Fungi, An Advanced Treatise*, vol. 3, p. 505, ed. G. C. Ainsworth and A. S. Sussman. New York: Academic Press.

Ainsworth, G. C., Sparrow, F. K. & Sussman, A. S. (eds.) (1973). *The Fungi, An Advanced Treatise*, vol. 4. New York: Academic Press.

Alexander, M. (1977). *Introduction to Soil Microbiology*, 2nd edn. New York: John Wiley.

Allen, M. B. (1963). Nitrogen fixing organisms in the sea. In *Symposium on Marine Microbiology*, ed. C. H. Oppenheimer, p. 85. Illinois: C. C. Thomas.

Anand, J. C. & Brown, A. D. (1968). Growth rate patterns of the so-called osmophilic and non-osmophilic yeasts in solutions of polyethylene glycol. *Journal of General Microbiology*, **52**, 205–12.

Balows, A. & Brenner, D. J. (1981). The genus *Legionella*. In *The Prokaryotes*, vol. I, ed. M. P. Starr, H. Stolp, H. G. Trüper, A. Balows and H. G. Schlegel, p. 1091. Berlin: Springer-Verlag.

Bassham, J. A. (1976). Clean fuels for biomass, sewage, urban refuse. In *Agricultural Wastes*, p. 205. Chicago: Institute of Gas Technology.

Batterton, J. C. Jr & Baalen, C. Van (1971). Growth responses of blue-green algae to sodium chloride concentration. *Archiv für Mikrobiologie*, **76**, 151.

Baumann, P. & Baumann, L. (1981). The marine gram-negative eubacteria:

Genera *Photobacterium*, *Beneckea*, *Alteromonas*, *Pseudomonas* and *Alcaligenes*. In *The Prokaryotes*, vol. II, ed. M. P. Starr, H. Stolp, H. G. Trüper, A. Balows and H. G. Schlegel, pp. 1302–31. Berlin: Springer-Verlag.

Ben-Amotz, A. & Avron, M. (1973). The role of glycerol in osmotic regulation of the halophilic alga *Dunaliella parva*. *Plant Physiology*, **51,** 875.

Ben-Amotz, A. & Avron, M. (1980). Osmoregulation in the halophilic algae *Dunaliella* and *Asteromonas*. In *Genetic Engineering of Osmoregulation. Impact on Plant Productivity for Food, Chemicals and Energy*. Basic Life Sciences, vol. 14, ed. D. W. Rains, R. C. Valentine and A. Hollaender, p. 91. New York and London: Plenum Press.

Bergey's Manual of Determinative Bacteriology (1974). Ed. R. E. Buchanan and N. E. Gibbons. Baltimore: Waverley Press.

Besnier, V., Bazin, M., Marchelidon, J. & Genevot, M. (1969). Étude de la variation du pool intracellulaire des acides amines libres d'une diatomée marine en fonction de la salinité. *Bulletin de la Société Chimique de France*, **51,** 1255.

Biebl, H. & Pfennig, N. (1981). Isolation of members of the family Rhodospirillaceae. In *The Prokaryotes*, vol. I, ed. M. P. Starr, H. Stolp, H. G. Trüper, A. Balows, and H. G. Schlegel, p. 267. Berlin: Springer-Verlag.

Billen, G. (1982). Modelling the processes of organic matter degradation and nutrients recycling in sedimentary systems. In *Sediment Microbiology. Special Publications of the Society for General Microbiology*, **7,** ed. D. B. Nedwell and C. M. Brown, p. 15.

Bisson, M. A. & Gutknecht, J. (1977). Osmotic regulation in the marine alga *Codium decorticatum*. II. Active chloride influx exerts negative feedback control on the turgor pressure. *Journal of Membrane Biology*, **37,** 85.

Bisson, M. A. & Kirst, G. O. (1979). Osmotic adaptation in the marine alga *Griffithsia monilis* (Rhodophyceae): role of ions and organic compounds. *Australian Journal of Plant Physiology*, **6,** 523.

Bold, H. C. & Wynne, M. J. (1978). *Introduction to the Algae*. Englewood Cliffs, New Jersey: Prentice-Hall.

Borowitzka, L. J. & Brown, A. D. (1974). The salt relations of marine and halophilic species of the unicellular green alga, *Dunaliella*. *Archives of Microbiology*, **96,** 37.

Borowitzka, L. J., Kessly, D. S. & Brown, A. D. (1977). The salt relations of *Dunaliella*. *Archives of Microbiology*, **113,** 131.

Borowitzka, L. J., Demmerle, S., Mackay, M. A. & Norton, R. S. (1980). Carbon-13 nuclear magnetic resonance study of osmoregulation in a blue-green alga. *Science*, **210,** 650.

Bottomley, P. J. & Van Baalen, C. (1978a). Dark hexose metabolism by phototrophically and heterotrophically grown cells of the blue-green alga (cyanobacterium) *Nostoc* sp. strain Mac. *Journal of Bacteriology*, **135,** 888.

Bottomley, P. J. & Van Baalen, C. (1978b). Characteristics of heterotrophic growth in the blue-green alga *Nostoc* strain Mac. *Journal of General Microbiology*, **107,** 309.

Bowen, G. D. (1981). Misconceptions, concepts and approaches in rhizosphere biology. In *Contemporary Microbial Ecology*, ed. D. C. Ellwood, J. N. Hedger, M. J. Latham, J. M. Lynch and J. H. Slater, p. 283. London: Academic Press.

Brown, A. D. (1974). Microbial water relations: Features of the intra-cellular composition of sugar tolerant yeasts. *Journal of Bacteriology*, **118,** 769.

Brown, A. D. (1976). Microbial water stress. *Bacteriological Reviews*, **40,** 803.

Brown, A. D. (1978). Compatible solutes and extreme water stress in eukaryotic microorganisms. *Advances in Microbial Physiology*, **17,** 181.

Brown, A. D. & Simpson, J. R. (1972). Water relations of sugar-tolerant yeasts: the role of intracellular polyols. *Journal of General Microbiology,* **72,** 589.

Brown, C. M. & Stanley, S. O. (1972). Environment-mediated changes in the cellular content of the 'pool' constituents and their associated changes in cell physiology. In *Environment Control of Cell Synthesis and Function*, ed. A. C. R. Dean, S. J. Pirt and D. W. Tempest, p. 363. London and New York: Academic Press.

Brown, L. M. & Hellebust, J. A. (1978). Sorbitol and proline as intracellular osmotic solutes in the green alga *Stichococcus bacillaris. Canadian Journal of Botany*, **56,** 676.

Carpenter, E. J. (1972). Nitrogen fixation by a blue-green epiphyte on pelagic *Sargassum. Science*, **178,** 1207.

Carpenter, E. J. & Price, C. C. (1977). Nitrogen fixation, distribution and production of *Oscillatoria* (*Trichodesmium*) spp. in the western Sargasso and Caribbean Seas. *Limnology and Oceanography*, **22,** 60.

Castenholz, R. W. (1981). Isolation and cultivation of cyanobacteria. In *The Prokaryotes*, vol. I, ed. M. P. Starr, H. Stolp, H. G. Trüper, A. Balows and H. G. Schlegel, p. 236. Berlin: Springer-Verlag.

Castenholz, R. W. & Pierson, B. K. (1981). Isolation of members of the family Chloroflexaceae. In *The Prokaryotes*, ed. M. P. Starr, H. Stolp, H. G. Trüper, A. Balows and H. G. Schlegel, p. 290. Berlin: Springer-Verlag.

Cavanaugh, C. M., Gardiner, S. L., Jones, M. L., Jannasch, H. W. & Waterbury, J. B. (1981). Prokaryotic cells in the hydrothermal vent tube worm *Riftia pachyptila* Jones: possible chemoautotrophic symbionts. *Science*, **213,** 340.

Cohen-Bazire, G., Sistrom, W. R. & Stanier, R. Y. (1957). Kinetic studies of pigment synthesis by non-sulfur purple bacteria. *Journal of Cellular and Comparative Physiology*, **49,** 25.

Cohen, Y., Jorgensen, B. B., Padan, E. & Shilo, M. (1975). Sulphide dependent anoxygenic photosynthesis in the cyanobacterium *Oscillatoria limnetica. Nature,* **257,** 489.

Cohn, F. (1872). Untersuchungen über Bacterien. *Beiträge zur Biologie der Pflanzen*, **1,** 127.

Copeland, H. F. (1956). *The Classification of Lower Organisms*. Palo Alto: Pacific Books.

Coughlan, S. J. & Wyn Jones, G. W. (1982). Glycine betaine biosynthesis and its control in detached secondary leaves of spinach. *Planta* (*Berlin*), **154,** 6.

Craigie, J. A. (1969). Some salinity induced changes in growth pigments and cyclohexanetetrol content of *Monochrysis lutheri. Journal of the Fisheries Research Board of Canada*, **26,** 2959.

Cram, W. J. (1976). Negative feedback regulation of transport in cells. The maintenance of turgor volume and nutrient supply. In *Transport in Plants IIA Cells*, ed. U. Lüttge and M. G. Pitman. *Encyclopedia of Plant Physiology, New Series, Vol. 2A*, p. 284. Berlin: Springer-Verlag.

Csonka, L. N. (1980). The role of L-proline in response to osmotic stress in *Salmonella typhimurium*: selection of mutants with increased osmotolerance as strains which over-produce L-proline. In *Genetic Engineering of Osmoregulation. Impact on Plant Productivity for Food, Chemicals and Energy*, ed. D. W. Rains, R. C. Valentine and A. Hollaender, p. 35. New York and London: Plenum Press.

Dalton, H. (1980). Chemoautotrophic nitrogen fixation. In *Nitrogen Fixation*, ed. W. D. P. Stewart and J. R. Gallon, p. 177. London: Academic Press.

Dobereiner, J. & De-Polli, H. (1980). Diazotrophic rhizocoenosis. In *Nitrogen*

Fixation, ed. W. D. P. Stewart and J. R. Gallon, p. 301. London: Academic Press.

DUBININA, G. A. & GORLENKO, V. M. (1975). New filamentous photosynthetic green bacteria containing gas vacuoles. *Microbiology*, **44**,452.

DUGDALE, R. C. (1976). Nutrient cycles. In *The Ecology of Seas*, ed. D. H. Cushing and J. J. Walsh, p. 141. Oxford: Blackwell Scientific Publications.

EADY, R. (1981). Regulation of nitrogenase activity. In *Current Perspectives in Nitrogen Fixation*, ed. A. H. Gibson and W. E. Newton, p. 172. Canberra: Australian Academy of Science.

EINHUBER, G. & GIMMLER, R. (1980). The glycerol permeability of the plasmalemma of the halotolerant green alga *Dunaliella parva* (Volvocales). *Journal of Psychology*, **16,** 524.

EVANS, H. J., EMERICH, D. W., LEPO, J. E., MAIER, R. J., CARTER, K. R., HANUS, F. J. & RUSSELL, S. A. (1980). The role of hydrogenase in nodule bacteroids and free-living rhizobia. In *Nitrogen Fixation*, ed. W. D. P. Stewart and J. R. Gallon, p. 55. London: Academic Press.

EVANS, H. J., PUROHIT, K., CANTRELL, M. A., EISBRENNER, G., RUSSELL, S. A., HANUS, F. J. & LEPO, J. E. (1981). Hydrogen losses and hydrogenases in nitrogen-fixing organisms. In *Current Perspectives in Nitrogen Fixation*, ed. A. H. Gibson and W. E. Newton, p. 84. Canberra: Australian Academy of Science.

EVANS, M. C. W., BUCHANAN, B. B. & ARNON, D. I. (1966). A new ferredoxin-dependent carbon reduction cycle in a photosynthetic bacterium. *Proceedings of the National Academy of Sciences, USA*, **55,** 928.

FELBECK, H. (1981). Chemoautotrophic potential of the hydrothermal vent tube worm, *Riftia pachyptila* Jones (Vestimentifera). *Science*, **213,** 336.

FLOODGATE, G. D., FOGG, G. E., JONES, D. A., LOCHTE, K. & TURLEY, C. M. (1981). Microbiological and zooplankton activity at a front in Liverpool Bay. *Nature*, **290,** 132–6.

FLOWERS, T. J., TROKE, P. F. & YEO, A. R. (1977). The mechanism of salt tolerance in halophytes. *Annual Review of Plant Physiology*, **28,** 89.

FOGG, G. E., STEWART, W. D. P., FAY, P. & WALSBY, A. E. (1973). *The Blue-Green Algae*. London: Academic Press.

FULLER, R. C. (1978). Photosynthetic carbon metabolism in the green and purple bacteria. In *The Photosynthetic Bacteria*, ed. R. K. Clayton and W. R. Sistrom, p. 691. New York and London: Plenum Press.

GALINSKI, E. A. & TRÜPER, H. G. (1982). Betaine, a compatible solute in the extremely halophilic phototrophic bacterium *Ectothiorhodospira halochloris*. *FEMS Microbiology Letters*, **13,** 357–60.

GARLICK, S., OREN, A. & PADAN, E. (1977). Occurrence of facultative anoxygenic photosynthesis among filamentous and unicellular cyanobacteria. *Journal of Bacteriology*, **129,** 623.

GIBBONS, N. E. & MURRAY, R. G. E. (1978). Proposals concerning the higher taxa of bacteria. *International Journal of Systematic Bacteriology*, **28,** 1–6.

GIMMLER, H. & MÖLLER, E.-M. (1981). Salinity-dependent regulation of starch and glycerol metabolism in *Dunaliella parva. Pl. Cell. Env.* **4,** 367–375.

GOERING, J. J., DUGDALE, R. C. & MENZEL, D. W. (1966). Estimates of *in situ* rates of nitrogen uptake by *Trichodesmium* in the tropical Atlantic Ocean. *Limnology and Oceanography*, **11,** 614.

GOLUBIC, S. (1976). Organisms that build stromatolites. In *Stromatolites*, ed. M. R. Walter, p. 113. Amsterdam: Elsevier.

GORLENKO, V. M. (1975). Characteristics of filamentous phototrophic bacteria from freshwater lakes. *Microbiology*, **44,** 682.

GORLENKO, V. M. & PIVOVAROVA, T. A. (1977). On the belonging of blue-green alga *Oscillatoria coerulescens* Gicklhorn, 1921 to a new genus of Chlorobacteria *Oscillochloris* nov. gen. (In Russian with English summary.) *Isvestiya Akademii Nauk SSSR, Seriya Biologischeskaya*, **3,** 396.
GREENWAY, H. & SETTER, T. L. (1979a). Na^+, Cl^- and K^+ concentrations in *Chlorella emersonii* exposed to 100 and 335 mM NaCl. *Australian Journal of Plant Physiology*, **6,** 61.
GREENWAY, H. & SETTER, H. L. (1979b). Accumulation of proline and sucrose during the first hours after transfer of *Chlorella emersonii* to high NaCl. *Australian Journal of Plant Physiology*, **6,** 69.
HALL, J. L., HARVEY, D. M. R. & FLOWERS, T. J. (1978). Evidence for the cytoplasmic location of betaine in leaf cells of *Suaeda maritima*. *Planta* (*Berlin*), **140,** 59.
HANUS, F. J., MAIER, R. J. & EVANS, H. J. (1979). Autotrophic growth of H_2 uptake-positive strains of *Rhizobium japonicum* in an atmosphere supplied with hydrogen gas. *Proceedings of the National Academy of Sciences, USA*, **76,** 1788.
HELLEBUST, J. A. (1976a). Effect of salinity on mannitol synthesis in the green flagellate *Platymonas suecica*. *Canadian Journal of Botany*, **54,** 1735.
HELLEBUST, J. A. (1976b). Osmoregulation. *Annual Reviews of Plant Physiology*, **27,** 485.
HELLEBUST, J. A. (1980). Reactions to water and salt stress in plants. In *Plant Membrane Transport: Current Conceptual Issues*, ed. R. M. Spanswick, W. J. Lucas and J. Dainty, p. 147. North Holland, Amsterdam: Elsevier.
HERBERT, R. A. (1982). Nitrate dissimilation in marine and estuarine sediments. In *Sediment Microbiology. Special Publication of the Society for General Microbiology*, **7,** ed. D. B. Nedwell and C. M. Brown, pp. 53–71.
HERBERT, R. A., BROWN, C. M. & STANLEY, S. O. (1977). Nitrogen assimilation in marine environments. In *Aquatic Microbiology*, ed. F. A. Skinner and J. M. Shewan, p. 161. London: Academic Press.
HILLER, R. G. & GREENWAY, H. (1968). Effects of low water potentials on some aspects of carbohydrate metabolism in *Chlorella pyrenoidosa*. *Planta*, **78,** 49.
JANNASCH, H. W. & WIRSEN, C. O. (1979). Chemosynthetic primary production at East Pacific sea-floor spreading centers. *Bioscience*, **29,** 592.
JENNINGS, D. H. (1974). Cations and filamentous fungi: Invasion of the sea and hyphal functioning. In *Ion Transport in Plants*, ed. W. P. Anderson, p. 323. London and New York: Academic Press.
JOLIVET, Y., LARHER, F. & HAMELIN, J. (1982). Osmoregulation in halophytic higher plants: the protective effect of glycine betaine against the heat destabilization of membranes. *Plant Science Letters*, **25,** 193.
JONES, M. L. (1981). *Riftia pachyptila* Jones: observations on the Vestimentiferan worm from the Galapagos rift. *Science*, **213,** 333.
JØRGENSEN, B. B. (1981). Mineralization and the bacterial cycling of carbon, nitrogen and sulphur in marine sediments. In *Contemporary Microbial Ecology*, ed. D. C. Ellwood, J. N. Hedger, M. J. Latham, J. M. Lynch and J. H. Slater, p. 239. London and New York: Academic Press.
JOSHI, G., DOLAN, T., GEE, R. & SALTMAN, P. (1962). Sodium chloride effect on dark fixation of carbon dioxide for marine and terrestrial plants. *Plant Physiology*, **37,** 446.
KAPLAN, A., SCHREIBER, U. & AVRON, M. (1980). Salt-induced metabolic changes in *Dunaliella salina*. *Plant Physiology*, **65,** 810.
KARL, D. M., WIRSEN, C. O. & JANNASCH, H. W. (1980). Deep sea primary production at the Galapagos hydrothermal vents. *Science*, **207,** 1345.

KAUSS, H. (1969). Osmoregulation mit α-Galaktosylglyzeride bei *Ochromonas* und Rotalgen. *Berichte der Deutschen Botanischen Gesellschaft*, **82,** 115.
KAUSS, H. (1973). Turnover of galactosylglycerol and osmotic balance in *Ochromonas. Plant Physiology*, **52,** 613.
KAUSS, H. (1977). Biochemistry of osmotic regulation. In *Plant Biochemistry II*, vol. 13, ed. D. H. Northcote, p. 119. Baltimore: University Park Press.
KAUSS, H., LÜTTGE, U. & KRICHBAUM, R. M. (1975). Changes in potassium and isofloridoside content during osmoregulation in *Ochromonas malhamensis. Zeitschrift für Pflanzenphysiologie*, **76,** 109.
KELLY, D. P. (1981). Introduction to the chemolithotropic bacteria. In *The Prokaryotes*, vol. I, ed. M. P. Starr, H. Stolp, H. G. Trüper, A. Balows and H. G. Schlegel, p. 997. Berlin: Springer-Verlag.
KENDALL, C. G. St. C. & SKIPWITH, P. A. D'E. (1968). Recent algal mats of a Persian Gulf lagoon. *Journal of Sedimentary Petrology*, **38,** 1040.
KESSLY, D. S. & BROWN, A. D. (1981). Salt relations of *Dunaliella.* Transitional changes in glycerol content and oxygen exchange reactions on water stress. *Archives of Microbiology*, **129,** 154.
KIRST, G. O. (1975a). Wirkung unter schiedlicher Konzentrationen von NaCl und anderen osmotisch wirksamen Substanzen auf die CO_2-Fixierung der einzelligen Alge *Platymonas subcordiformis. Oecologia*, **20,** 237.
KIRST, G. O. (1975b). Ziehungen zwischen Mannitkonzentration und osmotischer Belastung bei der Brackwasseralge *Platymonas subcordiformis. Zeitschrift für Pflanzenphysiologie*, **76,** 316.
KIRST, G. O. (1977a). Composition of unicellular marine and freshwater algae with special reference to *Platymonas subcordiformis* cultivated in media with different osmotic strengths. *Oecologia*, **28,** 177.
KIRST, G. O. (1977b). Co-ordination of ionic relations and mannitol concentrations in the euryhaline, unicellular alga, *Platymonas subcordiformis* (Hasen) after osmotic shocks. *Planta*, **135,** 69.
KIRST, G. O. & BISSON, M. A. (1979). Regulation of turgor pressure in marine algae: Ions and low-molecular weight organic compounds. *Australian Journal of Plant Physiology*, **6,** 539.
KRUMBEIN, W. E., COHEN, Y. & SHILO, M. (1977). Solar Lake (Sinai) 4. Stromatolitic cyanobacterial mats. *Limnology and Oceanography*, **22,** 635.
KUSHNER, D. J. (1978). Life in high salt and solute concentrations: halophilic bacteria. In *Microbial Life in Extreme Environments*, ed. D. J. Kushner, p. 318. London: Academic Press.
LADYMAN, J. A. R., HITZ, W. D. & HANSON, A. D. (1980). Translocation and metabolism of glycine betaine by barley plants in relation to water stress. *Planta*, **150,** 191.
LANYI, J. H. (1974). Salt-dependent properties of proteins, from extremely halophilic bacteria. *Bacteriological Reviews*, **38,** 272.
LARSEN, H. (1967). Biochemical aspects of extreme halophilism. *Advances in Microbial Physiology*, **1,** 97.
LARSEN, H. (1981). The family Halobacteriaceae. In *The Prokaryotes*, vol. I, ed. M. P. Starr, H. Stolp, H. G. Trüper, A. Balows and H. G. Schlegel, p. 985. Berlin: Springer-Verlag.
LEWIN, R. A. (1977). *Prochloron*, type genus of the Prochlorophyta. *Phycologia*, **16,** 217.
LEWIN, R. A. (1981). The Prochlorophytes. In *The Prokaryotes*, ed. M. P. Starr, H. Stolp, H. G. Trüper, A. Balows and H. G. Schlegel, p. 257. Berlin: Springer-Verlag.

Li, W. K. W., Glover, H. E. & Morris, I. (1980). Physiology of carbon photo-assimilation by *Oscillatoria thiebautii* in the Caribbean Sea. *Limnology and Oceanography*, **25,** 447.

Liu, M. S. & Hellebust, J. A. (1976). Regulation of proline metabolism in the marine centric diatom *Cyclotella cryptica. Canadian Journal of Botany*, **54,** 949.

McCarthy, J. J. & Carpenter, E. J. (1979). *Oscillatoria* (*Trichodesmium*) *thiebautii* (Cyanophyta) in the central North Atlantic Ocean. *Journal of Phycology*, **15,** 75.

McDade, J. E., Shepard, C. C., Fraser, D. W., Tsai, T. R., Redus, M. A. & Dowdle, W. R. (1977). Legionnaires' disease. Isolation of a bacterium and demonstration of its role in other respiratory disease. *New England Journal of Medicine*, **297,** 1197–203.

MacRobbie, E. A. C. (1974). Ion uptake. In *Algal Physiology and Biochemistry*, ed. W. D. P. Stewart, p. 676. Oxford: Blackwell.

Madigan, M. T. & Brock, T. D. (1977). CO_2 fixation in photosynthetically grown *Chloroflexus aurantiacus. FEMS Microbiology Letters*, **1,** 301.

Mague, T. H., Mague, F. C. & Holm-Hansen, O. (1977). Physiology and chemical composition of nitrogen-fixing phytoplankton in the central North Pacific Ocean. *Marine Biology*, **41,** 213.

Mague, T. H., Weare, N. M. & Holm-Hansen, O. (1974). Nitrogen fixation in the North Pacific Ocean. *Marine Biology*, **24,** 109.

Malcolm, J. J. & Stanley, S. O. (1982). The sediment environment. In *Sediment Microbiology. Society for General Microbiology Special Publications*, **7,** p. 1.

Malik, K. A. & Schlegel, H. G. (1981). Chemoautotrophic growth of bacteria able to grow under N_2-fixing conditions. *FEMS Microbiology Letters*, **11,** 63.

Measures, J. C. (1975). Role of amino acids in osmoregulation of non-halophilic bacteria. *Nature* (*London*), **257,** 398.

Melack, J. M. & Kilham, P. (1972). Lake Mahega: a mesothermic, sulphate-chloride lake in Western Uganda. *African Journal of Tropical Hydrobiology and Fisheries*, **2,** 141.

Metting, B. (1981). The systematics and ecology of soil algae. *Botanical Reviews*, **47,** 195.

Miller, D. M., Jones, J. H., Yopp, J. H., Tindall, D. R. & Schmid, W. E. (1976). Ion metabolism in a halophilic blue-green alga, *Aphanothece halophytica. Archives of Microbiology*, **111,** 145.

Mirg, S. L. & Hellebust, J. A. (1976). Effect of salinity and osmolarity of the medium on amino acid metabolism in *Cyclotella cryptica. Candian Journal of Botany*, **54,** 938.

Mohammad, F. A. A., Reed, R. H. & Stewart, W. D. P. (1982). The halophilic cyanobacterium *Synechocystis* DUN52 and its osmotic responses. *FEMS Microbiology Letters* (in press).

Mortenson, K. E. & Thorneley, R. N. F. (1979). Structure and function of nitrogenase. *Annual Reviews of Biochemistry*, **48,** 387.

Nedwell, D. B. (1982). The cycling of sulphur in marine and freshwater sediments. In *Sediment Microbiology. Special Publication of the Society for General Microbiology*, **7,** ed. D. B. Nedwell and C. M. Brown, p. 73.

Norton, R. S., Mackay, M. A. & Borowitzka, L. J. (1982). The physical state of osmoregulatory solutes in unicellular algae. *Biochemical Journal*, **202,** 699.

Oesterhelt, D. (1976). Isoprenoids and bacteriorhodopsin in Halobacteriaceae. In *Progress in Molecular and Subcellular Biology*, vol. 4, ed. F. E. Hahn, p. 134, Berlin: Springer-Verlag.

Okamoto, H. & Suzuki, I. (1964). Intracellular concentration of ions in the halophilic strains of *Chlamydomonas*. I. Concentrations of Na, K and Cl in the cell. *Zeitschrift für Allgemeine Mikrobiologie*, **4,** 350.

Orla-Jensen, S. (1909). Die Hauptlinien des natürlichen Bakterien systems. *Zentralblatt für Bakteriologie 2*, **22,** 305.

Ostrom, B. (1976). Fertilization of the Baltic by nitrogen fixation in the blue-green alga *Nodularia spumigena. Remote Sensing of the Environment*, **4,** 305.

Paerl, H. W., Webb, K. L., Baker, J. & Wiebe, W. J. (1981). Nitrogen fixation in waters. In *Nitrogen Fixation*, vol. 1, ed. W. J. Broughton, p. 193. Oxford University Press.

Paul, E. A. & Voroney, R. P. (1981). Nutrient and energy flows through soil microbial biomass. In *Contemporary Microbial Ecology*. ed. D. C. Ellwood, J. N. Hedger, M. J. Latham, J. M. Lynch and J. H. Slater, p. 215. London and New York: Academic Press.

Pavlicek, K. A. & Yopp, J. H. (1982). Betaine as a compatible solute in the complete relief of salt inhibition of glucose-6-phosphate dehydrogenase from a halophilic blue-green alga. *Plant Physiology*, **69,** supplement no. 324.

Pedrosa, F. O., Dobereiner, J. & Yates, M. G. (1980). Hydrogen-dependent growth and autotrophic carbon dioxide fixation in *Derxia. Journal of General Microbiology*, **119,** 547.

Pfennig, N. (1969). Photosynthetic bacteria. *Annual Reviews of Microbiology*, **21,** 285.

Pfennig, N. (1977). Phototrophic green and purple bacteria. A comparative systematic survey. *Annual Reviews of Microbiology*, **31,** 275.

Pfennig, N. & Trüper, H. G. (1981). Isolation of members of the families Chromatiaceae and Chlorobiaceae. In *The Prokaryotes*, ed. M. P. Starr, H. Stolp, H. G. Trüper, A. Balows and H. G. Schlegel, p. 279. Berlin: Springer-Verlag.

Pierson, B. K. & Castenholz, R. W. (1974). A phototrophic gliding filamentous bacterium of hot springs, *Chloroflexus aurantiacus* gen. and sp. nov. *Archives of Microbiology*, **100,** 5.

Postgate, J. R. (1981a). Discussion to paper. *The Molecular and Genetic Manipulation of Nitrogen Fixation*, by J. R. Postgate and F. C. Cannon. *Philosophical Transactions of the Royal Society, London,* **B292,** 589.

Postgate, J. R. (1981*b*). Microbiology of free-living nitrogen-fixing bacteria, excluding cyanobacteria. In *Current Perspectives in Nitrogen Fixation*. ed. A. H. Gibson and W. E. Newton, p. 217. Canberra: Australian Academy of Science.

Potts, M. (1980). Blue-green algae (Cyanophyta) in marine coastal environments of the Sinai Peninsula; distribution, zonation, stratification and taxonomic diversity. *Journal of Phycology*, **19,** 60.

Potts, M. & Whitton, B. A. (1980). Vegetation of the intertidal zone of the lagoon of Aldabra, with particular reference to the photosynthetic prokaryotic communities. *Proceedings of the Royal Society, London*, **B208,** 13.

Preston, T., Stewart, W. D. P. & Reynolds, C. S. (1980). Bloom-forming cyanobacterium *Microcystis aeruginosa* overwinters on sediment surface. *Nature*, **288,** 365.

Quader, H. & Kauss, H. (1975). Die Rolle einiger Zwischenstoffe des Galactosylglyzerinstoffwechsels bei der Osmoregulation in *Ochromonas malhamensis*. *Planta* (*Berlin*), **124,** 61.

Rau, G. H. (1981). Low $^{15}N/^{14}N$ in hydrothermal vent animals: ecological implications. *Nature*, **289,** 484.

Raven, J. A. (1975). Algal cells. In *Ion Transport in Plant Cells and Tissues*, ed. D. A. Baker and J. L. Hall, p. 125. Amsterdam: North Holland.

Richardson, D. L., Mohammad, F. A. A., Reed, R. H. & Stewart, W. D. P.

(1982). Freshwater and halophilic cyanobacteria: osmotic responses in extreme environments. In *Abstracts of the IVth International Symposium on Photosynthetic Prokaryotes, Bombannes – Bordeaux, September 1982*. A36.

Rippka, R., Deruelles, J., Waterbury, J. B., Herdman, M. & Stanier, R. Y. (1979). Generic assignments, strain histories and properties of pure cultures of cyanobacteria. *Journal of General Microbiology*, **111,** 1.

Robson, R. L. & Postgate, J. R. (1980). Oxygen and hydrogen in biological nitrogen fixation. *Annual Reviews of Microbiology*, **34,** 183.

Round, F. E. (1981). *The Ecology of Algae*, Cambridge University Press.

Rowbotham, T. J. (1980). Preliminary report on the pathogenicity of *Legionella pneumophila* for freshwater and soil amoebae. *Journal of Clinical Pathology*, **33,** 1179.

Rubey, W. W. (1951). Geologic history of sea water. An attempt to state the problem. *Geological Society of America Bulletin*, **62,** 1111.

Sampaio, M. J. A. M., Da Silva, E. M. R., Dobereiner, J., Yates, M. G. & Pedrosa, F. O. (1981). Autotrophy and methylotrophy in *Derxia gummosa, Azospirillum brasilense* and *A. lipoferum*. In *Current Perspectives in Nitrogen Fixation*, ed. A. H. Gibson and W. E. Newton, p. 444. Canberra: Australian Academy of Science.

Shere, S. M. & Jacobson, L. (1970). Mineral uptake in *Fusarium oxysporum* f. sp. *vasinfectum*. *Physiologia Plantarum*, **23,** 51.

Schlegel, H. G. & Jannasch, H. W. (1981). Prokaryotes and their habitats. In *The Prokaryotes*, ed. M. P. Starr, H. Stolp, H. G. Trüper, A. Balows and H. G. Schlegel, p. 43. Berlin: Springer-Verlag.

Schobert, B. (1974). The influence of water stress on the metabolism of diatoms. I. Osmotic resistance and proline accumulation in *Cyclotella meneghiniana*. *Zeitschrift für Pflanzenphysiologie*, **74,** 106.

Schobert, B., Untner, E. & Kauss, H. (1972). Isofloridosid und die Osmoregulation bei *Ochromonas malhamensis*. *Zeitschrift für Pflanzenphysiologie*, **67,** 385.

Singh, R. N. (1961). *Role of Blue-Green Algae in Nitrogen Economy of Indian Agriculture*. New Delhi: Indian Council of Agricultural Research.

Slayman, C. L. & Slayman, C. W. (1968). Net uptake of potassium in *Neurospora*. Exchange for sodium and hydrogen ions. *Journal of General Physiology*, **52,** 424.

Smith, A. J. (1973). Synthesis of metabolic intermediates. In *The Biology of Blue-Green Algae*, ed. N. G. Carr and B. A. Whitton, p. 1. Oxford: Blackwell Scientific Publications.

Smith, A. J. & Hoare, D. S. (1977). Specialist phototrophs, lithotrophs and methylotrophs: a unity among a diversity of prokaryotes? *Bacteriological Reviews*, **41,** 419.

Söderlund, R. & Svensson, B. H. (1976). The global nitrogen cycle. In *Nitrogen, Phosphorus and Sulphur – Global Cycles*, ed. B. H. Svensson and R. Söderlund. *Ecological Bulletin (Stockholm)*, **22,** 23.

Stacey, G., Van Baalen, C. & Tabita, F. R. (1977). Isolation and characterisation of a marine *Anabaena* sp. capable of rapid growth on molecular nitrogen. *Archives of Microbiology*, **114,** 197.

Stanier, R. Y. & Van Niel, C. B. (1962). The concept of a bacterium. *Archiv für Mikrobiologie*, **42,** 17.

Stanley, S. O. & Brown, C. M. (1974). Influence of temperature and salinity on the amino acid pools of some marine pseudomonads. In *Effect of Ocean Environment on Microbial Activity*, ed. R. R. Colwell and R. Y. Morita, p. 92. Baltimore: University Park Press.

Stewart, W. D. P. (1962). Fixation of elemental nitrogen by marine blue-green algae. *Annals of Botany*, **26,** 439.

Stewart, W. D. P. (1964). Nitrogen fixation by Myxophyceae from marine environments. *Journal of General Microbiology*, **36,** 415.

Stewart, W. D. P. (1967). Nitrogen turnover in marine and brackish habitats. II. Use of ^{15}N in measuring nitrogen fixation in the field. *Annals of Botany*, NS, **31,** 385.

Stewart, W. D. P. (1975). Biological cycling of nitrogen in intertidal and supralittoral marine environments. In *Proceedings of the Ninth European Marine Biology Symposium*, ed. H. Barnes, p. 637. Aberdeen University Press.

Stewart, W. D. P. (1980). Some aspects of structure and function in nitrogen-fixing cyanobacteria. *Annual Reviews of Microbiology*, **34,** 497.

Stewart, W. D. P. (1982). Nitrogen fixation – its current relevance and future potential. *Israel Journal of Botany*, **31,** (in press).

Stewart, W. D. P., Preston, T., Peterson, H. G. & Christofi, N. (1982). Nitrogen cycling in eutrophic freshwaters. *Philosophical Transactions of the Royal Society, London, B*, **296,** 491.

Stewart, W. D. P., Rowell, P. & Rai, A. N. (1980). Symbiotic nitrogen-fixing cyanobacteria. In *Nitrogen Fixation*, ed. W. D. P. Stewart and J. R. Gallon, pp. 239–77. London: Academic Press.

Stewart, W. D. P., Sinada, F., Christofi, N. & Daft, M. J. (1977). Primary production and microbial activity in Scottish freshwater habitats. In *Aquatic Microbiology*, ed. F. A. Skinner and J. M. Shewan, p. 31. *Society for Applied Bacteriology Symposium Series 6*. London and New York: Academic Press.

Svensson, B. H. & Söderlund, R. (eds.) (1976). Nitrogen, phosphorus and sulphur global cycles. *Ecological Bulletin* (*Stockholm*), **22,** 23.

Talling, J. F., Wood, R. B., Prosser, M. V. & Baxter, R. M. (1973). The upper limit of photosynthetic productivity by phytoplankton: evidence from Ethiopian soda lakes. *Freshwater Biology*, **3,** 53.

Tempest, D. W., Meers, J. L. & Brown, C. M. (1970a). Influence of environment on the content and composition of microbial free amino acid pools. *Journal of General Microbiology*, **64,** 171.

Tempest, D. W., Meers, J. L. & Brown, C. M. (1970b). Synthesis of glutamate in *Aerobacter aerogenes* by a hitherto unknown route. *Biochemical Journal*, **117,** 405.

Tindall, D. R., Yopp, J. H., Miller, D. M. & Schmid, W. D. (1978). Physico-chemical parameters governing the growth of *Aphanothece halophytica* (Chroococcales) in hypersaline media. *Phycologia*, **17,** 179.

Tison, D. L., Pope, D. J., Cherry, W. B. & Fliermans, C. B. (1980). Growth of *Legionella pneumophila* in association with blue-green algae (cyanobacteria). *Applied and Environmental Microbiology*, **39,** 456.

Trüper, H. G. (1976). Higher taxa of the phototrophic bacteria: Chloroflexaceae fam. nov., a family for the gliding, filamentous, phototrophic, 'green' bacteria. *International Journal of Systematic Bacteriology*, **26,** 74.

Trüper, H. G. & Imhoff, J. F. (1981). The genus *Ectothiorhodospira*. In *The Prokaryotes*, ed. M. P. Starr, H. Stolp, H. G. Trüper, A. Balows and H. G. Schlegel, p. 274. Berlin: Springer-Verlag.

Trüper, H. G. & Pfennig, N. (1981). Characterization and identification of the anoxygenic phototrophic bacteria. In *The Prokaryotes*, ed. M. P. Starr, H. Stolp, H. G. Trüper, A. Balows and H. G. Schlegel, p. 229. Berlin: Springer-Verlag.

Waughman, G. J., French, J. R. J. & Jones, K. (1981). Nitrogen fixation in some terrestrial environments. In *Nitrogen Fixation*, vol. I, *Ecology*, ed. W. J. Broughton, p. 135. Oxford: Clarendon Press.

Wetherell, D. F. (1963). Osmotic equilibration and growth of *Scenedesmus obliquus* in saline media. *Physiologia Plantarum*, **16,** 82.

WHITE, D. E., HERN, J. D. & WARING, G. A. (1963). Chemical composition of sub-surface waters. *U.S. Geological Survey Professional Paper 440-F.*

WHITTENBURY, R. & KELLY, D. P. (1977). Autotrophy: A conceptual phoenix. In *Symposium of the Society for General Microbiology*, vol. 27, *Microbial Energetics*, p. 121. Cambridge University Press.

WILLERT, D. J. VON (1974). Der Einfluss von NaCl auf die Atmung und Activität der Malatdehydrogenase bei einigen Halophyten und Glycophyten. *Oecologia*, **14,** 127.

WOESE, C. R., MAGRUM, L. J. & FOX, G. E. (1978). Archaebacteria. *Journal of Molecular Evolution*, **11,** 245.

WYN JONES, R. G. (1980). An assessment of quaternary ammonium and related compounds as osmotic effectors in crops plants. In *Genetic Engineering of Osmoregulation. Impact on Plant Productivity for Food, Chemicals and Energy,* ed. D. W. Rains, R. C. Valentine and A. Hollaender, p. 155. New York and London: Plenum Press.

WYN JONES, R. G. & STOREY, P. (1981). Betaines. In *Physiology and Biochemistry of Drought Resistance in Plants*, ed. L. G. Paleg and D. Aspinall, p. 171. Sydney and New York: Academic Press.

YAMADA, T. & SAKAGUCHI, K. (1980). Nitrogen fixation associated with a hotspring green alga. *Archives of Microbiology*, **124,** 161.

ANALYSIS OF MICROORGANISMS IN TERMS OF QUANTITY AND ACTIVITY IN NATURAL ENVIRONMENTS

DAVID C. WHITE

310 Nuclear Research Building, Florida State University, Tallahassee, Florida 32306, USA

INTRODUCTION

The problem

Two major problems complicate the assessment of microbes in their natural environments: they are commonly found attached to surfaces and they occur in consortia usually containing different physiological and morphological types. An understanding of the role of surfaces began with the observation of 'bottle effects' where microbial growth and activity was stimulated by the interposition of surfaces (Zobell, 1943) particularly in oligotrophic environments (Jannasch & Pritchard, 1972; Filip, 1978). Such observations were confirmed and amplified by the publication of a monograph on the effects of surfaces by Marshall in 1976 and clearly summarized in two recent books (Ellwood *et al.*, 1979; Bitton & Marshall, 1980).

Microbial consortia, as morphologically diverse microcolonies on solid surfaces, can be detected using scanning electron microscopy (Sieburth, 1975). Specific elements in such consortia can be defined by fluorescent antibody staining (Ward & Frea, 1979); moreover zones of aerobiosis and anaerobiosis can be detected within these consortia by tetrazolium reduction (Paerl, 1980). The creation of anaerobic microzones in aerobic environments has been particularly well documented in studies of dental plaque formation (Gibbons & van Houte, 1975; Newman, 1980; Rutter, 1979).

Consortia, rather than single-species microcolonies, can be readily detected in the morphologically heterogeneous film which forms on a titanium surface exposed to rapidly flowing seawater (Fig. 1A) and is apparent in the diverse cell wall morphologies of a microcolony recovered from coral rubble sediment fixed with glutaraldehyde before dissolving away that coral sand in acid (Fig. 1B). In both electron micrographs the presence not only of microbes but their extracellular products is clearly seen.

How can microbial assemblies like those illustrated in Fig. 1 be

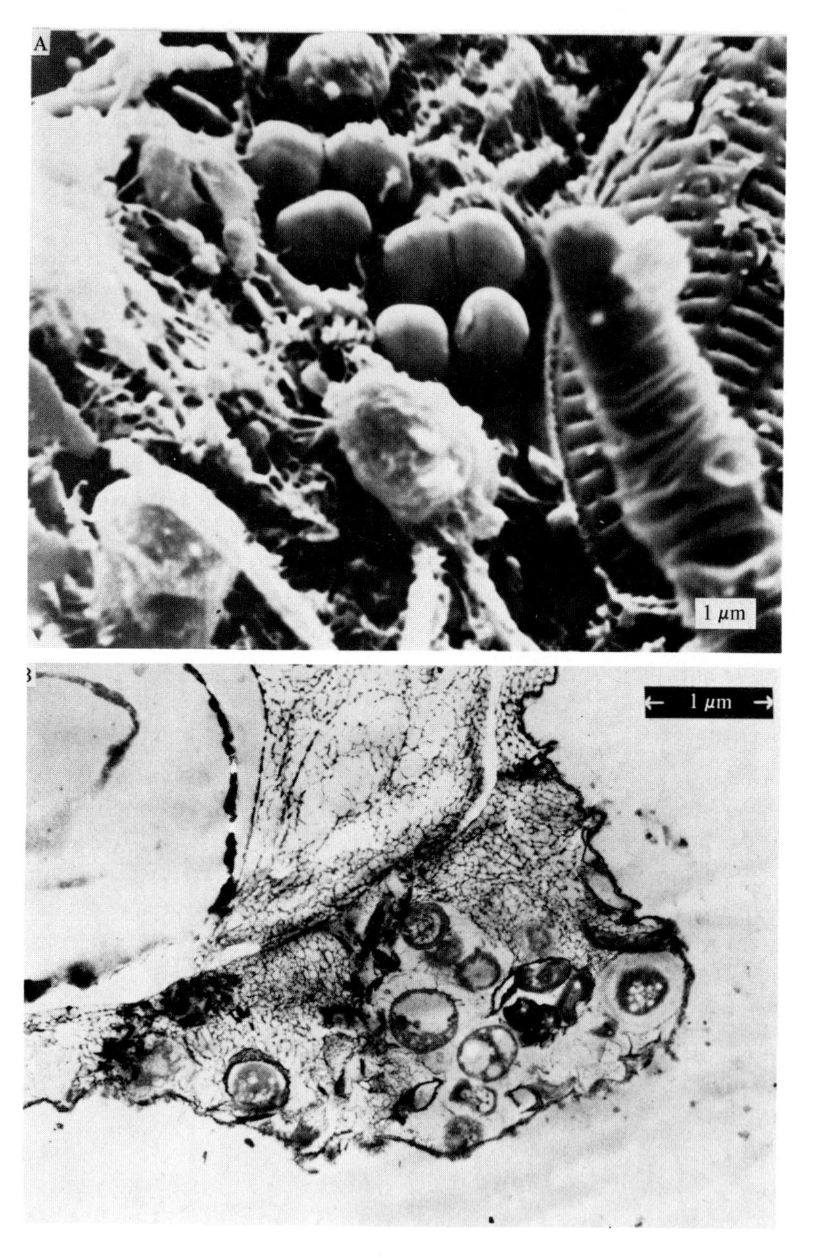

Fig. 1. (A) Scanning electron micrograph of the detrital microbiota on a pine needle incubated in the Apalachicola Bay estuary for 5 weeks (reproduced from White *et al.*, 1979a). The bar represents 1 μm. (B) Electron micrograph of a thin section of a microcolony attached to a diatom frustule stained with ruthenium red and osmium tetroxide (reproduced from D. J. W. Moriarty & A. C. Hayward (1982). Ultrastructure of bacteria and the proportion of gram negative bacteria in marine sediments. *Microbial Ecology* in press.) The bar represents 1 μm.

described quantitatively? Methods requiring the culture of microorganisms used so successfully in public health microbiology are not adequate. Plate counts of seawater organisms consistently underestimate the microbial community when compared with direct counting techniques (Butkevich, 1932; Zobell, 1946; Collins & Kipling, 1957; Jannasch & Jones, 1959; Perfil'ev & Gabe, 1969) or to numbers estimated by extracting adenosine triphosphate (ATP) (Holm-Hansen & Booth, 1966). Similarly, viable counts underestimate the number of bacteria present in the soil when compared with estimates of prokaryotic cell wall muramic acid (Millar & Casida, 1970). Methods requiring removal of microorganisms from surfaces have also not proved either reproducible or quantitative. Direct counts of acridine orange stained microbes using epifluorescent microscopy has proved useful in estimating bacteria in filtrates from a water column (Daley & Hobbie, 1975) but requires the quantitative removal of organisms from the substrate before filtration and counting when applied to sediments. The high-speed blending operation needed to remove the microbes from the sediment before staining and counting was neither quantitative nor reproducible when compared to specific assays for muramic acid (Moriarty, 1980). The quantitative release of microbes with their nuclear bodies intact is necessary if they are to be stained with acridine orange. The widespread occurrence of an extracellular glycocalyx which binds microbes to surfaces is characteristic of microbes found in nature (Costerton *et al.*, 1981) and makes the quantitative release of microbes from a solid substrate unlikely.

New methods were developed which required neither growth with its attendant problems of microbial selection nor removal of cells from the solid surface. Two major attributes define the ecology of microbial assemblies, these include on the one hand biomass which forms the basis for food chains and on the other metabolic activities of the microbiota that provide the driving force for vital biogeochemical cycles. The methods described below were designed to measure these two fundamental aspects of microbial ecology.

THE ANALYTICAL PROCEDURES

General comments

The requirement for a nonselective method for analysing intact microbial consortia can be met through an assessment of the biochemical properties of cells. If specific biochemical constituents are restricted to particular subsets of the microbial community then the analysis can give information about the community composition. These can be called the 'signature' components of specific microbial groups. A number of 'signature' chemicals have been identified and used in this way. Actually a continuum of biochemical compounds exists ranging from those typical of all cells which are useful as general biomass indicators, to the more or less specific 'signature' compounds described above.

The analysis of biomass and of community composition involves the extraction, isolation, and separation of the biochemical markers. Using similar techniques it is also possible to estimate metabolic activities of the consortia and various subsets of the latter. Metabolic activities can be estimated by measuring the incorporation of turnover of particular biochemical compounds after preincubation with labeled precursors. Cellular components that are known to turnover rapidly in nature on cell death can be used to estimate the 'viable' biomass.

In the past 5 years this laboratory has developed a suite of analytical methods based on an efficient one phase chloroform-methanol extraction (Bligh & Dyer, 1959) of environmental samples (White *et al.*, 1979c). Lipids, particularly phospholipids, are quantitatively extractable from microbial assemblies and are rapidly metabolized in nature (White *et al.*, 1977; King *et al.*, 1977). This procedure also liberates all the water-soluble components of the microbial cytosol including adenosine nucleotides and free amino acids (Davis & White, 1980). These water-soluble biochemicals can easily be recovered from the aqueous phase after lipid extraction. The residue remaining can then be hydrolyzed to release bound lipid constituents and other components which yield additional information on the biomass and community composition. A general scheme for the analytical procedure is illustrated in Fig. 2. The methods involve isolation of each compound, separation by various chromatographic techniques, followed by derivatization and analysis by gas–liquid chromatography (GLC). The application of splitless injection techniques with highly efficient glass open tubular

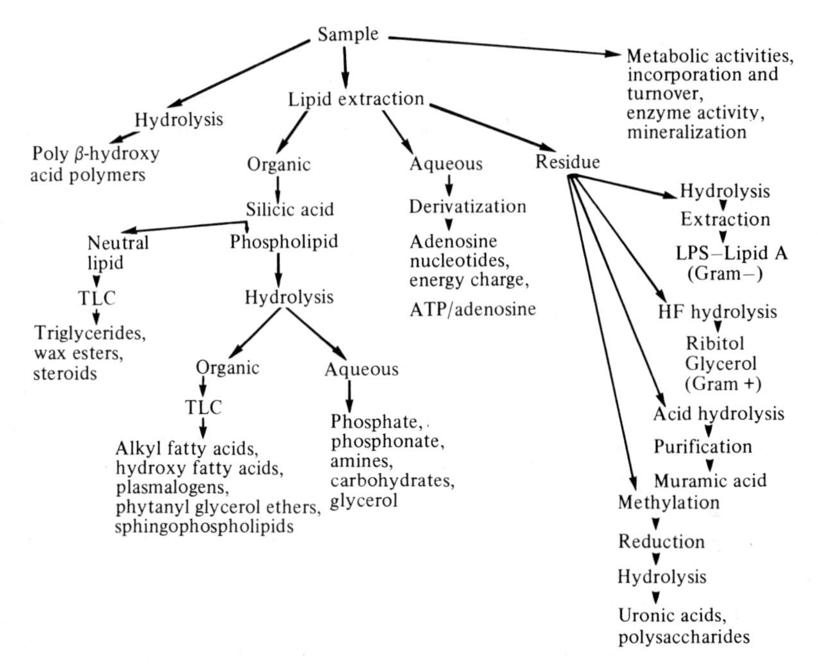

Fig. 2. Diagram of the analytical scheme for the biochemical analysis of microbial consortia.

capillary columns using newly designed electron capture or flame ionization detection devices makes sensitivities in the pico-molar range possible. Replacement of conventional detectors by a rapid scanning quadrupole mass spectrometer allows an identification of molecular structure from its fragmentation pattern as well as the efficient detection of the heavy isotope ^{13}C and ^{15}N in major ions from each component separated by the capillary GLC.

THE MEASURES

The estimation of biomass

As stated earlier, biomass is determined by measuring components common to all cells. The rapid metabolism of ATP in living cells and its easy determination using the very sensitive luciferin–luciferase assay led to its widespread usage in assessing the biomass of plankton in the water column (Holm-Hansen & Booth, 1966). The application of this assay to sediment ecosystems produced a spate of papers each describing an improved method for quantitative ATP recovery. A major problem was the need to inactivate ATPase enzymes whilst leaving the enzymatic assay system unaffected (Davis & White, 1980). Attempts to assay ADP and AMP in order to calculate adenylate energy charge values aggravated the problem since these assays required additional enzymes. The need to measure biomass in microbial films growing on metallic surfaces precluded the use of acid extraction techniques and a chemical procedure specific for adenine containing cell components was devised. This involved the formation of a 1-N^6 ethenoadenosine derivative using chloroacetaldehyde and allowed detection by fluorescence in the 10^{-12} molar range. Components were then separated by high pressure liquid chromatography (HPLC) using fluorescence detectors (Davis & White, 1980). This assay could measure all the adenine containing components including adenine, adenosine, AMP, cyclic-AMP, NAD, ADP and ATP from the aqueous phase after lipid extraction to measure other biomass components. Minor modifications in the extraction procedure were required to inactivate ATPase. When applied to detrital microbiota, total adenosine nucleotides were more closely related to microbial biomass as estimated by the phospholipid content than were ATP measurements alone (Davis & White, 1980). In sediments, the interpretation of results by this or the enzymatic assays may be complicated by the presence of extracellular adenosine nucleotides.

Phospholipids constitute a part of every cellular membrane and form a relatively constant proportion of the membranes of various microbes (White *et al.*, 1979a, c). These components are not found in storage lipids and have a relatively rapid turnover both in living cells and in killed cells added to the environment (King *et al.*, 1977; White *et al.*, 1977, 1979c). Phospholipids are quantitatively extractable and make up 98% of eubacterial membrane lipids and about 50% of the lipids of other organisms. Lipid phosphate is readily measured after perchloric acid digestion and colorimetric analysis to a sensitivity of 10^{-9} moles (White *et al.*, 1979c). This allows the detection of about 10^9 bacteria the size of *E. coli*. The successive application of acid and hydrofluoric acid (HF) hydrolysis using GLC to assay the glycerol, increased the sensitivity to 10^{-11} moles corresponding to phospholipid present in about 10^6 bacteria the size of *E. coli* (Gehron & White, 1982b). Palmitic acid is present in virtually all microbial lipids (Kates, 1964) and can be assayed readily by GLC in the 10^{-12} molar range corresponding to about 5×10^5 bacteria the size of *E. coli* (Bobbie & White, 1980).

Muramic acid (MA) is a unique component of prokaryotic cell walls. The value of MA as a biomass indicator that is neither selective nor requires the quantitative removal of cells from sediments has been discussed before (Moriarty, 1977; King & White, 1977). The lipid-extracted sediment residue can be hydrolyzed and the MA quantitatively extracted, purified and assayed by GLC at sensitivities down to 10^{-11} moles corresponding to about 10^8 bacteria the size of *E. coli* (Fazio *et al.*, 1979; Findlay *et al.*, 1982). MA present in cyanophytes, gram negative and gram positive bacteria are different (Millar & Cassida, 1970; Moriarty, 1980) so independent methods for assessing each are needed to investigate community composition in greater detail.

The biomass of gram negative bacteria in sediments has been estimated by the assay of ketodeoxyoctonate, beta-hydroxymyristate and anti-human complement activity present in lipopolysaccharides (LPS) after extraction by trichloroacetic acid or phenol water (Saddler & Wardlaw, 1980). Mild acid hydrolysis of the lipid-extracted residue followed by re-extraction yields 4 to 10 times more bound hydroxy fatty acids. These hydroxy fatty acids can be purified and assayed by GLC with sensitivities down to 10^{-12} (flame ionization detection) or 10^{-13} (electron capture detection) moles (Parker *et al.*, 1982). These sensitivities correspond to 10^7 or 10^6 *E. coli* cells. The recovery of lipids extracted from microorganisms added to sediments was quantitative and analysis of the hydroxy fatty acid patterns yielded insight into the composition of the gram negative bacterial community (Parker *et al.*, 1982).

Gram-positive bacteria forming teichoic acids can be assayed in sediments using cold concentrated HF to specifically and quantitatively release glycerol and ribitol from the lipid extracted residue (Gehron *et al.*, 1982). The sensitivity (down to 10^{-11} moles) corresponds to 5×10^6 bacteria resembling *Staphylococcus aureus*. This assay was used to confirm the increase in proportion of gram positive bacteria seen in deeper sediments by the analysis of transmission electron micrographs of cell wall structures (Moriarty, 1980).

It is important to record that adenosine nucleotides, phospholipids, MA, and the LPS of dead bacteria are quite rapidly lost from sediments (Davis & White, 1980; White *et al.*, 1979c, King *et al.*, 1977; Moriarty, 1977; Saddler & Wadlaw, 1980). Clearly these biomass assays can provide good estimates of the standing viable crop.

Community composition

By measuring biochemical components that are more or less restricted to subsets of the microbial flora, an assessment of community composition is possible. The use of 'signature' lipids has allowed our laboratory to investigate two major divisions of the microbial community. These are the separation of prokaryotic from eukaryotic microbes on the one hand and the ratio of aerobic to anaerobic organisms on the other.

The ratio of prokaryotes to microeukaryotes can be estimated by analysis of the phospholipid alkyl fatty acids. We have used short branched-chain 15 carbon saturated acids, cis-vaccenic acid, and the 17 and 19 carbon cyclopropane fatty acids as 'signature' lipids of various groups of bacteria (Bobbie & White, 1980). The 10-methyl palmitic, the monoenoic 15 and 17 carbon with w6 and w8 isomers (especially if they occur in pairs), the iso 15 monoenoic acids with w3 and w7 unsaturation, the trans monoenoic branched and straight chain 16 and 17 carbon fatty acids could all be added to the list of 'signature' lipids of bacteria according to the elegant work of R. B. Johns and his colleagues (Perry *et al.*, 1979).

Phospholipids with more glycerol than phosphate in the polar head groups (largely phosphatidyl glycerol) are found predominantly in bacteria. The phosphatidyl glycerol lipids may be estimated by measuring the acid-labile glycerol from the differential hydrolysis of phospholipids (Gehron & White, 1982b).

Polyenoic fatty acids with more than 20 carbon atoms are characteristic of eukaryotes and those with alpha linolenic unsaturation patterns (w3 series) are found in the eukaryotic microflora. Those with gamma linolenic unsaturation patterns (w6 series) are found in the microfauna (Erwin, 1973; Bobbie & White, 1980). The absence of 18 carbon polyenoic fatty acids in marine diatoms has been used to measure the effects of light on the detrital microbiota (Bobbie *et al.*, 1981). Triglycerides are readily assayed by measuring their glycerol content (Gehron & White, 1982a). The monohydroxy steroids, phosphatidyl inositol, and most of the sulfolipid synthesising capacity are found in microeukaryotes containing photosystem II and are additional 'signature' compounds for this group (White *et al.*, 1980a). The presence of phosphonate in microeukaryotes can be determined by selective hydrolysis (Kittredge & Roberts, 1969) and phosphonates can be detected in the detrital microbiota (White *et al.*, 1980a).

The balance between anaerobic and aerobic bacterial species can also be estimated with signature lipids. Boon *et al.* (1977) showed that *Desulfovibrio desulfuricans* contained high concentrations of ester-linked iso- and anteiso-branched 3-hydroxy-fatty acids. We have detected a high level of specific hydroxy fatty acids from phospholipid fractions and these are ester linked in microcosms enriched in sulfate-reducing bacteria (Fredrickson, 1981). The types of hydroxy fatty acids found in sediments resemble patterns from acetate-utilizing strains rather than those from classic *Desulfovibrio* species. This may reflect the central role of acetate metabolism in anaerobic sediments (Lovely & Klug, 1982).

Plasmalogens are amongst the few unique lipids found in anaerobic bacteria. These compounds yield fatty aldehydes on mild acid hydrolysis and are found both in specific organs of higher eukaryotes and in anaerobic fermentative bacterial species (Goldfine & Hagen, 1972; Kamio *et al.*, 1969). Plasmalogens are rarely found in higher plants (Mangold, 1972), in fish or in molluscs (Malins & Varanasi, 1972; Thompson, 1972).

Sphingolipids are very rare in bacteria (Karlsson, 1970) but unique phosphosphingolipids have been found among the anaerobic fermenters (LaBach & White, 1969; White & Tucker, 1970; Rizza *et al.*, 1970). The phosphosphingolipids can be isolated from these organisms and the branched sphinganines derived from them assayed by GLC (White *et al.*, 1969).

The methane-forming bacteria contain di- and tetra-phytanyl glycerol ether phospholipids (Tornabene & Langworthy, 1979). These compounds can be extracted, purified, derivatized and assayed by HPLC (Martz *et al.*, 1982).

Analytical methods can now distinguish three major groups of anaerobic bacteria by their signature lipids. Microcolonies of mixed cell types such as the example illustrated in Fig. 1B have revealed the signature lipids of anaerobic fermentative bacteria. Eukaryotic populations were shown to be very low in estuarine muds judged by the polyenoic fatty acid and monohydroxysteroid content of the samples. The presence of acid labile plasmalogen phosphate and fatty aldehydes in these sediments showed that the anaerobic species were concentrated in the aerobic portion of the sediment (White *et al.*, 1979b). The aerobic sediment horizon shows 3 to 5 times the biomass (measured as lipid phosphate), at least 10 times as much PHB and 10 to 50 times more fatty aldehydes derived from the plasmalogens of the anaerobes than are seen in the anaerobic horizons. This reinforces the suggestion that anaerobes exist in microcolonies surrounded by heterotrophic aerobic species.

Metabolic activity

The measurement of metabolic activities including phosphodiesterase, sulfatase, glycosidases and alkaline phosphatase enzymes, as well as the heterotrophic metabolism of simple substrates, rates of phospholipid and sulpholipid synthesis, the incorporation of thymidine into DNA, and oxygen utilization all correlate well with the biomass measurements of muramic acid, ATP and lipid phosphate from the microbiota of detrital surfaces (White *et al.*, 1979a, c, Bobbie *et al.*, 1981). However differences in enzymatic activities per unit biomass can be detected by changing the availability of inorganic phosphate (Morrison & White, 1980) or the biodegradability of the surface (Bobbie *et al.*, 1978). The incorporation of ^{3}H-thymidine into DNA is a measurement of prokaryotic activity because microeukaryotes lack the nucleotide salvage pathways (White *et al.*, 1980a). This assay has been greatly improved by Moriarty & Pollard (1981). Sulfolipids are rare in bacteria but high in photosynthetic microeukaryotes so the ratio of sulfolipid to phospholipid synthesis gives an indication of the proportion of metabolic activities due to each class of organism (White *et al.*, 1980a; Bobbie *et al.*, 1981).

Rates of incorporation of isotopes into lipids provide a reliable measurement of activity when applied to microbial film forming on metal surfaces or detritus. When applied to sediment systems, however, these methods can only give an indication of 'potential' activity since delicate interactions between organisms are certain to be disturbed by the addition of substrates.

Turnover of cellular components can allow estimations of growth rates to be made. However, organisms may be growing at different growth rates within the same microbial community. Muramic acids in the detrital microbiota show complex turnover patterns having both fast and slow components (King & White, 1977). Turnover of glycerol phosphate esters derived from phospholipids after a short exposure to labeled precursors gives a useful estimate of average growth rates. For example the lag in the saturation of the precursor pool after the termination of a short 'pulse' of cardiolipin synthesis is related both to the turnover of phosphatidyl glycerol, its direct precursor, and growth rate in monocultures (White & Tucker,

1969). This also holds true among the detrital microbiota (King *et al.*, 1977). Direct estimates of growth rates agree with values derived from estimates of rates of phospholipid synthesis, muramic acid turnover and respiratory activity (Morrison *et al.*, 1977). Lipids with very little metabolic activity such as the glycerol phosphoryl choline derived from phosphatidyl choline have proved accurate quantitative measurements of the rate of predation by grazing invertebrates (Morrison & White, 1980).

Isolation of cellular components using GLC analysis can be coupled with mass spectrometry to measure the enrichment of ^{13}C after pulse-chase experiments. Use of ^{13}C labeled precursors has a number of advantages over the radioactive ^{14}C isotope. No radioactivity is involved so experiments can be done more readily in the field. The specific activity is high (more than 99%) so that precursor concentrations can be used that are near to actual ambient values, and there is less likelihood of distortion in experimental results. Enrichment in ^{13}C can easily be detected using the selective ion detection mode in the mass spectrometer.

It is important to emphasize once more, that while these methods may be excellent when applied to thin films of microbes on easily manipulated surfaces such as detritus or fouling coupons, in the sediments the disturbances following exposure of the system to labeled substrates can induce increased metabolic activities.

Is it possible to develop an independent method which could accurately define the effects of sediment disturbance? Measurement of the nutritional status of the microbiota might achieve this aim since it gives information on recent metabolic activity.

Nutritional status

Our laboratory has developed both short and longer term methods for estimating nutritional status. Adenylate energy charge determination using a fluorometric assay can be applied to adenosine nucleotides derived from the aqueous fraction of the lipid extraction procedure (Davis & White, 1980). An even more sensitive indicator of nutritional status is provided by measuring the homeostatic mechanism which controls the adenylate energy charge. For example as the ATP synthetic rate declines with stress there is a compensatory decrease in AMP within the cell so that the energy charge ratio remains constant. Two mechanisms for decreasing the

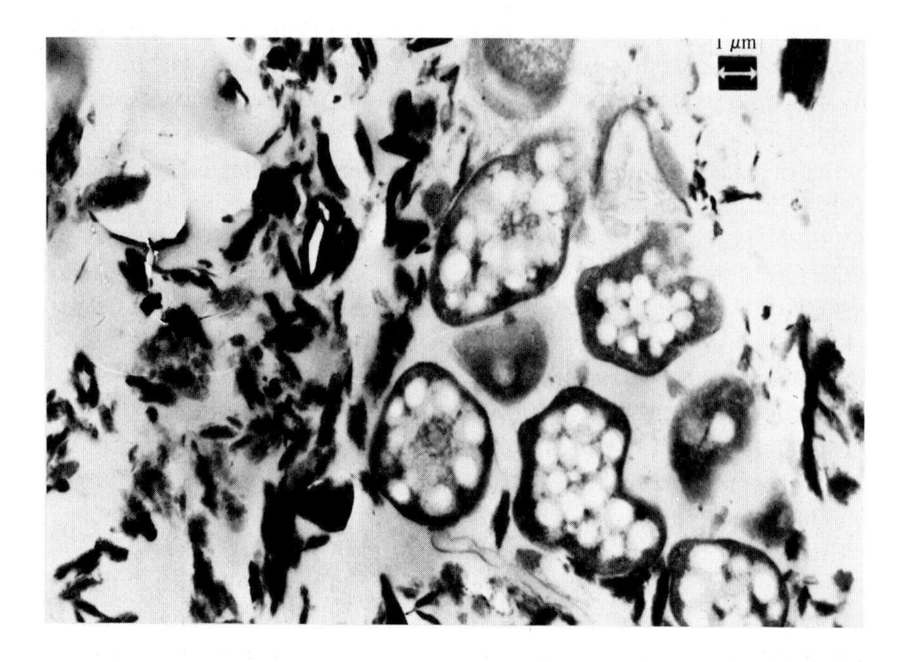

Fig. 3. Electron micrograph of a thin section of a microcolony from coral sediment showing the accumulations of poly-beta-hydroxy alconates (reproduced from D. J. W. Moriarty & A. C. Hayward (1982). Ultrastructure of bacteria and the production of gram negative bacteria in marine sediments. *Microbial Ecology* in press.) The bar represents 0.1 μm.

intracellular AMP pool are by the excretion of AMP or by its hydrolysis to adenosine. Very gentle filtration of exponentially growing *E. coli* did not affect the adenylate energy charge but did increase the adenosine/ATP ratio 10-fold (Davis & White, 1980).

Recent measurements with ^{14}C-adenine have shown that the stress of increasing salinity in an estuarine detrital microbiota induced a rapid excretion of ^{14}C-adenosine and ^{14}C-AMP which corresponded to the amount of ^{14}C-ATP lost. There is a marked increase in extracellular adenosine as stationary growth phase continues in *P. atlantica* monocultures (Uhlinger & White, 1982).

Longer term measurements of prokaryote nutritional status all based on poly-beta-hydroxy buyrate or more accurately multiple polymers of poly-beta-hydroxy alkonates (PHA), (Findlay & White, 1982b) and uronic acid containing extracellular glycocalyx. PHA accumulation has often been detected in microcolonies taken from the environment (Fig. 3). Rapid formation of PHA and a cessation of its turnover was reported in a detrital microbiota when the supply of carbohydrate and oxygen exceeded critical values for nitrogen and phosphate (Nickels *et al.*, 1979). With a proper balance of carbon, nitrogen and phosphate, a rapid turnover of PHA coincided with an increase in the microbial biomass. The hot chloroform-soluble polymer extracted from a sedimentary microbiota or from monocultures can be quantitatively recovered, hydrolyzed and the resulting short chain hydroxy-fatty acids analyzed by GLC (Findlay & White, 1982b). This results in an increase in sensitivity down to 10^{-11} moles hydroxy-fatty acid derived from multiple polymers of different short chain OH-fatty acids each subject to separate metabolic control mechanisms. Aeration of anaerobic sediments increases the PHA from 0.4 μg/μmole phospholipid by about ten times (White *et al.*, 1979b). The sparse ground water aquifer microbiota contains about twenty times more PHA/nmole MA than is found in surface organisms (White *et al.*, 1982a).

Extracellular glycocalyx accumulation follows nutrient deprivation in prokaryotes (Sutherland, 1972, 1977; Costerton *et al.*, 1981). This material stains with ruthenium red (Fig. 1B) due to its uronic acid content. The presence of uronic acids in the extracellular glycocalyx appears to be as characteristic in polymers from outside the cellular cytoplasmic membrane as is the presence of hydroxyproline in collagen (Fazio *et al.*, 1982). Such extracellular polymers can be assayed quantitatively by reducing the uronic acids to alcohols with sodium borodeuteride. Hydrolysis and analysis by GLC-mass spectrometry gives the carbohydrate composition and, by deuterium enrichment, the proportion of uronic acids in the original polymer (Fazio *et al.*, 1982). It is possible to correlate exopolymer formation and nutritional status as reflected by measurements of adenylate energy charge values and its control mechanisms (Uhlinger & White, 1982).

Eukaryotes synthesize endogenous storage lipid triglycerides that are readily lost when fungi, crustaceans, or the detrital microbiota are deprived of food (Gehron & White, 1982a). A convenient measurement of eukaryotic nutritional status is the ratio of triglyceride glycerol to phospholipid. This has been used to define the effects of confinement in aquaria on a natural population. A similar assay for the more slowly metabolized wax esters of marine eukaryotes based on the GLC analysis of fatty alcohols is also under development.

EXPERIMENTAL VALIDATION OF THE ANALYTICAL PROCEDURES

A series of experiments using the scheme shown in Fig. 2 was applied to a range of different microbial assemblies in an attempt to confirm the value of this approach.

Pure culture experiments

Mixtures with different proportions of lyophilized monocultures were analyzed for 'signature' components. As the amount of *Neurospora* was increased in mixtures of lyophilized *Escherichia coli* and *Neurospora crassa*, there were proportional increases in polyenoic and long chain (24 carbon atoms) fatty acids, lipid glucose, lipid serine, lipid inositol, triglyceride glycerol, cell-wall inositol and cell-wall glucosamine concentration. With increasing *E. coli* in the mixture cyclopropane and cis-vaccenic fatty acids, lipid ethanolamine and muramic acid, and the total lipid phosphate values all increased (White *et al.*, 1979a). This type of experiment is a good test of the analytical methodology.

Isolation of organisms from natural ecosystems

A second test is to isolate pure cultures or consortia from the environment and to examine cultures of these for specific 'signature' compounds or for characteristic patterns of the latter. The assumption must be made that growth in isolation induces no critical changes in cell composition compared with growth in nature. This approach has been successfully exploited by R. B. Johns and his colleagues. For example, they have identified a bacterium containing unusual trans monoenoic fatty acids from marine sediments (Gillan *et al.*, 1980). In addition they have reported unusual bacterial polyenoic fatty acids present in a marine *Flexibacter* (Johns & Perry, 1977), as well as characteristic polyenoic fatty acids and neutral lipids from marine diatoms (Volkman *et al.*, 1980). In this laboratory lipids from detrital fungal monocultures were compared to lipids from the detritus itself and similarities and differences reported (White *et al.*, 1979b).

Manipulation of communities

A convincing test of the biochemical methodology applied to a surface microbiota is to relate scanning electron micrographs to the chemical analyses, particularly if the community structure can be manipulated. The rich detrital microbiota that forms on strings of Teflon squares exposed to an estuarine environment constitutes a convenient analytical substrate which can easily be manipulated. In experiments designed to generate a fungal 'heaven' or 'hell' using antibiotics, nutrient supplements and pH changes, morphological differences were demonstrated which were clearly reflected by the chemical composition of lipids, cell walls, nucleotides and by the metabolic activities of microeukaryotes in 'heaven' and prokaryotes in 'hell' (White *et al.*, 1980a). The second type of manipulation involved exposure to light which induced the growth of a sheet of diatoms on the substrate (Bobbie *et al.*, 1981). Once more clear morphological differences were accompanied by a light-induced 2- to 4-fold rise in bacterial and microfaunal constituents, a 10- to 15-fold increase in microfloral-fungal components and a 50- to 100-fold increase in the fatty acid and lipid carbohydrates found in diatoms.

Responses to specific inhibitors

Inhibitors, especially if they are reasonably specific, can be used to manipulate the surface growth. Nutrient supplementation and addition of the inhibitors molybdate and chloroform altered the activities of sulfate reducing and methane generating organisms responsible for the redox reactions in anaerobic microbial consortia (Oremland & Taylor, 1978). Inhibition or stimulation of sulfate reducing activities in sediments is reflected in the recovery of certain extractable, ester-linked, phospholipid hydroxy-fatty acids (Fredrickson, 1981).

Alterations due to specific predation

A convincing confirmation of the value of the analytical scheme shown in Fig. 2 is the effects of predation on a microbiota due to grazers having a recognized specificity. The sand dollar, *Mellita quinquiesperforata*, selects a specific fraction of the sediment including non-photosynthetic microeukaryotes for its diet (Lane, 1977).

Viable diatoms have been detected in its faeces. Comparison of a sedimentary microbiota before and after sand dollar predation showed that components characteristic of bacteria or photosynthetic microeukaryotes were not affected. Sand dollar grazing produced a drastic reduction in foraminifera (White *et al.*, 1980b; Findlay, 1981). Grazing of detrital microbiota by the relatively non-specific gammaridean amphipods increased the proportion of bacteria and their activities in the microbiota (Morrison & White, 1980). In contrast, the *Mellita* species with highly developed mouth parts selectively grazed diatoms. This is reflected in the morphology of the residual community and in the biochemical assays (Smith *et al.*, 1982a).

SOME APPLICATIONS OF CHEMICAL ANALYSIS

Sampling

To understand the role of microbes in the ecology of sediments, many samples must be examined. Often it is not possible to process these samples in the field. It was decided to investigate the effects of sample preservation and storage on biomass and community composition (Federle & White, 1982). It is often necessary to remove for further examination, by sieving sediment samples through 500 nm mesh screens with 2.5% saline. This yields a dilute sediment suspension which requires a larger volume of methanol and chloroform for field extraction. If available, centrifugation at 4 °C can be used to reduce the extraction volumes. If sediments are held for 5 days before sieving and extraction at 4 °C or −70 °C, phospholipid losses are 20% and 40% respectively. Refrigeration or freezing leads to a fall in many polyenoic fatty acids and the addition of methanol:chloroform mixtures or replacing the air with argon did not prevent this loss. Storage of sieved samples in 10% formalin led to a rise in phospholipid values and increases in polyenoic fatty acids due to lysis of macroinvertebrate organisms. The most satisfactory preservation of both phospholipids and polyenoic fatty acids was by sieving in the field directly into 10% formalin, although this still caused a 17% loss in phospholipid content. No changes in fatty acid composition were seen when sieved and field extracted controls were compared.

Successions in surface colonization

Microbial successions can be demonstrated on the surface of oak leaves exposed to an estuarine environment (Morrison *et al.*, 1977). Oxygen utilization rates and phosphodiesterase activity rise to their highest levels within 2 weeks and remain relatively constant. Galactosidase and glucosidase activities and ^{14}C-acetate incorporation into lipids show a steady increase in activity. The heterotrophic potential, alkaline phosphatase activity and phospholipid synthesis reach a maximum at the fourth week and then decline. Initial colonization is by organisms with a high ratio of muramic acid to ATP, however these are succeeded by organisms with a progressively lower ratio. Scanning electron microscopy, shows that the earliest colonisers with high muramic acid to ATP values are mainly bacteria. After 6 weeks diatoms and fungal mycelia that contain ATP but no muramic acid, dominate the population. A similar succession occurs on metal surfaces exposed to rapidly flowing seawater. An increase in heat transfer resistance across the metal surfaces follows the typical exponential kinetics of a bacterial growth curve and also parallels the observed increase in microbial biomass (Nickels *et al.*, 1981a). The community measured by the fatty acid composition of extracted microbial lipids and also by scanning electron microscope morphology shows early bacterial colonization followed by a complex community of bacteria and microeukaryotes.

Effect of the chemical nature of the surface

There are differences in the morphology of microbial film attracted to aluminum, titanium, stainless steel, and copper surfaces from seawater (Berk *et al.*, 1981). The particular succession observed depends more on the chemical nature of the surface than on its texture. Titanium surfaces rapidly accumulate a complex film with greater numbers of bacteria and microeukaryotes than does aluminium exposed to the same stream of rapidly flowing seawater. Copper stimulates the formation of an extracellular glycocalyx in which the microbes proliferate. Titanium attracts organisms of a highly unusual morphology that are extremely rare in seawater.

The film that accumulates rapidly on titanium has a lower heat flow resistance than does the film that forms on aluminum: it is also more readily removed by simple manual brushing (Nickels *et al.*,

1981a). The film forming on aluminum surfaces contains a much greater proportion of bacteria forming an extracellular glycocalyx than does the film appearing on titanium. The degree of corrosion correlates closely with the microbial biomass on aluminum but not on titanium surface.

A comparison of the microbiota colonizing azoic marine sands with the same sand to which sterile bentonite clay had been added after an 8 week exposure to flowing unfiltered pelagic seawater showed stimulation of bacterial growth (increased short branched and cis-vaccenic fatty acids) accompanied by increases in the microfauna (increases in the polyenoic fatty acids) (Smith *et al.*, 1982a). Similar experiments with barite (barium sulfate), a major component of oil and gas well drilling fluids, induced no significant change in biomass or community composition.

Effects of biodegradability of the surface on the microbial film

The microbiota colonizing the surface of non-degradable polyvinylchloride Christmas tree fragments contained one-half to one-tenth the amount of biomass measured as ATP, oxygen utilization, alkaline phosphatase activity, phosphodiesterase activity and muramic acid than that found on flash pine needles (*Pinus elliotti*) during a 14 week incubation period in the estuary (Bobbie *et al.*, 1978). The community on the biodegradable substrate was also morphologically and biochemically more diverse. The glycolytic enzyme activities associated with polysaccharide degradation were not found in the microbiota growing on the artificial substrate. The community on the artificial substrate was enriched in bacteria and was more attractive to gammaridean amphipods than the more diverse community from the biodegradable substrate.

Effect of differences in microtopology of the same surface

The microtopography of silica surfaces of the same grain size and pore space markedly affects the microbial surface film (Nickels *et al.*, 1981b). Loss of microdiscontinuities in the surface decreases the microbial biomass measured as lipid phosphate and total palmitic acid, decreases the microflora measured as the alpha linolenic series of polyunsaturated fatty acids but increases the grazing microfauna estimated as the gamma linolenic series of polyunsaturated fatty acids.

Effects of shifts in nutrient composition and light

Manipulation of the microbiota by exposing it to different nutrients in seawater induced changes in its morphology as detected by scanning electron microscopy. Biochemical analyses for biomass, community composition, and metabolic activity also changed significantly (White *et al.*, 1980a; Bobbie & White, 1980). These experiments were used as a test of the validity of the methodology.

The effect of light on the detrital microbiota exposed to the same flowing seawater system was to cause different rates of increase in a range of biochemical signature compounds (Bobbie *et al.*, 1981). Smallest increases were seen in the markers for bacteria and the microfauna. There were intermediate increases in the microfloral-fungal signatures, whilst the greatest increases were in compounds associated with diatoms. This experiment was also used to help validate the signature concept.

Effects of mechanical disruption

Manual brushing of the microfouling films forming on titanium and aluminum in rapidly flowing seawater removed most of the microbiota but left a residual film that stimulated even more rapid recolonization when they were returned to the seawater stream (Nickels *et al.*, 1981a). The residual communities retained on titanium or aluminum after brushing were different in detailed community composition but both were enriched in bacteria compared to the communities present before brushing. The community remaining on aluminum was more resistant to removal and was enriched in bacteria forming cyclopropane fatty acids compared to the unbrushed microbiota. These bacteria form filaments that were seen by scanning electron microscopy to penetrate in and under the corrosion gel. These communities were associated with twice the extracellular glycocalyx estimated as the ratio of total organic carbon to palmitic acid compared with the unbrushed community.

The application of continuous mechanical brushing accentuated differences between the microbial films growing on titanium and aluminum in running seawater (Nickels *et al.*, 1981c). Continuous brushing induced the formation of a film having greater heat flow resistance compared with the unbrushed controls. The titanium surface film maintained a low heat flow resistance as well as a low extracellular glycocalyx content with continuous brushing but showed a diverse cellular biomass made up of bacteria and microeukaryotes an order of magnitude more than that appearing on the aluminum. This conclusion is based on the analysis of lipid extractable ester-linked palmitic acid. The film that formed on aluminum with continuous brushing contained about 35 times more extracellular glycocalyx than did a similar film on titanium. Films on titanium and aluminum had quite different compositions. For example the ratio of galactose to galacturonic acid in the glycocalyx was 3.5 in the film from titanium but only 0.85 in that from aluminum (Fazio *et al.*, 1982).

Marine sands dried in the sun for several weeks were placed in plexiglas troughs at the site of origin of the sand in 32 meters of seawater. Similar trays containing the same sand were placed on a platform above the water over the sampling site. Both sets of sand were exposed to 200 ml min^{-1} unfiltered seawater obtained from a depth of 26 metres. This corresponded to 0.1 volumes of water $(\text{volume of sediment})^{-1}$ min^{-1}. The experiment continued for a period of 8 weeks with both samples being exposed to light of the same intensity and spectral quality as that reaching seafloor sediments. Sediment samples from both sets of trays were compared with the cores taken from the site (Nickels *et al.*, 1981a). The biomass, and both bacterial and macrobiotic community composition, were greater and more diverse in the systems placed under water than in the surface troughs. Both surface and underwater trough sediments had a lower total biomass and a less diverse community structure than that detected in sedimentary cores from the same area.

Effects of exposure to xenobiotic compounds

Azoic marine sands exposed as described above showed sensitive responses to xenobiotics in the $\mu g\ l^{-1}$ range (Smith *et al.*, 1982a). Paraformaldehyde infusions at 1 $\mu g\ l^{-1}$ caused an increase in the cellular biomass (lipid phosphate) but depressed the numbers of microeukaryotes (detected in the extracted lipids as linoleic, oleic, dodecanoic and polyenoic fatty acids with chain lengths longer than 20 carbon atoms); dichlorophenols at 4 $\mu g\ l^{-1}$ stimulated bacterial biomass having short branched and cis-vaccenic fatty acids in their lipids as well as microeukaryotes with polyenoic fatty acids having more than 20 carbon atoms; pentachlorophenol increased bacteria using the anaerobic fatty acid desaturating pathway and therefore having cis-vaccenic acid in their extractable lipids but depressed the eukaryotic microflora containing $\omega 3$ polyenoic fatty acids.

The application of biochemical analyses to the reef building coral *Montastrea annularus*, showed dose–response related effects to the exposure to oil and gas well drilling muds in the part per million range (White *et al.*, 1982b). Exposure induced the following changes: shifts in the free amino acid pools, a drop in phospholipid content with a shift from saturated to polyunsaturated fatty acids, a fall in triglycerides and an increase in bacterial contamination indicated by an increase in cis-vaccenic acid in the lipids.

Effects of predation

Grazing amphipods remove the detrital microbiota without affecting the delicate morphology of the ventral surface of oak leaves (Morrison & White, 1980). Different detrital surfaces consisting of mixed hardwood leaves collected from flood plains, *Vallisneria* blades, and polyethylene films shaped like hardwood leaves were incubated in litter baskets at different stations in an estuary. The respiratory activity, total phospholipid and PHB of the microbiota correlated with the nature of the surface and not with its position in the estuary (White *et al.*, 1979d). The microbiota attracted to the litter baskets correlated with the station and not with gross estimates of microbial biomass. Amphipod distribution however, showed a significant correlation with concentrations of some bacterial fatty acids in the detrital microbiota. Amphipod grazing of a detrital microbiota prelabeled with ^{14}C showed a fall in ^{14}C associated with the detritus, and an increase in ^{14}C associated with the water column, amphipod fecal pellets and specific lipids of the amphipods rose with increasing grazing density (Morrison, 1980). Amphipod grazing at natural densities caused a rise in the microbial biomass (measured as ATP and lipid phosphate), and increased rates of oxygen utilization, PHB synthesis, lipid synthesis, and $^{14}CO_2$ release from simple substrates by detrital microorganisms (Morrison & White, 1980). Alkaline phosphatase activity decreases with grazing suggesting increased availability of inorganic phosphate. The community structure shifted from a diatom-fungal-bacterial community to one dominated by bacteria. Using a pulse-chase labeling technique, the metabolically stable glycerol phosphoryl choline derived from the phosphatidyl choline of the detrital microbiota disappeared and the metabolically active glycerol phosphorylglycerol derived from phosphatidyl glycerol showed a more rapid turnover rate in the grazed population.

In experiments with J. R. Robertson of the University of Georgia, Sapelo Island Marine Institute, exclusion of snails from tidal mud flats for 24 days led to the conclusion that deposit feeding snails normally reduce the microbial biomass and decrease the relative proportions of bacterial branched pentadecanoic fatty acids by half. In addition they cause a ten fold drop in cyclopropane fatty acids in the sedimentary horizon which is 6 cm deep; however they do not affect the tidally disrupted surface layer. Cyclopropane fatty acids tend to accumulate in older non-growing bacterial cultures (Knivett & Cullen, 1965) adding weight to the suggestion that grazing stimulated microbial growth.

Using these quantitative biochemical methods it was possible to explain how two sympatric amphipods with different mouthparts could divide the detrital microbiota (Smith *et al.*, 1982b). *Melita appendiculata* grazes the non-photosynthetic microeukaryotes allowing a preferential rise in bacteria and diatoms. *Gammarus mucronatus* is less fastidious, reflecting the possession of a less specialized feeding apparatus. It reduces the population of both the bacteria (MA) and the microeukaryotes (ratio of lipid glycerol to lipid phosphate). *Gammarus* feeding stimulated growth and metabolic activity from a smaller microbial population. This was shown by higher total adenosine nucleotides, ATP, adenylate energy charge, and the ratio of adenosine to ATP.

Grazing by either amphipod significantly increases the bacterial and microeukaryotec biomass and their metabolic activities. Grazing at 5 times the usual field density further increased the bacterial and diatom biomass at the expense of slower growing algae and fungi. Ungrazed detrital microbiota showed metabolic stress as evidenced by elevated adenosine to ATP ratios.

It is clear from these and unpublished observations with both detrital and sedimentary communities that any study which excludes natural grazing will lead to considerable distortions in the analysis.

Description of a new microbiota

Application of the analytical methods discussed in this paper to ground water aquifer sediments, recovered with extreme care to prevent contamination with surface organisms, has revealed the existence of a unique microbial community (White *et al.*, 1982a).

The microbial biomass assessed by MA, LPS, total extractable fatty acids, and phospholipid determination was a hundred times sparser than communities located at the surface. There was no evidence of microeukaryotes in this habitat. The microbiota was enriched in extractable ester linked, phospholipid hydroxy-fatty acids characteristic of the sulfate reducing bacteria and in plasmalogen derived fatty aldehydes characteristic of fermentative anaerobic heterotrophs. The microbiota contained a higher proportion of gram positive bacteria than that found at the surface. A higher proportion of linoleic and a lower proportion of cis-vaccenic fatty acids were present in the extractable phospholipids than normally seen in surface organisms from estuarine or deep sea sediments. The microbiota are nutritionally stressed as is clear from the 20-fold increase in PHA to MA ratio over that seen in the surface microbiota.

CONCLUSIONS

In most environments microbes are concentrated on surfaces where they exist as multicomponent prokaryotic and microeukaryotic assemblies which have been shown by electron microscopy to be morphologically diverse. Methods that distort or destroy interactions between members of these complex microbial communities have not yielded much insight into their dynamics. Numerical estimates based on isolation and viable count determination consistently underestimate the total amount of biomass present.

At least two major parameters are needed to define the ecology of these microbial assemblies. These are the microbial biomass, which forms the basis of natural food chains, and the metabolic activity, which is responsible for many of the biogeochemical processes vital to the operation of the biosphere. Non-selective techniques for determining biomass rely on the measurement of components found in all living cells. To gain an insight into community structure, assays for cellular components found in specific subsets of the general population are needed. These are 'signature' components of specific microbial groups.

Measurements of microbial activity generally involve the incorporation or metabolism of a precursor compound. The metabolic activity of the community could be altered during manipulation of the system needed to expose it to the labeled precursor. If an estimate of the nutritional status of some subsets of the community is made at the time of exposure to the labeled precursors, the possible consequences of any disturbances can be assessed.

The biochemical assays described in this study estimate the microbial biomass without any need for microbial growth or for the quantitative recovery of microbes from surfaces on which they grow. These methods throw light on community structure by analysing the 'signatures' of its various subsets. By correlating measurements of nutritional status with metabolic activity this type of biochemical analysis can start to unravel problems associated with the accurate estimation of true *in situ* microbial activity.

The application of sensitive analytical methods to natural microbial communities is only just beginning. As methods improve both in sensitivity and in selectivity and the catalogues of 'signature' lipids increase, a deeper insight into the complex microbial worlds that drive the biogeochemical cycles of this planet will be possible. With this insight, the system that spawned the biosphere and maintains it today will appear ever more beautiful.

ACKNOWLEDGMENTS

This work would have been impossible without the dedicated collaboration of S. J. Morrison, J. D. King, J. S. Herron, C. W. Taylor, J. S. Nickels, J. H. Parker, S. D. Fazio, H. L. Fredrickson, R. F. Martz, G. A. Smith, R. H. Findlay, W. M. Davis, D. J. Uhlinger, T. W. Federle, M. J. Gehron, R. J. Bobbie and W. I. Miller III of the Florida Sate University, Department of Biological Science. Thanks are due also to D. J. W. Moriarty and A. C. Hayward of CSIRO, Cleveland and the University of Queensland, St Lucia respectively in Queensland, Australia for their electron micrographs of microbial consortia.

This work was supported by contract 31-109-38-4502 from the Department of Energy (Argonne National Laboratory, Argonne, IL), contract N00014-75-C0201 from the Department of the Navy, Office of Naval Research, Ocean Science and Technology Detachment (NSTL Station, MS) grants R-0806143010 and R-0806143020 from the US Environmental Protection Agency, Gulf Breeze, FL, grants OCE 80-19757 and DEB-7818401 from the Biological Oceanography Program of the National Science Foundation, grant NAG2-149 from the Advanced Life Support Office, National

Aeronautics and Space Administration, and grants NA-81AA-D-00091 and 04-7-158-4406 from the National Oceanic and Atmospheric Administration Office of Sea Grant.

REFERENCES

Berk, S. G., Mitchell, R., Bobbie, R. J., Nickels, J. S. & White, D. C. (1981). Microfouling on metal surfaces exposed to seawater. *International Biodeterioration Bulletin,* **17,** 29–37.

Bitton, G. & Marshall, K. C. (1980). *Adsorption of Microorganisms to Surfaces.* New York: J. Wiley & Co.

Bligh, E. G. & Dyer, W. J. (1959). A rapid method of lipid extraction and purification. *Canadian Journal of Biochemistry and Physiology,* **35,** 911–17.

Bobbie, R. J., Morrison, S. J. & White, D. C. (1978). Effects of substrate biodegradability on the mass and activity of the associated estuarine microbiota. *Applied and Environmental Microbiology,* **35,** 179–84.

Bobbie, R. J., Nickels, J. S., Smith, G. A., Fazio, S. D., Findlay, R. H., Davis, W. M. & White, D. C. (1981). Effect of light on the biomass and community structure of the estuarine detrital microbiota. *Applied and Environmental Microbiology*, **42,** 150–8.

Bobbie, R. J. & White, D. C. (1980). Characterization of benthic microbial community structure by high resolution gas chromatography of fatty acid methyl esters. *Applied and Environmental Microbiology,* **39,** 1212–22.

Boon, J. J., deLeeuw, J. W., van de Hoek, G. J. & Vosjan, J. H. (1977). Significance and taxonomic value of iso and anteiso monoenoic fatty acids and branched beta-hydroxy acids in *Desufovibrio desulfuricans. Journal of Bacteriology,* **129,** 1183–91.

Butkevich, V. S. (1932). Zur Methodik der Bakteriologischen Meeresunterstachungen und einige Angaben uber die Verteilung der Bakterien im Wasser und in dem Buden des Barents Meeres. *Transactions of the Oceanographic Institute, Moscow,* **2,** 5–39.

Collins, V. G. & Kipling, C. (1957). The enumeration of waterborne bacteria by a new direct count method. *Journal of Applied Bacteriology,* **20,** 257–64.

Costerton, J. W., Irvin, R. T. & Cheng, K.-J. (1981). The bacterial glycocalyx in nature and disease. *Annual Review of Microbiology,* **35,** 299–324.

Daley, R. J. & Hobbie, J. E. (1975). Direct counts of aquatic bacteria by a modified epifluorescence technique. *Limnology and Oceanography*, **20,** 875–82.

Davis, W. M. & White, D. C. (1980). Fluorometric determination of adenosine nucleotide derivatives as measures of the microfouling, detrital and sedimentary microbial biomass and physiological status. *Applied and Environmental Microbiology,* **40,** 539–48.

Ellwood, D. C., Melling, J. & Rutter, P. (eds.) (1979). *Adhesion of Microorganisms to Surfaces*. New York: Academic Press.

Erwin, J. A. (1973). Fatty acids in eukaryotic microorganisms. In *Lipids and Biomembranes of Eukaryotic Microorganisms,* pp. 41–143. New York: Academic Press.

Fazio, S. A., Uhlinger, D. J., Parker, J. H. & White, D. C. (1982). Estimations of uronic acids as quantitative measures of extracellular polysaccharide and cell wall polymers from environmental samples. *Applied and Environmental Microbiology,* **43** (in press).

FAZIO, S. D., MAYBERRY, W. R. & WHITE, D. C. (1979). Muramic acid assay in sediments. *Applied and Environmental Microbiology,* **38,** 349–50.

FEDERLE, T. W. & WHITE, D. C. (1982). Preservation for biochemical analysis of the biomass and community composition of the microbiota. *Applied and Environmental Microbiology,* **44** (in press).

FILIP, Z. (1978). Effect of solid particles on growth and metabolic activities of microorganisms. In *Microbial Ecology,* ed. M. W. Loutit & J. A. R. Miles, pp. 102–4. New York: Springer-Verlag.

FINDLAY, R. H. (1981). The effect of the sand dollar *Mellita quinquiesperforata* on the benthic microbial community. Thesis, Florida State University, Tallahassee, Florida.

FINDLAY, R. H., MORIARTY, D. J. W. & WHITE, D. C. (1982). Improved method of determining muramic acid from environmental samples. *Geomicrobiology Journal,* **4** (in press).

FINDLAY, R. H. & WHITE, D. C. (1982a). The effects of the sand dollar *Mellita quinquiesperforata* on the benthic microbial community. *Journal of Experimental Marine Biology and Ecology,* **25** (in press).

FINDLAY, R. H. & WHITE, D. C. (1982b). Assay of polymeric beta-hydroxy alkylonates from environmental samples. *Applied and Environmental Microbiology,* **44** (in press).

FREDRICKSON, H. L. (1981). Lipid characterization of sedimentary sulfate-reducing communities. *Abstracts of American Society of Microbiology,* **1981,** 205.

GEHRON, M. J., MORIARTY, D. J. W., SMITH, G. A. & WHITE, D. C. (1982). Determination of the gram negative gram positive bacterial content of sediments. *Applied and Environmental Microbiology,* **44** (in press).

GEHRON, M. J. & WHITE, D. C. (1982a). Quantitative determination of the nutritional status of detrital microbiota and the grazing fauna by triglyceride glycerol analysis. *Journal of Experimental Marine Biology and Ecology,* **25** (in press).

GEHRON, M. J. & WHITE, D. C. (1982b). Sensitive measurements of phospholipid in environmental samples. *Applied and Environmental Microbiology,* **44** (in press).

GIBBONS, R. J. & VAN HOUTE, J. (1975). Bacterial adherence in oral microbial ecology. *Annual Review of Microbiology,* **29,** 19–44.

GILLAN, F. T., JOHNS, R. B., VERHEGEN, T. V., VOLKMAN, J. K. & BAVOR, H. J. (1980). Trans-monounsaturated acids in a marine bacterial isolate. *Applied and Environmental Microbiology,* **41,** 849–56.

GOLDFINE, H. & HAGEN, P. O. (1972). Bacterial plasmalogens. In *Ether Lipids: Chemistry and Biology,* ed. F. Snyder, pp. 329–50. New York: Academic Press.

HOLM-HANSEN, O. & BOOTH, C. R. (1966). The measurement of adenosine triphosphate in the ocean and its ecological significance. *Limnology and Oceanography,* **11,** 510–19.

JANNASCH, H. W. & JONES, G. E. (1959). Bacterial populations in seawater as determined by different methods of enumeration. *Limnology and Oceanography,* **4,** 128–39.

JANNASCH, H. W. & PRITCHARD, P. H. (1972). The role of inert particulate matter in the activity of aquatic microorganisms. *Memorie dell 'Instituto Italiano di Idrobiologia,* **29,** 289–308.

JOHNS, R. B. & PERRY, G. T. (1977). Lipids of the marine bacterium *Flexibacter polymorphus. Archives of Microbiology,* **14,** 267–71.

KAMIO, Y., KANEGASAKI, S. & TAKAHASHI, H. (1969). Occurrence of plasmalogens in anaerobic bacteria. *Journal of General Microbiology,* **15,** 439–51.

KARLSSON, K. A. (1970). Sphingolipid long chain bases. *Lipids,* **5,** 878–91.

KATES, M. (1964). Bacterial lipids. *Advances in Lipid Research,* **2,** 17–90.
KING, J. D. & WHITE, D. C. (1977). Muramic acid as a measure of microbial biomass in estuarine and marine samples. *Applied and Environmental Microbiology,* **33,** 777–83.
KING, J. D., WHITE, D. C. & TAYLOR, C. W. (1977). Use of lipid composition and metabolism to examine structure and activity of estuarine detrital microflora. *Applied and Environmental Microbiology,* **33,** 1177–83.
KITTREDGE J. S. & ROBERTS, E. (1969). A carbon phosphorous bond in nature. *Science,* **164,** 27–42.
KNIVETT, V. A. & CULLEN, J. (1965). Some factors affecting cyclopropane acid formation in *Escherichia coli. Biochemical Journal,* **96,** 771–6.
LABACH, J. P. & WHITE, D. C. (1969). Identification of ceramide phosphorylethanolamine and ceramide phosphorylglycerol in the lipids of an anaerobic bacterium. *Journal of Lipid Research,* **10,** 528–34.
LANE, J. M. (1977). Bioenergetics of the sand dollar *Mellita quinquiesperforata* (Leske 1778). Thesis, University of South Florida, Tampa, Florida.
LOVLEY, D. R. & KLUG, M. J. (1982). Intermediary metabolism of organic matter in the sediments of a eutrophic lake. *Applied and Environmental Microbiology,* **43,** 552–60.
MALINS, D. C. & VARANASI, U. (1972). The ether bond in marine lipids. In *Ether Lipids: Chemistry and Biology*, ed. F. Snyder, pp. 313–20. New York: Academic Press.
MANGOLD, H. K. (1972). The search for alkoxylipids in plants. In *Ether Lipids: Chemistry and Biology*, ed. F. Snyder, pp. 297–312. New York: Academic Press.
MARSHALL, K. C. (1976). *Interfaces in Microbial Ecology*. Cambridge, Massachusetts: Harvard University.
MARTZ, R. F., FINDLAY, R. H. & WHITE, D. C. (1982). Measurement of the biomass of methane forming bacteria in environmental samples. *Applied and Environmental Microbiology,* **44** (in press).
MILLAR, W. N. & CASIDA, L. E. (1970). Evidence for muramic acid in the soil. *Canadian Journal of Microbiology,* **18,** 299–304.
MORIARTY, D. J. W. (1977). Improved method using muramic acid to estimate biomass of bacteria in sediments. *Oecologia,* **26,** 317–23.
MORIARTY, D. J. W. (1980). Problems in the measurement of bacterial biomass in sandy sediments. In *Biogeochemistry of Ancient and Modern Sediments*, ed. P. A. Trudinger, M. R. Walter & B. J. Ralph, pp. 131–39. Canberra, Australia: Australian Acadamy of Science.
MORIARTY, D. J. W. & POLLARD, P. C. (1981). DNA synthesis as a measure of bacterial growth rates in seagrass sediments. *Marine Ecology Progress Series,* **5,** 151–6.
MORRISON, S. J. (1980). Trophic interactions between detrital microbiota and detritus-feeding estuarine gammaridean amphipods. Thesis, Florida State University, Tallahassee, Florida.
MORRISON, S. J. & WHITE, D. C. (1980). Effects of grazing by estuarine gammaridean amphipods on the microbiota of allochthonous detritus. *Applied and Environmental Microbiology,* **40,** 659–71.
MORRISON, S. J., KING, J. D., BOBBIE, R. J., BECHTOLD, R. E. & WHITE, D. C. (1977). Evidence of microfloral succession on allochthonous plant litter in Apalachicola Bay, Florida, U.S.A. *Marine Biology,* **41,** 229–40.
NEWMAN, H. N. (1980). Retention of bacteria in oral surfaces. In *Adsorption of Microorganisms to Surfaces*, ed. G. Bitton & K. C. Marshall, pp. 207–051. New York: John Wiley & Co.
NICKELS, J. S., KING, J. D. & WHITE, D. C. (1979). Poly-beta-hydroxybutyrate

metabolism as a measure of unbalanced growth of the estuarine detrital microbiota. *Applied and Environmental Microbiology,* **37,** 459–65.

Nickels, J. S., Bobbie, R. J., Lott, D. F., Martz, R. F., Benson, P. H. & White, D. C. (1981a). Effect of manual brush cleaning on the biomass and community structure of the microfouling film formed on aluminum and titanium surfaces exposed to rapidly flowing seawater. *Applied and Environmental Microbiology,* **41,** 1442–53.

Nickels, J. S., Bobbie, R. J., Martz, R. F., Smith, G. A., White, D. C. & Richards, N. L. (1981b). Effect of silicate grain shape, structure and location on the biomass and community structure of colonizing marine microbiota. *Applied and Environmental Microbiology,* **41,** 1262–8.

Nickels, J. S., Parker, J. H., Bobbie, R. J., Martz, R. F., Lott, D. F., Benson, P. H. & White, D. C. (1981c). Effect of cleaning with flow-driven brushes on the biomass and community composition of the marine microfouling film on aluminum and titanium surfaces. *International Biodeterioration Bulletin,* **17,** 87–94.

Oremland, R. S. & Taylor, B. F. (1978). Sulfate reduction and methanogenesis in marine sediments. *Geochimica et Cosmochimica Acta,* **42,** 209–14.

Paerl, H. W. (1980). Attachment of microorganisms to living and detrital surfaces in freshwater systems. In *Adsorption of Microorganisms to Surfaces*, ed. G. Bitton & K. C. Marshall, pp. 375–402. New York: John Wiley & Co.

Parker, J. H., Smith, G. A., Fredrickson, H. L., Vestal, J. R. & White, D. C. (1982). A sensitive assay for gram negative bacteria in sediments based on hydroxy-fatty acids from lipopolysaccharide lipid A. *Applied and Environmental Microbiology,* **44** (in press).

Perfil'ev, R. V. & Gabe, D. R. (1969). *Capillary Methods of Investigating Microorganisms.* Toronto: University of Toronto Press.

Perry, G. J., Volkman, J. K., Johns, R. B. & Bavor, H. J. (1979). Fatty acids of bacterial origin in contemporary marine sediments. *Geochimica et Cosmochimica Acta*, **43,** 1715–25.

Rizza, B., Tucker, A. N. & White, D. C. (1970). Lipids of *Bacteroides melaninogenicus. Journal of Bacteriology*, **101,** 84–91.

Rutter, P. (1979). The accumulation of organisms on teeth. In *Adhesion of Microorganisms to Surfaces*, ed. D. C. Elwood, J. Melling, & P. Rutter, pp. 137–64. New York: Academic Press.

Saddler, J. N. & Wardlaw, A. C. (1980). Extraction, distribution and biodegradation of bacterial lipopolysaccharides in estuarine sediments. *Antonie van Leeuwenhoek Journal of Microbiology*, **46,** 27–39.

Sieburth, J. McN. (1975). *Microbial Seascapes*. Baltimore: University Park Press.

Smith, G. A., Nickels, J. S., Bobbie, R. J., Richards, N. L. & White, D. C. (1982a). Effects of oil and gas well drilling fluids on the biomass and community structure of the microbiota that colonize sands in running seawater. *Archives of Environmental Contamination and Toxicology*, **11,** 19–23.

Smith, G. A., Nickels, J. S., Davis, W. M., Martz, R. F., Findlay, R. H. & White, D. C. (1982b). Perturbations of the biomass, metabolic activity, and community structure of the estuarine detrital microbiota: resource partitioning by amphipod grazing. *Journal of Experimental Marine Biology and Ecology*, **25,** (in press).

Sutherland, I. W. (1972). Bacterial exopolysaccharides. *Advances in Microbial Physiology*, **8,** 143–214

Sutherland, I. W. (1977). Bacterial expolysaccharides – their nature and production. In *Surface Carbohydrates of Prokaryotic Cell*, ed. I. W. Sutherland, pp. 27–96. New York: Academic Press.

THOMPSON, G. A. (1972). Ether-linked lipids in molluscs. In *Ether Lipids: Chemistry and Biology*, ed. F. Snyder, pp. 321–40. New York: Academic Press.

TORNABENE, T. G. & LANGWORTHY, T. A. (1979). Diphytanyl and tetraphytanyl glycerol ether lipids of methanogenic Archaebacteria. *Science,* **203,** 51–3.

UHLINGER, D. J. & WHITE, D. C. (1982). Correlation between extracellular polymer production and physiological status in a marine Pseudomonad. *Abstracts of the American Society for Microbiology,* **1982,** 222.

VOLKMAN, J. K., JOHNS, R. B., GILLAN, F. T. & PERRY, G. J. (1980). Microbial lipids of an intertidal sediment. I. Fatty acids and hydrocarbons. *Geochimica et Cosmochimica Acta,* **44,** 1133–43.

WARD, T. E. & FREA, J. I. (1979). Determining the sediment distribution of methanogenic bacteria by direct fluorescent antibody methodology. In *Methodology for Biomass Determinations and Microbial Activities in Sediments ASTM STP 673*, ed. C. D. Litchfield & P. L. Seyfried, pp. 75–86. Philadelphia: American Society for Testing Materials.

WHITE, D. C. & TUCKER, A. N. (1969). Phospholipid metabolism during bacterial growth. *Journal of Lipid Research,* **10,** 220–33.

WHITE, D. C., TUCKER, A. N. & SWEELEY, C. C. (1969). Characterization of the *iso*-branched sphinganines from the ceramide phospholipids of *Bacteroides melaninogenicus. Biochimica et Biophysica Acta,* **187,** 527–32.

WHITE, D. C. & TUCKER, A. N. (1970). Ceramide phosphoryl-glycerol phosphate, a new sphingolipid found in bacteria. *Lipids,* **5,** 56–62.

WHITE, D. C., BOBBIE, R. J., MORRISON, S. J., OOSTERHOF, D. K., TAYLOR, C. W. & MEETER, D. A. (1977). Determination of microbial activity of estuarine detritus by relative rates of lipid biosynthesis. *Limnology and Oceanography,* **22,** 1089–99.

WHITE, D. C., BOBBIE, R. J., HERRON, J. S., KING, J. D. & MORRISON, S. J. (1979a). Biochemical measurements of microbial mass and activity from environmental samples. In *Native Aquatic Bacteria: Enumeration, Activity and Ecology ASTM STP 695*, ed. J. W. Costerton & R. R. Colwell, pp. 69–81. Philadelphia: American Society for Testing and Materials.

WHITE, D. C., BOBBIE, R. J., KING, J. D., NICKELS, J. S. & AMOE, P. (1979b). Lipid analysis of sediments for microbial biomass and community structure. In *Methodology for Biomass Determinations and Microbial Activities in Sediments, ASTM STP 673*, ed. C. D. Litchfield & P. L. Seyfried, pp. 87–103. Philadelphia: American Society for Testing and Materials.

WHITE, D. C., DAVIS, W. M., NICKELS, J. S., KING, J. D. & BOBBIE, R. J. (1979c). Determination of the sedimentary microbial biomass by extractible lipid phosphate. *Oecologia,* **40,** 51–62.

WHITE, D. C., LIVINGSTON, R. J., BOBBIE, R. J. & NICKELS, J. S. (1979d). Effects of surface composition, water column chemistry and time of exposure on the composition of the detrital microflora and associated macrofauna in Apalachicola Bay, Florida. In *Ecological Processes in Coastal and Marine Systems*, ed. R. J. Livingston, pp. 83–116. New York: Plenum Press.

WHITE, D. C., BOBBIE, R. J. NICKELS, J. S., FAZIO, S. D. & DAVIS, W. M. (1980a). Nonselective biochemical methods for the determination of fungal mass and community structure in estuarine detrital microflora. *Botanica Marina,* **23,** 239–50.

WHITE, D. C., FINDLAY, R. H., FAZIO, S. D., BOBBIE, R. J., NICKELS, J. S., DAVIS, W. M., SMITH, G. A. & MARTZ, R. F. (1980b). Effects of bioturbation and predation by *Mellita quinquiesperforata* on the sedimentary microbial community structure. In *Estuarine Perspectives*, ed. V. S. Kennedy, pp. 163–71. New York: Academic Press.

WHITE, D. C., NICKELS, J. S., GEHRON, M. J., PARKER, J. H., MARTZ, R. F. & RICHARDS, N. L. (1982a). Biochemical measures of coral metabolic activity, nutritional status, and microbial infection with exposure to oil and gas well drilling fluids. In *Wastes In the Ocean*, vol. 4, *Energy Wastes in the Ocean*, ed. I. W. Duedall, D. R. Kester, P. K. Park, B. H. Ketchum. New York: John Wiley and Co. (In press.)

WHITE, D. C., NICKELS, J. S., PARKER, J. H., FINDLAY, R. H., GEHRON, M. J., SMITH, G. A. & MARTZ, R. F. (1982b). Biochemical measures of the biomass, community structure and metabolic activity of the ground water microbiota. In *Proceedings of the First International Conference of Ground Water Quality Research*, ed. C. H. Ward. New York: Wiley-Interscience Pub. Co. (In press.)

ZOBELL, C. E. (1943). The effect of solid surfaces upon bacterial activity. *Journal of Bacteriology,* **46,** 39–56.

ZOBELL, C. E. (1946). *Marine Microbiology: A Monograph of Hydrobacteriology*, pp. 41–58. Waltham, Massachusetts: Chronica Botanica Co.

LABORATORY MODEL SYSTEMS FOR THE INVESTIGATION OF SPATIALLY AND TEMPORALLY ORGANISED MICROBIAL ECOSYSTEMS

JULIAN W. T. WIMPENNY, ROBERT W. LOVITT AND J. PHILIP COOMBS

Department of Microbiology, University College, Newport Road, Cardiff CF2 1TA, UK

INTRODUCTION

Some of the most interesting questions relevant to microbial ecology concern the physico-chemical composition of aqueous solutions and a knowledge of how they change with time and in space. To many microbial physiologists these questions may not be immediately relevant, since in one of the commonest laboratory microcosms, the chemostat, the liquid is normally well-mixed and its composition is known.

In all natural ecosystems the physico-chemical composition of liquids determines the types and numbers of microorganisms that can grow in their presence. Equally important is a knowledge of the relative motion of the solution with respect to microbe and the relative flow of solutes compared both to the transporting solution and to the organism. If this view is accepted, it raises many questions which are really in the realms of hydrodynamics since the uptake of substrates and the removal of products are all dictated by the motion of solute molecules in a small space adjacent to the outer cell boundary.

There exists in Nature a large family of microbial ecosystems where microbes and the solutions which bathe them are more or less fixed at particular coordinates in space. Here molecular diffusion is the dominant solute transfer mechanism and is likely to be the chief growth-limiting factor in a particular habitat. Some of these are listed in Table 1 and are discussed in more detail by Wimpenny (1981). Such ecosystems are found in a wide range of habitats and vary in scale from a few micrometres in some microbial films, to tens

Table 1. *Examples of spatially heterogeneous microbial ecosystems*

Class of ecosystem	Example	Scale	Comment
Vertically organized (stratified)	Microbial film	μm to mm	Almost any surface exposed to suitable nutrient solution can be colonized by microbes. Examples include marine fouling, trickling filter media, disc fermenters, the 'quick vinegar' process, neuston, phylloplane colonization, dental plaque etc.
	Microbial colonies	μm to mm	Colonies show spatial heterogeneity and also exhibit limited morphogenetic potential
	Sediment systems	mm to cm	Sediment systems found at the base of all types of water: oceans, lakes, estuaries, rivers, streams, ponds, flooded soils. Systems contain mineral and organic matter and is homogeneously water saturated.
	Soil	cm to m	Irregularly stratified complex mixture of organic and inorganic, living and non-living matter containing variable ratios of gaseous and aquatic phases.
	Water	cm to m	Stratified water bodies including thermally stratified fresh water lakes, permanently stratified meromictic seas and lakes such as the Black Sea, Lake Gek-Gel and Solar Lake
Radially or cylindrically organized	Soil crumbs	mm	Irregular structures containing mineral and organic matter and living organisms. The array often held together with biologically generated polymers. May have steep oxygen and substrate gradients across them.
	Microbial floc	mm	Irregular clumps of living microorganisms found for example in activated sludge plants.
	Mycelial balls	mm	Formed commonly in industrial fermentations as tightly intertwined mycelia. Can have steep solute gradient across them.
	Rhizosphere	μm to mm	A loosely organized microbial film surrounding plant roots. Can reduce oxygen levels near them.
Horizontally organized (zonation)	Thermal gradients	cm to m	Hot sulphur springs: show changes in characteristic population down a thermal gradient.
	Organic gradient	m	Region at or below sewage outfall in a river or stream.
	Salt gradients	cm to m	Gradients in salinity at marine shorelines or at the edges of salt lakes or brine pools lead to changes in populations.

or even hundreds of metres in stratified meromictic lakes. Because organisms are located at fixed positions in space, we refer to these ecosystems as 'spatially organized' or 'heterogeneous' structures. It is worth discussing the difference between homogeneity and heterogeneity in more detail.

Homogeneity versus heterogeneity

Liquid culture systems have been, and will continue to be, essential in determining the relationship between the composition of the natural environment and the physiology of microbes. Such systems form a primary tool in the analysis of single and mixed communities in the laboratory and have been used effectively since the time of Winogradsky as an enrichment system from which many of the major groups of bacteria were isolated.

However, it was not until the invention of open systems such as the chemostat (Monod, 1950; Novick & Szillard, 1950) that the true potential of liquid cultures was realized. These systems allowed responses of microbial populations to changing environmental conditions to be studied in a systematic way. This approach has been extremely fruitful and has transformed our ideas of microbial growth in natural environments.

All liquid culture systems are essentially homogeneous in spatial terms. This means that chemical analyses performed at any point within the stirred container should give identical results. We must add a proviso to this definition. Such homogeneity exists only on a crudely microscopic scale since the existence of organisms acting as sinks and sources for particular solute molecules means that submicroscopic gradients exist very close to the cell surface. Homogeneity, which is the goal of the fermentation technologist, implies that solute exchange between sources and sinks can only be expressed in scalar terms.

Most natural ecosystems lack this homogeneity. Indeed, the statement that a particular group of organisms is spatially organized or heterogeneous indicates that solute transfer has not only magnitude but direction as well. Such vectored flow from sources to sinks must be characteristic of all structured ecosystems.

Liquid cultures are spatially homogeneous; but if they are operating as a batch culture, they are said to be 'closed' and they are temporally heterogeneous. Only the chemostat operating under steady-state conditions provides an environment which is spatially and temporally homogeneous. Even here, the early phases after inoculation, or oscillatory states, or the transient changes following perturbations of the culture, lead to temporal heterogeneity. Natural ecosystems are usually more complex. They are nearly always spatially heterogeneous and temporally heterogeneous as well. Stratification and zonation are evidence of spatial differentation whilst successions of microbial types lead to temporal differentiation. To complicate matters, definitions of spatial and temporal heterogeneity are governed by scale. Thus over long periods of time, certain habitats may be in a steady-state since, taken as a whole, their composition does not change significantly as new growth exactly balances death, decay and the recycling of nutrients. On a microscopic scale, however, the whole ecosystem may be undergoing numerous successional changes. Natural ecosystems are heterogeneous and highly complex structures.

Is there a case for relying on homogeneous liquid culture systems for the analysis of microbial behaviour in such complex ecosystems? There is no doubt that the most important advances in our understanding of microbial physiology have come from homogeneous culture systems. The chemostat has provided a sophisticated tool for furthering our knowledge of responses of pure and mixed cultures of bacteria to ecologically sensible concentrations of nutrients and it could be argued that everything we need to know about an organism, or about a mixed culture of bacteria, can be determined in carefully worked out continuous culture experiments. To accept such a view is to join the ranks of the reductionists who consider that any phenomenon can be explained if one fully understands the properties of its simplest elements. The opposite view, which is one we have taken, is that as a science proceeds to new levels of complexity, so new rules must be formulated to explain it. It is not that rules applying to simpler levels are wrong or even inapplicable, only that they may be irrelevant to an understanding of behaviour at a higher level of complexity.

Structural and temporal heterogeneity are vitally important properties of natural ecosystems for the following reasons:

(1) Solute transfer by diffusion plays a dominant part in determining the metabolic rates in particular ecosystems. Jørgensen (1981) has compared rates in natural ecosystems where anoxic and oxic zones couple organisms of the sulphur cycle and it is clear that such rates can vary over six orders of magnitude.

(2) Environmental heterogeneity allows interactions between organisms which may not be able to grow together in the same homogeneous habitat. Once again, organisms of the sulphur cycle illustrate this well since the metabolism of obligate aerobes and obligate anaerobes is coupled via the diffusion of sulphur compounds across the oxic-anoxic interface.

(3) Heterogeneity encourages diversity of microbial types since it

provides a wide range of niches for the development of different species.

(4) The more diverse the population, the more resilient it is to environmental stress; thus heterogeneity encourages stability.

(5) The frequency and extent of environmental perturbation and the degree of heterogeneity (which may imply the existence of environmental 'bolt holes' but is certainly related to the range of physiological types present) both determine the rate and range of evolutionary flexibility for particular ecosystems.

Accepting the importance of heterogeneity, our thesis is that heterogeneous models are needed to study heterogeneous phenomena.

The importance of model systems

There are two main approaches to solving the problems posed by the complex nature of microbial ecology. The first is to study them *in situ*, and as this volume suggests, the methodology is reaching a stage in its development where this is becoming more and more rewarding. The second route is to construct simplified laboratory model systems which can reveal important elements of the behaviour of the natural ecosystem. As Margalef (1967) has said, 'the purpose of laboratory analogues is not to duplicate or imitate nature, but to have simple situations that make it easy to identify operational mechanisms first in the experiment and then in nature'.

The expectation is that both approaches outlined above should be complementary leading at some point in the future to a complete understanding of all the interactions taking place in particular microbial habitats. Extreme views exist in microbial ecology; these range from a total condemnation of laboratory models as irrelevant artefacts, to those who have complete faith in the chemostat as capable of answering any question in microbiology. A model, therefore, is rather like a temperamental horse – it will almost certainly get you somewhere, but unless you keep it under tight control you may not be quite sure where. Treat the thing with deep suspicion, beat it often and it will more than likely serve you well!

The aim of this chapter is to collect together some of the available information on selected laboratory model systems which incorporate spatial heterogeneity, since the latter is a fundamental property of most natural ecosystems yet one that has been deliberately

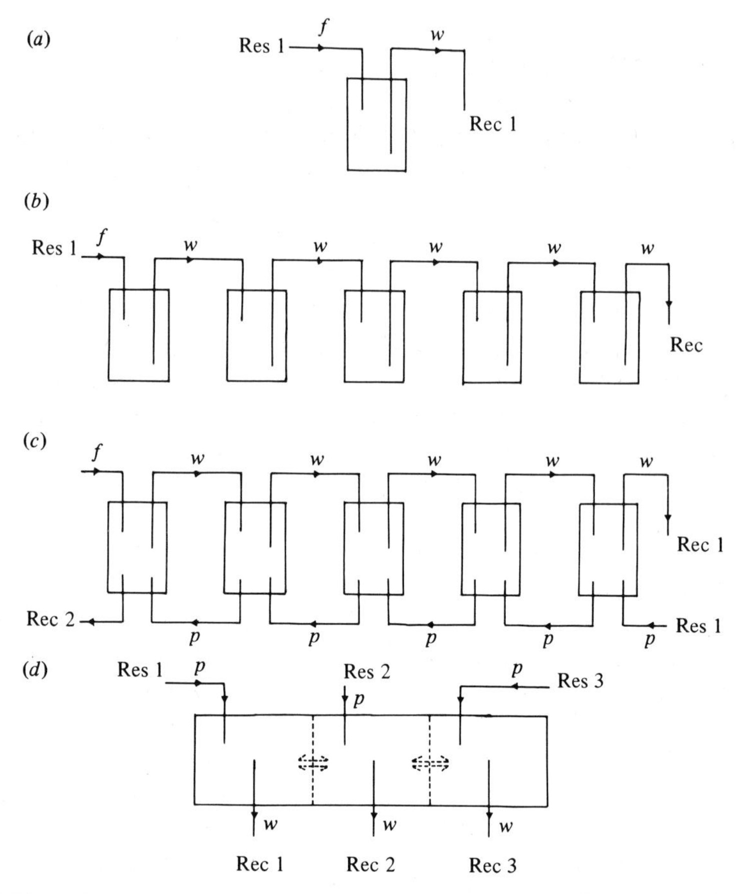

Fig. 1. The major homogeneous fermentation systems used to investigate microbial ecology in the laboratory. (*a*) the chemostat; (*b*) the multi-stage chemostat; (*c*) the gradostat and; (*d*) a semi-open gradostat (the Herbert model). f = flow, p = pump, w = weir, Res = reservoir, Rec = receiver.

excluded from the most widely used homogeneous laboratory microcosms.

LIQUID CULTURE SYSTEMS

The major liquid culture systems used in the laboratory are described in Fig. 1 and include the single-stage continuous culture vessel (Fig. 1*a*), the multi-stage continuous culture system (Fig. 1*b*) and bidirectionally linked systems such as the gradostat which will be described in more detail later (Fig. 1*c* & 1*d*). The chemostat at steady state is the only truly homogenous system where neither inputs to nor outputs from the system change with respect to time and space. The chemostat has a number of useful fundamental properties in the analysis of populations of single microorganisms or of communities. For reviews in this area see Jannasch & Mateles (1974), Pirt (1975), Slater (1981) and Kuenen & Harder (1982).

Single-stage systems

The simplest form of heterogeneity seen in single-vessel systems is principally concerned with temporal changes. Whilst this is obviously true of batch cultures it may be a characteristic of some continuous-flow systems. Temporal heterogeneity has been used to study various aspects of the physiology and growth of microbes in continuous culture; for example, the kinetics of nutrient uptake has been investigated in purple sulphur bacteria using the 'cyclostat', where nutrient input is varied over time in a regular fashion (Beeftink & van Gemerden, 1979). Van Gemerden (1974) has studied the coexistence of purple sulphur bacteria in the chemostat when light intensity was varied. Similarly Chisholm & Nobbs (1976) stressed the importance of the time dimension in defining the niche of an organism when discussing the stable coexistence of phytoplankton in phosphate-limited cultures with alternating light and dark cycles. Studies with mixotrophic thiobacilli have also shown that the interval between changes in medium nutrient composition can play an important part on the final outcome of competition experiments (Gottschall, 1980).

Multi-stage systems

Multi-stage systems have all the advantages of the chemostat except that they are physically more complex. These systems have been used for some time to observe the sequential degradation of complex wastes such as coke oven liquor (Abson & Todhunter, 1961) and diesel oil (Pritchard *et al.*, 1976) and in addition to investigate the nitrification and denitrification of effluents, (Hawkes, 1977). More recently D. Nedwell (personal communication) has used a five-stage system as a sediment model to observe the sequential use of electron acceptors, such as sulphate and carbon dioxide, when benzoate or glucose is degraded to methane. Fig. 2 shows the rates at which sulphate reduction and methanogenesis occur sequentially down the array. R. J. Parkes (personal

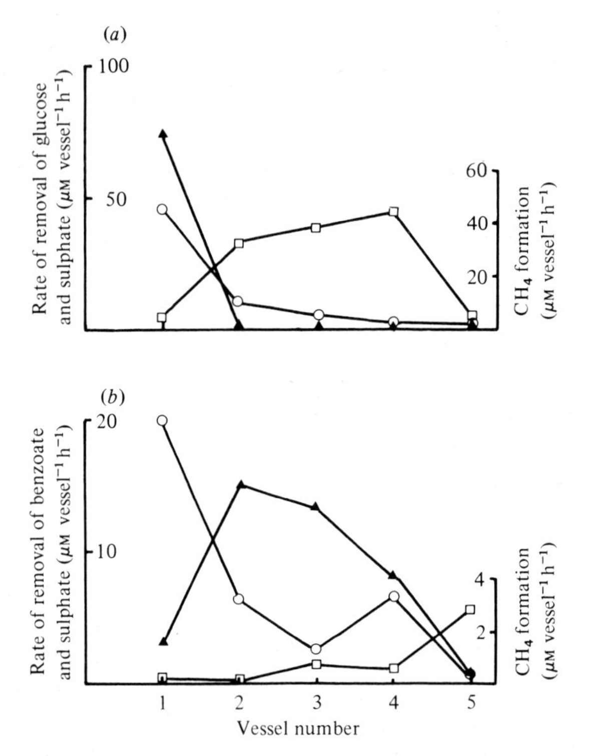

Fig. 2. Rates of sulphate reduction and methane formation in a five-vessel multi-stage system at steady state. Feeds contained: (*a*) 10 mM glucose, 10 mM sulphate; (*b*) 10 mM benzoate, 10 mM sulphate; $D = 0.002\ h^{-1}$ for the whole array. The figure shows the rate of removal of carbon substrate, (▲); rate of removal of sulphate, (○); rate of formation of methane, (□). (D. Needwell, personal communication.)

communication) has devised a similar system to observe the degradation of paper and paper effluents.

Multi-stage systems with bidirectional flow

The incorporation of bidirectional flow to a multi-stage system adds a further dimension to its heterogeneity and allows microbial communities to develop in a more realistic fashion since the system can now include physical spaces where growth, starvation and death can occur and where interactions between elements of the system are possible. A few such models have been described and these are briefly reviewed below.

The Cooper-Copeland model. Margalef (1967) first developed an analogue of an estuarine plankton system which incorporated some bidirectional flow through a series of interlinked vessels. The latter consisted of stirred 300 to 400 cm^3 glass flasks linked together in series. This basic model was further developed by Cooper & Copeland (1973) who constructed a system consisting of five 9 gallon containers linked to each other by two glass tubing couplers. Rates of exchange were controlled by varying the diameter of the connecting tubes. This system was used to generate a salinity gradient across the array allowing the development of a typical estuarine planktonic flora.

The gradostat model. The gradostat (Fig. 1*c*) was developed using similar principles (Lovitt & Wimpenny, 1979, 1981a). It normally consists of five glass vessels, each with a capacity of 500 to 1000 cm^3. Bidirectional exchange is facilitated by pumping medium up the system in one direction and by letting it flow downwards over weirs in the other. Some experiments illustrating properties of the gradostat are described later.

Semi-open systems (Herbert Model). R. Herbert (personal communication) has developed a diffusion coupled multi-stage system for investigating sediment ecosystems. It is conceptually similar to the 'Ecologen' (New Brunswick Scientific Co., Inc., New Jersey, USA) a device in which three vessels are linked to a common solute pool through membrane filters. The three used in the Herbert model are linked in series via membrane filters and the system is open to solute flow along the system and to cells across it (Fig. 1*d*). Using such a system *Clostridium butyricum* was grown on a glucose-salts medium in vessel 1 whilst *Desulfovibrio* species located in vessel 2 used the fermentation products generated by the clostridium as carbon and energy sources. Vessel 3 contained a *Chromatium* species which oxidized the sulphide generated by the sulphate reducer to produce sulphate which then diffused back to vessel 2 establishing a functional sulphur cycle.

Properties of the gradostat

Our experience of multi-stage continuous-flow systems is confined to the gradostat and its theoretical and experimental properties and applications will now be discussed.

Theory. Transfer of solutes in the gradostat is described by the following differential equation where x_n is the concentration of a solute in the nth vessel:

$$V\frac{dx}{dt}n = ux_{n-1} + vx_{n+1} - (u + v)xn \qquad (n = 1, 2 \ldots N) \ (1)$$

u and v are media flow rates in each direction through the array, and V is the volume of each vessel.

If $x_o = a$ and $x_{n+1} = b$, and are concentrations of particular solutes in each reservoir, and the initial conditions are such that $x_n = 0$ at time $t = 0$ ($n = 1,2, \ldots N$), then solving the differential equation gives:

$$x_n = \left(\frac{u}{v}\right)^{\frac{n}{2}} \frac{1}{N+1} \sum_{k=1}^{N} \frac{C_k}{\alpha_k} (e^{\alpha_k t} - 1) \sin\left(\frac{nk\pi}{N+1}\right) \tag{2}$$

where:

$$\alpha_k = 2\lambda \cos \frac{k\pi}{N+1} + \mu, \ C_k$$

$$= 2\lambda \left\{ a + b\left(\frac{v}{u}\right)^{\frac{N+1}{2}} (-1)^{k+1} \right\} \sin\left(\frac{k\pi}{N+1}\right),$$

$$\lambda = -\frac{\sqrt{(u+v)}}{V} \quad \text{and} \quad \mu = \frac{-u+v}{V}$$

Equation (2) defines the concentrations of a solute in a particular vessel at any time after pumping commences. However, as t tends to ∞ the system tends to a steady state given by:

$$x_n = E(u/v)^n + F \tag{3}$$

where:

$$E = (a - b)/\{1 - (u/v)^{N+1}\} \quad \text{and}$$

$$F = \{b - a(u/v)^{N+1}\}/\{1 - (u/v)^{N+1}\}$$

Where flow rates u and v are equal this simplifies to:

$$x_n = A + (n+1)B \tag{4}$$

where:

$$A = \{(N+2)a - b\}/(N+1) \quad \text{and} \quad B = (b - a)/(N+1)$$

Flow rates and solute distribution. Using a dye as a solute and keeping flow in each direction constant, we obtained a linear distribution of dye through the array of vessels as predicted mathematically. However, when flow rates are different in each direction the steady state distribution of solutes can be altered. The results of a computer simulation where flow rates vary from 1 : 1 to 10 : 1 are shown in Fig. 3.

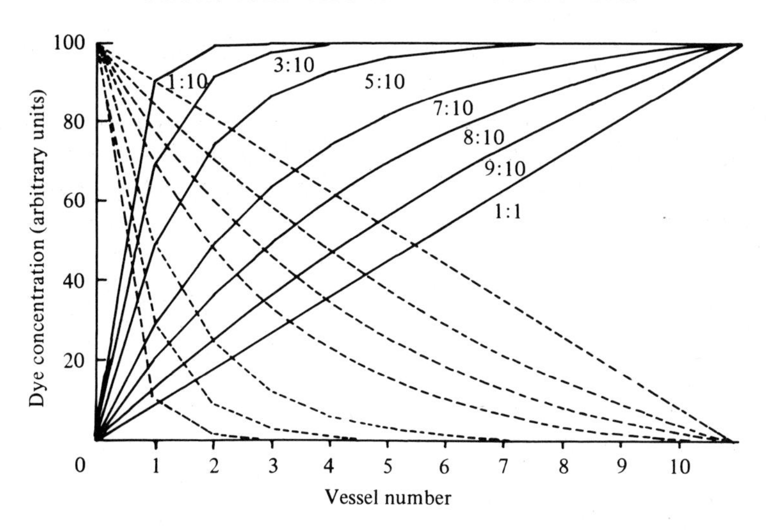

Fig. 3. The solute distribution obtained at steady state when asymmetric pumping is used in the gradostat. The ratios employed are shown in the figure and range from 1 : 1 to 10 : 1.

Residence times. The distribution of residence times in each vessel of each of the three main classes of culture system described above are compared at the same overall dilution rate (Fig. 4). Individual vessels of a unidirectional multi-stage system wash out sequentially whilst the system as a whole has approximately the same residence time as a single vessel of the same volume (Herbert, 1961). Residence times in the gradostat are quite different and are dependent on the position of the vessel in the array. Thus vessels at the periphery of the system have lower residence times than those near its centre. The average residence time for the whole system is approximately that of a single vessel running at the same dilution rate. This pattern becomes important when considering the use of these systems for enrichment cultures and for determining the kinetics of microbial growth within the system and introduces unique constraints on cells growing in such a system.

Growth in opposing gradients. The position at which growth takes place in the gradostat can be easily controlled by varying the direction of flow and the relative concentrations of two growth-limiting nutrients. In experiments with *Paracoccus dentrificans* growth was controlled in this manner by feeding the organism with various concentrations of succinate and nitrate from opposite ends of the system. The growth position could be predicted fairly satisfactorily using a simple Monod growth model in a computer simulation. The effect of Monod growth parameters on growth

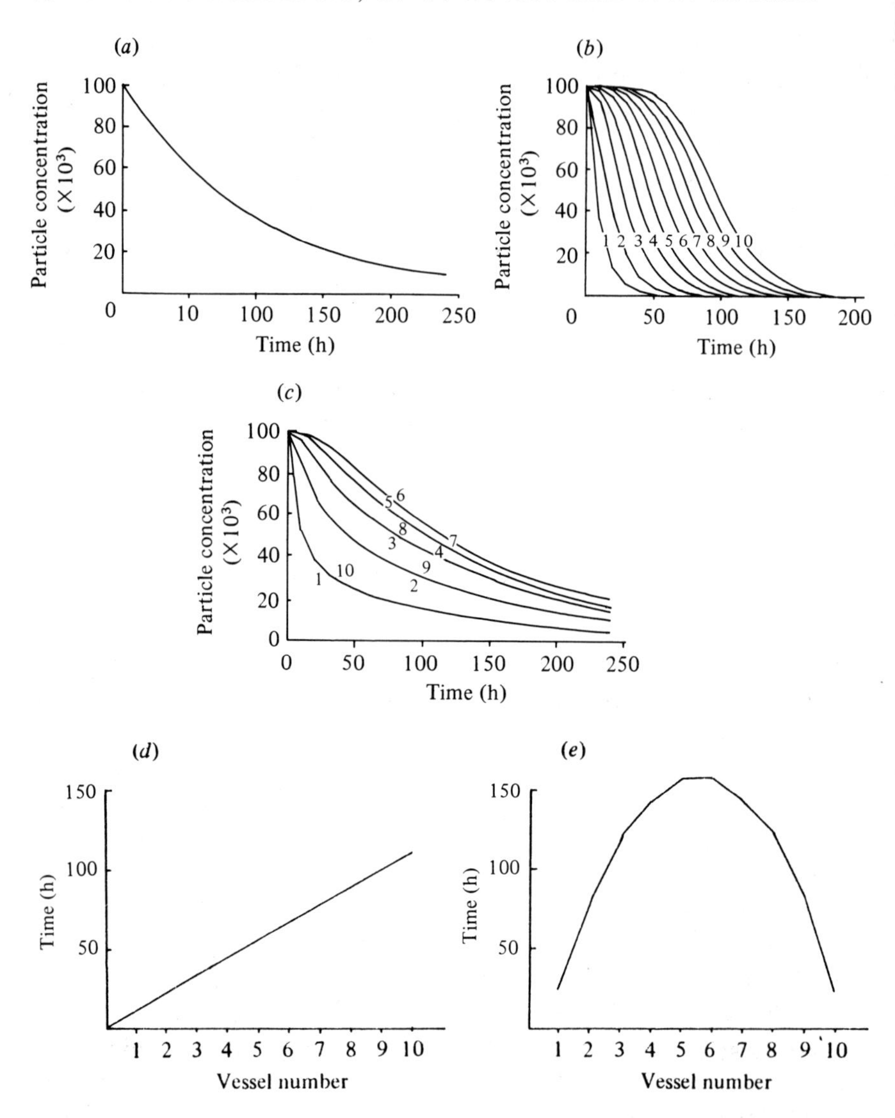

Fig. 4. Washout curves obtained from simulations of various fermenter conformations: (*a*) the chemostat, (*b*) a 10-element multi-state chemostat, (*c*) a 10-element gradostat. Each system operating at an overall dilution rate of 0.01 h^{-1}. The system contains uniformly distributed particles at the start of the simulation. Graphs (*d*) and (*e*) show the time required for cell concentrations to fall to 40% of their original value for the multi-stage chemostat and the gradostat respectively.

position was then studied in more general terms using the same computer model. The most dramatic effects were observed when the ratio of the yield coefficients for the two limiting substrates was altered. Other parameters such as the maximum specific growth rate

(μ_{max}) affected the rate at which steady states were achieved. The ratio of affinity constant for each of the two limiting substrates had little effect on the growth position suggesting that growth was substrate limited.

Changes in flow rate had significant and interesting effects. As the rate of nutrients supply increased, growth ceased to be nutrient limited where growth occurred. Increasing the flow rate further caused this region of unrestricted growth to spread outwards. This led to measurable nutrient gradients across the array. Only when the dilution rate exceeded a critical value for the system as a whole did the population washout. This value was higher than anticipated, presumably because of the variable distribution of residence times within the system.

The gradostat as a model for studying spatial and temporal interactions

Linked metabolism. One of the most interesting aspects of growth in heterogeneous environments is the possibility of linking metabolic processes that normally take place in mutually exclusive environments. The cyclic oxidation and reduction of nitrogen and sulphur are obvious examples. This was demonstrated in the gradostat using a lactate-sulphate enrichment in which sulphur compounds were cycled between sulphate reducing bacteria and thiobacilli. Community structure is, therefore, more diverse and complex than one would expect from a chemostat enrichment using the same medium and dilution rate (R. W. Lovitt & J. W. T. Wimpenny, unpublished observations).

Adaptation. When *Escherichia coli* was grown in opposing gradients of oxidant and reductant, adaptive changes took place across the whole array. Populations from each vessel showed major changes in their ability to oxidize substrates, and in the levels of oxidative enzyme activities and cytochrome pigments present (Lovitt & Wimpenny, 1981b).

Growth and viability. In several experiments with mixed cultures viable numbers of microorganisms were seen to drop in an exponential fashion away from the vessel in which growth took place. It is clear that inhibition and antagonism are possible in the gradostat and that some vessels in the system may not support or may actually be inimical to growth. The same regions however allow transfer of solutes between growth points, a situation common in many structured microbial ecosystems.

Competition. Competition can be studied in a number of ways in the gradostat. In an experiment demonstrating competitive exclusion, a steady-state population of *Esherichia coli* growing in glucose-oxygen counter gradients was reduced to very low levels by the addition first of *Pseudomonas aeruginosa* which occupied the aerobic end and second of *Clostridium acetobutyricum* which grew at the anaerobic end of the array (R. W. Lovitt & J. W. T. Wimpenny, unpublished observations). Varying the position of interacting populations within the system and therefore altering specific constraints to growth is also possible by manipulating relevant substrate concentrations.

Transients. Transient phenomena are complex especially in multistage systems such as the gradostat. The transition between steady states in a lactate-sulphate plus nitrate enrichment culture was followed after increasing the lactate and reducing the sulphate and nitrate concentrations. Fig. 5 shows that levels of sulphide increase markedly for a few days until new steady-state values are established. In this short period environmental conditions could select for quite different microbial populations.

Possible further developments

Multi-dimensional systems. Two- and three-dimensional systems have been suggested by Margalef (1967) and by Wimpenny (1981). However no published work has yet appeared on this subject, presumably because of the practical problems concerning system complexity. Investigations of multi-dimensional systems may be best approached using some of the other methods described in this paper. It is also theoretically possible to separate cells from solute flow by interposing filters between them. The cell compartment may itself be open or closed and raises the possibility of carrying out many ecologically relevant experiments.

GEL-STABILIZED MODELS

The addition of agar to a liquid medium prevents solute movement by convection currents or by mechanical mixing leaving molecular diffusion as the only mechanism for solute transfer. The presence of microorganisms at fixed points within or on the gel matrix intro-

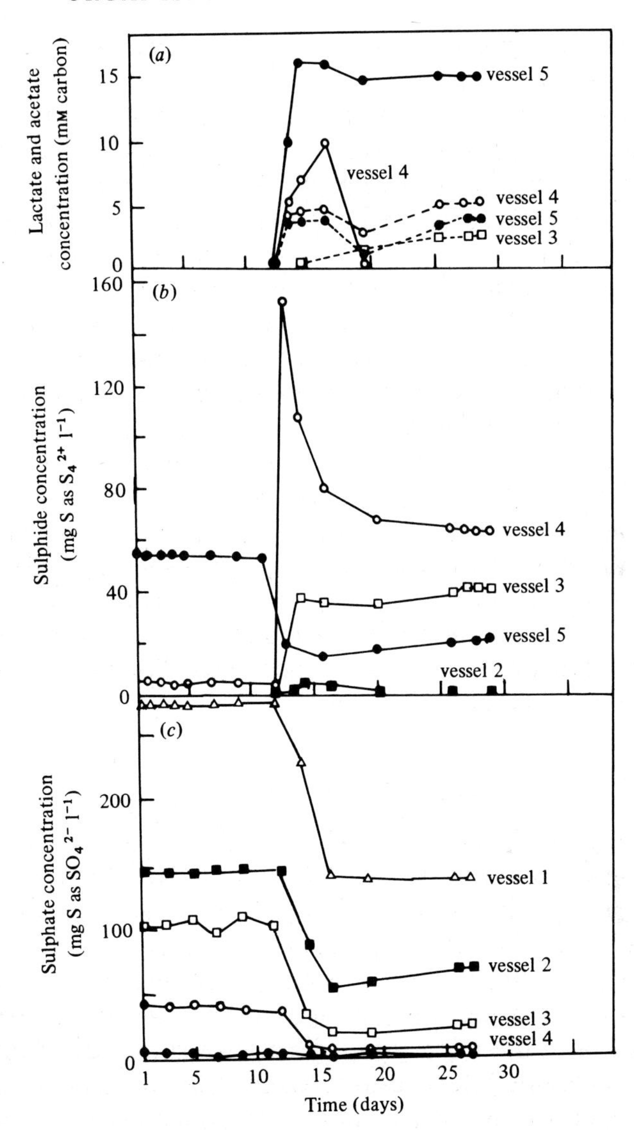

Fig. 5. Transient changes between two steady states in a lactate : sulphate + nitrate enrichment culture growing in the gradostat. The limiting substrates lactate, sulphate and nitrate were 27, 10 and 5 mM respectively and were changed on day 12 to 33 mM lactate, 5 mM sulphate and 1 mM nitrate. The figure shows the levels of (*a*) carbon compounds: lactate (solid line) and acetate (broken line) (*b*) sulphide and (*c*) sulphate in vessels 1 (△), 2 (■), 3 (□), 4 (○) & 5 (●).

duces an element of heterogeneity if they act as sinks for dissolved nutrients or as sources for metabolic products. Heterogeneity is increased if nutrients are supplied to such a gel matrix from spatially discrete sources.

History of gel systems

One of the first uses of a gelling agent as a means of subjecting microbes to solute gradients was by Beijerinck in 1889. Crystals of different compounds were placed on the surface of gelatin plates and the effects of these compounds on lawns of bacteria were examined. The technique was called 'auxonographie'. Another early use of semi-solid media was for the cultivation and identification of microbes, particularly anaerobes, since it was considered that the addition of sufficient agar to prevent convection currents would also prevent aeration of the medium. This was first proposed by Pringsheim in 1910 (cited by Spray, 1936) and subsequently developed by several workers so that in 1936, Spray published an identification key for clostridia based on their reactions in a variety of semi-solid media.

Semi-solid media have also been used for diagnostic purposes by Whittenbury (1963) in a study of the conditions affecting the utilization of fermentable substrates by lactic acid bacteria, and by de Vries & Stouthamer (1969) who used an agar/deep-tube technique to compare the oxygen sensitivity of twelve strains of *Bifidobacter*.

The work of Williams (1938a,b, 1939a,b) is particularly interesting. A number of bacteria were mixed with a semi-solid medium and incubated under pressure in oxygen. Under these conditions, the bacteria formed discrete bands of growth which Williams referred to as 'bacterial spectra'. The position and density of the bands could be varied by changes in oxygen or carbon dioxide tension, incubation temperature, inoculum size, nutrient and inhibitor levels, and by the addition of electron donors and acceptors.

Tschapek & Giambiagi (1954) also observed bands of growth of an *Azotobacter* when it was fixing nitrogen in a gel system and both they and Williams compared the bacterial growth bands to Liesegang ring formation. Liesegang rings occur comparatively frequently where diffusing solutes react to form a precipitate in a gel-stabilized matrix. Thus they appear when silver nitrate diffuses through a weak solution of potassium dichromate ions

(Liesegang, 1896; Hedges, 1932) or when unequal concentrations of antigen and antibody diffuse towards one another in an Ouchterolony plate (Crowle, 1973). Although Liesegang rings have been investigated widely in the past there is no generally accepted explanation for their formation.

A gel-stabilized laboratory model gradient system

A gel-stabilized system for observing microbial growth in solute gradients was originally described by Wimpenny & Whittaker (1979) and further developed by Wimpenny *et al.* (1981). The system is contained in a glass beaker in which a layer of semi-solid medium containing an initially homogeneous inoculum is poured over a source layer containing 1% (w/v) agar, the same basal medium plus a diffusible solute. If the source layer contains glucose and the gel is incubated aerobically, cells proliferate in opposing gradients of glucose and oxygen. As the system develops, a number of measurements may be made using needle electrodes and by removing a gel-core which can then be sliced and analysed. Microbial growth can be determined by optical density measurements on homogenized and resuspended slices or by using a gel scanner.

Band formation in the model system

Wimpenny *et al.* (1981) reported that a number of different bacteria generated multiple growth bands in a gel-stabilized system. Of these, *Bacillus cereus* consistently produced several bands and much of our detailed work has been concentrated on different strains of this organism. Fig. 6 shows the time course of band development in a system which was established in a test tube and which could be scanned directly.

Band formation is influenced by a number of factors. Anaerobic incubation abolishes the appearance of growth bands, hence oxygen is needed for their development. Interestingly, though, oxygen could not be detected in aerobically incubated gels after the first 2 to 3 h. High or zero concentrations of glucose throughout the growth layer abolished band formation and it seemed that glucose gradients were important. The position of the bands also depended on the concentration of glucose in the source layer; the higher the concentration, the nearer the surface bands occurred. Basal nutrient levels were also important. The lower the nutrient concentration, the

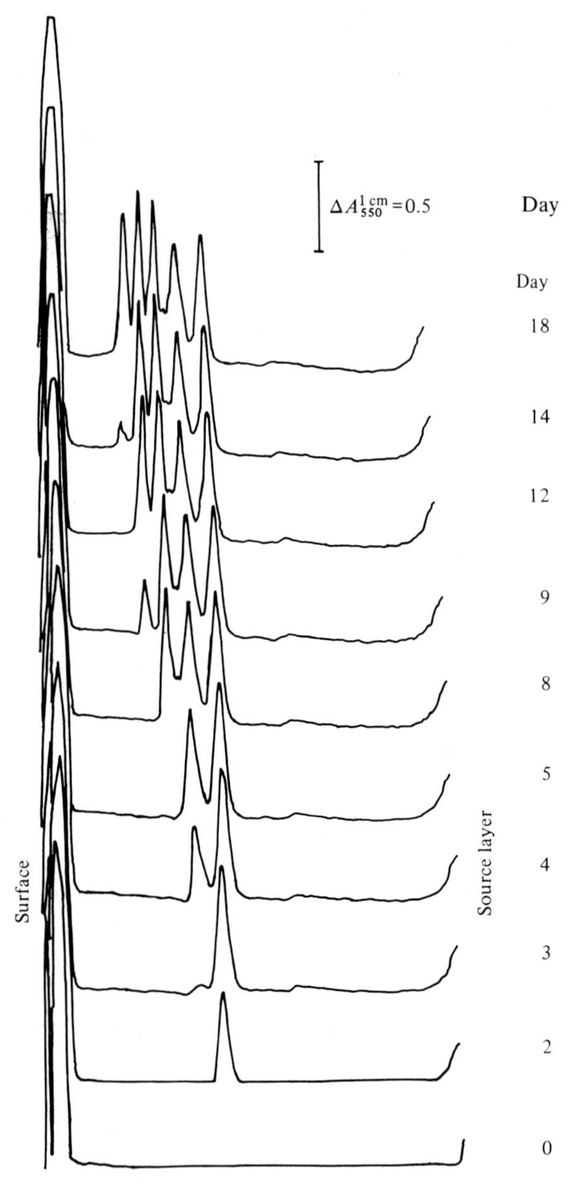

Fig. 6. The development of bands of *Bacillus cereus* C11 in a test-tube system. The basal medium was casamino acids/yeast extract/salts plus $20\,g\,l^{-1}$ glucose in the source layer. The gel was incubated at 30 °C and was scanned directly using a Unicam SP500 spectrophotometer with a Gilford gel scanning attachment. The gel was 49.7 mm long at day 0 and 45.7 mm long at day 18 as a result of slight shrinkage.

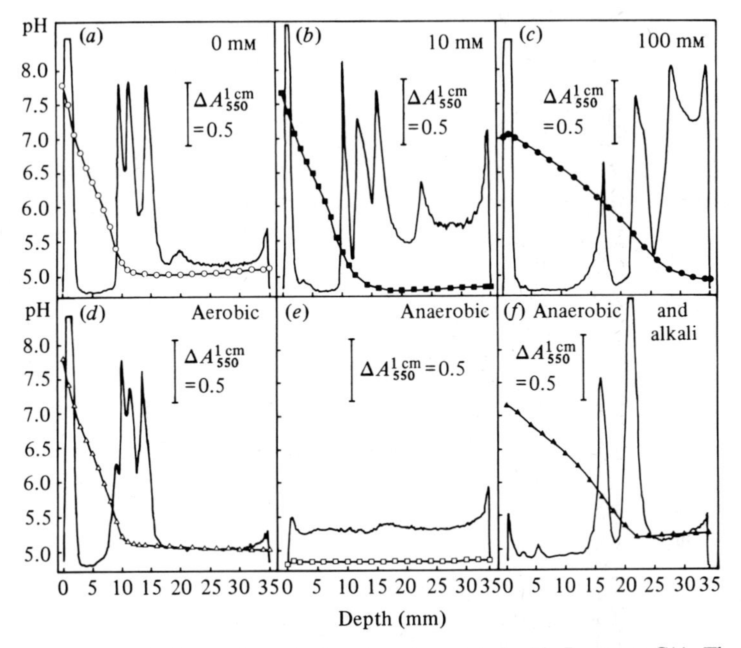

Fig. 7. Gel scans and pH profiles of gel systems inoculated with *B. cereus* C11. The basal medium was casamino acids/yeast extract/salts plus 20 g l^{-1} glucose in the source layer. In (*a*–*c*) the medium was initially at pH 7.0: (*a*) unbuffered; (*b*) plus 10 mM phosphate buffer; (*c*) 100 mM phosphate buffer. Measurements were made after 5 days incubation at 30 °C. In (*d*–*f*) gels were incubated under aerobic and anaerobic conditions: (*d*) aerobically; (*e*) anaerobically; (*f*) anaerobically with an agar overlayer containing 20 ml Tris-HC1, pH 8.0. pH profiles are indicated by the symbols ○, ■, ●, △, □, ▲. Measurements were made after 6 days incubation at 30 °C.

nearer the surface the bands appeared. Other variables, such as agar type and concentration and incubation temperature, appeared to have little effect; and band formation was independent of the growth phase of the inoculum.

pH gradients in the developing gels were determined with needle electrodes and revealed two processes: acidification presumably due to anaerobic fermentation reactions in most of the gel, and alkalization near the surface. A pH differential of about 2.5 units was maintained between the two regions over the incubation period.

The pH profile could be manipulated. For example, increasing the buffering capacity allowed heavier growth nearer the glucose source and also altered the band pattern (Fig. 7*a*–*c*). In separate experiments, *Bacillus cereus* was shown to be inhibited by pH values below 5.0 to 5.5, values which were measured in lower regions of the gel during incubation. The relevance of pH was established by investigating growth in anaerobic gels which became acidic throughout and in which no bands appeared. However, if a layer of gel containing alkali was poured over the surface of a complete system which was incubated anaerobically, then bands of growth appeared (Fig. 7*d–f*).

We consider that the gel system developed as follows: glucose diffusing from the source layer was fermented anaerobically generating acidic products which reduced the pH and inhibited growth. Near the surface, aerobic growth at the expense of amino acids led to the formation of alkaline products which diffused downwards. The first band occurred where the descending alkaline front met the ascending glucose and was the result of fermentative growth. This generated more acidic fermentation products which inhibited growth in the immediate vicinity of the first band. Glucose diffused past this band until it reached the next alkaline region and the process was repeated.

Computer simulations, however, suggested that this hypothesis alone was not sufficient to explain periodic growth since they predicted that a single moving growth front rather than multiple growth bands would result. It is suggested, therefore, that an asymmetric response to pH is involved where cells are inhibited by low pH, but do not begin to grow again until the pH reaches a higher value. Such a mechanism may involve sporulation and germination.

It is obvious that band formation by *Bacillus cereus* in a diffusion gradient system is a specific phenomenon which may depend on the ability of the organism to oxidize amino acids, to ferment glucose and to produce spores. However, as will be clear in the next section, periodic growth of this sort is actually very common. While we believe that an activation threshold model explains band formation, there may be many different biochemical and physiological reasons for such asymmetries in heterogeneous ecosystems.

Further examples of band formation

We have received personal communications from other workers who have observed but not formally reported the phenomenon of banded growth in gel systems. Examples of band formation are summarized in Table 2. H. Veldkamp has described a system in which alternate red and black bands of *Rhodopseudomonas palustris* and a sulphate-reducer were seen in gradients of sulphide and oxygen. W. D. Grant has observed multiple bands of an

Fig. 8. Bands of *Escherichia coli* in a test-tube gel system containing a mineral salts basal medium plus yeast extract in the source layer. The gel was photographed after 17 weeks incubation at 30 °C.

Table 2. *Examples of band formation by pure and mixed cultures of bacteria in gel-stabilized, liquid and natural systems*

Organism	Basal medium	Diffusing solute(s)	Comments	Reference
Gel-stabilized media				
Bacillus subtilis *Bacillus mycoides* *Proteus* sp. strain X19 *Bacillus mycoides* *Staphylococcus aureus*/*Bacillus subtilis* *Salmonella enteritidis*	Nutrient broth	Oxygen (under pressure)	Observed bands of a growth which were considered to be a form of 'growth spectrum' and were compared to Liesegang rings	Williams (1938a) Williams (1939a)
Azotobacter sp.	Mineral salts, mannitol	Oxygen, nitrogen	Bands produced under nitrogen-fixing conditions only; may be due to oxygen toxicity	Tschapek & Giambiagi (1954)
Serratia indica *Streptococcus faecalis* NCTC 775 *Escherichia coli* K12 *Bacillus subtilis* *Bacillus megatherium* *Lactobacillus confusus* NCIB 4037	Casamino acids, yeast extract, salts	Oxygen, glucose	In a survey of band formation was common in facultative anaerobes and aerotolerant anaerobes	Wimpenny *et al.* (1981)

B. cereus D11 *Bacillus* sp. C19 *B. cereus* NCTC 2599 *B. cereus* NCTC 6474 *B. cereus* NCTC 9947 *B. cereus* var. *mycoides* NCTC 9680	Casamino acids, yeast extract, salts	Oxygen, glucose	Band formation consistently noted; steep pH gradients appeared to be main trigger	Coombs & Wimpenny (1982)
Rhodopseudomonas palustris/ Sulphate reducer	Yeast extract, lactate	Oxygen	Alternate red/black bands	H. Veldkamp (personal communication)
Ectothiorhodospira sp.	Mineral salts, acetate, yeast extract	Sulphide, oxygen	Band formation noted; may be partly due to oxygen toxicity	W. D. Grant (personal communication)
Benzoate-degrading consortium	Artificial sea water, salts	Benzoate, oxygen		C. M. Brown (personal communication)
Escherichia coli K12	Mineral salts, oxygen	Yeast extract	Prolific multiple band formation	H. Abdolahi (personal communication)
Liquid media				
Escherichia coli	Galactose	Oxygen		Adler (1966)
Natural systems				
Gallionella	Estuarine mud	Not determined	About twelve 'well-defined narrow layers' using capillary peloscope.	Perfil'ev & Gabe (1969)

Ectothiorhodospira species in countergradients of oxygen and sulphide. C. M. Brown noticed bands of a benzoate-degrading consortium in gradients of benzoate and oxygen. H. Abdolahi has demonstrated that many bands of *Escherichia coli* K12 were formed when the organism was grown in a mineral salts medium in counter gradients of yeast extract and oxygen. Fig. 8 shows such a system established in a test tube and photographed after 17 weeks incubation at 30 °C.

Even these few examples show that banded growth can be seen in a variety of different systems and suggest that the periodic growth of bacteria in response to solute gradients may be a widespread phenomenon. Indeed, it has been reported in a naturally occurring system by Perfil'ev & Gabe (1969) who recorded the appearance of about twelve bands of the iron oxidizer *Gallionella* in estuarine mud, examined using a capillary peloscope.

Other uses for gel-stabilized systems

Although discussion has centred on band formation since this phenomenon is so intriguing, gel-stabilized models have other parts to play which may be more significant in practical microbial ecology. For example, it is well suited to investigating microbial interactions in horizontally stratified ecosystems. We have used such a model to investigate the ecology of oil storage tank water bases (Wimpenny *et al.*, 1981). This is an economically important habitat since the development of sulphate-reducing bacteria leads to corrosion of the steel base of the tank. Normal sampling techniques disturb the spatial organization of the system which almost certainly has both aerobic and anaerobic niches. The model system, established in semi-solid agar over a steel plate and located beneath a layer of gas oil, become spatially differentiated showing aerobic and anaerobic regions and gradients of oxygen, redox potential, sulphide and iron. Such a model demonstrates a 'perfect' answer to the water base problem; that is, it demonstrates how such an ecosystem **could** develop if molecular diffusion were the sole transport mechanism.

An interesting area for investigation is to discover which are the important determinants for growth at particular positions in space within heterogeneous ecosystems. Preliminary experiments suggest that diffusion coefficients and solute concentrations are more important than cellular kinetic parameters, in predicting growth

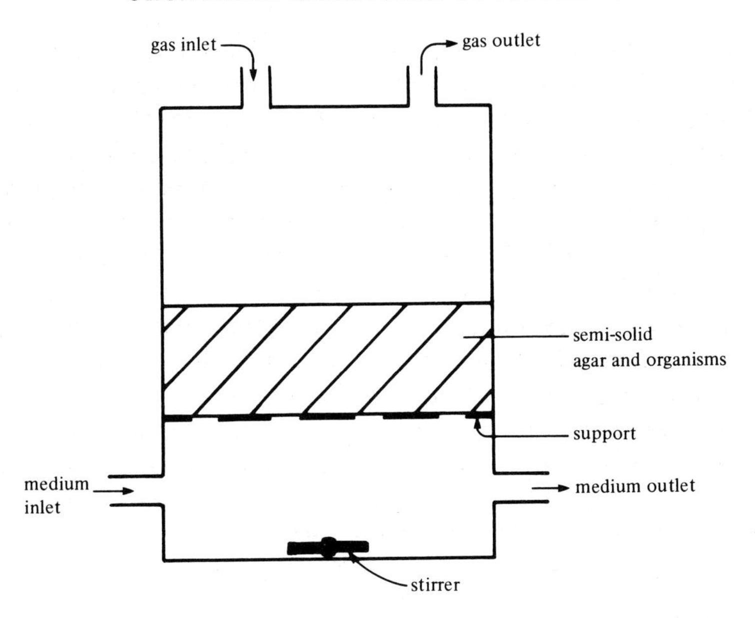

Fig. 9. Diagrammatic representation of a gel-stabilized system which is open with respect to solutes and gases. Such a system is proposed by the authors, but is not yet fully operational.

positions in gradient systems, since growth under these conditions is commonly diffusion-limited.

Gel-stabilized model systems suffer from a number of problems and limitations. They are closed systems with respect to solutes and not suitable for the investigation of steady states, although it is possible to devise a system which is open for solutes and gas exchange (Fig. 9). The generation of any quantity of gas can disrupt the system, and finally, the real world is not made up of semi-solid agar! One solution might be the addition of particulate matter to the gel to model a sediment for example. Provided that these points are borne in mind, these models have a potentially important part to play in our understanding of microbial ecology.

Two-dimensional gel-stabilized model systems

Gel systems described so far have incorporated single dimensional diffusion gradients. In addition, each system is closed so that solute concentrations at any one point vary with time. To overcome this, Caldwell and co-workers (Caldwell & Hirsch, 1973; Caldwell *et al.*, 1973) devised a two-dimensional agar plate system constructed from transparent plastic in which steady-state solute gradients were generated. The agar slab which was 2.5 cm square and 0.3 cm deep, was fed on two sides at right angles from chambers containing saturated solutions plus crystals of diffusible solutes. In addition, the surface was continuously washed with fresh medium providing a sink for each solute. The presence of solute crystals ensured that the source solution was always saturated and hence at a constant concentration, so that after equilibration which took about two days, steady-state gradients developed. The surface of the agar was then seeded with a thin layer of agar containing an inoculum of organisms, and the system incubated at a suitable temperature.

The mathematical model defining rates of diffusion in this system was verified by measuring the growth of *Escherichia coli* on radioactively labelled acetate. The growth of *Rhodomicrobium, Thiopedia* and *Hyphomicrobium* in gradients of methylamine and sodium sulphide was also investigated and segregation of these organisms was shown. The system was further used to study plankton from a forest pond, subjected to gradients of acetate and mud, the latter providing a natural source of nutrients. A major advantage of the Caldwell system, which has great potential in microbial ecology, is that it is capable of segregating 'stenobiotic' organisms, that is, those with narrow, well-defined habitat requirements, from the more versatile 'eurybiotic' organisms with less-exacting requirements.

The principle of a two-dimensional diffusion system has been pursued by Wimpenny and co-workers in an attempt to simplify the technique for routine use. The need to establish steady-state gradients was avoided by the construction of 'stopped time-dependent gradient plates' (STDGP). Diffusion gradients of pre-determined characteristics are established over a period of time in a square agar plate. The surface is inoculated with a thin layer of agar containing organisms. A metal grid capable of cutting the agar into a number of squares is then inserted into the agar preventing further diffusion. The system is incubated until visible growth is observed. Preliminary work with the system has shown a partial separation of *Bacillus subtilis* var. *niger* and *Bacillus cereus* in gradients of pH and salinity and has investigated the growth of a pseduomonad in a biocide/EDTA gradient system (Wimpenny, 1981).

A simplification of the system is to expose organisms to a solute gradient in one dimension only. Another variation is the use of a 'wedge-plate' technique originally devised by Szybalski (1951) in which agar containing acid, for example, is poured into a square

Petri dish set at an angle so that there is a greater depth of agar on one side. After this layer has set, the plate is put onto a level surface and a second agar layer containing alkali is poured over it. Opposing wedges of acid and alkali are generated. Diffusion vertically will distribute pH evenly in this dimension, but because diffusion is relatively slow horizontal gradients of acid and alkali are formed. The surface of the agar can then be inoculated with a lawn of organisms which are subject to different pH values depending on their relative position on the plate. To improve the system, it is possible to produce two-dimensional wedge plates using four instead of two layers of agar. This seems the most convenient way to construct large two-dimensional gradient plates with all the advantages these have in analysing responses of single organisms or microbial communities to changing environmental factors.

PERCOLATING COLUMNS

In many natural habitats a microbe is located in a fixed position and is subject to the flow of solutions. A clear example is that of rainwater flowing downwards through the soil, but it includes flow of water through pipes, channels, streams and rivers or the motion of tides and waves past organisms attached to fixed surfaces. In all these cases, the organism is constantly subject to 'fresh' medium. This may be especially significant in dilute environments where organisms need to scavenge low concentrations of available nutrients.

Investigations into such systems have concentrated on replicating characteristics of the soil using percolating columns, where microbes are exposed to a net downward flow of aqueous solutions. Downward flow leads to vertical stratification of microorganisms and their metabolites.

A simple percolating column was described by McLaren (1969) in which one species, dependent on a source of substrate, generated products which were used as a nutrient by another which, in turn, generated a second product. Analysis of such a system must be carefully made and Bazin *et al.* (1976) point out potential pitfalls. For example, assuming that the column is at steady state can lead to errors. If the system is assumed not to be in a steady state, then parameters of time and distance down the column must be taken into account. Thus, when a drop of nutrient solution is added to the

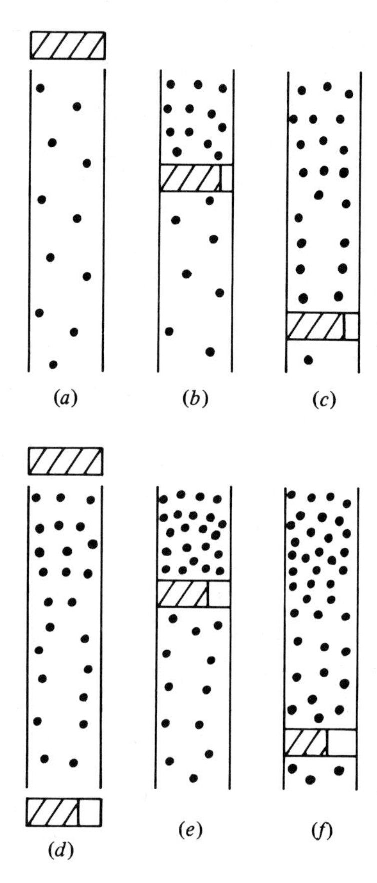

Fig. 10. Diagrammatic representation of the breakdown of a substrate in a percolating column. The relative numbers of microbes are indicated by the density of the dots and the concentration of substrate by the length of the shaded portion. (After Bazin *et al.*, 1976.)

column bacterial numbers increase and nutrient concentration falls (Fig. 10). However, as the nutrient concentration decreases lower down the column, the increase in bacterial numbers is smaller. When the first drop has left the column, the microbial population is no longer evenly distributed. When a second drop of nutrient is added, this disappears more rapidly in the region of increased biomass further increasing the change in distribution. The specific reaction rate of substrates and microbial biomass is therefore dependent on time and on spatial position within the column.

One of the first continuous-flow columns was devised by Macura & Malik (1958) and consisted of a tube packed with soil or other porous material through which nutrient solutions were passed. The system was open and quasi-steady-states could be achieved, if low input concentrations were used. The model suffered since microbial numbers could not be assessed without perturbing the system. Ardakani *et al.* (1973) used soil columns with sample ports located at intervals down the tube allowing changes to be followed as a function of depth as well as time. Bazin & Saunders (1973) replaced soil with glass beads and natural mixed populations with pure cultures of nitrifying bacteria growing on a completely defined medium. Macura & Kunck (1965a,b) showed that *Nitrosomonas* species predominated at the top of such columns whilst *Nitrobacter* species grew in the lower regions, indicating that the column was becoming spatially organized.

Whilst the use of glass beads has significant advantages for modelling interactions in the soil, they do not always predict the observed data accurately (see for example Prosser & Gray, 1977). Compared to a glass bead column, soil has a variety of organic and inorganic surfaces to which microbes can attach and in addition, competing microbes are also present. The glass bead does not reproduce the spatial heterogeniety of soil crumbs, which may have anaerobic centres even under aerobic conditions. This can be partially overcome by using hollow supports in which microbes may grow. These have been used successfully by Atkinson and co-workers in continuously stirred microbial film fermenters and this work is described elsewhere in this chapter (pp. 98–106).

THE BACTERIAL COLONY

The bacterial colony is included in this account because it is a good example of the growth of bacteria under heterogeneous conditions. It is in the strictest sense a 'natural ecosystem', although it appears in all its glory in laboratory microcosms called Petri dishes. Colonies are generally easy to grow, and their rates of growth and shape respond to environmental manipulation. Many colonies may be recognized by a characteristic form which implies that the developing structure, starting from a single organism, generates specific spatial patterns. The bacterial colony, therefore, is a model for film formation on the one hand and for diffusion-dominated morphogenesis on the other.

Bacterial colonies, in spite of their fundamental importance to microbiologists everywhere, have seldom been studied for their own sake. There are exceptions: Legroux & Magrou (1923) examined stained sections of old colonies of a vibrio and discerned changes in viability judged by the ability to take up vital stains. Only organisms near the surface of the colony or the tube-like folds that ran across the structure took up the stain. Other organisms were clearly anoxic and in the process of lysing. Perhaps the most detailed investigation of colonial growth was by Pirt (1967) who investigated radial growth rates and proposed a kinetic model which accounts for his own and other observations that radial growth is linear (Palumbo *et al.*, 1964, 1971; Cooper *et al.*, 1968). Pirt argued that if all the cells in the colony were dividing at the same exponential rate then the colony radius (assuming that the structure had the form of a flat disc) should increase exponentially. Pirt's kinetic model suggested that exponential growth occurred only in a narrow region at the edge of the growing colony and was capable of predicting linear growth rates. Pirt made the simplifying assumption that the colony remained substantially flat, and hence that no growth took place inside it. Although this assumption was not a bad one since most colonies are actually very flat, it was challenged by Wimpenny (1979) who determined colony profiles using an optical method. A number of colony types representing different cellular morphologies demonstrated an underlying common pattern of behaviour where the growing edge of the colony rose in a concave fashion to a plateau region which then increased in height more slowly to the domed or flattened centre of the colony. Wimpenny argued that growth in the leading edge region was indeed exponential at or near the μ_{max} for the organism; however, some growth also took place in the inner parts of the colony, but this was nutrient-limited since it depended on diffusion of solutes from the agar below. Growth here led to the observed increase in height.

Such investigations suggest a basic structural pattern for colony growth. However, they say little about the subtler elements of the surface topology. This is an area badly in need of further research. Photographs of the surface structure of various strains of *Bacillus cereus* show significant differences which are easily distinguished by the naked eye but are difficult to quantify (Fig. 11). Preliminary examination using optical transformation techniques suggests that these slight changes in pattern can be detected by Fourier analysis; however, the technique needs to be developed before it can be used reliably (G. Harburn & J. W. T. Wimpenny, unpublished observations).

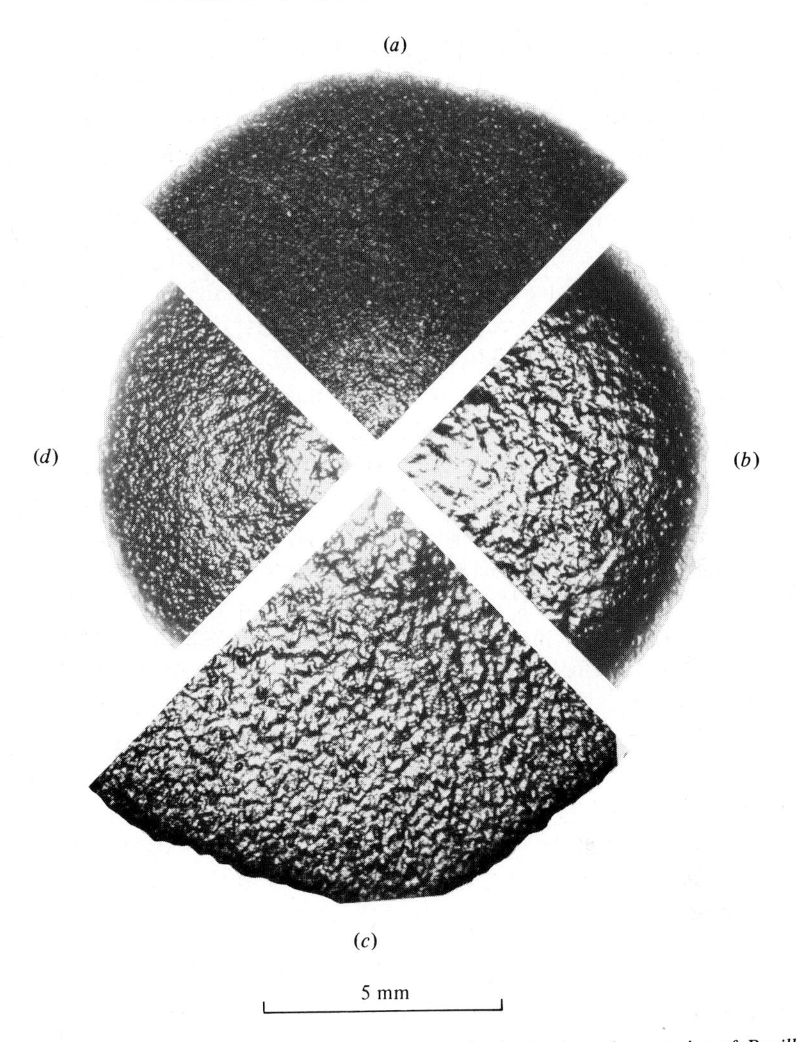

Fig. 11. Photographs showing the surface topology of colonies from four strains of *Bacillus cereus* grown for 48 h on tryptone-soya agar at 30 °C. Strain numbers: (*a*) NCTC 9680; (*b*) NCTC 9947; (*c*) C19; (*d*) D11.

The use of the scanning electron microscope has revealed a considerable amount of information on the structure of bacterial colonies (Whittaker & Drucker, 1970; Drucker & Whittaker, 1971a, b; Bibel & Lawson, 1972a, b; Todd & Kerr, 1972; Afrikian *et al.*, 1973; Fass, 1973; Bauer *et al.*, 1975). Examination of *Bacillus cereus* colonies using this technique shows clearly the orientation of individual organisms. Those near the periphery of the colony are very long and arranged parallel to one another in swirling loops.

Those nearer the centre of the structure are considerably shorter and appear to be orientated in random directions. It is probable that size variation is related to growth rate and these pictures confirm that cells near the colony periphery are dividing near their μ_{max}.

Additional information can be gained about conditions within the colony itself using microelectrode techniques. We have examined oxygen and pH profiles across colonies of bacteria including *Bacillus cereus*. Oxygen profiles indicate that below about 30 to 40 μm within the colony, conditions become anoxic. Oxygen dissolved in the agar layer is also depleted. Such values for oxygen penetration fit well with calculations made by Pirt (1967) and calculations based on measurements of actual and potential colony respiration rates made by Wimpenny & Lewis (1977). In old large colonies however, pO_2 measurements indicate that the whole structure is aerobic. Although this may seem surprising at first, it must be remembered that nearly all the organic reducing agents (carbon sources, etc.) have been exhausted by this time. pH profiles have so far not been extensively pursued, although early experiments indicate that colonies of *B. cereus* remained close to pH 7.0 in the upper layers and became slightly acidic in the lower anaerobic regions. This fits well with observations made on the growth of *B. cereus* in gel systems as described earlier.

Another approach to colony development is to examine the distribution of enzyme activities across the structure. We have obtained horizontal sections from frozen colonies of *Enterobacter cloacae* and assayed the latter for oxidative enzyme activities. Results demonstrated an increase in such activities near the aerobic surface layers of these colonies (Wimpenny & Parr, 1979).

So far, this work can only be regarded as a beginning and the construction of a numerical model of bacterial colony development must be regarded as a laudable goal in the near future. The architecture of the developing colony is complex as it depends not only on intrinsic factors including the shape of individual organisms, size range, method of reproduction, the production of extracellular macromolecules and motility, but on extrinsic factors, the most important of which is molecular diffusion.

MICROBIAL FILMS

Microbial film has been recognized as economically important for many years. For example, Zobell (1943) considered that the fouling of ships' hulls by higher animals was due to the initial colonization of the hull surface by films of microorganisms. It is probably safe to say that almost all surfaces exposed to aqueous solutions containing even small amounts of nutrients under appropriate physico-chemical conditions will allow the surface deposition of microbial cells followed by their proliferation and the appearance of a coherent microbial film. Biofilms, as they are sometimes called, can cause major problems in the industrial and public health sectors (Characklis, 1980). However, they are gainfully employed in effluent treatment plants, where they coat the surfaces of trickling filters and rotating disc aerators. Microbial film is common in aquatic environments where it forms at almost any solid–liquid interface and appears as neuston at the air–water interface. Microbes also form layers at liquid–liquid interfaces, for example between the oil and water layers at the base of oil storage tanks. Microbial film forms on plant surfaces including leaves, roots and germinating seeds and on animal surfaces including many internal epithelial tissues and, of course, as dental plaque on teeth.

Microbial films vary in thickness from a few micrometres to tens of millimetres and consist of a layer of microbes commonly held in a gelatinous or slimy matrix of extracellular polymer produced by one or more of the species present. Other organic and inorganic matter may be trapped in the matrix and the whole structure may become physically complex. At the film surface there is a stagnant layer of water which provides one barrier to solute transfer. As the film thickens it provides an increasing resistance to solute diffusion and this often leads to nutrient starvation in the deeper regions of the film. Because of its low solubility and importance as an electron acceptor oxygen is most likely to be one of these limiting substrates.

Film formation seems to take place as follows. An initially clean surface quickly becomes coated with adsorbed molecules, commonly proteins if they are present. Microbes suspended in the medium pass through the stagnant layer of water by one of a variety of hydrodynamic forces and attach to the solid surface. Attachment occurs in two stages. Weak or reversible attachment is governed by surface charges on the cell and on the substratum including London-van der Waals forces, interfacial tension and covalent bonding. Irreversible or permanent bonding depends on the production of specific polymers by the organism. In the presence of sufficient nutrients the cells proliferate and the film increases in thickness. In the presence of different organisms the developing film can be colonized by other species which might be unable on their own to colonize the solid surface but which might contribute to the organization of the film once it has started to form. The film differentiates spatially and temporally. If it is deep enough lower layers can become anoxic and where present, anaerobic species can develop. Anoxia, cell death and the generation of gases can destabilize the film which can detach or slough away from the solid surface. There is evidence that growth can take place in a cyclical fashion (Howell & Atkinson, 1976).

Dental plaque formation is one example of a film whose formation has been well documented. The clean enamel surface is rapidly coated with salivary glycoproteins and is then colonized by a variety of bacteria including the polysaccharide-producing streptococci, such as *Streptococcus mutans* and *S. mittior*. Relatively aerobic species such as neisseria are found early in plaque development and these are followed by a succession of facultative anaerobic or aerotolerant species including lactobacilli, actinomyces and coryneforms. Finally, obligate anaerobes such as *Veillonella* species appear.

Model microbial film fermenters

Examination of Table 3 shows that a very wide range of laboratory model film fermenters have been developed. However, rather specific questions have been asked of each. Many model film fermenters have been used to predict the kinetic properties of film involved in the waste water treatment industry in order to assess the consequences of loading these bioreactors with high concentrations of organic solutes. As predicted by many and demonstrated experimentally by Bungay and his colleagues using microelectrode techniques (Bungay *et al.*, 1969; Whalen *et al.*, 1969; Bungay & Chen, 1981; Chen & Bungay, 1981) the higher the organic loading the more rapidly the film respires and the more likely it is that the film will become anaerobic at some point. Anaerobiosis may be detrimental to the health of the film and may in the end cause it to detach from its support. Whilst there have been many numerical models devised to describe the kinetic properties of biofilm from the chemical engineering point of view, the microbiological structure and physiology of these films as ecosystems have scarcely been mentioned in the literature. Another group of *in vitro* models has been described by researchers into dental plaque as biofilm. Some

Table 3. *Model film fermenters*

Fermenters type	Organism	Nutrient source	Nutrient concentration (mg l^{-1})	Surface	Film thickness (mm)	Comment	Reference
Flat plate	mixed	glucose, mineral-salts	25–1000	plastic	0.48–1.4	simulation of trickling filter	Maier *et al.* (1967)
Flat plate	mixed	glucose, mineral-salts	up to 10 000	glass or aluminium	0.073 or 2.0	kinetics of fixed film reactors. Constant film thickness	Atkinson *et al.* (1968), Atkinson & Daoud (1970)
Rotating disc	mixed	sewage	N.D.	N.D.	0.194	4 foot diameter discs for effluent treatment	Borchardt (1971)
Rotating disc	mixed	various effluents	5.9–365 (BOD)	N.D.	0.210	sectored rotating discs. Describes model for substrate uptake kinetics	Farmularo *et al.* (1978)
Rotating drum	mixed	glucose, mineral-salts	27–440	plastic	up to 0.25	kinetics of fixed film reactors	Kornegay & Andrews (1968)
Rotating tube	mixed	sewage	200 (BOD)	perspex coated with kieselguhr	1.15–1.5	oxygen penetration measured at 0.2 mm	Tomlinson & Snaddon (1966)
Packed bed	mixed	glucose, mineral-salts	180–330	plastic	0.705	4 inch tube plus table tennis balls. Model trickling filter	Monadjemi & Behn (1971)

Table 3. *Model film fermenters—(contd.)*

Fermenters type	Organism	Nutrient source	Nutrient concentration (mg 1^{-1})	Surface	Film thickness (mm)	Comment	Reference
Packed column	mixed	sewage	250–300	plastic	up to 2.0	trickling filter model	Bruce & Merkens (1970)
Packed column	mixed	molasses	up to 3000	plastic tube or sheet	0.118–2.55	trickling Filter model	Rinche & Wolters (1971)
Packed column	mixed	acetate	2.2–7.2	glass	up to 0.25	solute gradient and film thickness	Rittman & McCarty (1980)
Glass slide or cover slip	mixed	sewage	50	glass	0.076–0.096	film reaction kinetics	Lee *et al.* (1976)
Glass slide or cover slip	mixed	nutrient broth	—	glass	0.021	limiting film thickness	Sanders (1966)
Glass slide or cover slip	mixed	nutrient broth	N.D.	glass	0.2	oxygen profiles in film using microelectrodes	Bungay *et al.* (1969), Whalen *et al.* (1969)
Glass slide or cover slip	30 pure species	various	N.D.	various glasses	0.001	survey of microbial adhesion to surfaces	Zvyagintsev (1959)
Glass slide or cover slip	mixed	seawater	4–5 (total organics)	glass	0.0012	microbial adhesion to surfaces	Zobell (1946)
Glass slide or cover slip	mixed	seawater	N.D.	glass, ground glass	0.00123–0.0056	role of bacteria in marine fouling	Wood (1950)
Glass or cover slips	mixed	seawater plus or minus 10 mg glucose or nutrient broth		glass	0.02–0.06	stack of cover slips in laminar flow of nutrient	Pederson (1982)
Glass	mixed or	various	5–100	glass	0.004–0.0122	film attachment	Dias *et al.*

Glass fermenter	8 species	nutrient broth	N.D.	glass	0.001	comparison of attachment to fermenter walls	Larsen & Dimmick (1964)
Continuously stirred or fluidized bed	mixed	methanol	16–57	glass	0.04–1.207	fluidized bed reactor	Mulcahy & LaMotta (1978)
Continuously stirred or fluidized bed	mixed or a pure yeast culture	various	N.D.	glass or steel filament balls or plastic	0.23–0.69	fluidized bed reactor for effluent treatment or for mass culture	Atkinson & Davies (1972) Atkinson & Knight (1975) Atkinson *et al.* (1968)
Artificial mouth	cariogenic streptococci	complex including synthetic saliva	N.D.	enamel (extracted teeth)		changes in pH and Eh in *in vitro* plaques	Russell & Coulter (1975)
Artificial mouth and test tubes	pure cariogenic streptococci	cycle of complex media	N.D.	enamel acrylic porcelain glass nichrome stainless steel	N.D.	cariogenicity and plaque formation in *in vitro* model systems	Jordan & Keyes (1966)
Miscellaneous	mixed	nutrient broth	500	rock (trickling filter)	0.4–1.5	pO_2 in natural film using microelectrodes	Chen & Bungay (1981)
Miscellaneous	mixed or pure *Nitrobacter*	tapwater plus 100 mg each of NH_4^+ NO_2^- or NO_3^-		membrane	0.2–0.5	film of filtered packed cells	Williamson & McCarty (1976)
Miscellaneous	pure cultures of oral bacteria	complex		PTFE	0.3	constant depth PTFE pan fermenter	Coombe *et al.* (1982)

N.D. = not determined.

Table 4. *Information sought concerning the formation, organization and function of microbial films*

1. What is the effect of substratum surface on (a) adsorption of a molecular film for example of glycoproteins? (b) colonization by bacteria?
2. What is the rate of film growth and its structural differentiation?
3. What are the numbers and types of the colonizing species?
4. How are these species distributed in space and time over the life of the film?
5. What is the spatial and temporal distribution of cell viability over over the life of the film?
6. What is the nature of and the part played by matrix material?
7. What are the dynamic properties of solute transfer between the film and its surrounding environment?
8. What is the distribution of physico-chemical elements such as pH, Eh, pO_2 and of physiologically important solutes across the film?
9. What is the distribution of respiratory capability, energy charge, oxidative enzymes, hydrolytic enzymes, polymer-synthesizing activities, etc, across the film?
10. What are the reasons for and the kinetics of film detachment?
11. What are the physical properties of microbial film?
12. What intracellular interactions take place in microbial film?
13. What is the ecological role and evolutionary significance of microbial film?

examples are included in Table 3; however, the approach taken here seems a whole world away from that taken by the chemical engineers. For example, there are few estimates of film thickness and little effort has been made to determine the rate constants for solute exchange between the liquid phase and the film itself. On the other hand plaque investigators have concentrated on:

(1) a description of the microbial population of plaque;

(2) the microscopic structure of plaque organization using optical, transmission and scanning electron microscopy;

(3) changes in pH and in redox potential and in concentration of certain solutes in plaque and;

(4) the composition and solute transfer properties of plaque fluids.

It seems worth setting out in Table 4 some of the questions that need answering before the biology of microbial film can be properly understood. If it is not always possible to ask these questions of natural microbial film, then film model systems ought to be capable of providing sensible answers.

The establishment of these aims determines the design of laboratory model fermenters. Thus, the ideal laboratory film fermenter ought, as a minimum, to have the following features:

(1) The ability to use any solid material as a surface for establishing film growth.

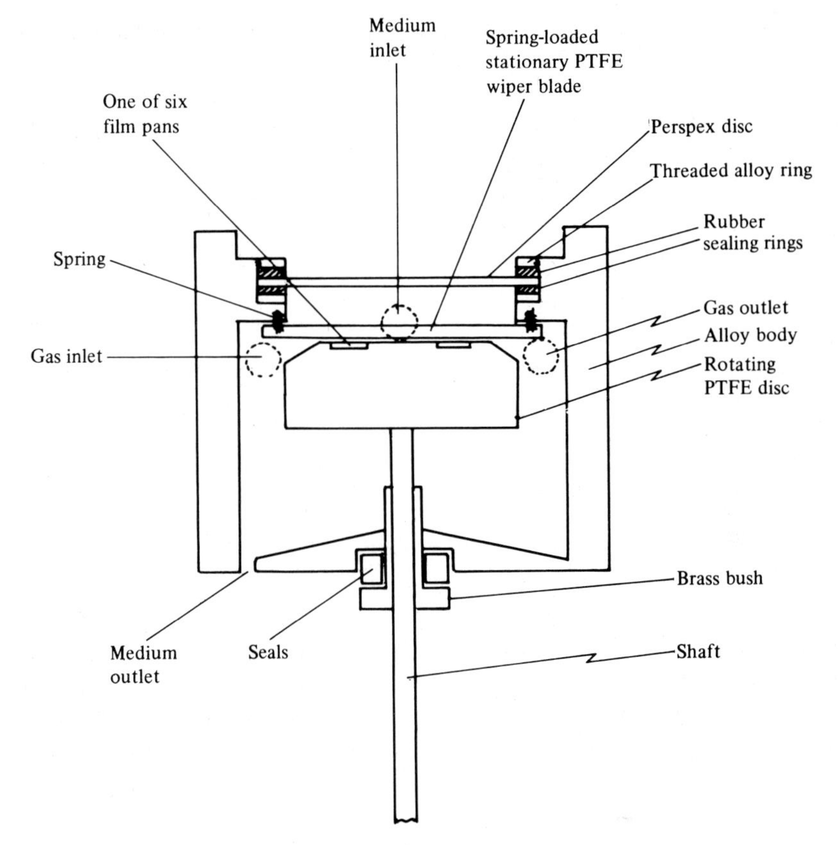

Fig. 12. Diagram of a laboratory model thin film fermenter developed in Cardiff by R. A. Coombe, A. Tatevossian & J. W. T. Wimpenny.

(2) A nutrient feed system that always allows any programmed regime of solute flow over the surface of the film.

(3) A sterile enclosure allowing the growth of pure single or mutiple species films.

(4) Facilities for producing 'steady state' or, at any rate, constant depth films to ensure reproducibility of film behaviour.

(5) Some system to allow the measurement of gradients in pH, Eh, pO_2 or pS^{2-} using microelectrode techniques.

(6) Facilities for fixing and sectioning the film for scanning and transmission electron microscopy and for electron microprobe analysis.

(7) Facilities for freeze sectioning the film in any dimension allowing an examination of the distribution of species, viability,

enzymes, substrates and metabolic products across the depth of the film.

The design developed by Coombe *et al.* (1982) goes some way to providing these facilities. The fermenter (Fig. 12) is enclosed in a sterile metal container and consists of a rotating PTFE disc containing six PTFE film pans. Each pan has a rim which is flush with the mother disc in operation. Each pan has a shallow recess, in our case this was 300 μm deep but could be any chosen depth. Film was grown on the PTFE base to the pan but the latter could easily be modified to include a disc of any material as its base. The pans were made from PTFE to facilitate freeze-sectioning of the entire pan plus its contents to minimize distortion to film sections. The rotating pan holder is swept by a stationary PTFE wiper blade to ensure that a constant depth film is produced. At the same time the disc is irrigated with sterile nutrient solution at any chosen rate. The fermenter itself can be gassed with any gas mixture of interest.

It is clear that, since the disc is rotating during operation microelectrodes cannot be deployed during film growth. The film pan may be removed at any time and replaced by a PTFE blank disc. The pan is now located in an inclined PTFE holder and nutrient solution is allowed to flow as a thin even laminar stream over its surface. A microelectrode holder attached to a micromanipulator is arranged normal to the surface of the film and measurements made as required. Clearly it is possible to alter the composition of the bathing solution to investigate transients in microelectrode output using this system. In short term experiments microbial growth and hence increases in film depth are likely to be small compared to the global compositions of the film itself.

DISCUSSION

To investigate the architecture of microbial ecosystems is to enter an exciting and sometimes beautiful new world, far removed from the featureless homogeneity of the chemostat! What information, for example, does the latter give us about the intricate pattern on the surface of the colony of *Bacillus cereus* or the strange spatial organization multiple band formation when this organism grows in gel-stabilized model systems?

Microbial ecology is about space and its occupation by organisms whose genotypes marshal the physiology of each individual so that its phenotypic behaviour is its best possible response to the physico-chemical constraints in its immediate environment. Microbial ecology is also about water, its presence or absence and the chemicals dissolved in it, since it is water itself that dominates the physico-chemical environment. Finally, microbial ecology is about time and the orderly unfolding of community organization with each second that passes.

The model systems briefly discussed in this chapter each incorporate elements of the heterogeneity found in natural habitats. The models and their most important properties are classified in Table 5. Of all the models described, only the linked stirred fermenter systems are capable of reaching true steady states and are hence open systems. The addition of bidirectionality to such models can allow growth in opposing gradients of solutes. It should be noted that in such systems, indeed in all natural systems incorporating spatial heterogeneity, a gradient of a limiting nutrient only exists between the source and the zone of growth. In other words, if a nutrient is growth-limiting it will not be transferred in measurable quantities beyond the organisms which need it. The linked homogeneous systems produce stepped changes in solute concentration rather than a true gradient in which solute concentrations alter continuously. This means that spatial resolution is poor, limited as it is to the number of vessels in the array. In practice five vessels seems a reasonable compromise between complexity and resolution. The major advantage of the bidirectional homogeneous systems is in their openness which allows the establishment of steady states, which in turn permits unequivocal interpretations on altering operating conditions.

Percolating columns as models are valuable analogues of ecosystems where there is a net flow of solutions in one direction. These naturally include soil systems, but could apply the motion of water in aquifers of all types. The percolating column is open to solutions and anything dissolved in them. They are not generally open to cell populations attached to surfaces, although in long-term experiments where net growth is exactly balanced by death and the sloughing off of cells, quasi-steady states can be achieved.

Gel-stabilized models are closed for cell populations and for most solutes as well, however, they may be open for gas transfer. Solute gradients are 'natural' and continuous. They therefore have the resolution necessary to distinguish between processes changing over very small regions in space as is made clear by the generation of

Table 5. *The main classes of laboratory model system used in the investigation of spatially or temporally organized microbial ecosystems*

Class of model	Characteristic properties	Examples of use	Reference
Single vessel homogeneous systems with cyclical environmental variations	Open, non-steady-state system showing temporal heterogeneity	Coexistence of competing species under variable light intensity	Van Gemerden (1974)
		Uptake kinetics	Beeftink & Van Gemerden (1979)
		Enrichment, competition and adaptation in temporally heterogeneous environments	Gottschall, (1980)
Linked homogeneous systems	Open systems, steady states usually possible, solute flow uni- or bi-directional. Each vessel homogenous	Estuarine model: growth in salinity gradients, Gradostat: opposing solute gradients established. Used to examine adaptation, competition, natural enrichments and spatial determinants in	Cooper & Copeland (1973), Lovitt & Wimpenny (1981 a,b), Wimpenny, (1981, 1982)

Percolating columns	Flow by gravity and capillarity through a column packed with solid substratum. Two orders of heterogeneity: (i) vertical segregation of species and solutes; (ii) microscopic heterogeneity at the level of soil crumb or film on glass beads	Commonly used to investigate soil ecosystems and nitrification	Lees & Quastel (1946), Macura & Malik, (1958), Bazin *et al.* (1976)
Gel-stabilized systems. (1) One dimensional	Aqueous ecosystem stabilized by the addition of gelling agent. Solute transfer by molecular diffusion only	Sensitivity of cells to oxygen	de Vries & Stouthamer (1969)
		Bacterial growth 'spectra' Liesegang ring formation	Williams, (1938a, 1939a) Tschapek & Giambiagi (1954), Wimpenny *et al.* (1981), Coombs & Wimpenny (1983)
Gel-stabilized systems. (2) Two dimensional	Two-dimensional gradients established along adjacent edges of square agar slabs	Analysis of microbial responses to gradients in two different solutes. Analysis of specific microbial habitats	Caldwell & Hirsch (1973), Wimpenny (1981)
Microbial film	Growth of layers of pure or mixed cultures of bacteria usually enclosed in polymer matrix. Models closely resemble natural ecosystem	Widely investigated in marine, effluent treatment and chemical-biological engineering including common health problems such as dental plaque	Atkinson & Fowler (1974), Characklis (1981), Russell & Coulter (1975), Coombe *et al.* (1981)
The bacterial colony	Natural multicellular array of cells dominated by diffusion gradients and having simple morphogenetic characteristics	Growth kinetics examined. Characteristic growth form. Oxygen penetration	Pirt (1967), Wimpenny (1979), Wimpenny & Lewis (1977), Wimpenny (1981)

multiple bands of growth. The transparency of gel systems offers a true 'window' on growth in the gel. The addition of particulate matter to gels may make the gel model closer to certain ecosystems such as marine sediments. Gels offer 'pure' solutions to ecological problems since they isolate molecular diffusion from other forms of solute transport. Gels are easy to manipulate, and allow extensive quantitative analysis. They can be investigated using computer simulations to verify experimental results or to make general predictions of the behaviour of microorganisms in ecosystems where solute flow is dominated by molecular diffusion.

Of all the model systems discussed in this article those established to investigate the behaviour of microbial film are the closest to reality. They are, in the short term, closed to cell growth though here again, in the longer term, sloughing away, death and nutrient recycling ensures that a sort of steady state develops. Whilst many model systems exist, not all of them are very useful and versions such as that described by Coombe *et al.* (1982), offer the best potential in understanding more about this ubiquitous and economically important ecosystem.

Finally the colony. This structure is a slightly specialized microcosm and it may be argued, not a model system at all. It is included in this discussion, because it is an entity whose spatial organization is due at least partly to solute diffusion gradients. It seems an appropriate model of other forms of morphogenesis and it is therefore worth detailed study.

To end with, we would like to consider microbial growth in structured ecosystems from a conceptually more abstract point of view. We must accept that rates of reaction can only be defined in scalar terms in homogeneous systems, whilst in spatially ordered ecosystems such phenomena have direction as well as magnitude. These vectorial reactions go some way to explaining the architecture of microbial communities.

If we consider a single organism as a discrete compartment, then it is surrounded by a region over which it has influence which we may call its 'domain'. Such domains may be source domains in which an organism is producing metabolites or they may be sink domains for substrates available to them in the environment. We may also consider the physico-chemical environment to be a domain whose composition governs the ability of particular phenotypes to proliferate in it. The 'habitat domain' of an organism and its multidimensional 'activity domains' are the two components making

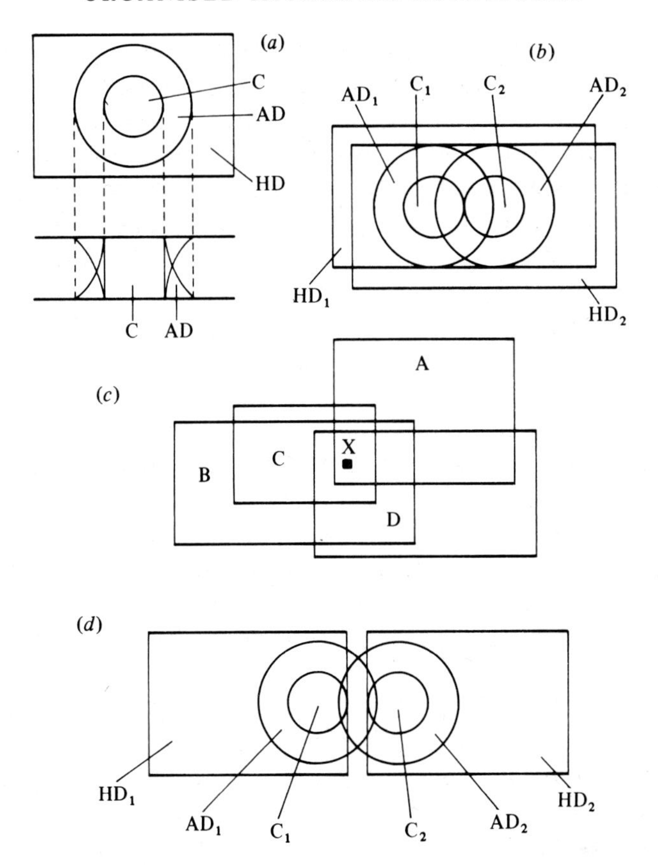

Fig. 13. Domains and compartments in microbial ecology. Compartment (C) and activity and habitat domains (AD and HD). The compartment may have source or sink domains for particular solutes as shown in the lower part of the figure. (*b*) Interacting compartments with overlapping habitat and activity domains. (*c*) A mixed microbial community growing under homogeneous conditions at one point (X) in their common habitat domains. Each organism shares a common activity domain. (*d*) Interacting compartments with exclusive habitat domains.

up its niche. Each type of domain has spatial boundaries to it. Thus the activity domain of a single organism is actually a small region surrounding the organism. A group of organisms organized perhaps as a stratum in a sedimentary ecosystem, may exert an influence over a considerable region above and below it. The activity domain of this stratum is then large.

Interactions between organisms or groups of organisms are of great importance to natural communities. Interactions may be beneficial or harmful but can only occur at significant rates if activity domains for interacting organisms or communities overlap. The degree of interaction or 'coupling' between compartments depends on the distance they are apart. Restrictions on this distance include the nature of the habitat domain or domains. Organisms with the same habitat domain can interact closely. Examples include syntrophic associations such as *Methanobacillus omelianski* (Bryant *et al.*, 1967) or *Chloropseudomonas ethylica* (Gray *et al.*, 1973). If habitat domains are exclusive, coupling can still occur if activity domains overlap. Thus, a sulphur cycle exists between thiobacilli located in an aerobic habitat domain and sulphate-reducing bacteria located in anaerobic habitat domains. Coupling occurs since sulphur compounds diffuse between the two habitat domains. Where two organisms compete for the same solute, the characteristics of their sink domain determine the outcome of the competition. Organisms with lower K_s values generate better sinks and hence have larger solute uptake domains than their competitors. Finally, solute flow is vectorial since it flows from sources to sinks and these occupy specific positions in spatially organized ecosystems. These ideas are illustrated in Fig. 13 and the whole approach (discussed more fully by Wimpenny, 1981) provides a conceptual framework for 'heterogeneous microbiology' and is at the same time an implicit condemnation of the use of homogeneous model systems to study it. We therefore strongly advocate the use of heterogeneous models such as those decribed here as a step towards a complete understanding of many microbial ecosystems.

Acknowledgements

We gratefully acknowledge help from many colleagues including H. Abdollahi, A. Tatevossian and M. W. A. Lewis. Special thanks are due to P. Waters who provided the colony photographs and to S. Jaffe for writing the computer simulation programmes. We would also like to thank the SERC, NERC and Esso Research Ltd., for grants that have supported many aspects of the work described in this article.

REFERENCES

Abson, J. W. & Todhunter, K. H. (1961). Plant for continuous biological treatment of carbonization effluents. *Society of the Chemical Industry Monograph,* **12,** 147–64.

Adler, J. (1966). Chemotaxis in bacteria. *Science,* **153,** 708–16.

Afrikian, E. G., St. Julian, G. & Bulla, L. A., Jr. (1973). Scanning electron microscopy of bacterial colonies. *Applied Microbiology,* **26,** 934–7.

Ardakani, M. S., Rehbock, J. T. & McLaren, A. D. (1973). Oxidation of nitrite to nitrate in a soil column. *Soil Science Society of America Proceedings,* **37,** 53–6.

Atkinson, B. & Daoud, I. S. (1970). The analogy between microbial 'reactions' and heterogeneous catalysis. *Transactions of the Institute of Chemical Engineers (London),* **48,** T245–T254.

Atkinson, B., Daoud, I. S. & Williams, D. A. (1968). A theory for the biological film reactor. *Transactions of the Institute of Chemical Engineers (London),* **6,** T245–T250.

Atkinson, B. & Davies, I. J. (1972). Completely mixed microbial film fermenter. Method for overcoming washout in continuous fermentation. *Transactions of the Institute of Chemical Engineers (London),* **50,** 208.

Atkinson, B. & Fowler, H. W. (1974). The significance of microbial film in fermenters. *Advances in Biochemical Engineering,* vol. 3, ed. T. K. Ghose, A. Fiechter & N. Blakebrough, pp. 224–77. Springer-Verlag, New York.

Atkinson, B. & Knight, A. J. (1975). Microbial film fermenters their present and future applications. *Biotechnology and Bioengineering,* **17,** 1245–67.

Bauer, H., Sigarlakie, E. & Faure, J. C. (1975). Scanning electron microscopy of three strains of *Bifidobacterium. Canadian Journal of Microbiology,* **21,** 1305–16.

Bazin, M. J. & Saunders, P. T. (1973). Dynamics of nitrification in a continuous flow system. *Soil Biology and Biochemistry,* **5,** 531–43.

Bazin, M. J., Saunders, P. T. & Prosser, J. I. (1976). Models of microbial interaction in the soil. *CRC Critical Reviews of Microbiology,* **4,** 463–99.

Beeftink, H. H. & van Gemerden, H. (1979). Actual and potential rates of substrate oxidation and product formation in continuous cultures of *Chromatium vinosum. Archives of Microbiology,* **121,** 161–7.

Beijerinck, M. (1889). Auxanography, a method useful in microbiological research, involving diffusion in gelatin. *Archives Neerlandaises des Sciences Exactes et Naturelles, Haarlem,* **23,** 367–72.

Bibel, D. J. & Lawson, J. W. (1972a). Scanning electron microscopy of L-phase Streptococci: Development of techniques. *Journal of Microscopy,* **95,** 435–53.

Bibel, D. J. & Lawson, J. W. (1972b). Development of streptococci L-form colonies. *Journal of Bacteriology,* **112,** 602–10.

Borchardt, J. A. (1971). Biological waste treatment using rotating discs. *Biotechnology and Bioengineering Symposium,* **2,** 131–40.

Bruce, A. M. & Merkens, J. C. (1970). Recent studies of high rate biological filtration. *Water Pollution Control,* **69,** 113–48.

Bryant, M. P., Wolin, E. A., Wolin, M. J. & Wolfe, R. S. (1967). *Methanobacillus omelianski,* a symbiotic association of two species of bacteria. *Archiv für Mikrobiologie,* **59,** 20–31.

Bungay, H. R. & Chen, Y. S. (1981). Dissolved oxygen profiles in photosynthetic microbial slimes. *Biotechnology and Bioengineering,* **23,** 1893–5.

Bungay, H. R., Whalen, W. J. & Sanders, W. M. (1969). Microprobe technique for determining diffusivities and respiration in microbial slime systems. *Biotechnology and Bioengineering*, **11,** 765–72.

Caldwell, D. E. & Hirsch, P. (1973). Growth of microorganisms in two-dimensional steady-state diffusion gradients. *Canadian Journal of Microbiology,* **19,** 53–8.

Caldwell, D. E., Lai, S. H. & Tiedje, J. M. (1973). A two-dimensional steady-state diffusion gradient for ecological studies. *Bulletin of Ecology Research Communication (Stockholm),* **17,** 151–8.

Characklis, W. G. (1980). Fouling biofilm development: a process analysis. *Biotechnology and Bioengineering* **23,** 1923–60.

Chen, Y. S. & Bungay, H. R. (1981). Microelectrode studies of oxygen transfer in trickling filter slimes. *Biotechnology and Bioengineering*, **23,** 781–92.

Chisholm, S. N. & Nobbs, A. A. (1976). Simulation of algal growth and competition in a phosphate limited cyclostat. In *Modelling biochemical processes in aquatic ecosystems*, ed. R. P. Canale, pp. 337–57. Ann Arbor Science, Michigan.

Coombe, R. A., Tatevossian, A. & Wimpenny, J. W. T.(1982). Bacterial thin films as *in vitro* models for dental plaque. In *Surface and Colloid Phenomena in the Oral Cavity: Methodological Aspects,* ed. R. M. Frank & S. A. Leach, pp. 239–49. IRL Press, London.

Coombs, J. P. & Wimpenny, J. W. T. (1982). The growth of *Bacillus cereus* in a gel-stabilised nutrient gradient system. *Journal of General Microbiology,* **128,** 3093–101.

Cooper, A. L., Dean, A. C. R. & Hinshelwood, D. C. (1968). Factors affecting the growth of bacterial colonies on agar plates. *Proceedings of the Royal Society of London, B,* **171,** 175–99.

Cooper, D. C. & Copeland, B. J. (1973). Responses of continuous-series estuarine microecosystems to point-source input variations. *Ecological Monographs,* **43,** 213–36.

Crowle, A. J. (1973). *Immunodiffusion.* Academic Press, London.

De Vries, W. & Stouthamer, A. H. (1969). Factors determining the degree of anaerobiosis of *Bifidobacterium* strains. *Archiv für Mikrobiologie,* **65,** 275–87.

Dias, F. F., Dondero, N. C. & Finstein, M. S. (1968). Attached growth of *Sphaerotilus* and mixed populations in a continuous flow apparatus. *Applied Microbiology,* **16,** 1191–9.

Drucker, D. B. & Whittaker, D. K. (1971a). Microstructure of colonies of rod-shaped bacteria. *Journal of Bacteriology,* **108,** 515–25.

Drucker, D. B. & Whittaker, D. K. (1971b). Examination of certain bacterial colonies by scanning electron microscope. *Microbios*, **4,** 109–13.

Famularo, J., Mueller, J. A. & Mulligan, T. (1978). Application of mass transfer to rotating biological contractors. *Journal of the Water Pollution Control Federation,* **50,** 653–71.

Fass, R. J. (1973). Morphology of ultrastructure of Staphylococcus L-colonies. Light, scanning and transmission electron microscopy. *Journal of Bacteriology,* **113,** 1049–53.

Gottschall, J. C. (1980). Mixotrophic growth of *Thiobacillus* A2 and its ecological significance. Ph.D. Thesis, The University of Groningen, Holland.

Gray, B. H., Fowler, C. F., Nugent, N. A., Rigopoulis, N. & Fuller, R. C. (1973). Reevaluation of *Chloropseudomonas ethylica* 2-K. *International Journal of Systematic Bacteriology,* **23,** 256–64.

Hawkes, H. A. (1977). Eutrophication of rivers, effects, causes and control. In *Treatment of Industrial Effluents*, ed. A. G. Callely, C. F. Forster & D. A. Stafford, pp. 159–92. Hodder & Stoughton, London.

Hedges, E. S. (1932). *Liesegang Rings and Other Periodic Phenomena.* Chapman & Hall, London.

Herbert, D. (1961). A theoretical analysis of continuous culture systems. In *Continuous Culture of Microorganisms.* Society for Chemistry and Industry Monograph, No 12, pp. 21–53. Society of Chemical Industry, London.

Howell, J. A. & Atkinson, B. (1976). Sloughing of microbial film in trickling filters. *Water Research*, **10,** 307–15.

Jannasch, H. W. & Mateles, R. I. (1974). Experimental bacterial ecology studied in continuous culture. *Advances in Microbial Physiology,* **11,** 161–209.

Jordan, H. V. & Keyes, P. H. (1966). *In vitro* methods for the study of plaque formation and carious lesions. *Archives of Oral Biology*, **11,** 793–801.

JØRGENSEN, B. B. (1981). Mineralisation and the cycling of carbon, nitrogen and sulphur in marine sediments. In *Contemporary Microbial Ecology*, ed. D. C. Ellwood, J. N. Hedger, M. J. Latham, J. M. Lynch & J. H. Slater, pp. 239–57. Academic Press, London.

KORNEGAY, B. H. & ANDREWS, J. F. (1968). Kinetics of fixed film biological reactors. *Journal of the Water Pollution Control Federation,* **40,** 460–8.

KUENEN, J. G. & HARDER, W. (1982). Microbial competition in continuous culture. In *Experimental Microbial Ecology*, ed. R. G. Burns & J. H. Slater, pp. 342–67. Blackwell Scientific Publications, Oxford.

LARSEN, D. H. & DIMMICK, R. L. (1964). Attachment and growth of bacteria on surfaces of continuous culture vessels. *Journal of Bacteriology,* **88,** 1380–7.

LEE, E. J., DE WITT, K. J., BENNETT, G. G. & BROCKWELL, J. L. (1976). Investigation of oxygen transfer to slime as a surface reaction. *Water Research,* **10,** 1011–17.

LEES, H. C. & QUASTEL, J. H. (1946). Biochemistry of nitrification in soil. I. Kinetics of and the effects of poisons on soil nitrification, as studied by a soil perfusion technique. *Biochemical Journal,* **40,** 803–15.

LEGROUX, R. & MAGROU, J. (1923). Etat organisé des colonies bacteriennes. *Annales de l'Institut Pasteur,* **34,** 417–33.

LIESEGANG, R. E. (1896). Ueber einige Eigenschaften von Gellerton. *Naturwissenschaften Wohenschrift,* **11,** 353–62.

LOVITT, R. W. & WIMPENNY, J. W. T. (1979). The gradostat: a tool for investigating microbial growth and interactions in solute gradients. *Society for General Microbiology Quarterly,* **6,** 80.

LOVITT, R. W. & WIMPENNY, J. W. T. (1981a). The gradostat: a bidirectional compound chemostat and its applications in microbiological research. *Journal of General Microbiology,* **127,** 261–8.

LOVITT, R. W. & WIMPENNY, J. W. T. (1981b). Physiological behaviour of *Escherichia coli* grown in opposing gradients of oxidant and reductant in the gradostat. *Journal of General Microbiology,* **127,** 269–76.

MCLAREN, A. D. (1969). Steady state studies of nitrification in soil: theoretical considerations. *Soil Society of America Proceedings,* **33,** 273.

MACURA, J. & KUNC, F. (1965a). Continuous flow method in soil microbiology. IV. Decomposition of glycine. *Folia Microbiologyia,* **10,** 115–24.

MACURA, J. & KUNCK, F. (1965b). Continuous flow method in soil microbiology. V. Nitrification. *Folia Microbiologica,* **10,** 125–35.

MACURA, J. & MALIK, I. (1958). Continuous flow method for the study of microbiological processes in soil samples. *Nature,* **182,** 1796–7.

MAIER, W. J., BEHN, V. C. & GATES, C. D. (1967). Simulations of the trickling filter process. *Journal of the Sanitation Engineering Division, American Society of Civil Engineering*, **93,** 91–112.

MARGALEF, R. (1967). Laboratory analogues of estuarine plankton systems. In *Estuaries: Ecology and Populations*, ed. G. M. Lauff, pp. 515–24. Hornshafer, Baltimore.

MONADJEMI, P. & BEHN, V. C. (1971). Oxygen uptake and mechanism of substrate purification in a model trickling filter. *Proceedings of the Fifth International Water Pollution Research Conference,* Paper number 11–12.

MONOD, J. (1950). La technique de culture continue; theorie et applications. *Annales de l'Institut Pasteur,* **79,** 390–410.

MULCAHY, L. T. & LAMOTTA, E. J. (1978). Mathematical model of the fluidised bed biofilm reactor. Report No. Env. E. 59–78–2, Environmental Engineering program, Department of Civil Engineering, University of Massachussetts.

NOVICK, A. & SZILLARD, L. (1950). Description of the chemostat. *Science,* **112,** 714–16.

Palumbo, S. A., Rieck, V. T. & Witter, L. D. (1964). Growth parameters of surface colonies of bacteria. *Bacteriological Proceedings,* G106.

Palumbo, S. A., Johnson, M. G., Rieck, V. T. & Witter, L. D. (1971). Growth measurements on surface colonies of bacteria. *Journal of General Microbiology,* **66,** 137–43.

Pedersen, K. (1982). Method for studying microbial biofilms in flowing-water systems. *Applied and Environmental Microbiology,* **43,** 6–13.

Perfil'ev, B. V. & Gabe, D. R. (1969). *Capillary Methods of Investigating Microorganisms.* (English translation). Oliver & Boyd, Edinburgh.

Pirt, S. J. (1967). A kinetic study of the mode of growth of surface colonies of bacteria and fungi. *Journal of General Microbiology,* **47,** 181–97.

Pirt, S. J. (1975). *Principles of Microbe and Cell Cultivation.* Blackwell Scientific Publications, Oxford.

Pritchard, P. M., Ventullo, R. M. & Suflita, J. M. (1976). The microbial degradation of diesel oil in a multistage continuous culture system. In *Proceedings of the 3rd International Biodegradation Symposium*, ed. J. Miles Sharply & A. M. Kaplan. Applied Science Publishers, London.

Prosser, J. I. & Gray, T. R. G. (1977). Use of finite difference method to study a model system of nitrification at low substrate concentrations. *Journal of General Microbiology,* **102,** 119–28.

Rincke, G. & Wolters, N. (1971). Technology of plastic medium trickling filters. *Proceedings of the Fifth International Water Pollution Research Conference,* Paper number 11–15.

Rittmann, B. E. & McCarty, P. L. (1980). Model of steady-state-biofilm kinetics. *Biotechnology and Bioengineering,* **22,** 2343–57.

Russell, C. & Coulter, W. A. (1975). Continuous monitoring of pH and Eh in bacterial plaque grown on a tooth in an artificial mouth. *Applied Microbiology,* **29,** 141–4.

Sanders, W. M. iii (1966). Oxygen utilisation by slime organisms in continuous culture. *Air and Water Pollution International Journal,* **10,** 253.

Slater, J. H. (1981). Mixed cultures and microbial communities. In *Mixed Culture Fermentations*, ed. M. E. Bushell & J. H. Slater, pp. 1–24. Academic Press, London.

Spray, R. S. (1936). Semisolid media for the cultivation and identification of the sporulating anaerobes. *Journal of Bacteriology,* **32,** 135–55.

Szybalski, W. (1952). Gradient plates for the study of microbial resistance to antibiotics. *Bacteriological Proceedings,* 36.

Todd, R. L. & Kerr, T. J. (1972). Scanning electron microscopy of microbial cells on membrane filters. *Applied Microbiology,* **23,** 1160–2.

Tomlinson, T. G. & Snaddon, D. H. M. (1966). Biological oxidation of sewage by films of microorganisms. *International Journal of Air and Water Pollution,* **10,** 865–81.

Tschapek, M. & Giambiagi, N. (1954). Die bildung von Liesegang'schen ringen durch Azotobakter bei O_2-hemmung. *Kolloid Zeitschrift,* **135,** 47–8.

Van Gemerden, H. (1974). Coexistence of organisms competing for the same substrate: an example among the purple sulfur bacteria. *Microbial Ecology,* **1,** 104–19.

Whalen, W. J., Bungay, H. R. & Sanders, W. M. (1969). Microelectrode determinations of oxygen profiles in microbial slime systems. *Environmental Science and Technology,* **3,** 1297–8.

Whittaker, D. K. & Drucker, D. B. (1970). Scanning electron microscopy of intact colonies of microorganisms. *Journal of Bacteriology,* **104,** 902–9.

WHITTENBURY, R. (1963). The use of soft agar in the study of conditions affecting the utilization of fermentable substrates by lactic acid bacteria. *Journal of General Microbiology,* **32,** 375–84.
WILLIAMS, J. W. (1938a). Bacterial growth 'spectrum' analysis. Methods and applications. *The American Journal of Medical Technology,* **4,** 58–61.
WILLIAMS, J. W. (1938b). Bacterial growth 'spectrums'. II. Their significance in pathology and bacteriology. *American Journal of Medical Technology,* **14,** 642–5.
WILLIAMS, J. W. (1939a). The nature of gel mediums as determined by various gas tensions and its importance in growth of microorgansims and cellular metabolism. *Growth,* **3,** 181–96.
WILLIAMS, J. W. (1939b). Growth of microorganisms in shake cultures under increased oxygen and carbon dioxide tensions. *Growth,* **3,** 21–33.
WILLIAMSON, K. & MCCARTY, P. L. (1976). Verification studies of the bio-film model for bacterial substrate utilization. *Journal of Water Pollution Research Federation,* **48,** 281–96.
WIMPENNY, J. W. T. (1979). The growth and form of bacterial colonies. *Journal of General Microbiology,* **114,** 483–6.
WIMPENNY, J. W. T. (1981). Spatial order in microbial ecosystems. *Biological Reviews,* **56,** 295–342.
WIMPENNY, J. W. T. (1982). Responses of microbes to physical and chemical gradients. *Philosophical Transactions of the Royal Society of London, B,* **297,** 497–515.
WIMPENNY, J. W. T., COOMBS, J. P., LOVITT, R. W. & WHITTAKER, S. G. (1981). A gel-stabilised model ecosystem for investigating microbial growth in spatially ordered solute gradients. *Journal of General Microbiology,* **127,** 277–87.
WIMPENNY, J. W. T. & LEWIS, M. W. A. (1977). The growth and respiration of bacterial colonies. *Journal of General Microbiology,* **103,** 9–18.
WIMPENNY, J. W. T. & PARR, J. A. (1979). Biochemical differentiation in large colonies of *Enterobacter cloacae. Journal of General Microbiology,* **114,** 487–9.
WIMPENNY, J. W. T. & WHITTAKER, S. (1979). Microbial growth in gel-stabilised nutrient gradients. *Society for General Microbiology Quarterly,* **6,** 80.
WOOD, E. J. F. (1950). Investigation on underwater fouling. II. The biology of fouling in Australia. *Australian Journal of Marine and Freshwater Research,* **1,** 85.
ZOBELL, C. E. (1943). The effect of solid surfaces upon bacterial activity. *Journal of Bacteriology,* **46,** 39–56.
ZOBELL, C. E. (1946). *Marine Microbiology: A Monograph of Hydrobacteriology.* Chronica Botanica Co., Waltham, Mass.
ZVYAGINTSEV, D. G. (1959). Adsorption of microorganisms by glass surfaces. *Microbiology,* **28,** 104–8.

PROPERTIES AND PERFORMANCE OF MICROORGANISMS IN LABORATORY CULTURE; THEIR RELEVANCE TO GROWTH IN NATURAL ECOSYSTEMS

DAVID W. TEMPEST, OENSE M. NEIJSSEL AND WANDA ZEVENBOOM

Laboratorium voor Microbiologie, Universiteit van Amsterdam, Nieuwe Achtergracht 127, 1018 WS Amsterdam, The Netherlands

INTRODUCTION

In our view, a serious misconception is engendered in the currently popular aphorism that 'what is true for *Escherichia coli* is also true for an elephant, only more so'.[1] For whereas this notion may gain support from the fact that the cells of an elephant are, like the *E. coli* cell, composed of the ubiquitous biopolymers (nucleic acids, proteins, carbohydrates and lipids) that are coded, synthesized and structured in a uniform fashion, nevertheless there are marked *physiological* differences between the two types of cells. These differences are not simply, nor principally, those that distinguish prokaryotes from eukaryotes, but derive from the environmental conditions that the cells of higher animals routinely experience, as compared with those of the free-living microbe. Thus, the cells that together make up an elephant (or a man) spend the whole of their existence in a closely regulated (internal) environment. Generally they have only poorly developed powers of adaptability and if the animals' internal environment is caused to shift beyond certain narrow limits (of, say, temperature, pH, oxygenation, salt balance, nutrient concentration, etc.) the cells cease to function and the animal dies. In contrast, the free-living microbial cells frequently must experience shifts of environment which they are powerless to control. Indeed, in a closed environment like that of a batch culture, microbes provoke through their own metabolism extensive shifts in the chemical environment; yet, seemingly, they can readily and rapidly accommodate to these changes, often without any perceptible effect on growth rate. This quick accommodation to marked shifts in the environment is effected, we now know, by organisms

[1]Attributed to J. Monod (see Koch, 1976).

changing themselves, structurally and functionally; and to such an extent can these free-living cells change *phenotypically*, as it is said, that it is quite impossible to specify any microbe in precise terms of structure and function without, at the same time, specifying the environmental conditions prevailing during its growth.

This capacity of microbial cells to modulate extensively their own structure and functioning accounts for the ease with which such cells can be cultivated in the laboratory under conditions (of batch or continuous culture) that may be far removed from those prevailing in their natural surroundings; but it raises the question of the relationship between this immensely plastic physiology and growth conditions extant in natural ecosystems. Does this mobile physiology, in fact, represent properties and processes that microbes necessarily must have acquired in the course of evolution to cope with the vicissitudes of life outside the laboratory culture? To what extent **are** the properties and performances of microorganisms in laboratory culture relevant to their growth in natural environments? Within these questions lie thoughts which circumscribe the scientific content of this chapter.

In the above connection, it is important to recognise that natural ecosystems generally are heterogeneous with respect to their indigenous microbial populations, and it is not easy, therefore, to assess the properties that are being expressed *in situ* by individual species within these populations. Hence, in order to come to some understanding of their behaviour in natural environments, it has become established practice to isolate and cultivate representative species *in vitro* and to examine their behaviour in laboratory culture. This classical approach to studies of microbial eco-physiology has unquestionably yielded a vast body of relevant information, although inherent in it is the tacit assumption that the properties and performances of microbes in axenic laboratory culture mirror closely those expressed in their natural habitat(s). In a broader, more general, context this is clearly true; methanogenic bacteria grow in the rumen of the cow and generate methane, in much the same way as they do in a laboratory culture. But can the same be said of the behaviour of *Escherichia coli* in the human intestine? Does it ever grow in the gut with a doubling time of 20 minutes? Perish the thought! So at what rate does *E. coli* grow in the gut or, for that matter, in some polluted body of water? What is the nature of the growth-constraining influence(s), and how do these affect the physiology of the cells?

Among the many factors constraining growth (like temperature, pH, etc.) it reasonably might be concluded that nutrient availability is the one most likely to limit overall cell proliferation. The appetites of most microbes are so voracious, and their mineralizing potential so vast, that even if the oceans were suddenly to be filled with nutrient broth they would not sustain a rapid growth of microorganisms for more than a few days. So other considerations apart, microbial populations frequently, and ubiquitously, must be limited in their growth (rate) by the availability of essential nutrient substances; and it is thus in their response to such extreme, nutrient-limiting conditions that we should expect to find expressed properties that are ecologically important (Veldkamp, 1977; Tempest & Neijssel, 1978).

A crucial question that follows from the above reasoning, and one that has not been adequately answered so far, is whether, under nutritionally restricting conditions, cellular proliferation continues (albeit at a grossly sub-maximal rate) or whether organisms opt for a period of rest to await a further influx of nutrient before resuming growth. Studies with chemostat cultures have shown convincingly that the vast majority of known microbial species can accommodate readily and rapidly to nutrient-limited environments, and can continue to grow at rates that are but 10% or less of the potential maximum. On the other hand, there are of course some species that not only form resting bodies (for example, spores) when starved of nutrient, but do so in nutrient-limited chemostat environments (Dawes & Mandelstam, 1970); so there is no uniformity in the strategy and tactics adopted by different microbial species in coping with extremes of environment. However, in attempting, as we do here, to assess the relevance to growth in natural environments of the properties and performances of organisms in laboratory culture, it is sensible, as a first approximation, to consider the behaviour of but a few microbial species when exposed to conditions that one might expect to be widespread in the biosphere.

Although one might imagine the biosphere to contain an almost infinite variety of ecosystems, certain types must predominate. For, as pointed out by Shilo (1980), 'seventy percent of the surface of the biosphere, and more than 90% of its volume is occupied by aquatic biotopes' and that 'most of these ecosystems can be considered extreme habitats since they approach in their nutrient content the lowest possible limits necessary to sustain organisms'. Hence, in this contribution, it is appropriate to concentrate attention upon low-nutrient environments and to consider principally the influence that specific nutrient-limiting conditions exert on the physiology of a few selected species (for example, *Klebsiella aerogenes, Bacillus subtilis, Candida utilis*), with special reference to structural and functional changes, regulation of nutrient fluxes and their energetic consequences. Within this body of information are to be seen changes in cellular physiology of a more general kind (for example, essentially quantitative changes in the content of some components), but some other changes are more specific and, indeed, sufficiently so as to be clearly (and, perhaps, unequivocally) indicative of particular growth conditions. It follows, therefore, that in those cases where it is possible to isolate in sizeable quantity organisms from a naturally growing population such as to allow of direct biochemical analysis, then certain *physiological indicators* may be used to assess the nature of the environmental forces acting upon the population *in situ*. This approach to studies on microbial ecology and ecophysiology is illustrated here by reference to the behaviour of *Oscillatoria agardhii* Gomont, a cyanobacterium that grows in some shallow Dutch lakes in concentrations sufficient to generate a 'bloom' during periods of the year.

PRIMARY NUTRITIONAL FACTORS OF ECOLOGICAL IMPORTANCE

Since all living matter seemingly contains nucleic acids and protein, it will require for its synthesis substantial amounts of the component elements – that is, carbon, hydrogen, oxygen, nitrogen, sulphur and phosphorus. Microbial cells also have an absolute need for potassium and magnesium; hence for growth to proceed at a rate that is not restricted nutritionally, the environment must contain enzyme-saturating concentrations of each of the elements specified above in forms that are readily assimilable, along with a suitable energy source. Carbon is, of course, available in combination with either oxygen alone (carbon dioxide) or hydrogen and oxygen (organic matter); and since the atmosphere generally contains not less than 0.04% (v/v) carbon dioxide, there will be present in virtually all air-equilibrated aqueous environments at least 10 μM inorganic carbon. The concentration of organically bound carbon may be substantially higher (average values for seawater and North American lake and river waters being 250 and 400 μM, respectively –

Brock, 1966), but bearing in mind that organic matter will be composed of a variety of individual compounds, each of which must be transported into the cell separately, there well may be present insufficient of any particular compound to saturate the appropriate uptake system(s) (Tempest & Neijssel, 1978).

With regards to other essential nutrients, natural aquatic environments generally contain only low concentrations of readily assimilable nitrogen (ammonia or nitrate) and phosphorus; and more specific environments may be depleted of either magnesium, potassium or sulphur. So it is reasonable to suppose that the growth of all types of microorganisms (autotrophs as well as heterotrophs) will, periodically, experience one of six basic nutrient limitations – that is, of carbon-, nitrogen-, sulphur-, phosphorus-, potassium- and magnesium-source – and one might expect these organisms, generally, to be adept at coping with each of these nutritional (growth-limiting) extremes.

Apart from these ubiquitously essential nutrients, some organisms may have a requirement for gases other than carbon dioxide. Among these, oxygen concentration frequently is crucial, both with respect to the growth of strict aerobes and obligate anaerobes; and for specialist organisms, so too are gases like molecular hydrogen and methane. In this connection, it should be remembered that only gases that are dissolved in the aqueous micro-environment surrounding the cell can enter into metabolism, and that the solubility coefficients of most of these gases are exceedingly low. So even though the partial pressures of some individual gases in the atmosphere are high, the concentrations available to the organisms will be low (Table 1). In the context of microbial growth in nutrient-limited environments, then, it is relevant to consider the effects that these extreme conditions exert on microbial structure and functioning, and to identify the kinds of physiological compromises that must be invoked in accommodating to these conditions. In this connection, it was pointed out by Tempest & Neijssel (1978) that organisms might be expected to modify their behaviour in four principal respects: (1) they would, where possible, synthesize high levels of enzymes appropriate to the uptake and assimilation of the growth-limiting nutrient and/or induce (or derepress) the synthesis of an alternative high-affinity uptake mechanism; (2) at the same time, they might be expected to modulate the rates of uptake and metabolism of excess (non-limiting) nutrients – particularly when one such nutrient is the carbon and energy source; (3) they would, where possible, re-

Table 1. *Influence of temperature on the concentrations of oxygen, nitrogen and carbon dioxide in pure water that is in equilibrium with normal air at atmospheric pressure*[a]

Temperature	Oxygen		Nitrogen		Carbon dioxide	
°C	μg ml^{-1}	mM	μg ml^{-1}	mM	μg ml^{-1}	mM
0	14.5	0.45	23.3	0.83	1.3	0.030
10	11.2	0.35	18.2	0.65	0.9	0.020
20	9.1	0.28	15.0	0.54	0.7	0.016
30	7.5	0.23	12.8	0.46	0.5	0.011
40	6.5	0.20	11.0	0.39	0.3	0.007
50	5.6	0.18	9.6	0.34	0.2	0.005
60	4.8	0.15	8.3	0.30	—	—
70	3.9	0.12	6.7	0.24	—	—
80	2.9	0.09	5.2	0.19	—	—
90	1.7	0.05	3.0	0.11	—	—

[a]Normal dry air is assumed to contain, by volume, 20.93% oxygen, 79.03% nitrogen and 0.04% carbon dioxide.

Note. The solubilities of oxygen and nitrogen will be markedly diminished by the presence of dissolved salts; for examples, the concentration of oxygen dissolved in saturated brine, at 20 °C and 750 mm air pressure will be only one-fifth that dissolved in pure water. The solubility of carbon dioxide is only slightly influenced by salt concentration and by pH value, but since it can react with water to produce bicarbonate, and since bicarbonate is non-volatile and very water soluble, the amount of material potentially convertible to carbon dioxide is markedly influenced by pH and salt concentration. Of the biologically important non-atmospheric gases, hydrogen and methane have about the same solubility as oxygen, and hydrogen sulphide is more than twice as soluble in water as is carbon dioxide.

arrange their metabolism so as to circumvent bottlenecks imposed by the specific growth limitation; and (4) organisms clearly would need to modulate coordinately the rates of synthesis of all their macromolecular components so as to allow 'balanced' growth to proceed at a grossly sub-maximal rate.

Evidence for the existence of each of these adaptive responses is to be found in the published literature; particularly in papers dealing with the behaviour of microorganisms in chemostat culture. Selected examples are presented in the next section.

INFLUENCE OF NUTRIENT-LIMITED ENVIRONMENTS ON THE PHYSIOLOGY OF MICROBIAL CELLS

Enzyme content

The uptake of nutrients and/or their primary reactions are mediated by enzymes; and therefore one might expect that the cellular content of these enzymes would be influenced by the level of availability of relevant nutrients. Thus it follows from Michaelis-Menten kinetics:

$$V = V_{max} \frac{s}{K_s + s} \tag{1}$$

that is, **other things being equal,** organisms can increase the rate of substrate uptake (V) at any limiting concentration of substrate (s) either by increasing the potential maximum uptake rate (V_{max}; that is, by increasing the cellular content of uptake enzyme(s)), or by lowering the half-saturation constant (K_s). This latter could be effected either by allosteric modulation of an existing enzyme or by synthesis of an alternative enzyme (or enzyme system) with a lower K_s value. Either way, the net effect would be to increase the scavenging capacity of the organism for the growth-limiting nutrient supply, and thus enhance its ability to compete with other species.

These types of response are routinely observed with nutrient-limited chemostat cultures of bacteria and other microorganisms (see list compiled by Tempest & Neijssel, 1978) and a few examples serve to make the point. Thus, cultures of *Saccharomyces cerevisiae* seemingly assimilate ammonia primarily by means of glutamate dehydrogenase, an enzyme with a relatively high K_m for this nitrogen source; and they respond to ammonia-limiting growth conditions by derepressing the synthesis of this enzyme some two to three fold (Table 2). In contrast, the Gram-negative bacterium *Klebsiella aerogenes* responds to low-ammonia environments by actually **repressing** the synthesis of its glutamate dehydrogenase whilst simultaneously derepressing the synthesis of a high-affinity system comprising glutamine synthetase and glutamate synthase (Table 2).

This form of regulation of enzyme synthesis (repression/derepression) is very common, as compared with induction processes, and clearly is the more relevant to microbial growth in natural nutrient-limited aquatic ecosystems. Not surprising, therefore, that similar changes in the uptake kinetics of inorganic nitrogen sources (nitrate and ammonia) have been found with the filamentous cyanobacterium *Oscillatoria agardhii* Gomont, a limnologically important species, when studied in laboratory culture (Fig. 1). Like *Klebsiella aerogenes*, this organism is unable to fix molecular nitrogen but can derive cellular nitrogen from either nitrate or ammonia (the nitrogen sources most commonly available *in situ*); and when grown

Table 2. *Influence of growth condition on the activities of enzymes of ammonia assimilation in microorganisms (activities in nmole min^{-1} (mg dry wt cells)$^{-1}$)*

	Low-affinity system	High-affinity system	
Organism and growth condition	Glutamate dehydrogenase	Glutamine synthetase[a]	Glutamate synthase
Saccharomyces cerevisiae			
Glucose-limited	620	1.0	n.d.
Phosphate-limited (excess glucose)	465	1.0	n.d.
Ammonia-limited (excess glucose)	1730	2.0	n.d.
Klebsiella aerogenes			
Glucose-limited	560	0.1	1
Phosphate-limited (excess glucose)	600	n.d	1
Ammonia-limited (excess glucose)	19	1.6	66

n.d. = not detectable.
[a]Arbitrary units.

under conditions of nitrate limitation, the cells were found to contain an uptake potential for nitrate (as well as for ammonia) that was some two to five fold higher than found with nitrate-sufficient, light (energy)-limited, cells. Interestingly, though somewhat confusingly, phosphate-limited cultures also possessed a high nitrogen uptake capacity (that is, $V_{max}(NO^{-}_{3}$ or $NH^{+}_{4}))$ when growing at low dilution rates, though only a low uptake capacity when growing at high dilution rates. Along with this high uptake potential expressed at low growth rates was a marked decrease in the cellular nitrogen content, closely similar to that observed with nitrogen-limited cells; but the culture still expressed properties characteristic of a phosphate-limitation (for example, the cells possessed a high alkaline phosphatase activity) although there were no grounds for supposing that these slowly growing cells were experiencing a primary nitrogen limitation. This observation will be discussed later.

The uptake of sugars by *Klebsiella aerogenes* and related organisms is mediated by a complex system of enzymes and proteinaceous carriers that collectively and simultaneously effect uptake and phosphorylation of the sugar; that is, a sugar phosphotransferase system, or PTS. But though possessing many components, the overall activity of this system can be fairly adequately and accurately

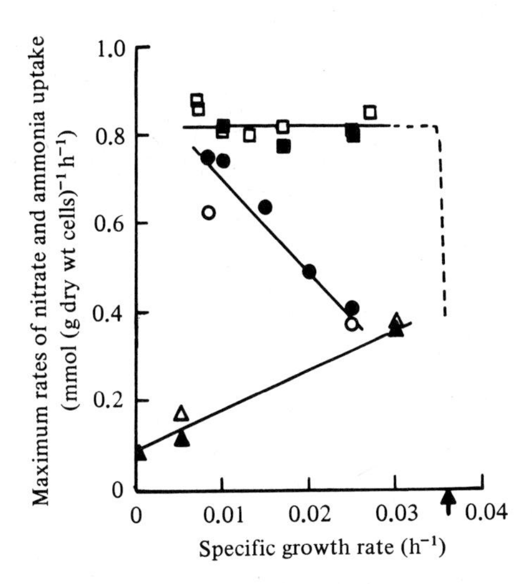

Fig. 1. Influence of the specific growth rate (μ) on the uptake capacity of *Oscillatoria agardhii* for nitrate ($V^{NO_3}_{max}$; closed symbols) and for ammonia ($V^{NH_4}_{max}$; open symbols). Organisms were grown in chemostat cultures that were, respectively, nitrate-limited (□), phosphate-limited (○), and light (energy)-limited (△). The arrow indicates the maximum specific growth rate expressed by this organism in a salts medium at 20°C, pH 8.0. Data of Zevenboom (1980).

assessed *in vitro* (Kornberg & Reeves, 1972; Hunter & Kornberg, 1979) thereby allowing the influence of growth-limiting conditions to be determined. Thus it was found, as expected, that the glucose-PTS of glucose-limited *K. aerogenes* cells was high, as compared with glucose sufficient cells, and increased substantially as the dilution rate was lowered and the stringency of the glucose limitation thereby increased (Fig. 2). At high dilution rates, and particularly under conditions of excess glucose, the glucose-PTS activity was low and seemingly insufficient to account for the actual rate of glucose uptake expressed by the growing culture. This suggested the existence of an alternative (presumably low-affinity) glucose uptake system, though definitive evidence could not be found (O'Brien *et al.*, 1980; Neijssel *et al.*, 1980).

At any fixed growth rate, the rate of uptake of essential nutrient(s) may vary substantially with the nature of the environment. Thus, glucose-sufficient cultures of *Klebsiella aerogenes* were found to take up glucose at a rate that was some two to five times higher than glucose-limited cultures growing at an identical rate (Neijssel & Tempest, 1975, 1979). Depending on the nature of the growth limitation, the excess glucose consumed was either polymerized to

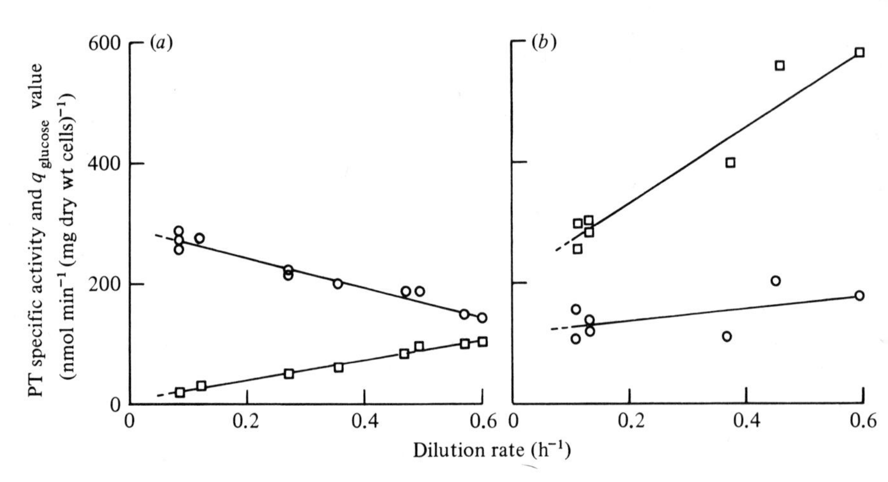

Fig. 2. Glucose phospho*enol*pyruvate phosphotransferase (PTS) activities and glucose uptake rates of *Klebsiella aerogenes* NCTC 318 chemostat cultures that were (*a*) glucose-limited and (*b*) potassium-limited. The squares represent the specific rates of glucose consumption by the growing cultures, and the circles represent the PTS activities of toluenized cells measured *in vitro*. Cultures were grown in a simple salts medium, with glucose as the sole carbon and energy source, at 35 °C and pH 6.8. Data of O'Brien *et al.* (1980).

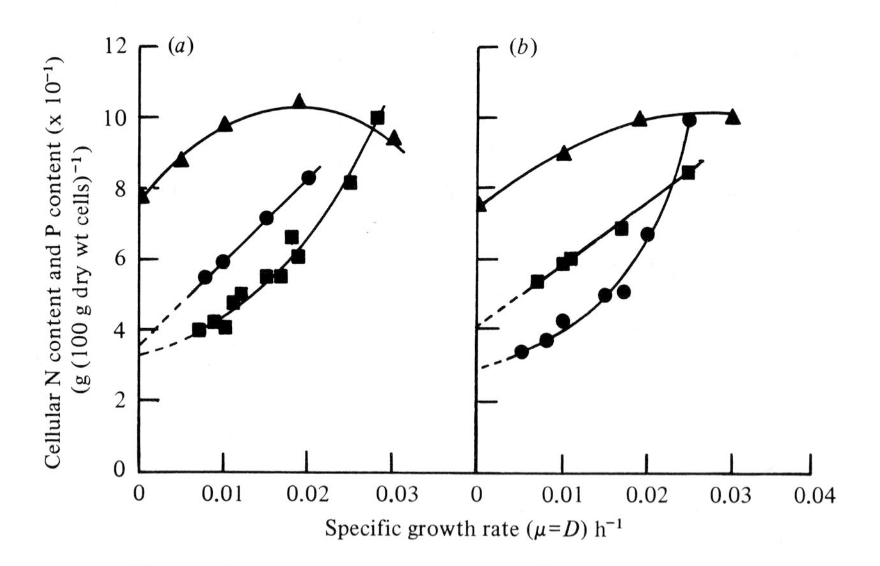

Fig. 3. Influence of the specific growth rate (μ) and (*a*) the cellular nitrogen content, and (*b*) the cellular phosphorus content of *Oscillatoria agardhii* Gomont grown in chemostat culture in simple salts media that were (■) nitrate-limited, (●) phosphate-limited, and (▲) light (energy)-limited. Cultures were grown at 20 °C, pH 8.0, at an incident irradiance of 2.3 to 3 W m^{-2} (light-limited) and 6 W m^{-2} (nitrate- or phosphate-limited). Data of Zevenboom (1980).

Table 3. *Influence of irradiance on culture biomass concentration and pigment content of* Oscillatoria agardhii, *growing at a constant rate (0.02 h^{-1})*

Incident irradiance (Wm^{-2})	Culture biomass concentration ($mg\ l^{-1}$)	Average irradiance (Wm^{-2})	Chlorophyll *a* content ($g(100\ g\ dry\ wt)^{-1}$)	C-phycocyanin content ($g(100\ g\ dry\ wt)^{-1}$)
6	200	3	1.35	9.0
13	430	3	1.10	5.5
25	680	5	0.95	5.0
40	650	7	0.50	2.5

Data of van Liere (1979).

polysaccharide (intracellular and extracellular) or partially oxidized to intermediary metabolites (such as gluconate, ketogluconate and acetate) that were excreted into the medium. Similarly, with a photosynthetic cyanobacterium, the cellular nitrogen content was found to vary markedly with the growth condition (hence its specific uptake rate varied); slowly growing nitrogen-limited cells, as well as phosphate-limited cells, contained markedly less nitrogen than did cells that were light-energy limited (Fig. 3*a*). And a similar relationship was found with respect to phosphorus content (Fig. 3*b*). These changes in nitrogen content correlated with, and were largely caused by, changes in the cellular content of chlorophyll *a* and C-phycocyanin (van Liere, 1979), and had associated with them marked changes in the kinetics of nitrate (and ammonia) uptake, particularly the uptake potential.

Whereas carbon-substrate metabolism in heterotrophs serves to generate both energy and intermediary metabolites required for cell synthesis, these two functions are served by light and carbon dioxide, respectively, in photoautotrophs such as the cyanobacteria. Therefore light might be considered an essential 'nutrient' for these photoautotrophs; and, just as with glucose-limited cultures of *Klebsiella aerogenes*, so one might expect that light-limited cultures of *Oscillatoria agardhii* would contain derepressed levels of the appropriate 'uptake' (light-trapping) system. And so they do (Table 3). Clearly, irradiance influences the synthesis of chlorophyll *a* and C-phycocyanin in a parallel manner. It is known that the regulatory systems of the different phycocyanobilins might involve a common precursor (Fogg *et al.*, 1973); and as the biosynthetic pathway of chlorophyll *a* is very similar to that of the phycocyanobilins, a

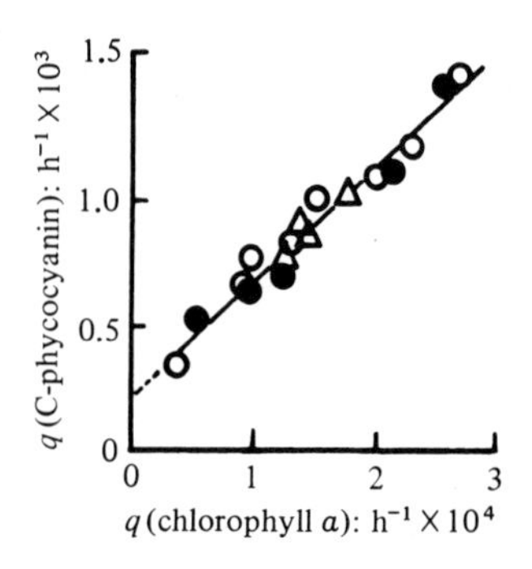

Fig. 4. Relationship between the specific rates of synthesis of C-phycocyanin and chlorophyll *a* found with light (energy)-limited cultures of *Oscillatoria agardhii* growing in a chemostat at different dilutions rates and with different incident irradiance values. These latter values were: (○) 6 W m^{-2}; (●) 13 W m^{-2}; and (△) 25 W m^{-2}. Data of van Liere (1979).

mechanism of coordinate regulation mediated through some common precursor seems an attractive hypothesis. However, though the rates of synthesis of both chlorophyll *a* and C-phycocyanin varied linearly with the growth rate of light-limited *O. agardhii* cultures, the ratio changed progressively in favour of the C-phycocyanin as the dilution rate was lowered (Fig. 4). Taken together, these data suggest that *O. agardhii* cells are provoked by progressive changes in light quantity and quality to synthesize the most useful pigments in optimum amounts so as to allow trapping and transfer of light energy with a high efficiency.

These changes in cellular pigment content lead to a more general consideration of the influence that environment exerts on the macromolecular content and composition of microbial cells.

Macromolecular composition

The mean cell volume of microbial cells is, in general, markedly affected by the growth environment and by the specific growth rate, and this is particularly evident with rod-shaped bacteria like *Bacillus megaterium* and *Klebsiella aerogenes* (Herbert, 1958), *Salmonella typhimurium* (Schaechter *et al.*, 1958), and with the filamentous cyanobacterium *Oscillatoria agardhii* (Fig. 5; van Liere, 1979). As expected, therefore, the amounts of DNA, RNA, protein and carbohydrate **per cell** increase progressively with growth rate (Maaløe, 1960). More significantly, however, there is (seemingly invariably) a marked differential increase in the cellular RNA content due to a stringent growth rate associated change in the

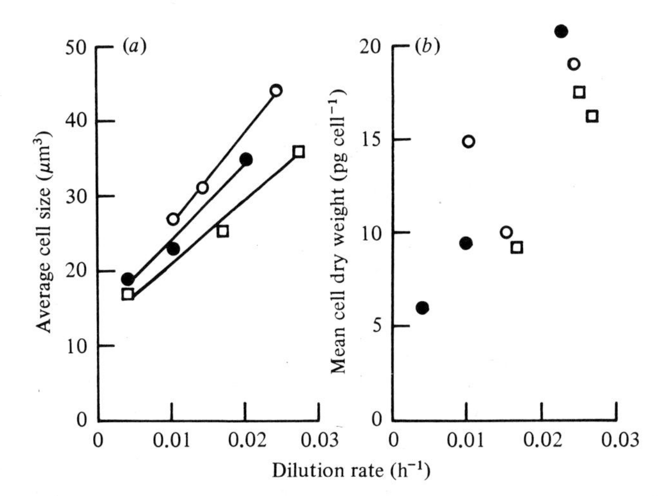

Fig. 5. Average cell volume (*a*) and dry weight per cell (*b*) of *Oscillatoria agardhii* Gomont growing at different rates in chemostat cultures that were light (energy)-limited with an incident irradiance of 40 W m^{-2} (○) and 6 W m^{-2} (●), and nitrogen (nitrate)-limited with an incident irradiance of 8–10 W m^{-2} (□). The temperature was controlled at 20 °C, and the pH at 8.0. Data of van Liere (1979).

cellular ribosome content (Kjeldgaard & Kurland, 1963). But this relationship also is greatly influenced by the growth temperature (Tempest & Hunter, 1965) such that the RNA content relates more closely to the relative growth rate (μ/μ_{max}; Tempest, 1976) than to the actual rate of cell proliferation.

As already mentioned, under some conditions organisms may take up nutrients at a rate that is substantially higher than that required to satisfy the basic requirements of cell synthesis, and lay down so-called 'energy-storage' compounds (Dawes & Senior, 1973). Principal among these compounds are polyphosphate, glycogen, poly-β-hydroxybutyric acid (PHB) and lipid, and these may accumulate within the cell in amounts that vary from a few per cent of the dry weight (in the case of polyphosphate in *Klebsiella aerogenes*) to as much as 86% of the dry weight in the case of PHB in *Hydrogenomonas eutropha* (Schlegel *et al.*, 1961). Again, although each of these polymers is known to accumulate under conditions of excess energy supply, and are further provoked in their syntheses by specific nutrient limitations (for example, sulphate limitation for polyphosphate accumulation, oxygen depletion in the case of PHB accumulation, and a high carbon:nitrogen ratio for glycogen storage), none of these conditions is absolutely specific and definitive.

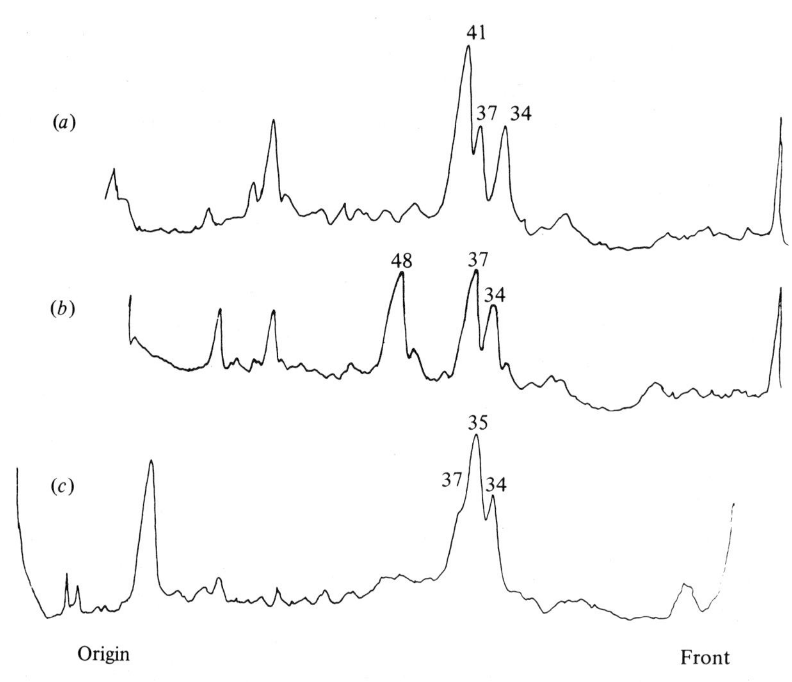

Fig. 6. Scans of SDS-polyacrylamide gels of membrane preparations from *Klebsiella aerogenes* that were grown in chemostat culture ($D = 0.3\,h^{-1}$; 37 °C; pH. 6.8.) under conditions of (*a*) sulphur (methionine) limitation, (*b*) glucose limitation, and (*c*) phosphate limitation. Preparations were made such as to extract specifically the outer wall proteins and, as can be seen, the spectrum of such proteins varied markedly with the growth condition. The numbers indicate the apparent molecular weights ($\times 10^{-3}$) of the different proteins. Data of Sterkenburg *et al.* (1981).

More clearly indicative of the growth environment, perhaps, is the polymer content and composition of the cells' envelope structure (that is, wall and membrane complex). Thus, with some Gram-positive bacteria, the presence of teichoic acid is characteristic of growth in the presence of an excess of phosphate whereas the presence of teichuronic acid in place of teichoic acid occurs principally, if not solely, under conditions of phosphate-limitation (Ellwood & Tempest, 1972).

The walls of Gram-negative bacteria are devoid of teichoic acid and teichuronic acid but contain much protein and lipid. Several distinct proteins are to be found in both the inner and outer components of the wall and the syntheses of a number of these proteins are extensively modulated in response to changes in the growth environment (Fig. 6) (Robinson & Tempest, 1973; Braun *et al.*, 1976; Sterkenburg *et al.*, 1981). It is reasonable to suppose, and in several instances known, that specific proteins function as porins or as translocators of certain nutrients. Hence the precise spectrum of proteins present in the inner and outer wall fractions well may prove to be highly characteristic of a particular growth condition.

Cell size and shape

The point has been made already that the mean cell volume and average dry weight of microbial cells generally increase progressively with growth rate. But these changes may not always be as uniform as shown in Fig. 5 (see, for example, Herbert, 1958) and lowering the dilution rate below some critical value may lead to an apparent impairment of the cell division process and the formation of a variety of aberrant cellular forms. Thus, when a glycerol-limited culture of *Klebsiella aerogenes* was grown at a series of diminishing dilution rates below $0.05\,h^{-1}$ (37 °C, pH 6.5) there was observed to be a marked change in cell morphology with a preponderance of elongated cells in the culture (Fig. 7). There was also found to be a significant decrease in the apparent viability of slowly growing steady state cultures; but the non-viable fraction did not appear to contain any of the filamentous cell types (Tempest *et al.*, 1967).

A fascinatingly different change in cell morphology, linked with growth rate, has been reported to occur with species of *Arthrobacter* (Luscombe & Gray, 1971). When growing in a glucose-limited chemostat culture at dilution rates above about $0.1\,h^{-1}$ the organisms were rod-shaped; below this critical dilution rate, however, they were almost coccoid. Of relevance, in an eco-physiological context, is the frequent finding of predominantly coccoid forms of this organism in soil samples, indicating probable low growth rates *in situ* with respect to these natural populations.

Energetics

Apart from the question of nutrient availability for microbial growth, the supply of biologically usable energy is equally important; for just as the environment might contain either a surfeit or a deficiency of some essential nutrient(s), so it might contain either an abundance of energy source(s) or very little. Heterotrophs and chemoautotrophs derive energy from the chemical transformation or modification of substances present in the biosphere and as

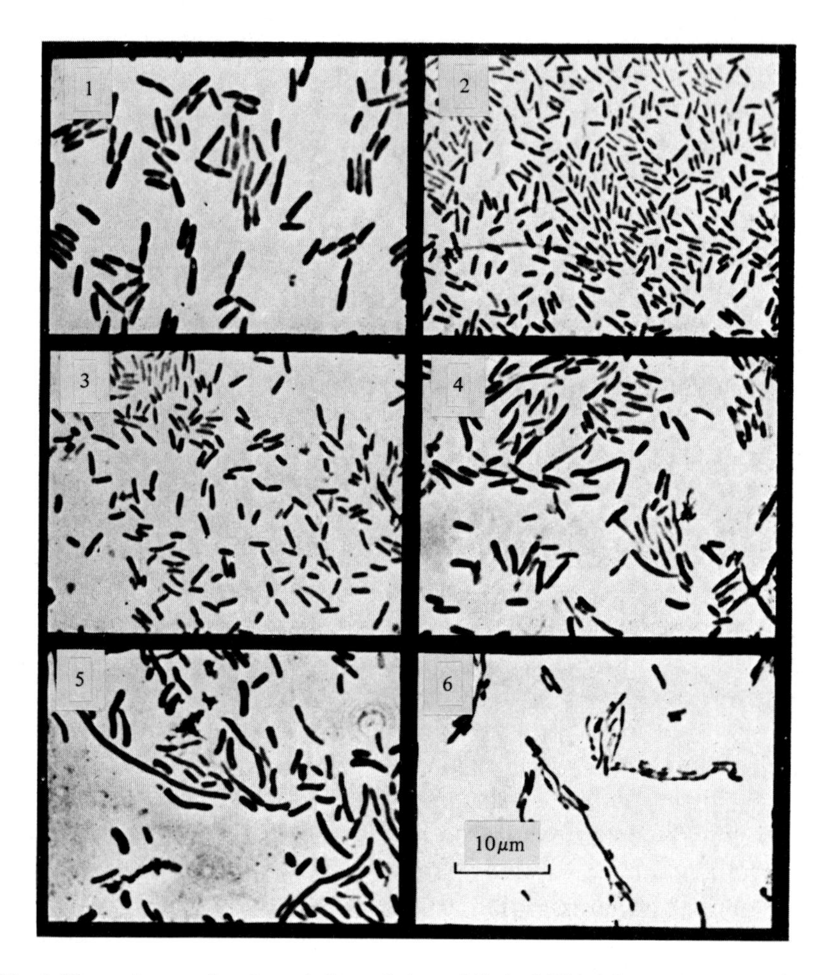

Fig. 7. Photomicrographs of populations of glycerol-limited *Klebsiella aerogenes*, growing at steady state rates of (1) 1.0 h^{-1}, (2) 0.24 h^{-1}, (3) 0.12 h^{-1}, (4) 0.015 h^{-1}, (5) 0.008 h^{-1} and (6) 0.004 h^{-1}. Organisms were photographed using phase contrast optics; the magnification was uniform throughout. Data of Tempest *et al.* (1967).

these become depleted they must be replenished. Hence, in any natural ecosystem the amount of usable energy source is likely to vary markedly. Similarly, phototrophs depend upon an energy source (sunlight) that again is available only periodically. So a common problem that faces all microorganisms is how to cope with both an excess and a growth-limiting supply of energy. It is relevant to consider, therefore, how organisms might accommodate to a fluctuating supply of energy source and how, in particular, they modulate the rates of energy-generating reactions to match precisely those energy-consuming reactions of biosynthesis and cell maintenance.

Under conditions of energy limitation, one might expect organisms to adjust automatically the rates of their anabolic (energy-consuming) reactions. No additional control mechanisms would seem to be necessary apart from those required to coordinately regulate synthesis of the different cellular polymers and thereby sustain balanced cell synthesis, albeit at a sub-maximal rate. There are, however, generally invoked changes in the cells' physiology with a depression of synthesis of (or modulation of) enzymes providing alternative mechanisms of energy generation. Thus, in the red halophilic bacteria, the synthesis of purple membrane components is promoted by a decrease in oxygen availability. And in aerobic and facultatively anaerobic heterotrophs, first an oxygen limitation provokes enhanced synthesis of terminal oxidases (some, seemingly, with an increased affinity for oxygen) and synthesis of enzymes providing potential alternative electron acceptors (for example, nitrate reductase); then, in the absence of oxygen, an alternative method of energy conservation (substrate-level phosphorylation) is invoked, where possible (see Harrison, 1976). With phototrophic organisms, a light limitation frequently is found to effect a derepression of synthesis of light-trapping pigments as already discussed (van Liere, 1979).

Problems of regulation become more complex, one might imagine, when the energy source is present in abundance since then the organism must either modulate the rate of ATP generation to match the growth requirement, or else invoke energy-spilling (ATPase-like) reactions. The respiratory control apparent with isolated mitochondria is one example of a fine control over the rate of ATP formation from ADP, but few types of **whole** cells, if any, have been shown to exhibit a similar respiratory control. It seems that whole cells generally possess mechanisms for dissipating excess energy as heat by invoking energy-spilling reactions (Hoogerheide, 1975; Neijssel & Tempest, 1976). With a heterotrophic bacterium, utilizing as energy source a carbon substrate that enters the cell by an active transport process, one might imagine regulatory mechanisms to operate at this uptake level. With some organisms, and with some substrates, this clearly is the case. Thus, whereas with *Klebsiella aerogenes* growing in a glucose-sufficient chemostat culture, the rate of carbon substrate catabolism was vastly in excess of that expressed by a glucose-limited culture growing at the same rate, the difference

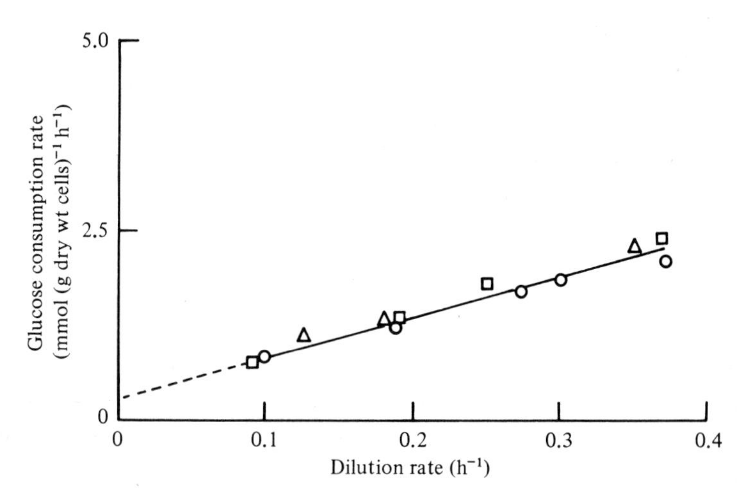

Fig. 8. Specific rate of glucose consumption as a function of dilution rate expressed by chemostat cultures of *Paracoccus denitrificans* that were, respectively (○) glucose-limited, (△) potassium-limited and (□) phosphate-limited. The culture temperature was maintained at 35 °C and the pH value adjusted automatically to 6.8. Previously unpublished data.

found was much less when mannitol was used as the carbon substrate (Neijssel & Tempest, 1976). And again, the differences found between glucose-sufficient and glucose-limited cultures of either *Paracoccus denitrificans* or *Bacillus subtilis* were much less than those observed with *K. aerogenes* (Fig. 8).

In contrast to substrates like glucose and mannitol, methanol can enter the cell freely by diffusion. And since the first enzyme of methanol catabolism (methanol dehydrogenase) is a membrane-associated enzyme that donates electrons to the respiratory chain (via its cofactor, PQQ; Duine & Frank, 1981), there would seem to be only limited possibilities for regulating the rates of substrate catabolism and respiratory energy generation (that is, excluding processes of respiratory control). In fact, though the cellular content of methanol dehydrogenase seemingly is modulated to some extent by the level of availability of its substrate, thereby effecting a coarse control over the rate of methanol catabolism, there does not appear to be any fine control mechanisms operative in these methanotrophic bacteria. Hence, when a methanol-limited chemostat culture of *Pseudomonas* (*Methylophilus*) *methylotrophus* is provided with a fluctuating supply of methanol there is seen to be a marked decrease in the carbon conversion efficiency (Brooks & Meers, 1973). Similarly, with batch cultures of *Pseudomonas extorquens*, the initial methanol concentration markedly influenced the yield of organisms obtained (that is, g organisms formed (g methanol consumed)$^{-1}$; Harrison, 1976).

With phototrophic organisms, light intensity is a factor of primary importance over which they have no immediate control. As already mentioned, some of these organisms can accommodate to low light intensities by derepressing the synthesis of their light-harvesting apparatus; but short-term changes in light intensity, and particularly exposure to high light intensities require other solutions. So far as we are aware, there are no reports published of phototrophic organisms being able to dissipate directly (for example, by some short-circuiting device) the excess membrane-associated energy generated as a consequence of excess light irradiation. The only well-defined process that might serve an energy-spilling function is photorespiration (Fig. 9). Here, it is clear that glycolic acid, generated from ribulose 1,5-bisphosphate by an oxygenase reaction, must either be excreted or converted into phosphoglycerate (the sole product of the ribulose 1,5-bisphosphate carboxydismutase reaction) by a sequence of enzyme-mediated reactions that concomitantly consume energy (ATP) and reductant. However, an excessive diversion of carbon through glycolic acid would drain the intracellular pool of key intermediates of carbon assimilation, thus inhibiting growth; and photoinhibition is not uncommon among phototrophic microorganisms – most particularly with the cyanobacteria (Fig. 10).

In summary, then, a problem of energy limitation is solved by organisms coordinately modulating the on-going rates of their anabolic processes, and slowing growth rate appropriately. And an energy overplus is equally readily overcome by cells invoking energy-spilling reactions (Neijssel & Tempest, 1976). The nature of these latter reactions are, in general, not well defined; though their existence is surely beyond doubt (Tempest, 1978; Tempest & Neijssel, 1980; but see, for a contrasting view, Hempfling & Rice, 1981).

PHYSIOLOGICAL INDICATORS OF NUTRIENT-LIMITED GROWTH CONDITIONS IN NATURAL ECOSYSTEMS

In the foregoing sections, attention has been drawn repeatedly to the nature of the physiological compromises that are invoked by microbial cells in accommodating to nutrient-limited (growth-

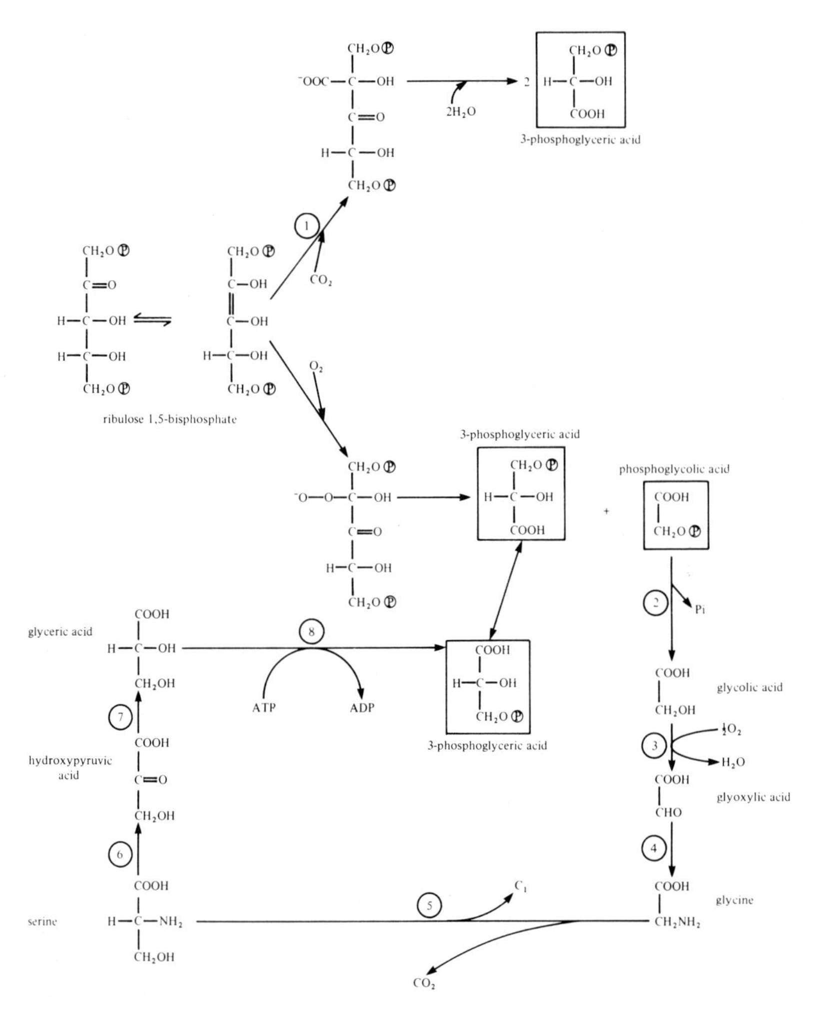

Fig. 9. Photorespiratory pathway in algae; glycolate synthesis and metabolism. Reactions are carried out by the following enzymes (1) ribulose 1,5-biphosphate carboxydismutase, (2) phosphoglycolate phosphatase, (3) glycolate dehydrogenase, (4) glutamate-glyoxylate amino transferase, (5) serine hydroxymethyl transferase, (6) serine-glyoxylate aminotransferase, (7) hydroxypyruvate reductase, (8) glycerate kinase. Taken from Tolbert *et al.* (1971).

restricting) laboratory culture conditions. Thus, specific nutrient limitations seemingly pose metabolic problems that require specific (possibly unique) solutions. In other words, phenotypes (that is, structural and functional variants) are synthesized that are appropriate to, and characteristic of, the dominant growth-limiting conditions; and it follows, then, that a knowledge of the actual phenotype expressed in any natural circumstance should allow

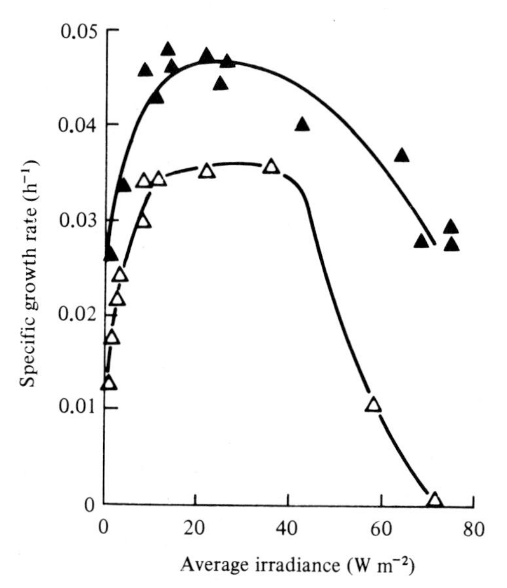

Fig. 10. Influence of the average incident irradiance value on the specific rate of growth of *Oscillatoria agardhii* Gomont. Upper plot (blocked triangles) are data for cultures growing at 28 °C; lower plot (open triangles) are data for cultures growing at 20 °C. From van Liere (1979).

conclusions to be drawn as to the nature of the environmental conditions that circumscribed its genesis. This, in a nutshell, is what is implied by 'physiological indicators'; and it is clear that if it was possible to isolate from some natural ecosystem a sufficient quantity of a particular species to allow for direct physiological analysis, then one ought to be able to identify the prominent environmental influences that were acting upon it during its growth. A preliminary exercise along these lines has been undertaken with respect to naturally occurring populations of *Oscillatoria agardhii* (Zevenboom, 1980); but before proceeding to detail these studies, it is necessary to discuss very briefly the problems inherent in adopting alternative (analytical and bioassay) approaches to studies of the growth-constraining influences that are operative in some natural ecosystem.

As pointed out by Brock (1966), freshwater environments frequently are depleted of readily assimilable nitrogen and phosphorus, and eutrophication with regards to these two elements often provokes vigorous growth of the phytoplankton. But as the phytoplankton community increases, so the availability of light in the water column decreases to a point where conditions become favourable for the growth of cyanobacteria (Gons, 1977; van Liere, 1979).

It may thus be concluded that in highly eutrophic lakes, phosphorus-, nitrogen- or light-availability may become overall growth-rate limiting.

The method used most widely so far for assessing the nature of the growth-limiting factor in natural environments has been the bioassay measurement (for example, Gerhart & Likens, 1975; Goldman, 1978). Briefly, this assay is performed as follows: after nutrient enrichment of a sample contained in a closed system, the increase in biomass is followed over several days and the degree of response used as a measure of nutrient deficiency. However, reliable though this method might seem, care is needed in interpreting the results as the following example will indicate. A sample of lakewater harbouring *Oscillatoria agardhii* as the predominant organism was found to contain 7.2 μM nitrate (that is, about six times the value of the half-saturation constant for nitrate uptake) and 0.1 μM phosphate (that is, about three times the half-saturation constant for phosphate). Thus, since a Monod-type relationship (Herbert *et al.*, 1956) was known to be operative here, with respect to both nitrate and phosphate uptake and growth, then one can calculate that the measured nitrate concentration would support a growth rate of $0.85\,\mu_{max}$, and that of phosphate a growth rate of only $0.75\,\mu_{max}$. Hence the concentration of each nutrient is potentially growth rate limiting though in reality growth is constrained by the availability of phosphate. However, let us consider what one might conclude from a bioassay study. Clearly, adding nitrate to a sample would not cause any increase in growth rate nor increase the amount of biomass produced over some extended period of time. Similarly, adding phosphate would cause a transient increase in growth rate, to $0.85\,\mu_{max}$, but the residual nitrate soon would be exhausted and there would be only a slight increase in biomass synthesized after an extended period of time. Finally, adding both nitrate and phosphate to the sample would cause both a sustained increase in growth rate and a marked increase in biomass synthesis, thus suggesting, erroneously, that the growth of the organisms *in situ* was limited by the availability of both nitrate and phosphate. We propose, therefore, that physiological indicators may provide a better basis upon which to assess the nature of the **primary** nutrient deficiency constraining microbial proliferation in natural ecosystems – provided, that is, that the indicators used are sufficiently specific and have been adequately investigated and defined.

Oscillatoria agardhii Gomont is the dominant cyanobacterium

Table 4. *Physiological indicators used in studies of* Oscillatoria agardhii *growing in laboratory culture and in Lake Wolderwijd*

Parameter	Growth condition		
	Energy limitation	Nitrogen limitation	Phosphorus limitation
Nitrate uptake[a]			
V_{max}	l	h	m
K_s	l	m	h
Ammonia uptake[a]			
V_{max}	l	h	m
K_s	l	l	l
Cell N content	h	v	m
Phosphate uptake			
V_{max}	l	l	h
K_s	l	l	h
Phosphorus content	h	m	l
Alkaline phosphatase	m	m	h
Chlorophyll *a* content	h	v	m
C-phycocyanin content	h	v	m

[a]The enzymes of nitrate uptake and assimilation are not constitutive. Thus, $N(NH_3)$-limited cells are unable to utilize nitrate immediately; in contrast, $N(NO_3)$-limited cells possess a highly active low-K_s (high affinity) ammonia uptake system.
h, high; l, low; m, moderate; v, variable.
Data of Zevenboom (1980) and Zevenboom *et al.* (1982).

present in Lake Wolderwijd, a hypertrophic body of water that is one of the 'Randmeren' of Lake IJsselmeer. This organism has been studied extensively in the laboratory, both in batch and chemostat culture, and the influence of a variety of appropriate environmental parameters on the cells' macromolecular composition and metabolic activities determined (Table 4). From these studies it became clear that a number of properties (like pigment content) were influenced by too many conditions to be used alone as indicators of the dominant growth-limiting conditions; however, others were more characteristic and, when used in combination with ancillary parameters, could provide good grounds upon which to evaluate the likely conditions prevailing *in situ*. Thus, using these physiological indicators, a detailed investigation was made, over a four year period, into the properties of natural populations in the above-mentioned lake with a view to identifying the major seasonal growth-restricting influences.

It is not necessary to go into all the details of this four year study since analysis of the data gathered over a single year (1977) serves to make the point. For nine months of this year (from March to November, inclusive), the concentration of utilizable nitrogen

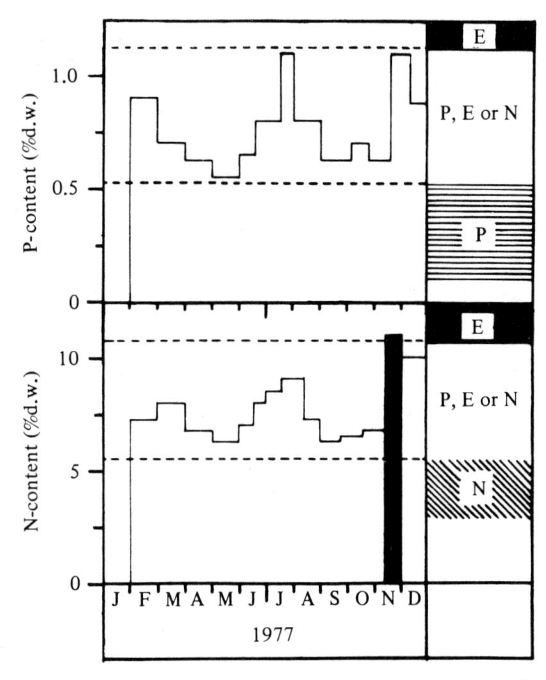

Fig. 11. Total cell-bound phosphorus and nitrogen contents of natural cyanobacterial populations in Lake Wolderwijd, examined throughout 1977. Data derived from variously limited chemostat cultures of *Oscillatoria agardhii* Gomont (the predominant cyanobacterium in this highly eutrophic lake) are shown in the right-hand column. The solid block shows the range of nitrogen and phosphorus contents indicative of a light (energy) limitation; the horizontal striped block shows the cellular contents relevant to a phosphate limitation; and the diagonal hatching indicates the range of nitrogen contents characteristic of nitrate-limited cells. The blank area labelled P, E or N is an 'overlap' region where no clear distinctions can be drawn between the differently limited cells. After Zevenboom (1980).

sources (nitrate and ammonia) in the lake water remained below the levels required to saturate the growth processes (nominally taken as ten times the value of the half-saturation constant). And similarly, from January to June (inclusive) as well as in December 1977, the level of available phosphate remained below that necessary to saturate the growth processes for this essential nutrient. But measurement of the cellular phosphorus and nitrogen contents (Fig. 11), as well as the contents of chlorophyll *a* and C-phycocyanin (Fig. 12) provided no firm basis upon which to conclude whether a nitrogen-, phosphate- or light-limitation prevailed during these months. However, a study of the uptake potential for nitrate, ammonia and phosphate of these natural populations was more revealing (Fig. 13) and gave clear indications of a phosphate limitation prevailing in early June and early November, a nitrogen

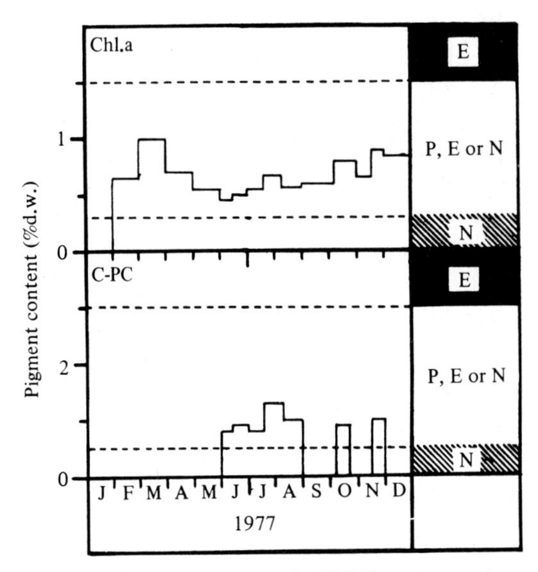

Fig. 12. Chlorophyll *a* (Chl.a) and C-phycocyanin (C-PC) contents of natural cyanobacterial populations taken from Lake Wolderwijd at successive times throughout 1977 (data of Zevenboom, 1980). For comparison, data obtained with variously limited chemostat cultures of *Oscillatoria agardhii* Gomont are shown in the right-hand column. The key is the same as that described for Fig. 11.

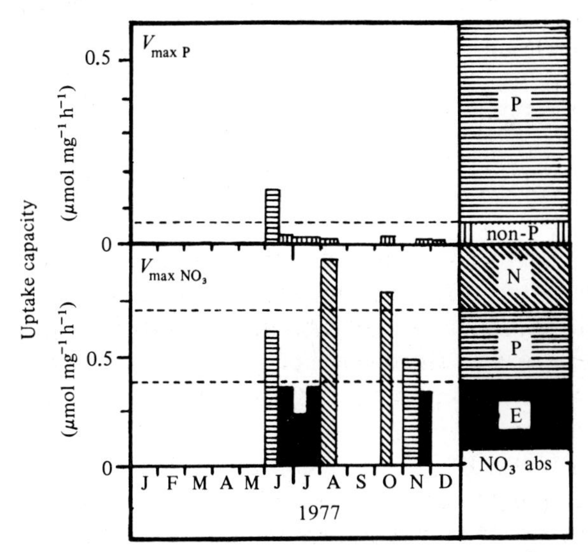

Fig. 13. Uptake capacity for phosphate (V_{maxP}) and for nitrate (V_{maxNO3}) of natural cyanobacterial populations occurring in Lake Wolderwijd throughout 1977). Comparative data obtained with variously limited chemostat cultures of *Oscillatoria agardhii* Gomont are shown in the right-hand column; the key being the same as that used in Figs. 11 and 12 (data of Zevenboom, 1980). The additional notation (NO_3abs) indicates the nitrate uptake capacity of cells grown in the absence of nitrate (e.g. on ammonia); the value (which represents the initial rate of nitrate uptake) is virtually zero.

limitation being extant in early August and mid-October, and of an energy limitation being the dominant growth constraining influence throughout the second half of June and all July, and again in late November. Moreover, though definite conclusions could not be drawn for the other periods of this year, with respect to the growth-limiting factors, it was possible to state justifiably that in January and February, and again in late December the natural population was not suffering a depletion of nitrogen source, nor was the availability of phosphate limiting growth during the second half of August and all of September. Indeed, only for one-quarter of this year (from March to May, inclusive) could nothing regarding conditions prevailing in this natural ecosystem be derived from a study of the physiological properties being expressed *in situ* by the organisms.

During those periods where it was not possible to identify a characteristic phenotype, there well may have been imposed upon the natural population some growth-constraining influence that has not so far been investigated. Alternatively, it is possible that some secondary, potentially growth-limiting, condition might have exerted a modifying effect. An example of the latter was apparent in the manner in which both average irradiance ($\bar{E}$) and growth rate (μ) influenced the cellular nitrogen content (Q) of a nitrate-limited culture of *Oscillatoria agardhii* (Fig. 14). Here it is clear that both a lowering of the irradiance value and an increase in growth rate potentiate an increase in the cellular nitrogen content, and that only at high irradiance combined with a low growth rate is a nitrogen-depleted phenotype expressed.

Similarly, when a culture of *Candida utilis* was grown in a potassium-limited chemostat environment with a small excess of glucose, and the input potassium concentration was then selectively increased, there was found to be no measurable increase in the extracellular potassium concentration (which was maintained at about 20 μM) until the input potassium concentration exceeded 3 mM, even though no free glucose could be detected after the input potassium concentration exceeded 0.9 mM (Fig. 15). Throughout the whole of this transition phase, the **cellular** potassium concentration progressively increased from that characteristic of a fully potassium-limited culture (0.15%, w/w, at a dilution rate of 0.1 h^{-1}; 30 °C; pH 5.5) to that routinely observed with a fully glucose-limited culture (that is, some 2.0%, w/w, irrespective of the growth rate); and there was a concomitant progressive increase in the growth

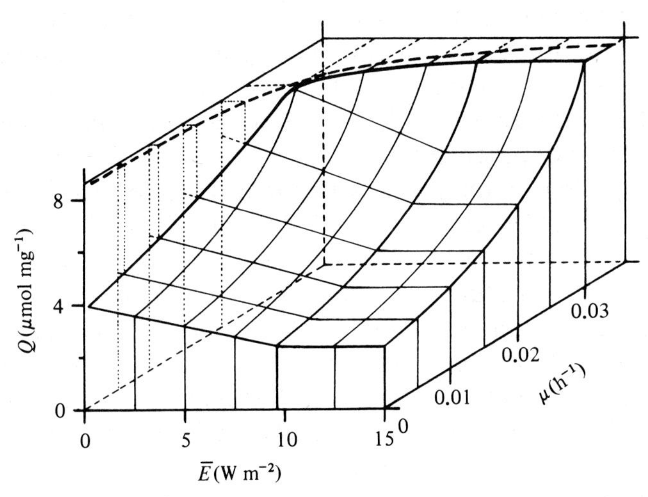

Fig. 14. Dependence of the nitrogen content (Q) of *Oscillatoria agardhii* Gomont on the specific growth rate (μ) and on the average light irradiance ($\bar{E}$), calculated according to the equations:

$$Q = \frac{-0.0066(\bar{E}) + 0.1633}{0.042 - \mu} \qquad \text{for } 0 < \bar{E} < 9.4 \text{ W m}^{-2}$$

$$Q = \frac{0.101}{0.042 - \mu} \qquad \text{for } \bar{E} > 9.4 \text{ W m}^{-2}$$

(From Zevenboom *et al.*, 1980).

efficiency with respect to both oxygen and glucose (Aiking & Tempest, 1976).

It is to be expected, then, that other complex inter-relationships will exist and that, in natural ecosystems, intermediate phenotypes frequently will be expressed. In our opinion, however, this does not diminish the utility of the 'physiological indicators' approach to studies of natural populations; it just means that more effort is needed to identify those cellular properties that are highly characteristic of specific growth conditions. As already mentioned, substrate uptake kinetics provide one such reliable indicator, and wall composition may well prove to be another.

EPILOGUE: MICROBIAL PHYSIOLOGY, ECOLOGY AND EVOLUTION

We began by questioning the generality of a currently popular aphorism, and we wish to conclude by offering another which we

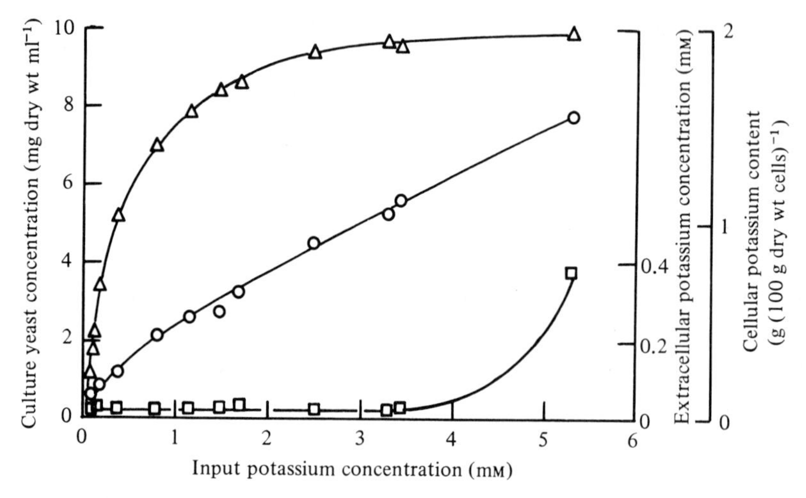

Fig. 15. Effect of varying the input potassium concentration on the steady state concentration of (□) extracellular potassium, (△) yeast biomass, and on the cellular potassium content (○) of *Candida utilis* organisms. No phosphate could be detected in the culture extracellular fluid after the input potassium concentration exceeded 0.8 mM, and no glucose after it exceeded 0.9 mM. Nevertheless, the extracellular potassium concentration did not rise until the input potassium concentration exceeded 3 mM ($D = 0.1$ h^{-1}; 30 °C; pH 5.5). Data of Aiking & Tempest (1976).

feel better accords with conclusions to be drawn from the studies reported herein: that is, 'life is a compromise'. There can be little doubt that such is the case, and that compromises are just as evident with respect to the elephant as with the *Escherichia coli* cell. But there is a difference between the two, and this lies in the level of organization at which that compromise is expressed: with the elephant it is expressed at the level of the whole multicellular organism, whereas with *E. coli* it necessarily is expressed at the level of the individual cell. In both cases, however, the compromise that is struck is dictated by the conditions to which these living creatures routinely are exposed in their natural ecosystems; conditions not only of environmental and nutritional constraint, but of competition. In this connection, it is relatively easy to assess those features of, say, the pachyderm that in the past have militated in favour of its survival; but to what extent can the same be said of *E. coli* or *Bacillus subtilis* or *Oscillatoria agardhii*? One can, of course, appreciate the advantage that accrues to methanogenic bacteria in being able to derive energy for growth from the transfer of electrons from hydrogen to carbon dioxide; and one can readily understand that this specialist physiology is not compatible with life in aerobic environment. Similarly, one might plausibly reason that much is to be gained by heterotrophic organisms in being able to grow both aerobically and anaerobically (that is, in being facultatively anaerobic in their metabolism). But wherein lies the advantage in possessing a strictly aerobic metabolism? This question is particularly relevant in the case of organisms like pseudomonads that commonly occupy aquatic biotopes and where rapid aerobic metabolism quickly strips the environment of oxygen. There would seem, to us, no ready explanation for the preponderance of strictly aerobic microbial species in the biosphere; particularly bearing in mind that the possession of but a few additional enzymes would allow these species to accommodate to a much wider range of growth conditions.

But since, with free-living microorganisms, compromises necessarily must be effected at the molecular level one should, perhaps, seek a resolution of the apparent paradox, mentioned above, in some of the general physiological and biochemical features of the microbial cells. Thus, we have already pointed to the special features of microbial behaviour relevant to their growth and survival in nutrient-depleted environments, and these now can be reassessed in the context of microbial evolution.

Here, we would first point once more to the overwhelming need of free-living microbes to be highly flexible, physiologically; but then add that this clearly will require organisms to carry all the relevant (and very considerable) extra genetic information. According to Koch (1976), perhaps 80% of *Escherichia coli*'s genetic content is committed to maintaining physiological flexibility.

Next we would draw attention to the special physiological requirements for scavenging traces of growth-constraining nutrient from the environment. This, we suggest, is facilitated by organisms possessing a vast nutrient uptake potential, and growth potential, relative to that which generally can be expressed in natural ecosystems. Thus, all the reactions leading to cell synthesis are held far away from their thermodynamic equilibrium values; the cells, consequently, are highly reactive and responsive to transient changes in the environment. But a high growth rate potential may well be incompatible with a large genome size (Cavalier-Smith, 1981), and hence the prokaryotic genome size is relatively small. It thus follows that if there are large demands upon the cells' genetic content imposed by a need to be physiologically plastic, and the genome size necessarily must be kept to a minimum, then it cannot permanently carry much redundant (spare) DNA. Microorganisms, one might expect, always will seek to dispose of genetic information that is not necessary for, or compatible with, growth in an ever-changing environment (Neijssel, 1980).

Now if this analysis is correct, and eco-physiological considerations require the genome size to be kept to a minimum, yet functionally replete, then the possibilities for evolutionary development will be severely restricted since this process seemingly requires the presence of spare mutable material (Clarke, 1974; Hartley, 1974). It may well be, therefore, that faced with these molecular constraints, evolution in prokaryotic species could not proceed towards the selection of a few super-competent 'generalist' organisms, but had to move divergently towards a spreading of the genetic information among many types of more specialist organisms. And these latter types are maintained **not** by virtue of some 'fitness' to compete continuously with other microbial species in the biosphere, but by being able to take advantage of occasional transient changes in the environment to which they are suited, and then hanging on, so to speak, during periods of adversity. This was the conclusion drawn by Konings & Veldkamp (1980) from studies reported by Sepers (1979) of ammonifying bacteria present in the Hollands Diep (a eutrophic body of water in the Delta area of the Netherlands). Such a conclusion also could adequately account for the presence of many obligately aerobic microbes in the biosphere as well as for the ubiquitous distribution of thermophiles and psychrophiles (that is, even in places where their growth is not particularly favoured).

However, if, as suggested, physiological considerations serve to impede evolutionary processes, then one might question how specialist organisms first arose. In this connection, the widespread occurrence of extrachromosomal elements among microbial species may be viewed as a strategy adopted by these organisms to retain within populations, rather than within individual cells, genetic properties that only rarely may be called into play. And these elements provide ample scope for evolutionary development. Moreover, the smaller plasmids seemingly do not influence markedly the maximum rate of genome replication (Dale & Smith, 1979; Zünd & Lebek, 1980) and therefore, in carrying additional genetic information, they act to reconcile the otherwise irreconcilable.

In conclusion, we suggest that a detailed understanding of the physiological properties expressed by microorganisms in laboratory culture is relevant not only to a fuller appreciation of the roles these organisms play in their natural surroundings, but also offers some rationale for an evolutionary process that seemingly has proceeded towards extreme species diversity and specialization.

Acknowledgements

We thank the Netherlands Organisation for Pure Scientific Research (ZWO) for financial support.

REFERENCES

AIKING, H. & TEMPEST, D. W. (1976). Growth and physiology of *Candida utilis* NCYC 321 in potassium-limited chemostat culture. *Archives of Microbiology,* **108,** 117–24.

BRAUN, V., HANCOCK, R. E. W., HANTKE, K. & HARTMANN, A. (1976). Functional organization of the outer membrane of *Escherichia coli.* Phage and colicin receptors as components of iron uptake systems. *Journal of Supramolecular Structure,* **5,** 37–58.

BROCK, T. D. (1966). *Principles of Microbial Ecology.* Englewood Cliffs, New Jersey: Prentice-Hall.

BROOKS, J. D. & MEERS, J. L. (1973). The effect of discontinuous methanol addition on the growth of a carbon-limited culture of pseudomonas. *Journal of General Microbiology,* **77,** 513–19.

CAVALIER-SMITH, T. (1981). The origin and early evolution of the eukaryotic cell. In *Molecular and Cellular Aspects of Microbial Evolution. Symposia of the Society for General Microbiology 32*, ed. J. F. Collins & B. E. B. Moseley, pp. 33–84. Cambridge University Press.

CLARKE, P. H. (1974). The evolution of enzymes for the utilization of novel substrates. In *Evolution in the Microbial World. Symposia of the Society for General Microbiology 24*, ed. M. J. Carlile & J. J. Skehel, pp. 183–217. Cambridge University Press.

DALE, J. W. & SMITH, J. T. (1979). The effect of a plasmid on growth and survival of *E. coli. Antonie van Leeuwenhoek,* **45,** 103–11.

DAWES, E. A. & SENIOR, P. J. (1973). The role and regulation of energy reserve polymers in micro-organisms. *Advances in Microbial Physiology,* **10,** 135–266.

DAWES, I. W. & MANDELSTAM, J. (1970). Sporulation of *Bacillus subtilis* in continuous culture. *Journal of Bacteriology,* **103,** 529–35.

DUINE, J. A. & FRANK, J. (1981). Methanol dehydrogenase: a quinoprotein. In *Microbial Growth on* C_1 *Compounds*, ed. H. Dalton, pp. 13–41. London: Heyden.

ELLWOOD, D. C. & TEMPEST, D. W. (1972). Effects of environment on bacterial wall content and composition. *Advances in Microbial Physiology,* **7,** 83–117.

FOGG, G. E., STEWART, W. D. P., FAY, P. & WALSBY, A. E. (1973). *The Blue-Green Algae.* London: Academic Press.

GERHART, D. Z. & LIKENS, G. E. (1975). Enrichment experiments for determining nutrient limitation: four methods compared. *Limnology and Oceanography,* **20,** 649–54.

GOLDMAN, C. R. (1978). The use of natural phytoplankton populations in bioassay. *Mitteilungen Internationale Vereinigung für Theoretische und Angewandte Limnologie,* **21,** 364–71.

GONS, H. J. (1977). On the Light-limited Growth of *Scenedesmus protuberans* Fritsch. Thesis: University of Amsterdam, The Netherlands.

HARRISON, D. E. F. (1976). The regulation of respiration rate in growing bacteria. *Advances in Microbial Physiology,* **14,** 243–313.

HARTLEY, B. S. (1974). Enzyme families. In *Evolution in the Microbial World. Symposia of the Society for General Microbiology 24*, ed. M. J. Carlile & J. J. Skehel, pp. 151–82. Cambridge University Press.

HEMPFLING, W. P. & RICE, C. W. (1981). Microbial bioenergetics during continuous culture. In *Continuous Cultures of Cells,* vol. 2, ed. P. H. Chalcott, pp. 99–125. Boca Raton, Florida: CRC Press.

HERBERT, D. (1958). Some principles of continuous culture. In *Recent Progress in Microbiology, VIIth International Congress of Microbiology,* pp. 381–96.

HERBERT, D., ELSWORTH, R. & TELLING, R. C. (1956). The continuous culture of bacteria; a theoretical and experimental study. *Journal of General Microbiology,* **14,** 601–22.

HOOGERHEIDE, J. C. (1975). Studies on the energy metabolism during the respiratory process by baker's yeast. *Radiation and Environmental Biophysics,* **12,** 281–90.

HUNTER, I. S. & KORNBERG, H. L. (1979). Glucose transport of *Escherichia coli* growing in glucose-limited continuous culture. *Biochemical Journal,* **178,** 97–101.

KJELDGAARD, N. O. & KURLAND, C. G. (1963). The distribution of soluble and ribosomal RNA as a function of the growth rate. *Journal of Molecular Biology,* **6,** 341–50.

KOCH, A. L. (1976). How bacteria face depression, recession and derepression. *Perspectives in Biology and Medicine,* **20,** 44–63.

KONINGS, W. N. & VELDKAMP, H. (1980). Phenotypic responses to environmental change. In *Contemporary Microbial Ecology*, ed. D. C. Ellwood, J. N. Hedger, M. J. Latham, J. M. Lynch & J. H. Slater, pp. 161–91. London: Academic Press.

KORNBERG, H. L. & REEVES, R. E. (1972). Correlation between hexose transport and phosphotransferase activity in *Escherichia coli. Biochemical Journal,* **126,** 1241–3.

LUSCOMBE, B. M. & GRAY, T. R. G. (1971). Effect of varying growth rate on the morphology of *Arthrobacter. Journal of General Microbiology,* **69,** 433–4.

MAALØE, O. (1960). The nucleic acids and the control of bacterial growth. In *Microbial Genetics. Symposia of the Society for General Microbiology 10*, ed. W. Hayes &. R. C. Clowes, pp. 272–93. Cambridge University Press.

MUR, L. R., GONS, H. J. & VAN LIERE, L. (1977). Some experiments on the competition between green algae and blue-green bacteria in light-limited environments. *FEMS Microbiology Letters,* **1,** 335–8.

NEIJSSEL, O. M. (1980). A microbiologist's view of genetic engineering. *Trends in Biochemical Sciences,* **5,** III–IV.

NEIJSSEL, O. M. & TEMPEST, D. W. (1975). The regulation of carbohydrate metabolism in *Klebsiella aerogenes* NCTC 428 organisms, growing in chemostat culture. *Archives of Microbiology,* **106,** 251–8.

NEIJSSEL, O. M. & TEMPEST, D. W. (1976). The role of energy-spilling reactions in the growth of *Klebsiella aerogenes* NCTC 418 in aerobic chemostat culture. *Archives of Microbiology,* **110,** 305–11.

NEIJSSEL, O. M. & TEMPEST, D. W. (1979). The physiology of metabolite overproduction. In *Microbial Technology. Symposia of the Society for General Microbiology 29*, ed. A. T. Bull, D. C. Ellwood & C. Ratledge, pp. 53–82. Cambridge University Press.

NEIJSSEL, O. M., HARDY, G. P. M. A., LANSBERGEN, J. C., TEMPEST, D. W. & O'BRIEN, R. W. (1980). Influence of growth environment on the phosphoenolpyruvate: glucose phosphotransferase activities of *Escherichia coli* and *Klebsiella aerogenes*: a comparative study. *Archives of Microbiology,* **125,** 175–9.

O'BRIEN, R. W., NEIJSSEL, O. M. & TEMPEST, D. W. (1980). Glucose phosphoenolpyruvate phosphotransferase activity and glucose uptake rate of *Klebsiella aerogenes* growing in chemostat culture. *Journal of General Microbiology,* **116,** 305–14.

ROBINSON, A. & TEMPEST, D. W. (1973). Phenotypic variability of the envelope proteins of *Klebsiella aerogenes. Journal of General Microbiology,* **78,** 361–70.

SCHAECHTER, M., MAALØE, O. & KJELDGAARD, N. O. (1958). Dependency on medium and temperature of cell size and chemical composition during balanced growth of *Salmonella typhimurium. Journal of General Microbiology,* **19,** 592–603.

SCHLEGEL, H. G., GOTTSCHALK, G. & VON BARTHA, R. (1961). Formation and utilization of poly-β-hydroxybutyric acid by Knallgas bacteria (*Hydrogenomonas*). *Nature (London),* **191,** 463–70.

SEPERS, A. B. J. (1979). De aerobe mineralisatie van aminozuren in natuurlijke aquatische mileus. Ph.D. thesis: University of Groningen, The Netherlands.

SHILO, M. (1980). Strategies of adaptation to extreme conditions in aquatic microorganisms. *Naturwissenschaften,* **67,** 384–9.

STERKENBURG, A., VLEGELS, P. A. P. & WOUTERS, J. T. M. (1981). Phenotypic variability of the major membrane proteins of *Klebsiella aerogenes. The Society for General Microbiology Quarterly,* **8,** 132.

TEMPEST, D. W. (1976). The concept of 'relative growth rate': its theoretical basis and practical application. In *Continuous Culture 6: Applications and New Fields*, ed. A. C. R. Dean, D. C. Ellwood, C. G. T. Evans & J. Melling, pp. 349–52. Chichester, England: Ellis Horwood.

TEMPEST, D. W. (1978). The biochemical significance of microbial growth yields: a reassessment. *Trends in Biochemical Sciences,* **3,** 180–4.

TEMPEST, D. W. & HUNTER, J. R. (1965). The influence of temperature and pH value on the macromolecular composition of magnesium-limited and glyceral-limited *Aerobacter aerogenes*, growing in a chemostat. *Journal of General Microbiology,* **41,** 267–73.

TEMPEST, D. W. & NEIJSSEL, O. M. (1978). Eco-physiological aspects of microbial growth in aerobic nutrient-limited environments. *Advances in Microbial Ecology,* **2,** 105–53.

TEMPEST, D. W. & NEIJSSEL, O. M. (1980). Growth yield values in relation to respiration. In *Diversity of Bacterial Respiratory Systems 2*, ed. C. J. Knowles, pp. 1–31. Boca Raton, Florida: CRC Press Inc.

TEMPEST, D. W., HERBERT, D. & PHIPPS, P. J. (1967). Studies on the growth of *Aerobacter aerogenes* at low dilution rates in a chemostat. In *Microbial Physiology and Continuous Culture*, ed. E. O. Powell, C. G. T. Evans, R. E. Strange & D. W. Tempest, pp. 240–56. London: HMSO.

TOLBERT, N. E., NELSON, E. B. & BRUIN, W. J. (1971). Glycolate pathway in algae. In *Photosynthesis and Photorespiration*, ed. M. D. Hatch, C. B. Osmond & R. O. Slayter, pp. 506–13. London: Wiley-Interscience.

VAN LIERE, L. (1979). On *Oscillatioria agardhii* Gomont; experimental ecology and physiology of a nuisance bloom-forming cyanobacterium. Ph.D. thesis: University of Amsterdam, The Netherlands.

Veldkamp, H. (1977). Ecological studies with the chemostat. *Advances in Microbial Ecology,* **1,** 59–94.

ZEVENBOOM, W. (1980). Growth and nutrient uptake kinetics of *Oscillatoria agardhii.* Ph.D. thesis: University of Amsterdam, The Netherlands.

ZEVENBOOM, W., DE GROOT, G. J. & MUR, L. R. (1980). Effects of light on nitrate-limited *Oscillatoria agardhii* in chemostat cultures. *Archives of Microbiology,* **125,** 59–65.

ZEVENBOOM, W., BIJ DE VAATE, A. & MUR, L. R. (1982). Assessment of factors limiting growth rate of *Oscillatoria agardhii* in hypertrophic Lake Wolderwijd, 1978, by use of physiological indicators. *Limnology and Oceanography,* **27,** 39–52.

ZÜND, P. & LEBEK, G. (1980). Generation time-prolonging R plasmids: correlation between increases in the generation time of *Escherichia coli* caused by R plasmids and their molecular size. *Plasmid,* **3,** 65–9.

ENERGY TRANSDUCTION AND SOLUTE TRANSPORT MECHANISMS IN RELATION TO ENVIRONMENTS OCCUPIED BY MICROORGANISMS

WIL N. KONINGS AND HANS VELDKAMP

Department of Microbiology, University of Groningen, The Netherlands

INTRODUCTION – THE MICROBE'S ENVIRONMENT

Although the nutritional requirements of many bacteria have been established in laboratory studies, little is known about the chemical composition of the microenvironments in which bacteria occur naturally. This is due to the fact that microbes are very small and that hardly any methods have been devised thus far to measure local concentrations of environmental components which are essential for their growth. One promising development is the application of microelectrodes which allows the measurement of the concentrations of oxygen (Revsbech, 1982) or of particular ions such as sulphide (Blackburn, Kleiber & Fenchel, 1975) in the vicinity of bacterial microcolonies.

One of the main differences between bacteria thriving in liquid laboratory cultures and those exposed to natural conditions is that the latter occur in a heterogeneous environment. Only part of the nutrients they use is dissolved in the water phase and freely available to living cells. Another part, and in many cases the larger one, is in some way or other associated with solid particles of varying complexity, with colloids or with chelating compounds. Some essential solutes such as iron may occur mainly in an insoluble form such as $Fe(OH)_3$ in sea water, as FeS in sediments, or in chemically more complex material. Altogether in nature, the solutes needed by bacterial cells are present in a variety of chemical species. The exact composition of complexes containing chemicals which are not directly available is largely unknown, as are the equilibria which exist between these and the water phase. A complicating factor of course is that release of solutes from these complexes often is largely or entirely due to microbial activities.

In their heterogeneous natural environment bacteria are often exposed to chemical and physical gradients which means that microbes in close proximity to each other may show different phenotypic responses to their microenvironment (Wimpenny, 1981, 1982). These gradients often are not stable for prolonged periods of time. Transient states both win respect to environmental conditions and metabolism are the rule rather than the exception in nature. In addition to concentration gradients of solutes needed for growth microbes are also often exposed to gradients in pH and redox potential, which are of particular importance since they affect solute uptake. The latter gradients are particularly common in soil and in the surface of the sediments of natural waters (Jørgensen, 1977). Anaerobic pockets in which production of reducing agents such as sulphide and organic acids occurs are often surrounded by capillaries filled with water having considerably higher pH and redox potential values. On the other hand the oxygen tension in the vicinity of micro-colonies may be zero in 'aerobic' environments (Wimpenny, 1981).

The concentrations of solutes needed for microbial growth in natural ecosystems are very much lower than those employed in batch cultures in the laboratory. Because of this they are often very difficult to determine. This was shown for example by Jøorgensen, Lindroth & Mopper (1981) in a study on excretion and distribution of free amino acids and ammonium in an intertidal sediment. The methods used to extract pore water from the sediment yielded quite different results in terms of the concentration of these solutes in the water phase. The use of one particular method even, such as centrifugation of the sediment at low speeds, showed considerable qualitative and quantitative differences which depended on the centrifugation time employed. Christensen & Blackburn (1980) concluded that much of the dissolved alanine present in a marine sediment estimated at 800 nM by chemical methods was biologically unavailable and that a more realistic value was of the order of 10 nM.

Williams (1975) reviewed the biological and chemical aspects of dissolved organic matter in sea water and concluded that the gross organic content of sea water lies in the range of 0.5–2 mg C l^{-1}. Table 1 (Williams, 1975) shows a synopsis of data collected for sea water.

Despite the considerable effort to collect the data presented in Table 1 it is evident that even in this relatively well-studied environment the item 'total carbohydrates', of which the actual

Table 1. *Average concentration of dissolved organic compounds in sea water*

Component	Concentration in sea water (as μg C l^{-1})
Vitamin B_{12}	0.0005
Thiamine	0.005
Biotin	0.001
Total fatty acids	5
Urea	5
Total free sugars	10
Total free amino acids	10
Total carbohydrates	200
Combined amino acids	50

After Williams (1975).

composition is little known, represents the major component of organic matter in sea water.

With respect to carbon and energy sources for heterotrophic bacteria in natural waters most attention has been given to sugars and amino acids. Mopper *et al*. (1980) determined the monosaccharide spectra of a variety of sea water samples and identified at least twelve sugars. Quantitatively glucose and fructose were by far the most important. The high fructose concentrations might be explained partly by epimerization of glucose to fructose. The total concentration of free sugars varied between 0.015 μM (Sargasso Sea) and 2 μM in near shore Baltic Sea water above the sediment-water interface.

As already indicated, an important point is not only the nutrient concentrations present in natural environments but also their availability for use by microbial cells. Chemical assay methods generally do not discriminate between available and non-available solutes. Gocke, Dawson & Liebezeit (1981) studied the availability of dissolved free glucose to heterotrophic microorganisms from aquatic environments by comparing concentrations determined chemically with the ($k_t + S_n$) values (sum of transport 'constant' (k_t) and the natural concentration of glucose (S_n)) obtained using a microbiological method. The latter values were considerably smaller and the conclusion was drawn that not all of the 'free' glucose determined chemically was actually available to heterotrophic microorganisms. These workers presented a tentative scheme for the equilibrium between 'bound free' and 'truly free' low molecular weight compounds. Similar schemes varying in complexity can be made for any natural environment. All schemes of this type share the principle that concentrations of bacterial nutrients estimated chemically in natural environments must be differentiated between a fraction which is directly available and the remainder which is present in various kinds of reservoir whose chemical composition is generally unknown. This holds for sugars as well as for amino acids (Dawson & Gocke, 1978) and also for inorganic ions essential for bacterial growth such as iron. Solutes needed for growth generally become available at a rate which does not allow growth to proceed at its maximal value. Although growth rate may be limited by any essential solute it seems that in many environments carbon and energy limitation is most commonly encountered by chemoorganotrophic microbes.

These considerations can be summarized as follows:

(1) Solutes needed for microbial growth are often present as different species in their natural environment. The part available to microorganisms is often small compared with the total reservoir which can be detected by chemical methods.
(2) Available solutes are often present in nanomolar concentration and the rate at which they are supplied to microbial cells does not generally allow their growth at maximal rates.
(3) The most common growth rate-limiting solute for heterotrophs is the carbon and energy source. This also holds for 'nutrient-rich' environments such as the sediments of natural waters.
(4) Microbes in nature often occur in gradients of solute concentrations and of such factors as pH, redox potential, light intensity, and sometimes also of temperature.
(5) Environmental conditions are changing continuously, which means that the physiological properties of cells and their membranes are always in transient states.
(6) In nature the chemical environment of a particular microbial cell is often determined at least in part by the activities of other cells of its own or of different species.

Although natural environments are highly complex and little is known about the dynamics of the unavailable solute fraction, it seems possible to circumvent these difficulties in laboratory studies. The nanomolar concentrations of solutes available in nature can easily be obtained in the laboratory in nutrient-limited chemostats run at low dilution rates ($0.01\ h^{-1}$). What is more bacteria can be exposed to fluctuation in solute type or concentration or to fluctuations in other environmental factors in such continuous culture systems. Studies of this kind are, however, quite rare, and we are still far from having a complete picture of the way in which naturally occurring bacteria translate solute flows available at extremely low concentrations and in ratios quite different from that which is optimum for orderly multiplication into the more concentrated and harmonious flow required.

Solute transport studies are still mainly at the stage of revealing general principles concerning energy generation and the uptake of single substrates. Due to the present state of our knowledge, the following part of this chapter is concerned mainly with general principles. The last section will focus on the role of the cytoplasmic membrane in interactions of the cell with its environment and describes how bacteria organized in different ways deal with the problems of translating solute flow into flows of energy generation and biosynthesis.

MECHANISMS OF ENERGY TRANSDUCTION AND SOLUTE TRANSPORT

The bacterial cell envelope

The point of contact between a bacterial cell and its environment is at the cell envelope. In most bacteria this is formed by the cell wall (peptidoglycan layer) and the cytoplasmic membrane located on the inner side of the cell wall. In Gram negative organisms a third layer, the outer membrane, is found outside the cell wall.

These three layers differ both in structure and function. Solutes with molecular weights of up to 600 can pass the outer membrane and the cell wall freely (Lugtenberg, 1981). The cytoplasmic membrane, on the other hand, does not allow free passage of solutes with small molecular weights. Most solutes (substrates and ions) can only pass through the cytoplasmic membrane via specific translocation systems. The cytoplasmic membrane, therefore, governs to a large extent the specific composition of the cytoplasm. A cytoplasmic composition differing drastically from that outside the cell is essential for cell metabolism and growth.

The cytoplasmic membrane consists of a liquid-crystalline bilayer made up of phospholipids in which proteins are embedded. The hydrophobic properties of its components make the membrane impermeable for many solutes. This feature is an essential prerequisite for the function of the energy transducing mechanisms located in the cytoplasmic membrane. Many different proteins are embedded in the phospholipid bilayer. The function of most of these proteins is not yet understood. A large number of them seem to be involved in the translocation of solutes across the cytoplasmic membrane. These proteins (carriers) perform the controlled influx of metabolizable substrates and ions and the efflux of endproducts of metabolism. Another large group of membrane proteins play a role in the conversion of chemical or light energy into a form which can be used by the intracellular metabolic machinery. Among these proteins are components of the ATPase complex and of electron transfer systems such as the respiratory chain.

From the above it should be clear that the cytoplasmic membrane plays an essential role in interactions between bacteria and their environment (Konings & Veldkamp, 1980). These interactions and phenotypic responses to changes in environmental conditions can only be understood when detailed information is available about metabolic processes located in the cytoplasm and about energy transducing and solute translocating processes operating in the cytoplasmic membrane.

The generation of metabolic energy

The metabolic energy-generating processes which are found in bacteria can be classified into two groups: processes coupled to substrate level phosphorylation on the one hand and chemiosmotic energy generating processes on the other (Thauer, Jungermann & Dekker, 1977). In substrate level phosphorylations the energy released by dehydrogenation or lyase reactions is used for the synthesis of ATP. In several of these reactions the energy released is initially conserved in 'energy rich' compounds such as acetyl CoA or acetylphosphate and subsequently transferred to ATP by kinase reactions. This ATP functions as a universal molecular carrier of biological energy.

The second group of metabolic energy generating processes has been recognized since the discovery of oxidative phosphorylation (Slater, 1981). The form of energy generated by these processes has been a matter of considerable controversy until experimental evidence in favour of the chemiosmotic hypothesis (Mitchell, 1966,

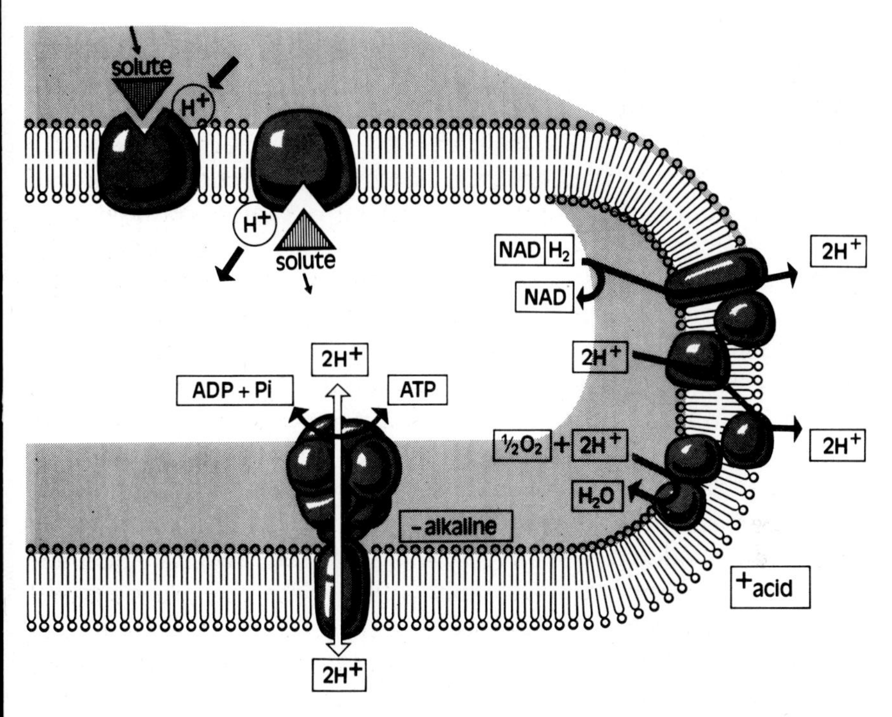

Fig. 1. Some chemiosmotic energy transducing systems in the cytoplasmic membrane of bacteria.

1972) became available. According to this hypothesis energy transducing systems, such as the electron transfer chains and the Ca^{2+}–Mg^{2+} stimulated ATPase complex, which are incorporated into the cytoplasmic membrane, act as electrogenic proton pumps and translocate protons across the cytoplasmic membrane from the cytoplasm to the external medium (Fig. 1). Since the cytoplasmic membrane is practically impermeable to ions and in particular to protons and hydroxyl ions this translocation results in the generation of an electrical potential ($\Delta\psi$) and a pH-gradient (ΔpH) across the membrane. An electrical potential is formed because with the protons positive charges are removed from the cytoplasm which accumulate in the external medium. A pH-gradient (a chemical proton gradient) is formed as a result of alkalinization of the cytoplasm due to removal of protons and acidification of the external medium due to accumulation of protons. Both gradients exert an inwardly-directed force on the protons, the proton motive force ($\Delta\tilde{\mu}_{H^+}$) is expressed as:

$$\Delta\tilde{\mu}_{H^+} = \Delta\psi - Z\Delta\text{pH (mV)} \quad (1)$$

in which $Z = 2.3\,RT/F$, R is the gas constant, T, the absolute temperature and F, the Faraday constant. The factor Z converts the pH-gradient into mV. Z has a numerical value of about 60 mV per pH unit at 25 °C.

It is important to realize that both components of the proton motive force are related but not equivalent. We will see below (see p. 28) that bacterial cells can interconvert a $\Delta\psi$ into a ΔpH or vice versa without changing the total proton motive force. The rates at which both components are generated by the proton pumps differ significantly. Since the electrical capacity of the membrane is low only a few charges have to be translocated for the generation of a $\Delta\psi$ of 1 mV. For the generation of a pH-gradient of 1 mV many more protons are required. Due to the cell's internal buffer capacity removal of protons will gradually increase the cytoplasmic pH while ejection of protons in the relatively large external volume will only slightly decrease the external pH. Consequently, the energy stored in the pH-gradient exceeds by far the energy content of an electrical potential of the same magnitude.

According to the chemiosmotic hypothesis the energy-transducing systems in the membrane convert chemical energy or light energy into electrochemical energy. This electrochemical energy can subsequently be used to drive energy-consuming processes by a reversed flow of protons and/or positive charges (Fig. 1). In this way the energy of the electrochemical gradient can be converted into chemical energy such as in the synthesis of ATP or into electrochemical energy as substrate or ion gradients or it can be used to drive mechanical work such as flagellar movement.

The central energy intermediates formed by each of the metabolic energy generating processes are thus distinctly different. For substrate level phosphorylation the intermediate is ATP whilst for chemiosmotic energy generating processes the intermediate is the electrochemical proton gradient, $\Delta\tilde{\mu}_{H^+}$. Each energy intermediate can drive certain energy-requiring processes in a cell but the processes driven by ATP are different from those driven by the proton motive force. ATP is a general energy intermediate in biosynthetic processes, it supplies energy for the transport of some solutes and it can also be used to generate a proton motive force. The proton motive force supplies the energy for flagellar movement, solute transport (secondary transport), reverse electron flow, the transhydrogenase reaction and it plays an essential role in many

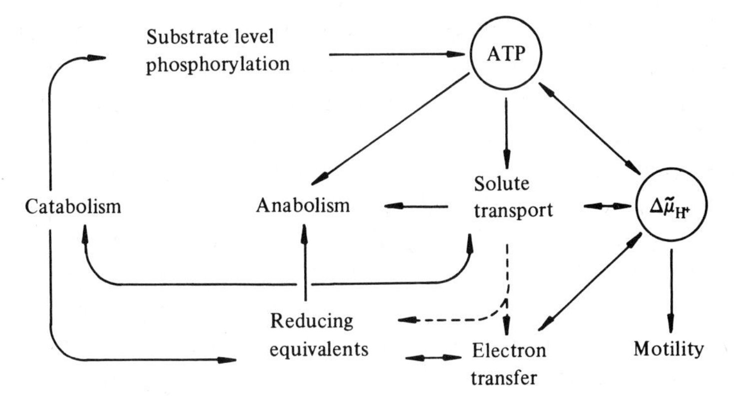

Fig. 2. Interaction between energy-generating and energy-consuming processes in bacteria.

other processes such as nitrogen fixation and DNA transport (for review see Konings, Hellingwerf & Robillard, 1981).

In a bacterial cell neither energy generating process functions independently but are both closely interlinked (Fig. 2). It will be shown below that the proton motive force drives the accumulation of catabolic substrates which are subsequently metabolized leading to the generation of ATP by substrate level phosphorylation and to the production of reduced intermediates like NADH or succinate. These reduced intermediates can be oxidized by proton pumping electron transfer systems whilst ATP can be hydrolysed by the proton pumping membrane-bound ATPase complex, each leading to the generation of a proton motive force.

The systems behave like connected vessels. When many reducing equivalents are available electron transfer will lead to the generation of a high proton-motive force and subsequently to the synthesis of ATP. On the other hand when the ATP content of a cell is high and the proton motive force is low the proton motive force can be generated by ATP hydrolysis and less ATP becomes available for biosynthesis.

Solute transport systems in the cytoplasmic membrane

The energy-transducing systems of cytoplasmic membranes are in essence transport systems for solutes (metabolizable substrates and ions). They can be classified into four groups according to the energy-transducing processes catalysed: (i) The primary transport systems convert chemical or light energy into electrochemical energy. These transport systems comprise the electrogenic proton pumps: the electron transfer systems, the Ca^{2+}–Mg^{2+} stimulated ATPase and in halobacteria the light-driven proton pump bacteriorhodopsin. (ii) The secondary transport systems driven by electrochemical energy. These are the main solute transport systems found in bacteria. (iii) The group translocation systems which translocate solutes by an enzymatic reaction. In this process the solute is chemically modified and the product is released in the cytoplasm. (iv) The ATP-driven transport systems translocate solutes through specific membrane proteins. The energy for translocation is supplied by phosphate-bond energy directly.

Primary transport systems

The main primary transport systems in bacteria are the electron transfer systems and the Ca^{2+}–Mg^{2+} stimulated ATPase complex. The electrogenic extrusion of protons by these proton pumps leads to the generation of a proton motive force which is usually internally alkaline and negative (Mitchell, 1976; for review see Konings, Hellingwerf and Robillard, 1981).

The driving force for proton extrusion in electron transfer systems is the redox potential difference across a proton translocation site ($\Delta E'_0$) and in the ATPase complex the free energy change of ATP-hydrolysis, the phosphate potential $\Delta G'_{ATP}$. When thermodynamic equilibrium between the proton extruding and the proton-motive force is reached the following equations hold:

$$n\Delta\tilde{\mu}_{H^+} = \Delta E'_0 \tag{2}$$

or

$$n\Delta\tilde{\mu}_{H^+} = \Delta G'_{ATP} \tag{3}$$

in which n is the number of protons translocated per electron transferred or per ATP hydrolysed.

These equations show that the electron transfer systems and the ATPase complex are coupled via the proton motive force. Both systems can also catalyse the reverse reactions: reverse electron flow leading to the generation of reduced equivalents and ATP-synthesis. When a bacterial cell has a high concentration of reducing equivalents and a low phosphate potential (low[ATP]/[ADP]ratio) electron transfer will generate a proton-motive force which will subsequently drive the synthesis of ATP. In the reverse situation

ATP-hydrolysis will generate a proton-motive force which will then drive the synthesis of reducing equivalents.

The proton motive force is a driving force for several energy-consuming processes in the cytoplasmic membrane. Since the activity of these energy-consuming processes will cause a dissipation of the proton-motive force continuous proton extrusion by the proton pumps is required to maintain high values. The rate at which protons can be translocated across the membrane by proton pumps, therefore, sets a limit to the rate at which work can be done. Bacteria have solved the problem of increasing the rate of proton translocation by increasing the cytoplasmic membrane surface area, thus allowing the incorporation of more proton pumps. Examples include the chromatophore membranes in phototrophic organisms and the stacked membranes found in methanogenic bacteria.

The proton-motive force has been measured in *E. coli* grown under aerobic conditions and under anaerobic conditions with fumarate as electron acceptor. Under aerobic conditions a proton-motive force of −200 mV is generated (Zilberstein, Schuldiner & Padan, 1979). Anaerobically in the presence of fumarate the proton-motive force is much lower and around −105 mV (Hellingwerf, Bolscher & Konings, 1981). Since, as has already been indicated, many cell metabolic processes are directly driven or influenced by proton-motive force, their activity will vary with the prevailing growth conditions. Clearly, the complex process of growth proceeds with proton-motive force values ranging between −100 mV and −200 mV, indicating that a balanced regulation of energy transducing processes must occur.

Secondary transport systems

The translocation of most solutes across the cytoplasmic membrane is driven by the proton-motive force or one of its components. Some compounds like undissociated weak acids or bases and lipophilic ions can cross the membrane passively without the involvement of specific carrier proteins. The transport of most solutes, however, is facilitated by specific carrier proteins. Often such proteins are also found for membrane permeable compounds, apparently to allow higher rates of solute translocation. The energy-dependent carrier-mediated translocation process is usually termed active transport.

Facilitated secondary transport can occur by three different mechanisms: (i) 'uniport', only one solute is translocated by the carrier protein; (ii) 'symport', two or more different solutes are

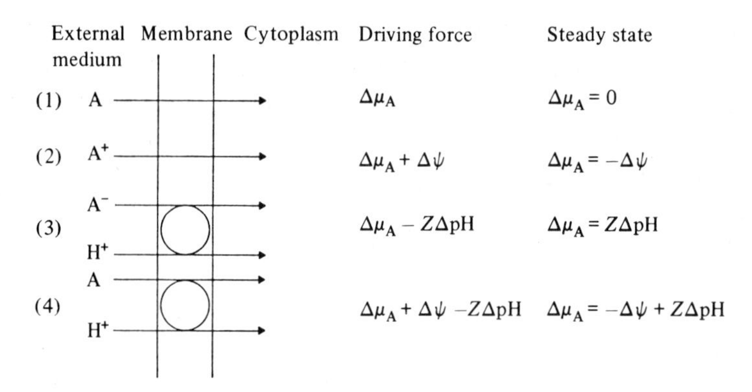

Fig. 3. Schematic presentation of four secondary transport processes: (1) passive transport of a neutral solute; (2) passive transport of a cation; (3) facilitated transport of an anion in symport with one proton; (4) facilitated transport of a neutral solute in symport with one proton. For each process the driving force and the steady state level of accumulation are given.

translocated by one carrier in the same direction; (iii) 'antiport', two or more different solutes are translocated by one carrier in opposite directions.

The driving force for passive and facilitated solute transport is supplied by the electrochemical gradients of the translocated solutes. The driving forces for different solute translocation processes and the steady state levels for solute accumulation when the driving force is zero are shown in Fig. 3.

These examples demonstrate that an equilibration of the internal concentration with the external concentration will be achieved at steady state only in the passive transport of a neutral solute (example 1). In the presence of a $\Delta\psi$, passive transport of a positively charged solute (example 2) can lead to an accumulation of this solute internally. On the other hand, in the absence of a $\Delta\psi$ the uptake of A^+ will generate a $\Delta\psi$, internally positive, and net uptake will stop before equilibration has been reached. In example 3 a symport system is shown for a negatively charged solute A^- which is translocated together with a proton. In this translocation process only the ΔpH component of the proton-motive force contributes to the driving force. In example 4 a neutral solute is transferred with a proton through a symport. In this translocation process the total proton-motive force contributes to the driving force for solute uptake.

In a similar way the driving forces and steady state levels of transport can be derived for more complex transport processes (Konings & Michels, 1980). These examples clearly demonstrate that the driving force for solute transport is dependent on the total charge and the number of protons which are transported. Consequently when during transport the composition and the magnitude of the proton-motive force is kept constant by primary transport systems, different levels of accumulation can be achieved for different solutes. This can be illustrated by the examples given in Fig. 3. With a proton-motive force of -180 mV consisting of a $\Delta\psi$ of -120 mV and a $Z\Delta pH$ of -60 mV the internal concentration of solute A at steady state will exceed the external concentration by a factor of 100 in example 2 ($60 \log[A]_{in}/[A]_{out} = +120$); by a factor of 10 in example 3 ($60 \log[A]_{in}/[A]_{out} = 60$) and by a factor of 1000 in example 4 ($60 \log[A]_{in}/[A]_{out} = 180$). It should also be obvious from these examples that changes in the composition of the proton-motive force will have different effects on the transport of different solutes. A dissipation of the pH gradient will strongly affect the ΔpH dependent transport systems. Many bacteria which grow at neutral pH values maintain an internal pH around 7.5 (see p. 168). Consequently, transport processes in which only the ΔpH component contributes to the driving force can only occur in media with external pH's below 7.5. In order to make accumulation of these solutes possible above pH 7.5 bacteria have developed symport systems with variable proton-solute stoichiometry (number of protons translocated per molecule of solute) (Konings & Booth, 1981). As a result the driving force for solute translocation varies. For instance in example 3 transport of A^- in symport with one proton can occur at pH 5.5. For transport at pH 7.5 an increase of the H^+-solute stoichiometry is required. With $2 H^+$ the driving force will be $\Delta\mu_A + \Delta\psi - 2 Z\Delta pH$ and since the ΔpH is zero the actual driving force will be $\Delta\mu_A + \Delta\psi$. Hence transport of A is coupled to the ΔpH at pH 5.5 and to the $\Delta\psi$ at pH 7.5. A consequence of this increase in H^+-solute stoichiometry is an increase of the energy cost of this solute transport process (see below). The H^+-solute stoichiometry has been measured only for a few transport processes. Evidence for a variable stoichiometry has been presented for organic acids such as lactate in *E. coli* (Ten Brink & Konings, 1980) and *S. cremoris* (Ten Brink & Konings, 1982) and for glucose-6-P, amino acids and lactose in *E. coli* (Le Blanc, Rimon & Kaback, 1980; Ramos & Kaback, 1977a, b, c; Ten Brink, Lolkema, Hellingwerf & Konings, 1981, Ahmed & Booth, 1981).

The information which is currently available about the driving forces for secondary solute transport indicates that at neutral pH values most solutes are taken up by the mechanisms shown in examples 2, 3 and 4 of Fig. 3 and that only a limited number of solutes are taken up in symport with more than one proton or with more than one positive charge. It seems, therefore, fair to state that on average the extrusion of about one proton is required to pay for every solute taken up. This energy requirement for solute transport can be expressed in ATP-equivalents by assuming that hydrolysis of ATP by the ATPase leads to the exclusion of two protons. The average cost for secondary solute transport is then around 0.5 ATP-equivalent.

Group translocation

The distinctive feature of group translocation types of solute transport is that, concomitant with transport, a chemical modification of the solute occurs resulting in the appearance of the product in the cytoplasm. At present the only thoroughly studied group translocation system in bacteria is the phosphoenolpyruvate-dependent sugar phosphotransferase system (PTS) (Postma & Roseman, 1976). This system catalyses the transport of various sugars according to the reaction:

$$\text{sugar}_{\text{out}} + \text{PEP} \xrightarrow[\text{HPr, Mg}^{2+}]{\text{E}_{\text{I}},\text{E}_{\text{II}}} \text{sugar P}_{\text{in}} + \text{pyruvate}$$

The overall transport which is catalysed can be broken down into several independent phosphoryl-group transfer steps as is schematically shown in Fig. 4. Several proteins are involved in this process. The soluble proteins Enzyme I and the Histidine-protein HPr are active in the transfer of phosphoryl groups from PEP to the sugar-specific membrane-bound enzymes II. Enzyme I is phosphorylated by PEP to form phospho-Enzyme I and pyruvate. Phospho-Enzyme I transfers the phosphoryl group to HPr. The Enzymes II contain sugar-specific binding sites and are active in the translocation and subsequent phosphorylation of the substrate. In several cases a third soluble protein, factor III, was shown to be involved in the transfer of the phosphoryl group from HPr to the membrane bound Enzyme II.

The PTS has been shown to be involved in the transport of several hexoses and hexitols almost exclusively in facultative and obligate

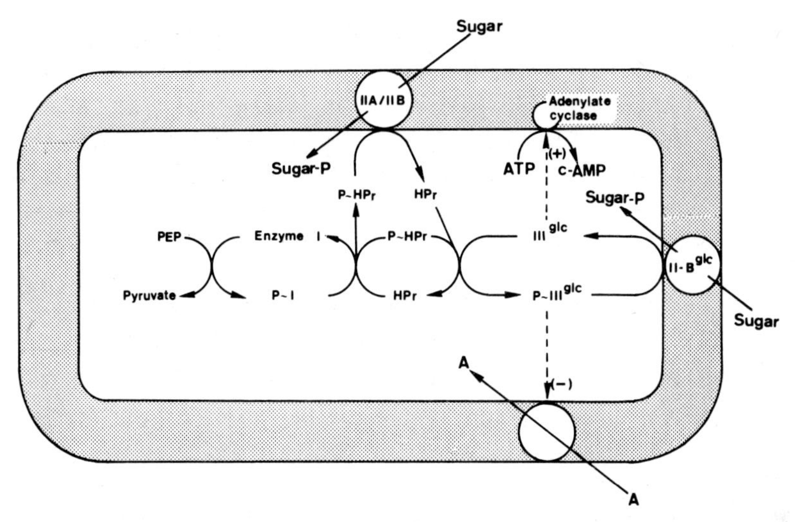

Fig. 4. Schematic representation of the PEP-dependent sugar phosphotransferase system in bacteria. I, enzyme I; IIA & IIB, enzymes II, III & factor III; HPr, histidine protein; A, solute.

anaerobes. In addition the PTS is linked to other cellular processes such as the regulation of the activity of several secondary transport systems (Saier, 1977). This regulation can occur by a direct inhibition of carrier proteins by the unphosphorylated form of factor III and by a stimulatory effect of the phosphorylated form of factor III on adenylate cyclase, the enzyme which synthesizes cAMP. The activated adenylate cyclase produces increased cAMP levels. This cAMP binds to the receptor protein and activates the synthesis of catabolic enzyme systems for non-PTS substrates (Postma & Scholte 1979; Meadow *et al.*, 1982). The result of this regulation is that in the presence of PTS-sugars the synthesis of carrier proteins for certain non-PTS substrates is repressed, whilst in their absence it is initiated.

ATP-dependent transport systems

In Gram-negative bacteria osmotic-sensitive transport systems are found in which periplasmic binding proteins play an essential role. The direct energy source for these transport systems appears to be acetyl-phosphate (Hong *et al.*, 1979). Such binding-protein-dependent transport systems are found for some amino acids and for phosphate (Boos, 1974).

There is also evidence for the existence of 'ATP'-dependent transport systems without the involvement of binding proteins.

These systems are found in Gram-positive and Gram-negative bacteria. Examples are sodium transport in *Streptococcus faecalis* (Heefner & Harold, 1980; Heefner, Kobayashi & Harold, 1980) and the potassium-ion TrkA transport system in *E. coli* (Rhoads & Epstein, 1977; Epstein & Laimins, 1979).

THE ROLE OF THE CYTOPLASMIC MEMBRANE IN INTERACTIONS OF BACTERIA WITH THEIR ENVIRONMENT

The cytoplasmic membrane separates the inner compartment, the cytoplasm, of a bacterial cell from its environment. The energy-transducing processes in the cytoplasmic membrane translocate solutes between compartments and generate electrochemical solute gradients between the two bulk phases. Solute concentrations on both sides of the cytoplasmic membrane will influence, therefore, the activities of the energy-transducing processes. It should be obvious that these activities will also be influenced by other cytoplasmic and environmental parameters such as specific regulating factors and physical parameters such as temperature and redox potential. Since all these energy transducing processes are closely interlinked via the proton motive force and/or the phosphate potential, altering one process will affect all the others. From these and other considerations it can be concluded that the proton-motive force, the phosphate potential and their related processes will strongly depend on conditions in the cytoplasm and in the environment. The following sections will focus on the environmental influences and will describe some well-documented studies on the effects of environmental factors on energy transducing processes in a few bacteria.

pH homeostasis

Bacteria are found in aquatic environments with pH values ranging from pH 1 to 2 in acidic sulphur springs up to 11 in soda lakes (Langworthy, 1978; Brown, Mayer, & Lelieveld, 1980; Grant & Tindall, 1980). Most bacteria live in environments at near-neutral pH values but a few acidophilic and some alkalophilic species have been described. Many bacteria are also able to survive in habitats in which the pH fluctuates drastically as a result of biological activity.

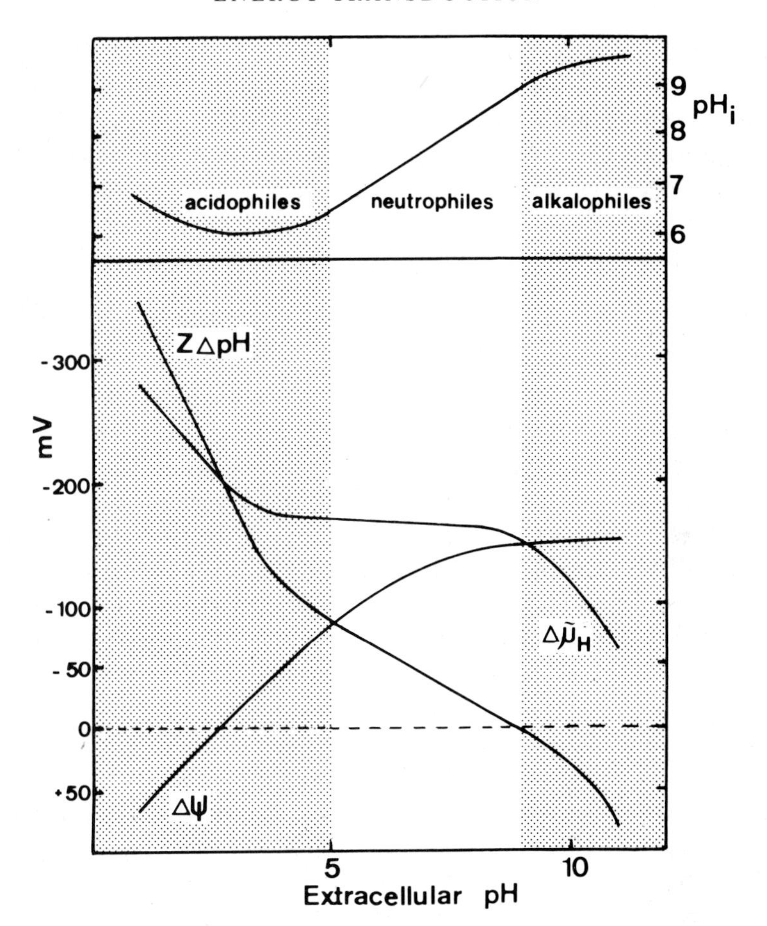

Fig. 5. $Z\Delta$pH, $\Delta\psi$, $\Delta\tilde{\mu}_{H+}$ and the internal pH (pH_i) in various bacteria as a function of the external pH. (Modified after Padan, Zilberstein & Schuldiner, 1981.)

The pH of the environment will influence the pH of the cytoplasm. Since most proteins, even those isolated from alkalophiles and acidophiles have a narrow pH range for optimum activity and/or stability these bacteria must have mechanisms that regulate the cytoplasmic pH values close to neutral (Padan, Zilberstein & Rottenberg, 1976).

In several acidophiles, neutrophiles and alkalophiles components of the proton-motive force have been measured and the results of these studies have been combined by Padan, Zilberstein & Schuldiner (1981). From these data a general picture of the magnitude of electrical potentials and the pH gradients in different bacteria at different external pH-values emerged. These relationships are presented in Fig. 5 without the experimental data. The reader is referred to the review of Padan, Zilberstein & Schuldiner (1981) for more detailed information. The sum of $\Delta\psi$ and $Z\Delta$pH gives the total proton-motive force. This figure shows that the magnitude and the composition of the proton-motive force varies drastically. In the acidophiles at pH 1 the proton-motive force is as high as −270 mV and consists of a large ΔpH and an *oppositely* directed (inside positive) $\Delta\psi$. In the neutrophiles the proton-motive force is around − 170 mV made up of both a $\Delta\psi$ and a ΔpH. In the alkalophiles the proton-motive force decreases dramatically with increasing external pH and is around −50 mV at pH 11. This proton-motive force contains a large $\Delta\psi$ component together with an oppositely directed (inside acid) ΔpH.

From the breakdown of proton-motive forces in acidophilic, neutrophilic and alkalophilic bacteria it can be concluded that the contribution of the $\Delta\psi$ and the ΔpH to the driving forces for uptake of solutes via secondary transport systems varies significantly. In acidophiles at pH values below 2.5 only the ΔpH can operate as a driving force, in the alkalophiles only $\Delta\psi$ is involved whilst between pH 2.5 and pH 8 both components of the proton-motive force can play a role. To make the conclusions clearer electroneutral and electrogenic transport can occur up to a pH value of 8. Above pH 8 all secondary transport occurs by electrogenic mechanisms.

From the ΔpH and the external pH the internal pH can be estimated (see Fig. 5). It is remarkable that a variation of the external pH from 1 to 11 leads to a variation of only 3 pH units in the cytoplasmic pH. In acidophiles the internal pH is poised at around pH 6.5, in neutrophiles at around pH 7.5 and in alkalophiles at around pH 9.5. The mechanisms which control the cytoplasmic pH have been studied extensively and these have been reviewed recently by Padan, Zilberstein & Schuldiner (1981) and Schuldiner & Padan (1982). Three different mechanisms have been suggested as playing a role in this process: (i) the primary proton pumps which always lead to an extrusion of protons and thus to an increase of the internal pH; (ii) Na^+/H^+ antiporter which extrudes Na^+ and returns protons into the cell; (iii) K^+/H^+ antiporter which extrudes K^+ in exchange for protons. Where present this antiporter appears to be electroneutral.

At this moment it is far from clear what the contribution of each of these systems is to the regulation of the internal pH in different bacteria.

Proton-motive force and the phosphate potential in growing and starved bacterial cultures

By now it should be clear that the proton-motive force and the phosphate potential largely determine the physiological condition of bacterial cells and consequently control the complex process of bacterial growth. Knowledge about the composition and magnitude of the proton-motive force, the magnitude of the adenosine nucleotide pools and in particular about the phosphate potential of bacterial cells under physiological conditions is therefore required for a proper understanding of the energy status and consequently the growth properties of bacteria. Since this energy status is affected by virtually all environmental factors like the composition and concentration of the carbon and/or energy source(s), sources of nitrogen and other essential nutrients, the ion composition, the environmental pH and temperature, an understanding of the phenotypic responses of bacteria to environmental changes can only be achieved if the effect of the environmental factors on the energy status is understood. However, the experimental procedures for measuring the different parameters, especially the components of the proton motive force in growing cells have only been developed recently. Consequently such information about growing bacteria is hardly available. Recently, procedures have been designed for measuring the $\Delta\psi$ and the ΔpH in bacteria growing in batch and continuous cultures and these procedures have been applied to cultures of *Streptococcus cremoris* (Otto, Lageveen, Veldkamp & Konings, 1982; Ten Brink & Konings, 1982). This organism was chosen because its metabolism is very simple: ATP is produced by substrate level phosphorylation only and a proton-motive force is generated by H^+ extrusion via the membrane-bound Ca^{2+}-Mg^{2+} stimulated ATPase or by efflux of lactate (see below).

In batch cultures which were not pH-controlled and which grow with lactose as sole energy source, the proton motive force at the start of the experiment (external pH = 6.8) was composed solely of a $\Delta\psi$ of about -90 mV. During logarithmic growth this $\Delta\psi$ was maintained at a rather constant value but as soon as lactose became depleted the $\Delta\psi$ dropped to zero within 90 min (Fig. 6).

During logarithmic growth the pH of the external medium decreases as a result of lactic acid production. The organism attempted to maintain the internal pH at around pH 6.6 and gradually a ΔpH was formed which reached a maximum value of

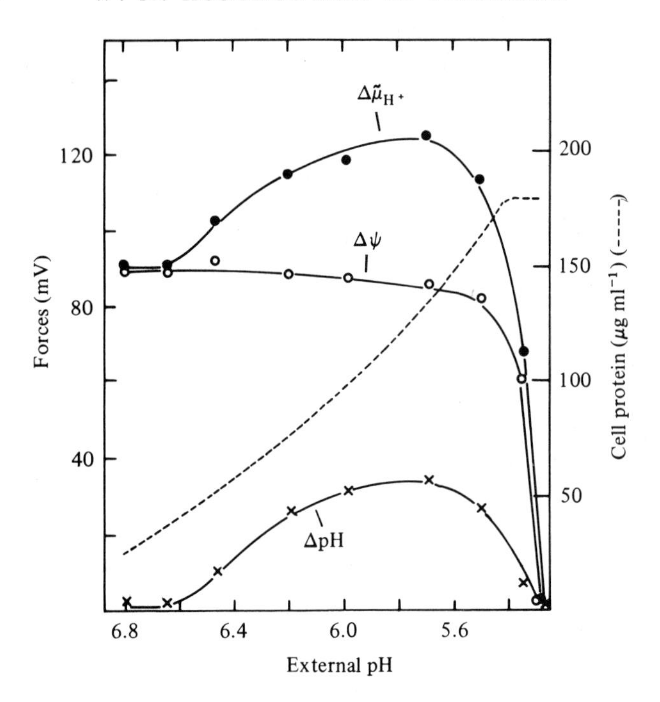

Fig. 6. The electrochemical proton gradient in *Streptococcus cremoris* during growth in batch culture on a complex medium with lactose ($2\ g\ l^{-1}$) as sole energy source. (×) $Z\Delta pH$, (○) $\Delta\psi$mV (●) $\Delta\tilde{\mu}_{H^+}$ ---, time course of cell protein synthesis.

−35 mV. In the final stage of logarithmic growth the internal pH could not be kept constant and decreased gradually with the external pH. As soon as growth ceased a rapid equilibration of the internal and external pH values occurred and the ΔpH collapsed. As a result of this variation of ΔpH, the proton-motive force increased during growth from −90 mV to −122 mV.

In batch cultures controlled at pH 6.8 the organism is not faced with the problem of maintaining its cytoplasmic pH value and no ΔpH was recorded. Also in these cultures the $\Delta\psi$ and thus the total proton motive force was kept at a constant value of around −90 mV during logarithmic growth and decreased rapidly to zero as soon as lactose was consumed (Fig. 7).

In these cultures the phosphate potential was also followed (Otto, Klont, Ten Brink & Konings, unpublished results). During logarithmic growth the $\Delta G'_{ATP}$ was rather constant at around −260 mV. When lactose had been consumed the internal concentration of ATP decreased from 1.6 mM to 0.025 mM within 90 min. Since the internal concentration of ADP also decreased (from 0.3 mM to

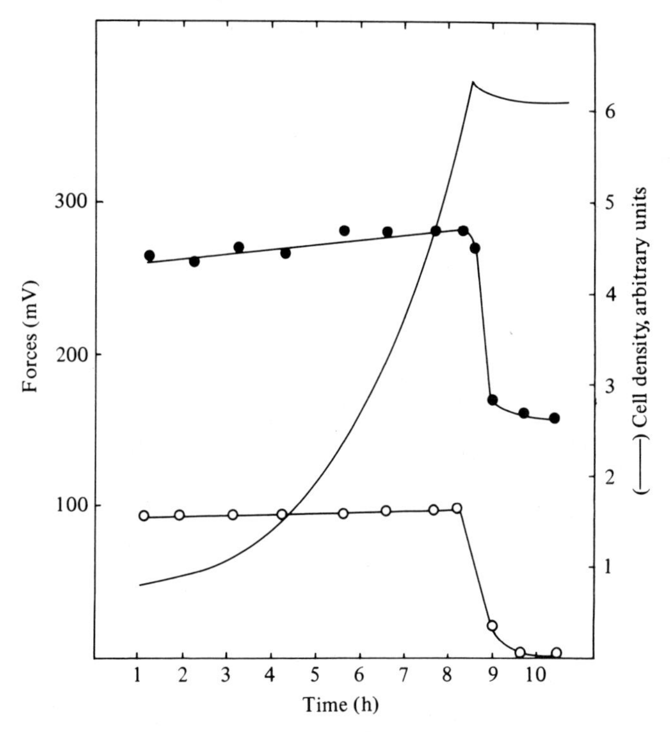

Fig. 7. $\Delta\tilde{\mu}_{H+}$ (°) and $\Delta G'_{ATP}$ (•) and cell density in *Streptococcus cremoris*, during growth in a pH-controlled batch culture at pH 6.8 on a complex medium with lactose (2 g l^{-1}) as sole energy source.

0.1 mM) whilst the internal concentration of organic phosphate increased (from 10 mM to 30 mM) the phosphate potential levelled off at −160 mV. These studies clearly demonstrate that *Streptococcus cremoris* maintains high values of the proton-motive force and $\Delta G'_{ATP}$ as long as growth supporting substrate is available, however both energy reservoirs dwindle rapidly when the energy supply stops. As a result, during logarithmic growth the organism is capable of accumulating essential solutes and of performing many energy requiring processes, but all these activities cease when the bacteria enter the starvation phase. Furthermore, since a proton-motive force is required to maintain the intracellular pools of metabolites, in the starvation phase these pools will gradually melt away. This is shown in Table 2 for the intracellular amino acid pools of *Staphylococcus epidermidis*. Already after two hours of starvation the intracellular pools of several amino acids are completely exhausted. Similar observations have been made for *Streptococcus*

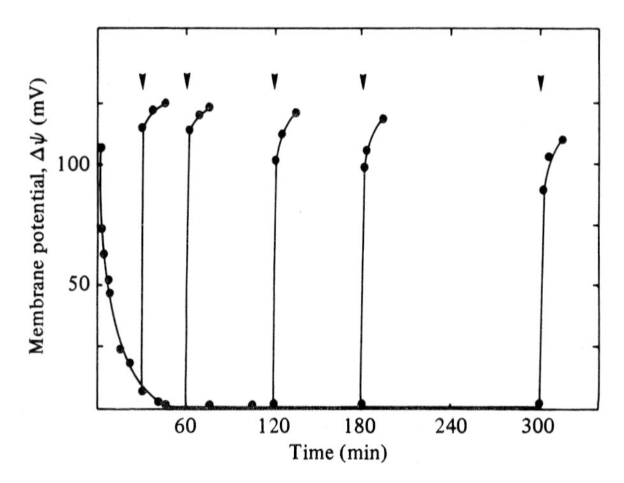

Fig. 8. Generation of a proton-motive force in energy-starved *S. cremoris* upon addition of lactose (indicated by the arrows) at different times after the start of starvation ($t = \circ$). Culture conditions are given in the legend to Fig. 7.

lactis (Thomas & Batt, 1969). Dissipation of energy reservoirs and the metabolite and ion pools will inevitably lead to a loss of viability of the cells. Cells will remain viable so long as essential cell structures (cell membrane, DNA, RNA, enzymes) are intact and the proton-motive force can be built up when energy is supplied. Fig. 8 shows that *S. cremoris* can still generate a high proton-motive force when lactose is added after a prolonged period of starvation. The gradual decrease in proton-motive force levels during starvation might indicate a gradual increase in the fraction of non-viable cells in the culture.

Mechanisms have been found in bacteria which will slow down the rate of energy or metabolite pool dissipation on starvation. The number of protons translocated per solute molecule transferred has been shown to be variable (see above). This enables an organism to maintain a high rate of solute uptake and a high internal concentration of metabolites or ions when the proton-motive force decreases by increasing the H^+/solute stoichiometry.

Furthermore, several transport and energy-transducing systems have been shown to be gated by the proton-motive force. When the proton-motive force decreases below certain levels such systems become blocked. Gating phenomena have been shown for the ATPase complex in *Streptococcus lactis* (Maloney, 1982), for amino acid transport in *Halobacterium halobium* (Lanyi & Silverman,

Table 2. *Effect of anaerobic starvation on the amino acid composition of the intracellular pool of* Staphylococcus epidermidis

Amino acid	Intracellular amino acid concentration (nmol (mg dry wt organisms)$^{-1}$) Time from start of harvesting (h):				
	0.75	2	5.75	7.25	24
Glutamate	40.0	6.9	2.6	2.1	0
Aspartate	33.6	31.5	10.2	5.1	0
Glycine + cysteine	4.6	0.6	0.6	0	0
Alanine	4.3	1.2	0.9	0	0
Lysine	12.2	6.6	0.6	0.5	0
Threonine	7.6	0	0	0	0
Serine	7.0	0	0	0	0
Proline	4.7	0	0	0	0
Valine	1.7	0.4	0	0	0
Leucine	2.9	0	0	0	0
Isoleucine	8.0	2.7	0.5	0	0
Tyrosine	1.9	0	0	0	0
Phenylalanine	2.9	0	0	0	0
Histidine	3.5	0	0	0	0
Total	134.6	49.9	15.4	7.7	0
NH_4^+	29.0	29.3	43.0	44.0	44.2

After Horan, Midgley & Dawes (1981) with permission.

1979) and for amino acid and K^+ transport in *Rhodopseudomonas sphaeroides* (Elferink *et al.*, 1982). As a result of this gating, ATP cannot further be hydrolysed by the ATPase complex, the proton-motive force cannot drive carrier-mediated solute uptake and accumulated solutes will not leave the cells via transport carriers. Dissipation of the pools will then occur only by passive transport processes.

Although information about such phenomena in different bacteria is still very scarce it is evident that some bacteria manage in this way to survive very long periods of starvation.

The generation of a proton-motive force by endproduct efflux

Strictly speaking, fermentative bacteria have no electron transfer systems which function as proton pumps. In these organisms a proton-motive force can be generated by the membrane-bound ATPase complex. Such a generation of proton-motive force by ATP hydrolysis would consume a considerable fraction of the ATP formed by substrate level phosphorylation meaning that less ATP is available for biosynthetic purposes. In order to avoid this ATP drain these organisms have developed transport systems which allow the efflux of endproducts of metabolism in symport with protons. In this way the reverse of the uptake process occurs: in the uptake process proton-motive force energy is converted into the energy of a solute gradient; in the efflux process the energy of the endproduct gradient is converted into a proton-motive force (Michels, Michels, Boonstra & Konings, 1979). Efflux of endproduct in symport with proton(s) will lead to the generation of an electrical potential if net charge is translocated, to the generation of a pH gradient if protons are translocated and to the generation of both electrical potential and a ΔpH if net charge and protons are translocated (see Fig. 3).

Since metabolic endproducts are continuously produced internally during fermentation a continuous efflux of endproducts occurs leading to the continuous generation of a proton-motive force. As a result, carrier-mediated efflux of endproducts can contribute significantly to metabolic energy during fermentation.

It is important to realize that the maximum value of the proton-motive force generated by this efflux process will be in thermodynamic equilibrium with the chemical potential of the endproduct gradient as has been described for the uptake processes. An increase of endproduct concentration in the medium will thus lead to a reduction in the endproduct gradient (unless the internal concentration becomes extremely high) and consequently the proton-motive force generated by this efflux process will fall. At high external concentrations of endproduct the organism depends on ATP-hydrolysis for the generation of a proton-motive force. Since under these conditions the inwardly directed proton-motive force would prevent carrier-mediated efflux of endproducts, a continuous excretion can proceed only if the H^+/endproduct stoichiometry is reduced for instance by translocating the endproduct in an electroneutral form by a passive diffusion process.

Experimental evidence for such a proton-motive force-generating mechanism has been supplied for *Escherichia coli* (Ten Brink & Konings, 1980) and for the homolactic fermentative bacteria *Streptococcus cremoris* (Otto, Sonnenberg, Veldkamp & Konings, 1980; Otto, Hugenholtz, Konings & Veldkamp, 1980; Otto, Lageveen, Veldkamp & Konings, 1982; Ten Brink & Konings, 1982). In this organism glucose and lactose are taken up by a PEP-dependent phosphotransferase system and subsequently metabolized quantitatively to lactate. The internal concentration of lactate can reach levels as high as 200 mM. The efflux of lactate is a carrier-mediated

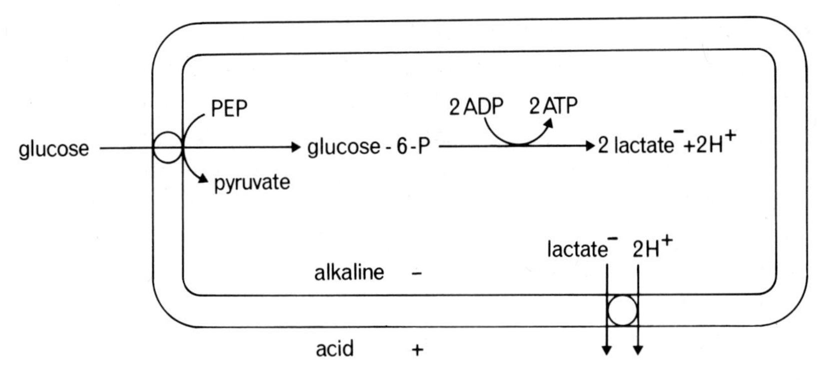

Fig. 9. Synthesis of ATP by substrate level phosphorylation and the generation of a proton-motive force by lactate efflux during homolactic fermentation.

process which occurs in symport with protons. Under optimal conditions (external pH above 6.7 external lactate concentrations below 10 mM) lactate leaves the cell with 2 protons (see Fig. 9). Per mole of glucose, therefore, 2 moles of ATP are formed by substrate level phosphorylation and 4 protons are excreted together with the lactate. Since 2 protons per glucose are formed during lactate production, lactate efflux leads to the net translocation of 2 protons per glucose consumed. Assuming that 2 protons are taken up per ATP synthesized by the ATPase complex the contribution to the metabolic energy of the cell by lactate efflux will be under optimal conditions 1 ATP-equivalent per glucose consumed, an additional energy gain of 50%.

During growth conditions in the medium will vary and so will the contribution of lactate efflux to metabolic energy. This is shown in Fig. 10 in which the proton-motive force and the lactate gradient ($\Delta\mu_{\text{lactate}}$) during growth of *S. cremoris* in batch cultures are given (Ten Brink & Konings, 1982). When growth proceeds and the external pH falls the ratio $\Delta\mu_{\text{lactate}}/\Delta\tilde{\mu}_{\text{H}^+}$ also drops, indicating a reduction in H^+/lactate stoichiometry and a consequent decrease in the contribution to metabolic energy by lactate efflux. Recently it was shown that not only the external pH determines the H^+/lactate stoichiometry in lactate efflux but also the external lactate concentration (Ten Brink & Konings, unpublished results). When the external lactate concentration increases the H^+/lactate stoichiometry falls.

A beneficial effect of lactate excretion on growth yield can, therefore, be expected at pH values above 6.3 under conditions in which the external lactate concentration is low. This could be

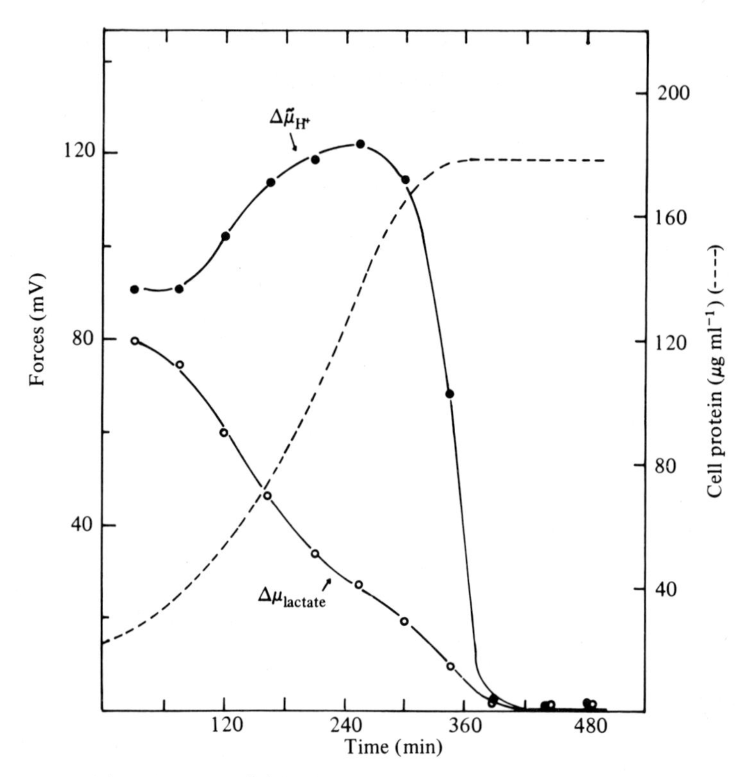

Fig. 10. $\Delta\tilde{\mu}_{H+}$(●) and $\Delta\mu_{lactate}$ (○) in *Streptococcus cremoris* during growth in batch culture on a complex medium with lactose ($2\,gl^{-1}$) as sole energy source. ---, time course of cell protein formation (modified after Ten Brink & Konings, 1982).

achieved by growing *S. cremoris* in a pH-controlled chemostat at pH 7, in the presence of the lactate-consuming *Pseudomonas stutzeri* (Otto, Hugenholtz, Konings & Veldkamp, 1980). In such a system the molar growth yield for *S. cremoris* increased, due to lactate removal by the pseudomonad, by 60–70%. This experiment shows that the relationship between a lactate producer and a lactate consumer can represent a case of mutualism since the interaction is profitable for both organisms.

It remains to be seen whether efflux of other metabolic endproducts can also lead to the generation of a proton-motive force and contribute to the energy budget of the bacterial cell.

Redox potentials in the environment and in the cytoplasm as a controlling factor for bacterial growth

An environmental parameter known for a long time to affect the growth properties of many microorganisms is the redox potential.

Hardly any information, however, is available about the number and nature of the cell functions which are affected by redox potential and how redox potential exerts its effect. In the last decade it has become evident that many important cell functions in bacteria, such as the reductive pentose phosphate cycle are catalysed by enzymes which contain redox-active disulphide groups, and that the redox state of these groups determines the activity of these enzymes (Buchanan, Wolosink & Schürmann, 1979). In several organisms a small hydrogen-carrier protein, thioredoxin, has been implicated in the regulation of enzyme activity by thiol-redox control. Hydrogen for the reduction of thioredoxin is supplied by NADPH in an NADPH-thioredoxin reductase-catalysed reaction, or by ferredoxin in a ferredoxin-thioredoxin reductase-catalysed reaction in fermentative and phototrophic bacteria. The reduced form of thioredoxin can effectively reduce the disulphide groups to dithiols in different proteins. It thus functions as a specific cellular equivalent of dithiothreitol (Holmgren, 1981).

Recent studies (Robillard & Konings, 1981; Konings & Robillard, 1982; Robillard & Konings, 1982) in *E. coli* have shown that the membrane-bound protein (EII) of the PEP-dependent hexose transport system (PTS) and the carrier proteins of lactose and proline contain redox-active disulphide groups. The redox state of these groups is determined by the redox potential of the environment (see Table 3) and by the proton-motive force. In the reduced dithiol form the proteins have a high affinity for their solute, in the oxidized, disulphide from the affinity is at least a factor of one hundred lower. Recently evidence has been presented that two such redox-sensitive disulphide groups are present in the transport proteins and that these groups are located on opposite sides of the membrane. Since dithiol-disulphide interconversions have been reported to play an essential role in many solute transport proteins and energy transducing enzymes, it is anticipated that the redox potential of the cytoplasm and/or the environment as well as the proton-motive force across the cytoplasmic membrane control the activity of important membrane functions.

The gating effects seen in proton-motive force driven transport processes are easier to understand when the activities of these systems are coupled to redox conversions. A finite proton-motive force is necessary to raise the redox potential past the midpoint potential of a given site before oxidation can occur.

Table 3. *Effect of redox reagents on the PEP-dependent phosphorylation of α-methyl-glucoside in inside-out membrane vesicles of* Escherichia coli ML 308–225

Redox agents	E_0' (mV)	Inhibition
Methyl viologen	−440	−
Benzyl viologen	−360	−
NAD^+	−320	+−
Riboflavin	−208	+
Plumbagin	−150	+
Methylene Blue	+ 11	+
PMS	+ 80	+
1,2-Naphthoquinone	+143	+
DCPIP	+217	+
Fe^{3+} cyt c	+254	+
1,4-Benzoquinone	+293	+
$Fe(CN)_6^{3-}$	+360	+

After Robillard & Konings (1981).

The effects of environmental ionic composition on the proton-motive force

The magnitude and the composition of the proton-motive force in bacteria depends largely on the ionic composition of the environment. This can be attributed to uncontrolled passive ion fluxes and to the activity of specific translocation systems. Both kinds of translocation processes can lead to either a decrease of the $\Delta\psi$, the ΔpH or both components of the proton-motive force or to an interconversion of $\Delta\psi$ and ΔpH.

As discussed above the activity of primary proton pumps leads to the extrusion of protons from the cytoplasm into the environment and will proceed until the proton extruding force is equal and opposite to the proton-motive force. Further extrusion of protons can only occur if the proton-motive force is dissipated, for instance by the uptake of positive charge as is the case in the uptake of a cation by passive movement or through a uniport (example 2 in Fig. 3). This process is driven by the $\Delta\psi$ only and leads to a fall in $\Delta\psi$ and consequently in proton-motive force. Further proton translocation by the primary proton pumps will increase both components of the proton-motive force until this has again reached its original value. The final result of these actions will be an increase in ΔpH at the expense of $\Delta\psi$. In a similar way it can be argued that electroneutral uptake of a solute in symport with protons (example 3 in Fig. 3) will lead to an increase of $\Delta\psi$ at the expense of ΔpH. No change in proton-motive force is expected when an equal number of positive charges and protons are taken up in the solute uptake process (see example 4 in Fig. 3). In this uptake process both components of the proton-motive force are equally involved in the driving force.

However, when more charges than protons are translocated the $\Delta\psi$ plays a more important role than the ΔpH as a driving force and solute uptake will lead to an increase in ΔpH. On the other hand when more protons than charges are translocated the ΔpH plays the dominant role as driving force and uptake of the solute will lead to an increase of $\Delta\psi$.

These considerations indicate that changes in composition of the proton-motive force will occur on solute uptake. In all these examples it has been assumed that the rate of proton-motive force generation using the primary proton pumps exceeds the rate of proton-motive force dissipation by solute uptake. This is not always so. In the presence of high concentrations of solutes which are accumulated rapidly in response to a $\Delta\psi$ or a ΔpH, a complete dissipation of one of the components and a decrease of the proton-motive force can occur (Michels *et al.*, 1981).

Properties of the cytoplasmic membrane in relation to conditions found in the natural environment

The currently available information concerning energy-transducing and solute transport systems and about proton-motive force can be used to discuss the function of bacterial cytoplasmic membranes in their natural environment.

It has been stated earlier that the concentration of solutes, available for uptake generally lies in the nanomolar range in most environments. The affinity of almost all solute transport systems lies in the micromolar range. This means that most solute transport systems function far below their maximum capacity. This property may be useful under transient conditions in order to allow high rates of uptake during short periods of abundance. This emphasis on rapid responses to environmental changes is also reflected in the surprisingly high number of specific carrier proteins present constitutively in the cytoplasmic membrane of most bacteria.

We have seen above that upon exhaustion of the energy source, a rapid dissipation of the proton-motive force can occur. If uptake of the energy source is mediated by a secondary transport system driven by proton-motive force, these bacteria would never be capable to refuel. This is possibly the main reason why bacteria have developed transport systems, especially for important energy sources such as sugars, which do not depend on the proton-motive force, but on other sources of energy, such as PEP. In several bacteria it has been shown that the level of PEP increases at the onset of starvation and remains high through prolonged starvation periods (Otto & Klont, unpublished results).

Regulatory mechanisms for the different transport systems seem to ensure uptake of the right solute at the right time. In the absence of a proton-motive force the redox state of the membrane-bound proteins of the PEP-dependent transport system (PTS) is such that the affinity for their substrate is high. During full activity of the PTS several secondary transport systems are inhibited. As a result of metabolism of the accumulated sugars a proton-motive force will be generated, the PTS becomes inhibited whilst the secondary transport systems become fully active. In this way a balanced uptake of energy source and essential solutes can be achieved.

The mechanisms by which bacteria manage to regulate a balanced uptake of individual solutes is not well understood at all. Possibly regulation is achieved by varying the number of copies of each transport carrier in the cytoplasmic membrane and/or by regulating the activity of the carriers by proton-motive force (via gating or redox state).

Most natural environments are energy-limited and the proton-motive force of bacteria in these environments will usually be very small. During short periods of energy supply the rate of proton-motive force generation on dissimilation of the energy source will largely determine the uptake of essential solutes and consequently the rate of biomass formation. The rate of proton-motive force generation thus determines to a large extent the reactivity of bacteria in natural environments.

Acknowledgements

We are grateful to F. B. van Es, B. Ten Brink, R. Otto and H. van Gemerden for stimulating discussions and critical comments. The help of R. Otto and B. Ten Brink in preparing the figures and Marry Pras and Mieke Broens in preparing the manuscript is greatly appreciated.

REFERENCES

AHMED, S. & BOOTH, I. R. (1981). The effects of partial and selective reduction in the components of the proton motive force on lactose uptake in *Escherichia coli. Biochemical Journal*, **200,** 583–9.

BLACKBURN, T. H., KLEIBER, P. & FENCHEL, T. (1975). Photosynthetic sulfide oxidation in marine sediments. *Oikos*, **26,** 103–8.

BOOS, W. (1974). Bacterial transport. *Annual Review of Biochemistry*, **43,** 123–47.

BROWN, M. H., MAYER, T. & LELIEVELD, H. L. M. (1980). The growth of microbes at low pH-values. In *Microbial Growth and Survival in Extremes of Environment*, ed. G. W. Gould & J. E. L. Corry, pp. 71–98. London: Academic Press.

BUCHANAN, B. B., WOLOSINK, R. A. & SCHÜRMANN, P. (1979). Thioredoxin and enzyme regulation. *Trends in Biochemical Sciences*, 93–5.

CHRISTENSEN, D. & BLACKBURN, T. H. (1980). Turnover of tracer (^{14}C, ^{3}H-labelled) alanine in in-shore marine sediments. *Marine Biology*, **58,** 97–103.

DAWSON, R. & GOCKE, K. (1978). Heterotrophic activity in comparison to the free amino acid concentrations in Baltic sea water samples. *Oceanologica Acta*, **1,** 45–54.

ELFERINK, M. G. L., FRIEDBERG, J., HELLINGWEY, K. J. & KONINGS, W. N. (1982). The role of the proton-motive force and electron flow in light-driven solute transport in *Rhodopseudomonas sphaeroides. European Journal of Biochemistry* (in press).

EPSTEIN, W. & LAIMINS, L. (1980). Potassium transport in *Escherichia coli.* Diverse systems with common control by osmotic forces. *Trends in Biochemical Science*, **5,** 21–3.

GOCKE, K., DAWSON, R. & LIEBEZEIT, G. (1981). Availability of dissoved free glucose to heterotrophic microorganisms. *Marine Biology*, **62,** 209–16.

GRANT, W. D. & TINDALL, B. J. (1980). The isolation of alkalophilic bacteria. In *Microbial Growth and Survival in Extremes of Environment*, ed. G. W. Gould & J. E. L. Corry, pp. 27–38. London: Academic Press.

HEEFNER, D. L. & HAROLD, F. M. (1980). ATP-linked sodium transport in *S. faecalis.* I. The sodium circulation. *Journal of Biological Chemistry*, **255,** 11396–402.

HEEFNER, D. L., KOBAYASHI, H. & HAROLD, F. M. (1980). ATP-linked sodium transport in *S. faecalis.* II. The energy coupling in everted membrane vesicles. *Journal of Biological Chemistry*, **255,** 11403–7.

HELLINGWERF, K. J., BOLSCHER, J. G. M. & KONINGS, W. N. (1981). The electrochemical proton gradient generated by the fumarate reductase system in *Escherichia coli* and its bioenergetic implications. *European Journal of Biochemistry*, **113,** 369–74.

HOLMGREN, A. (1981). Thioredoxin: structure and functions. *Trends in Biochemical Sciences*, **6,** 26–9.

HONG, J. S., HUNT, A. G., MASTERS, P. S. & LIEBERMAN, M. A. (1979). Requirement of acetyl-phosphate for the binding protein dependent transport systems in *Escherichia coli. Proceedings of the National Academy of Sciences, USA*, **76,** 1213–7.

HORAN, N. J., MIDGLEY, M. & DAWES, E. A. (1981). Effect of starvation on transport membrane potential and survival of *Staphylococcus epidermis* under anaerobic conditions. *Journal of General Microbiology*, **127,** 223–30.

JØRGENSEN, B. B. (1977). Bacterial sulfate reduction within reduced microniches of oxidized marine sediments. *Marine Biology*, **41,** 7–17.

JØRGENSEN, N. O. G., LINDROTH, P. & MOPPER, K. (1981). Extraction and distribution of free amino acids and ammonium in sediment interstitial waters from the Limfjord, Denmark. *Oceanologica Acta*, **4,** 465–74.

KONINGS, W. N. & BOOTH, I. R. (1981). Do the stoichiometries of ion-linked transport systems vary? *Trends in Biochemical Sciences*, **2,** 257–62.
KONINGS, W. N., HELLINGWERF, K. J. & ROBILLARD, G. T. (1981). Transport across bacterial membranes. In *Membrane Transport*, ed. S. L. Bonting & J. J. H. H. M. de Pont, pp. 257–83. Amsterdam: Elsevier/North-Holland.
KONINGS, W. N. & MICHELS, P. A. M. (1980). Electron-transfer driven solute accumulation across bacterial membranes. *Diversity of Bacterial Respiratory Systems*, ed. C. J. Knowles, pp. 33–87. Boca Raton Fl.: CRC Press.
KONINGS, W. N. & ROBILLARD, G. T. (1982). The physical mechanism for regulation of proton-solute symport in *Escherchia coli. Proceedings of the National Academy of Sciences, USA*, (in press).
KONINGS, W. N. & VELDKAMP, H. (1980). Phenotypic responses to environmental changes. In *Contemporary Microbial Ecology*, ed. D. C. Ellwood, J. N. Hedger, M. J. Latham, J. M. Lynch & J. H. Slater, pp. 161–91. London: Academic Press.
LANGWORTHY, T. A. (1978). Microbial life in extreme pH values. In *Microbial Life in Extreme Environments*, ed. D. J. Kushner, pp. 279–315. London: Academic Press.
LANYI, J. K. & SILVERMAN, M. P. (1979). Gating effects in *Halobacterium halobium* membrane transport. *Journal of Biological Chemistry*, **254,** 4750–5.
LEBLANC, G., RIMON, G. & KABACK, H. R. (1980). Glucose-6-phosphate transport in membrane vesicles isolated from *Escherichia coli:* effect of imposed electrical potential and pH gradient. *Biochemistry*, **19,** 2522–8.
LUGTENBERG, B. (1981). Composition and function of the outer membrane of *Escherchia coli. Trends in Biochemical Sciences*, **6,** 262–6.
MALONEY, P. C. (1982). Coupling between H^+ entry and ATP synthesis in bacteria. In *Current Topics in Membranes and Transport*, vol. 16, ed. C. L. Slayman, pp. 175–93. New York: Academic Press.
MEADOW, N. D., SAFFEN, D. W., DOTTIN, R. P. & ROSEMAN, S. (1982). Molecular cloning of crr gene and evidence that it is the structural gene for III^{Glc}, a phospho carrier protein of the bacterial phosphotransferase system. *Proceedings of the National Academy of Sciences, USA*, **79,** 2533–6.
MICHELS, P. A. M., HELLINGWERF, K. J., LOLKEMA, J. S. FRIEDBERG, I. & KONINGS, W. N. (1981). Effects of the medium composition of the components of the electrochemical proton gradient in *Rhodopseudomonas sphaeroides. Archives of Microbiology*, **130,** 357–61.
MICHELS, P. A. M., MICHELS, J. P. J., BOONSTRA, J. & KONINGS, W. N. (1979). Generation of an electrochemical gradient in bacteria by the excretion of metabolic endproducts. *FEMS Microbiology Letters*, **5,** 357–64.
MITCHELL, P. (1966). Chemiosmotic coupling and energy transduction. Glynn Research Ltd., Bodmin, England.
MITCHELL, P. (1972). Chemiosmotic coupling in energy transduction: a logical development of biochemical knowledge. *Bioenergetics*, **3,** 5–24.
MITCHELL, P. (1976). Possible molecular mechanism of the proton motive function of cytochrome systems. *Journal of Theoretical Biology*, **62,** 327–68.
MOPPER, K., DAWSON, R., LIEBEZEIT, G. & ITTEKKOT, V. (1980). The monosaccharide spectra of natural waters. *Marine Chemistry*, **10,** 55–66.
OTTO, R., HUGENHOLTZ, J., KONINGS, W. N. & VELDKAMP, H. (1980). Increase of molar growth yield of *Streptococcus cremoris* for lactose as a consequence of lactate consumption by *Pseudomonas stutzeri* in mixed cultures. *FEMS Microbiology Letters*, **9,** 85–8.
OTTO, R., LAGEVEEN, R. G., VELDKAMP, H. & KONINGS, W. N. (1982). Lactate efflux-induced electrical potential in membrane vesicles of *Streptococcus cremoris. Journal of Bacteriology*, **149,** 733–8.

OTTO, R., SONNENBERG, A. S. M., VELDKAMP, H. & KONINGS, W. N. (1980). Generation of an electrochemical proton gradient in *Streptococcus cremoris* by lactate efflux. *Proceedings of the National Academy of Sciences, USA*, **77,** 5502–6.

OTTO, R., TEN BRINK, B., VELDKAMP, H. & KONINGS, W. N. (1982). The relation between growth rate and electrochemical proton gradient of *Streptococcus cremoris. FEMS Microbiology Letters*, (in press).

PADAN, E., ZILBERSTEIN & ROTTENBERG, H. (1976). The proton electrochemical gradient in *Escherichia coli* cells. *European Journal of Biochemistry*, **63,** 533–41.

PADAN, E., ZILBERSTEIN & SCHULDINER, S. (1981). pH Homeostasis in bacteria. *Biochimica et Biophysica Acta*, **650,** 151–66.

POSTMA, P. W. & SCHOLTE, B. J. (1979). Regulation of sugar transport in *Salmonella typhimurium*. In *Function and Molecular Aspects of Membrane Transport*, ed. E. Quagliarullo *et al.*, pp. 249–57. Amsterdam: Elsevier/North-Holland Biomedical Press.

POSTMA, P. W. & ROSEMAN, S. (1976). The bacterial phosphoenol pyruvate: sugar phosphotransferase system. *Biochimica et Biophysica Acta*, **457,** 213–57.

RAMOS, S. & KABACK, H. R. (1977a). The electrochemical proton gradient in *Escherichia coli* membrane vesicles. *Biochemistry*, **16,** 848–54.

RAMOS, S. & KABACK, H. R. (1977b). The relationship between the electrochemical proton gradient and active transport in *Escherichia coli* membrane vesicles. *Biochemistry*, **16,** 854–9.

RAMOS, S. & KABACK, H. R. (1977c). pH-dependent changes in proton: substrate stoichiometry during active transport in *Escherichia coli* membrane vesicles. *Biochemistry*, **16,** 427–5.

REVSBECH, N. P. (1982). In situ measurement of oxygen profiles of sediments by use of oxygen microelectrodes. In *Handbook on Polarographic Oxygen Sensors: Aquatic and Physiological Applications*, ed. E. Gnaiger & H. Forstner, Berlin, Heidelberg, New York: Springer-Verlag.

RHOADS, D. B. & EPSTEIN, W. (1977). Energy coupling to net K^+ transport in *Escherichia coli* K-12. *Journal of Biological Chemistry*, **252,** 1394–401.

ROBILLARD, G. T. & KONINGS, W. N. (1981). Physical mechanisms for regulation of phosphoenolpyruvate-dependent glucose transport activity in *Escherichia coli. Biochemistry*, **20,** 5025–32.

ROBILLARD, G. T. & KONINGS, W. N. (1982). The role of dithiol-disulfide interchange in solute transport and energy transducing processes. *European Journal of Biochemistry*, (in press).

SAIER, M. H. (1977). Bacterial phosphoenolypyruvate: sugar phosphotransferase systems: structural, functional and evolutionary interrelationships. *Bacteriological Reviews*, **41,** 856–71.

SCHULDINER, S. & PADAN, E. (1982). How does *E. coli* regulate internal pH? In *Membranes and Transport*, ed. A. Martonosi. Plenum Publishing Co. (in press).

SLATER, E. C. (1981). The discovery of oxidative phosphorylation. *Trends in Biochemical Sciences*, **6,** 226–7.

TEN BRINK, B. & KONINGS, W. N. (1980). Generation of an electrochemical proton gradient by lactate efflux in *Escherichia coli* membrane vesicles. *European Journal of Biochemistry*, **111,** 59–66.

TEN BRINK, B. & KONINGS, W. N. (1982). The electrochemical proton gradient and lactate concentration gradient in *Streptococcus cremoris* grown in batch culture. *Journal of Bacteriology*, (in press).

TEN BRINK, B., LOLKEMA, J., HELLINGWERF, K. J. & KONINGS, W. N. (1981). Variable stoichiometry of proton: lactose symport in *Escherchia coli* cells. *FEMS Microbiology Letter*, **12,** 237–40.

Thauer, R. K., Jungermann, K. & Decker, K. (1977). Energy conservation in chemotrophic anaerobic bacteria. *Bacteriological Reviews*, **41,** 100–80.

Thomas, T. D. & Batt, R. D. (1969). Degradation of cell constituents by starved *Streptococcus lactis* in relation to survival. *Journal General Microbiology*, **58,** 347–62.

Williams, P. J. leB. (1975). Biological and chemical aspects of dissolved organic material in seawater. In *Chemical Oceanography*, ed. J. P. Riley & G. Skirrow, London, New York, San Francisco: Academic Press.

Wimpenny, J. W. T. (1981). Spatial order in microbial ecosystems. *Biological Reviews*, **56,** 295–342.

Wimpenny, J. W. T. (1982). Responses of microorganisms to chemical and physical gradients. *Philosophical Transactions of the Royal Society, London, B*, **297,** 497–515.

Zilberstein, D., Schuldiner, S. & Padan, E. (1979). Proton electrochemical gradient in *Escherichia coli* cells and its relation to active transport of lactose. *Biochemistry*, **18,** 669–73.

THE 'SHUT DOWN' OR 'GROWTH PRECURSOR' CELL – AN ADAPTATION FOR SURVIVAL IN A POTENTIALLY HOSTILE ENVIRONMENT

CRAWFORD S. DOW*, ROGER WHITTENBURY* AND NOEL G. CARR†

**Department of Biological Sciences, University of Warwick, Coventry CV4 7AL, UK*
†Department of Biochemistry, University of Liverpool, Liverpool L69 3BX, UK

INTRODUCTION

The concept we wish to explore and develop is that of the 'shut down' or 'growth precursor' cell – an entity first proposed by Dow & Whittenbury (1980). A number of bacterial species (see Dow & Whittenbury, 1980) give rise, at division, to a cell which differs from its parent (or sibling) in that it does not immediately embark upon a cell cycle, but only does so in response to an environmental trigger of some sort. An obvious implication of this proposition is that the 'shut down' cell is in a different physiological state to that of the parent or sibling cell also formed at division, a cell which is committed to growth and reproduction, regardless of environmental conditions, and therefore at risk in that death may ensue as a consequence. The 'shut down' cell, we would argue, is a survival mechanism especially suited for organisms whose habitat is aquatic environments where nutrient supply and growth conditions may be such that long periods of time elapse between periods favourable for growth. Indeed, it could well be that many aquatic bacteria spend most of their lives in the 'shut down' state, being able to carry out a restricted range of metabolism devoted solely to the preservation of the cell. In that active net biosynthesis is not involved, such a cell therefore, seems ideally equipped to prolong its life in unfavourable environments which would prove lethal to cells committed to growth and reproduction.

Prima facie evidence for the existence of 'shut down' cells, which we will now refer to subsequently as *growth precursor cells* (gp cells) will be drawn from research on the swarmer cells of *Rhodomicro-*

bium vannielii, *Hyphomicrobium*, and *Caulobacter crescentus.* All three species undergo complex morphogenetic cell cycles, in comparison, say, to *Escherichia coli* and, therefore, permit studies on recognizable entities within their cycles, entities which may have their parallels in cell cycles of morphologically less complicated bacteria, but which are not readily recognizable.

Some of the key features which characterize the gp cells of these three species are drawn together as a preliminary to the next phase of this article which is concerned with determining how widespread the gp cell phenomenon is amongst bacteria. Clearly the evidence will be of a more haphazard and circumstantial nature than we believe the case to be with the first three species considered; however, we hope to point to useful lines of investigation with other bacteria which (a) will elaborate on the nature of gp cells, (b) determine the universality of such a cell-cycle stage and (c) indicate that the 'aquatic environment' we alluded to earlier is not an exceptional environment in this context.

Of the other bacteria to be discussed, species of cyanobacteria and *E. coli*, in so far as available evidence allows, will be looked at in detail. Finally, other environmental adaptive mechanisms involving morphological responses will also be commented upon within the perspective of the examples of bacteria discussed.

RHODOMICROBIUM VANNIELII

Rhodomicrobium vannielii, a member of the Rhodospirillaceae, is an obligate polarly growing organism in which cell division is asymmetric. Three distinct cell types may be formed depending upon environmental conditions (Fig. 1). These are (a) ovoid cells which are joined by integral cellular filaments which may be branched, so forming ramifying groups of cells, (b) motile peritrichously flagellated swarmer cells which are formed at the filament tips and released by binary fission, (c) angular exospores formed at filament tips and which are also released by binary fission. The formation, subsequent growth and reproductive cycles of these cell types have been described by Whittenbury & Dow (1977). In addition *R. vannielii* expresses two vegetative cell cycles; one leads to the formation of characteristic multicellular complexes in which cell separation is by the formation of a plug in the filament, the other is a 'simplified' cell cycle in which the reproductive budding

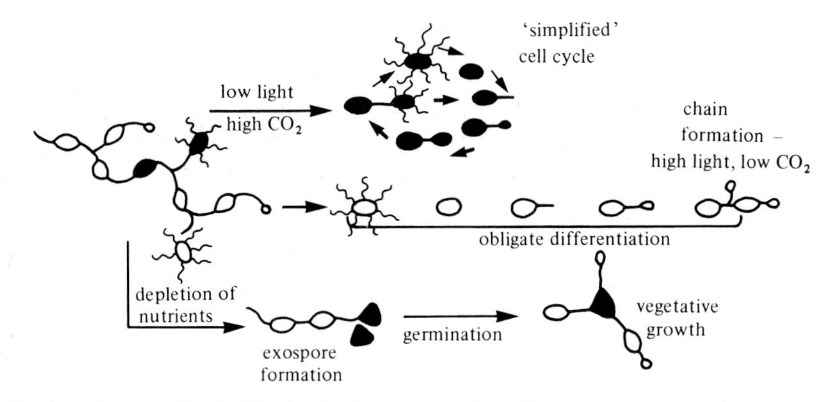

Fig. 1. Cell expression in *Rhodomicrobium vannielii.* Cell type expression can be correlated with specific environmental stimulii. Low light/high CO_2 concentrations are conducive to constitutive swarmer cell formation whilst high light/low CO_2 concentrations result in chain formation. Nutrient depletion leads to the formation of exospores which have resistance properties akin to those of *Bacillus* endospores but which germinate to form up to four vegetative cells (Whittenbury & Dow, 1977).

cell releases motile swarmer cells (Fig. 1) by binary fission (Dow & France, 1980). Each vegetative cycle is characterized therefore not only by the cell type to which it gives rise, but also by the mode of cell division.

Outstanding features of this organism, from an experimental viewpoint, are:

(a) it forms at least three functionally distinct cell types;
(b) it undergoes two vegetative cell cycles, depending upon growth conditions, which differ in cell types expressed and in the mode of progeny cell separation;

(c) that the precursor cell – from which all subsequent cell types are synthesized – is the motile swarmer cell. An additional, and very important, experimental bonus is the ease with which large quantities of homogeneous, and synchronized, swarmer cell populations can be obtained by selective filtration (Whittenbury & Dow, 1977). This makes *R. vannielii* particularly attractive both for the study of the control and regulation of the expression of the different types of cell and as a model to study the temporal and spatial control of gene expression.

R. vannielii swarmer cells are apparently formed in response to a diminution of light and an increase in CO_2 concentration. In the natural environment this organism exists primarily as multicellular complexes. From laboratory studies, it has been observed that as the colonies increase in biomass, the light available to cells within

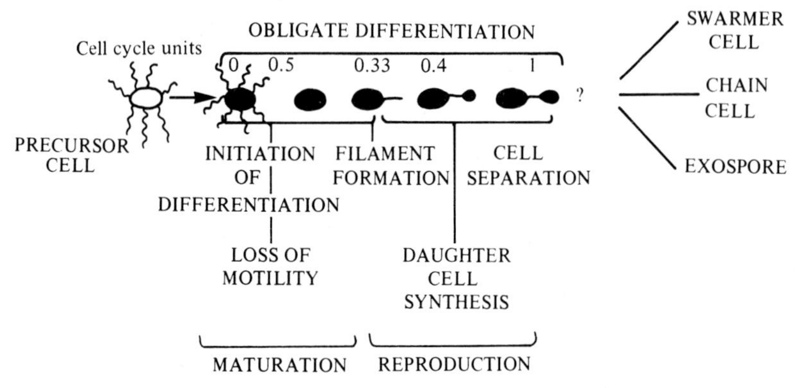

Fig. 2. *R. vannielii* swarmer cell cycle.

the complex is reduced and the dissolved CO_2 concentration rises (the rate being dependent on the nature of the organic C source available), so triggering the formation of the motile swarmer cells which only initiate development beyond the swarmer cell stage when they are relocated in a favourable growth site (or in fresh medium). This sequence of events poses several questions with respect to the nature and role of the swarmer cell.

(a) What is its physiological status when in the motility mode?
(b) How long can it remain motile?
(c) Is this a fixed or variable time period in relation to the rest of the cell cycle?
(d) How and when does the reproductive cell arising from the swarmer cell become committed to giving rise to the particular type of cell (one of three) subsequently formed? The ready availability of large quantities of homogeneous swarmer cell populations, displaying a high degree of synchrony, have enabled us to probe the molecular biology relating to these questions.

Selectively synchronized *R. vannielii* swarmer cells incubated phototrophically develop through an obligate series of well defined morphological stages (Fig. 2). These characteristic changes in cell morphology enable the temporal sequence of the differentiation cycle to be followed very precisely, not only by microscopy, but by cell volume distribution analysis (Fig. 3). This readily permits the correlation of biochemical and physiological characteristics with specific morphologies and provides quantitative data on the degree of synchrony, i.e. on the heterogeneity or otherwise of the culture.

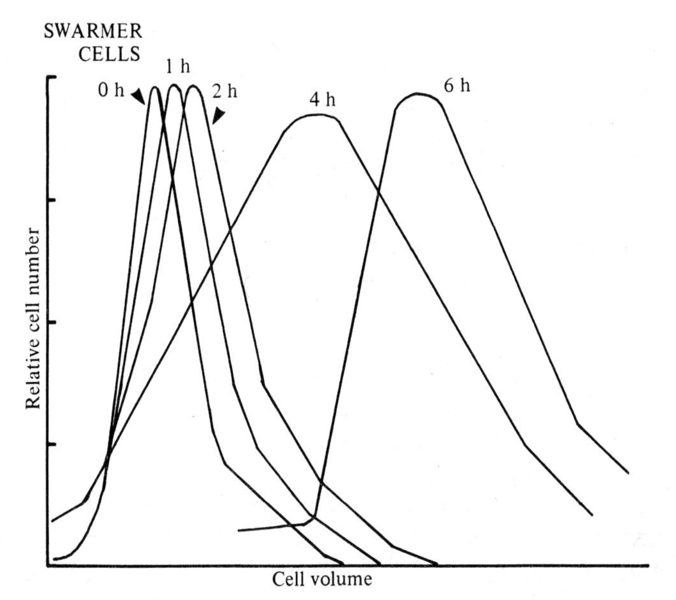

Fig. 3. Cell volume distribution profiles through the synchronous growth of *R. vannielii* swarmer cells. Cell volume distribution profiles generated from a ZBI Coulter Counter and C1000 Channelyzer linked to a Commodore Pet 8032.

The obligate differentiation sequence followed by the swarmer cells consists of two phases, maturation and reproduction. The maturation phase, which occurs only once in the lifetime of a cell, leads to the formation of the reproductive unit, the prosthecate cell, which enters permanently into the repetitive reproductive phase.

Exploiting this fact has led to the following observations.

Swarmer cell cycle of R. vannielii

Protein synthesis

The comparative soluble protein profiles representative of the swarmer cell and the reproductive cell are given in Fig. 4. From these autoradiographs it is clearly apparent that there are many quantitative distinctions between the two cell types. Qualitative differences are also evident. There are at least ten proteins which are unique to the swarmer cell and three which are uniquely characteristic of the reproductive cell, leading to the conclusion that some protein synthesis is specific to each of the two functionally distinct cell types.

The swarmer cell at time 0, although readily labelled with ^{35}S-methionine, presumably as a consequence of protein turnover,

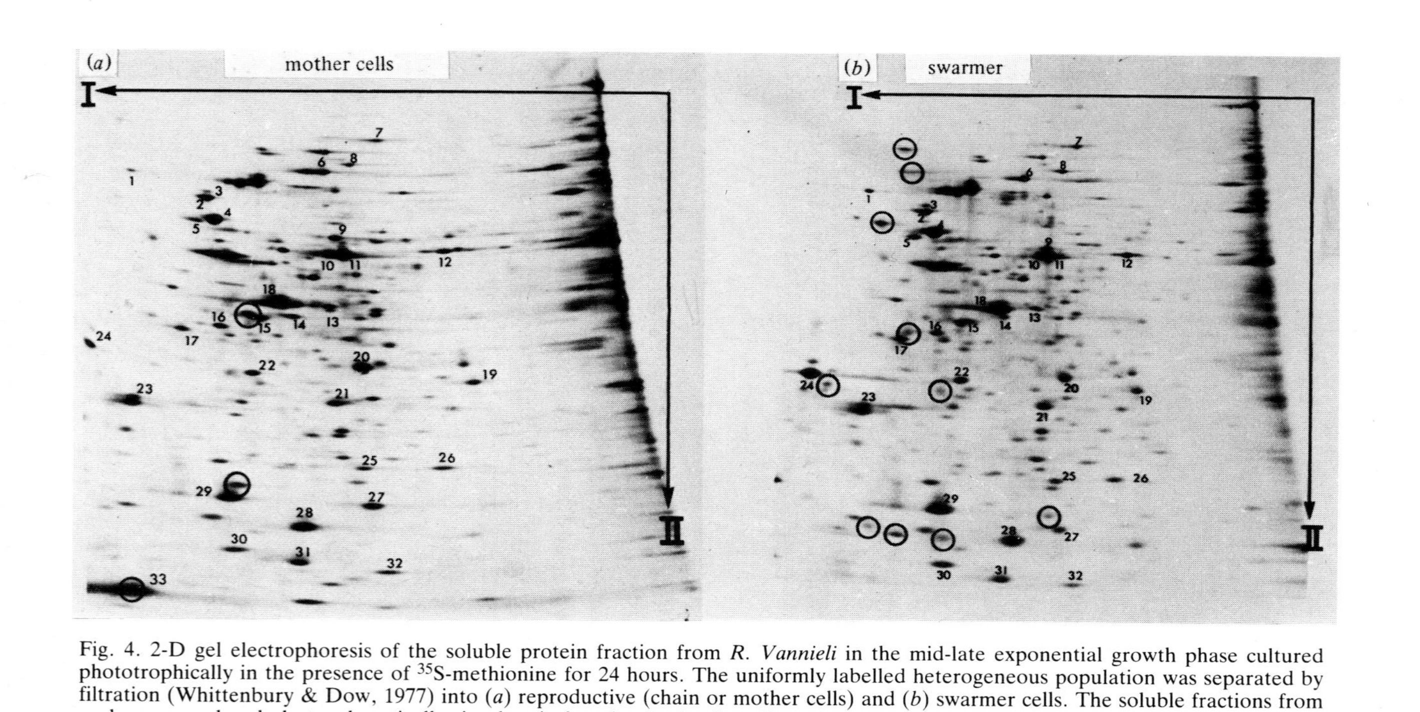

Fig. 4. 2-D gel electrophoresis of the soluble protein fraction from *R. Vannieli* in the mid-late exponential growth phase cultured phototrophically in the presence of ^{35}S-methionine for 24 hours. The uniformly labelled heterogeneous population was separated by filtration (Whittenbury & Dow, 1977) into (*a*) reproductive (chain or mother cells) and (*b*) swarmer cells. The soluble fractions from each were analysed electrophoretically: isoelectric focusing in the first dimension and SDS-polyacrylamide gradient electrophoresis (10%–3% acrylamide) in the second (O'Farrell, 1975). After drying the gels were autoradiographed.

To facilitate comparison the major peptides have been identified numerically. The resolution of the system is such that approximately 350 separate spots can be identified. Those which are cell type specific are circled.

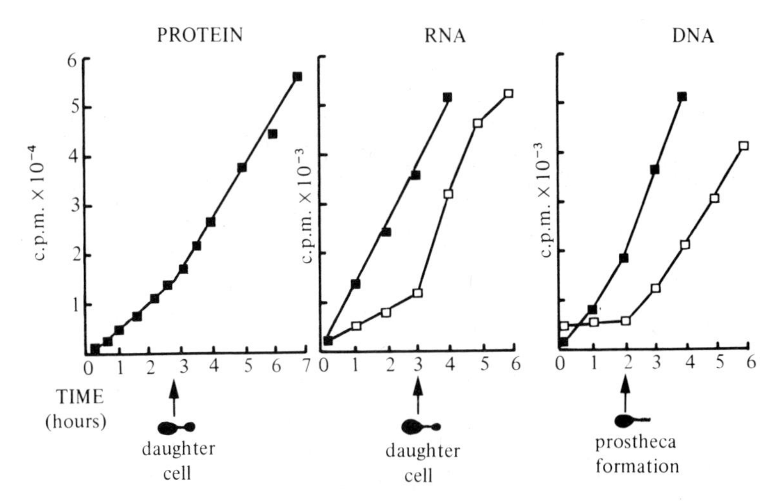

Fig. 5. Protein, RNA and DNA synthesis during the *R. vannielii* swarmer cell differentiation cycle. Protein synthesis was followed as the incorporation of ^{35}S-methionine (2 μM methionine + 2 μCi ml^{-1} ^{35}S-methionine) into total cell material.

Incorporation of ^{3}H-adenosine into DNA was estimated as counts incorporated into alkali-stable, trichloroacetic acid precipitable material, and incorporation into RNA as total TCA precipitable counts less those attributable to DNA (Potts & Dow, 1979). ■, control; □, synchronous culture.

Heterogeneous, early exponential cultures were used as controls.

shows *no net increase in biomass.* With the initiation of the differentiation cycle, however, net protein synthesis increases gradually and linearly until there is a sudden change in rate which directly correlates with the initiation of daughter cell synthesis (Fig. 5). In addition, by pulse labelling with ^{35}S-methionine (Fig. 6), it can be seen that sequential protein synthesis, one measure of sequential gene expression, occurs during synchronous growth. Whilst quantitative and qualitative changes in peptide synthesis are obvious, it can also be seen, particularly from the microdensitometer traces (Fig. 7) of the autoradiograph (Fig. 6), that the vast majority of these are associated with the maturation phase of the cycle and not with the reproductive phase. It is also possible to correlate the onset of the synthesis of specific proteins with specific stages in the morphogenetic cycle.

RNA synthesis

Studies with the specific RNA inhibitor rifampicin have shown that RNA synthesis occurs throughout the differentiation cycle. However, of particular interest is the observation that the maturation phase, i.e. swarmer cell to prosthecate cell stage, is inhibited by

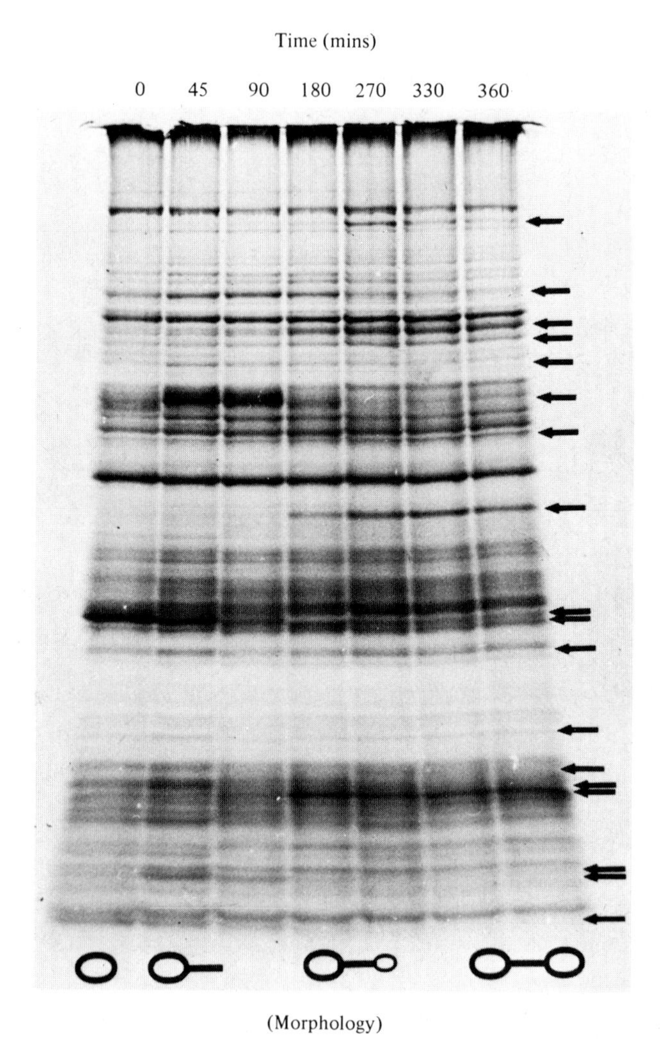

Fig. 6. Sequential protein synthesis during *R. vannielii* swarmer cell differentiation. A 20 litre heterogeneous culture in the mid-exponential growth phase was used to prepare a homogeneous swarmer cell population which was incubated phototrophically. At regular intervals 500 ml samples were removed and pulse labelled, under an identical cultural regime, with ^{35}S-methionine for fifteen minutes. Soluble protein extracts were subsequently analysed on a 10%–30% (w/v) gradient polyacrylamide gel (Laemmli, 1970) which was stained with Coomassie blue, dried and autoradiographed. Sample loading per track was standardised with respect to cpm.

The culture was monitored throughout by light and electron microscopy and by electronic cell volume distribution analysis.

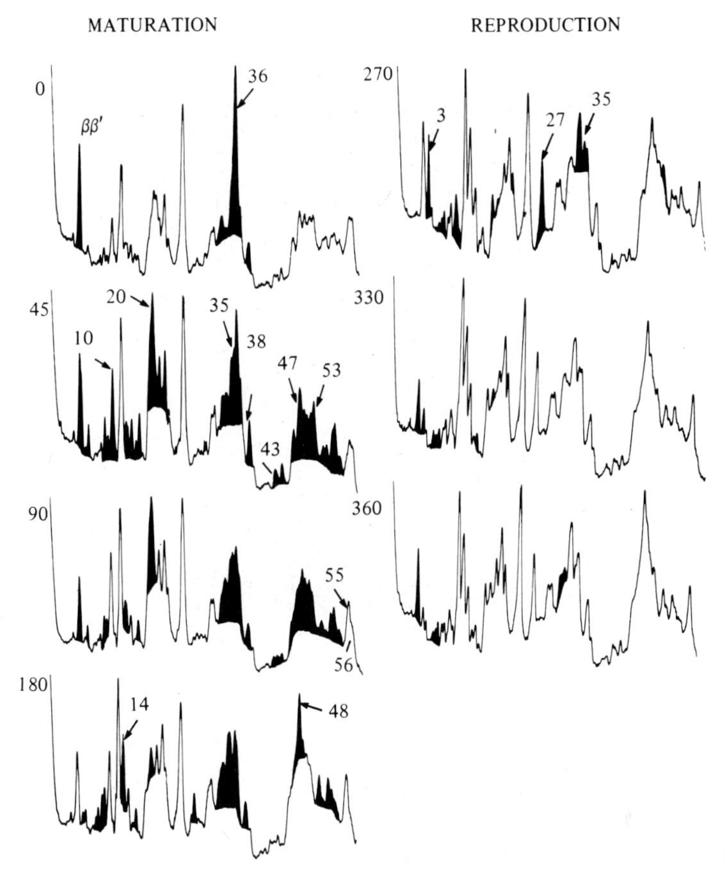

Fig. 7. Microdensitometer traces of the tracks from the autoradiograph shown in Fig. 6 with the quantitative and qualitative changes highlighted. The protein peaks have been numbered sequentially from the top of the gel. The time is given in minutes.

10–20 μg ml^{-1} whilst the reproductive phase is sensitive to 1 μg ml^{-1}. This may be indicative of a change in the DNA-dependent RNA polymerase holoenzyme – a point which will be discussed later on.

Incorporation of the radioactive nucleotide (^{3}H-adenosine) into RNA indicates a very low level of synthesis in the swarmer cell, the rate of synthesis being constant and linear until the initiation of daughter cell formation. At that stage (Fig. 5) there is a very marked increase in RNA synthesis (Potts & Dow, 1979). The kinetics of RNA synthesis can therefore be related to protein synthesis. On completion of the maturation phase there is a very marked increase in the rate of synthesis of RNA and protein heralding the initiation of the reproductive phase. More detailed analysis of ^{32}P-orthophosphate labelled RNA yields data which

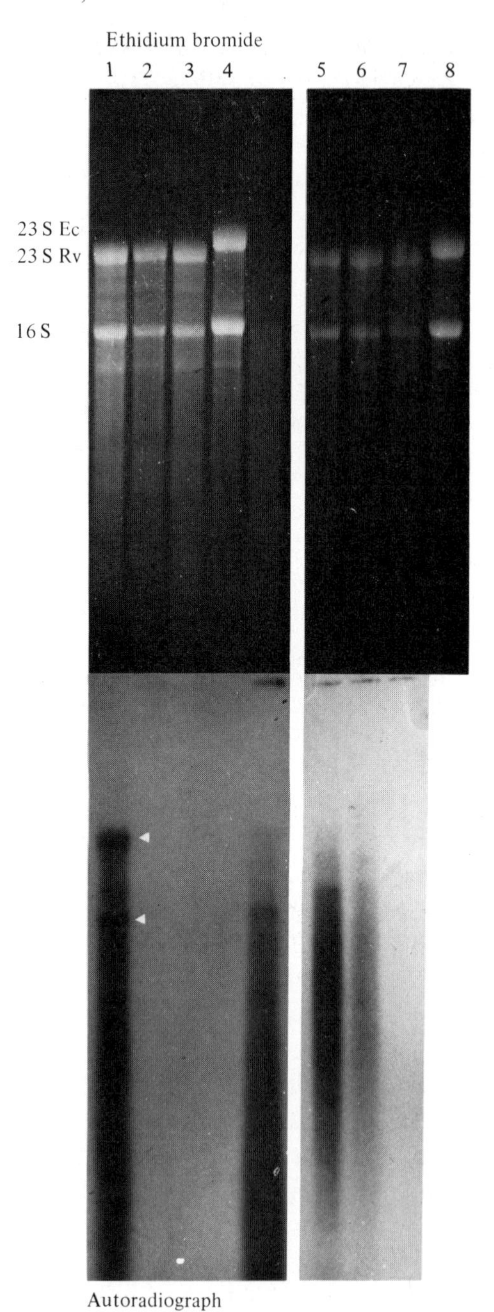

Fig. 8. Electrophoretic analysis of ^{32}P labelled RNA. Synchronized swarmer cells were grown for the times and under the light regimes stated in the presence of ^{32}P-labelled orthophosphate. Total RNA was isolated and analysed by agarose gel electrophoresis followed by autoradiography. Tracks: 1, 0–1 h light; 2, 0–1 h dark; 3, 0–2 h dark; 4, *E.coli*; 5, 0–0.25 h light; 6, 0–0.5 h light; 7, 0–0.5 h dark; 8, *E. coli*. (*a*) 1.5% agarose: 50% formamide flat bed gel stained with ethidium bromide, (*b*) autoradiograph of *a*.

indicates that the ribosomal gene(s) are not being transcribed during the first 30 minutes of the differentiation cycle, i.e. *rRNA synthesis is switched off in the swarmer cell* (Fig. 8).

DNA dependent RNA polymerase

The fundamental split of the swarmer cell cycle into two well defined stages (maturation and reproduction) and in particular the large difference in rifampicin sensitivity (an inhibitor which specifically interacts with the β subunit of the enzyme) of each phase, suggests major changes in the DNA dependent RNA polymerase, i.e. does the RNA polymerase play an active role in the recognition of specific promoter sites during differentiation, a process leading, irreversibly, to a particular cell type?

Using a slightly modified version of the micropurification technique of Gross *et al.* (1976), the holoenzyme has been purified from various stages of the differentiation cycle. The structure of the core enzyme is similar to that of *E. coli* (Burgess, 1969), i.e. β,β' subunits of molecular weight approaching 160 K, α subunits of molecular weight approaching 45 K. However, it can be seen from Fig. 9 that the *R. vannielii* β subunits are diffcult to resolve from one another and that the α subunits are of a lower molecular weight than those of *E. coli*.

The major changes occurring in core-associated proteins are that there is a quantitative decrease in a 95 K protein and a significant increase in a 70 K protein during the differentiation cycle. Two further proteins are consistently associated with the core enzyme, one of 80 K and another of 110 K. The 110 K protein appears to be a contaminant, whereas the 80 K protein is functionally associated with the core, as judged by 2-dimensional gel electrophoresis (Fig. 10).

Although these core-associated protein changes can be readily demonstrated analytically, there is, as yet, no direct evidence as to their functional significance. It is, however, tempting to suggest that the sequential gene expression observed during the differentiation cycle is, at least in part, regulated at the level of transcription by modification of the RNA polymerase and hence by modification of template specificity (Doi, 1977).

DNA synthesis

DNA synthesis, as measured by the incorporation of ^{3}H-adenosine into alkali-stable, TCA precipitable material, is not detectable in

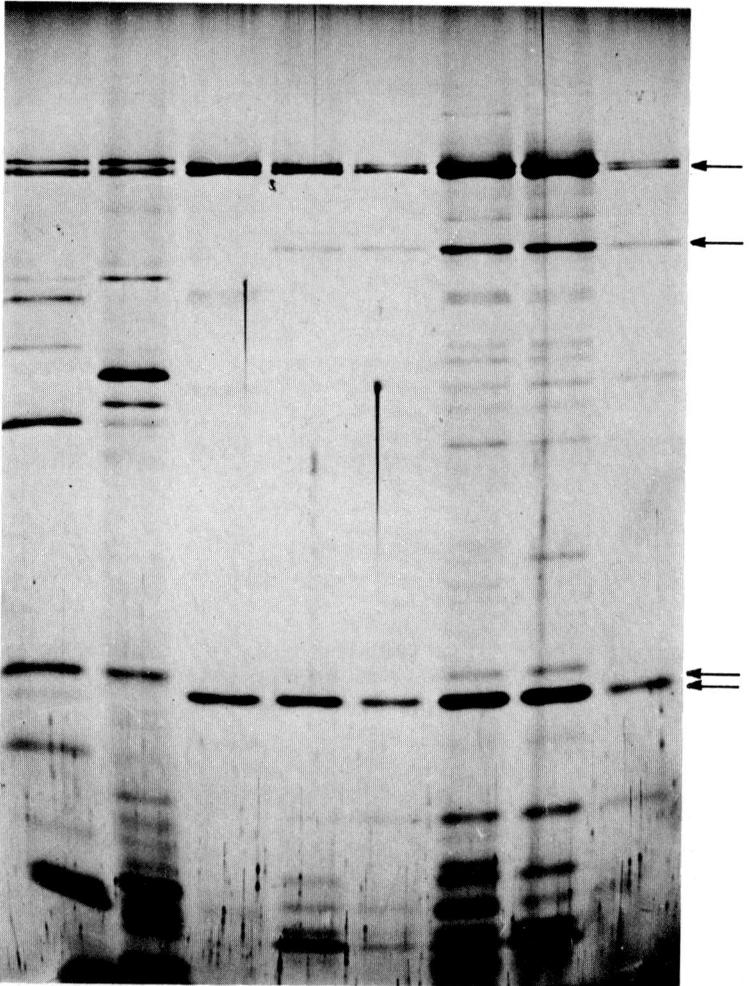

1. *E. coli* K 12
2. *E. coli* C
3. *R. vannielii*

Fig. 9. Comparative SDS-polyacrylamide gel of *E. coli* C, *E. coli* K12 and *R. vannielii* RNA polymerase preparations. The micropurification procedure employed was essentially that devised by Gross *et al.* (1976). This procedure, basically purification via a DNA cellulose column followed by a DEAE sephadex column, yields a significant number of additional DNA binding proteins. However, it is comparatively easy to demonstrate those which are associated with the RNA polymerase by 2-dimensional gel electrophoresis (Fig. 10). Samples were analysed on a 7.5–10% SDS polyacrylamide gel which was stained by the silver nitrate method of Wray *et al.* (1981). As can be readily appreciated from this gel considerable caution must be exercised in the interpretation of the data.

the swarmer cell (Potts & Dow, 1979). Its synthesis is initiated at the onset of, or immediately after, the initiation of prostheca formation, i.e. it is a characteristic solely of the reproductive phase (Fig. 5). The fact that the swarmer cell does not, or cannot, replicate its genome is obviously of importance when discussing the functional role of such a cell. However, it is the configurational state of the genome, or more correctly, of the nucleoid (Pettijohn, 1976; Hirschbein & Guillen, 1981) which is of considerable interest in this context.

It is well established that the *in vivo* structure of prokaryotic DNA displays several levels of organization (Pettijohn & Hecht, 1973; Drlica & Worcel, 1975; Guillen, 1979) which may play a significant role during replication (Gellert *et al.*, 1976), recombination (Mizuuchi *et al.*, 1978), transcription (Guillen, 1979) and transformation (Bohin *et al.*, 1982). It has consequently been suggested (Pettijohn, 1976; Hirschbein & Guillen, 1981) that there is a close correlation between DNA configuration and gene expression.

Giemsa staining of the *R. vannielii* nucleoid during the differentiation cycle (Whittenbury & Dow, 1977) has shown that in the swarmer cell the nucleoid is spherical, located centrally and appears to be in a condensed format. On the initiation of differentiation, before the start of DNA synthesis, the nucleoid assumes an elongated state and apparently moves towards the pole of the cell at which the prostheca is being formed. On completion of the daughter cell the nucleoid returns to the centre of the cell but retains the elongated format. These preliminary observations indicate that the nucleoid configuration can be correlated with distinct stages within the differentiation cycle, and, therefore, raises the question of the role of the nucleoid – in its different configurational states – in the regulation of gene expression.

Isolation of envelope-associated nucleoids, using the 'gentle lysis' procedure of Stonington & Pettijohn (1971) necessary for their isolation, proved difficult with *R. vannielii* cells and was associated with the fact that they can only be ruptured at the growing point (Fig. 11). However, sufficient lysate material was obtained from swarmer cells to reveal that the nucleoids within them had a sedimentation value greater than that of cells post the swarmer cell stage (Fig. 12) indicating that nucleoids in the swarmer cell are in the condensed state.

A consequence of the salt concentration (0.1 M NaCl) used in the

(a)

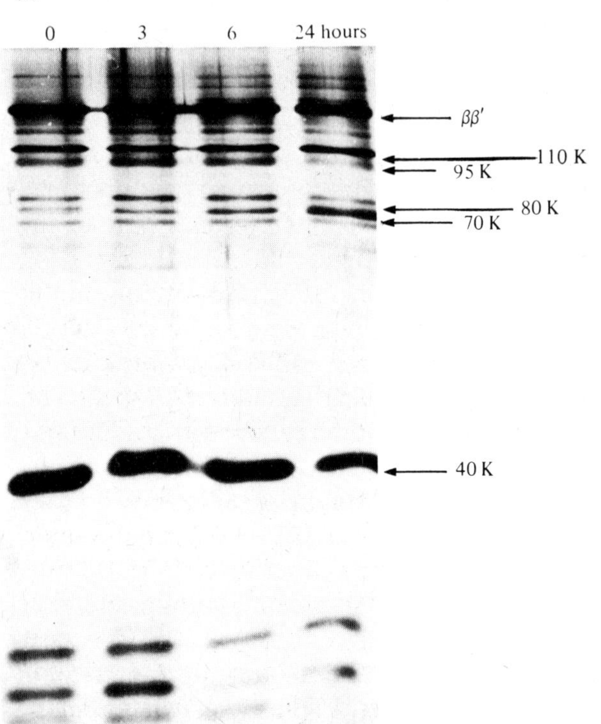

0
3
6
24 hours
ββ′
110 K
95 K
80 K
70 K
40 K

(b)

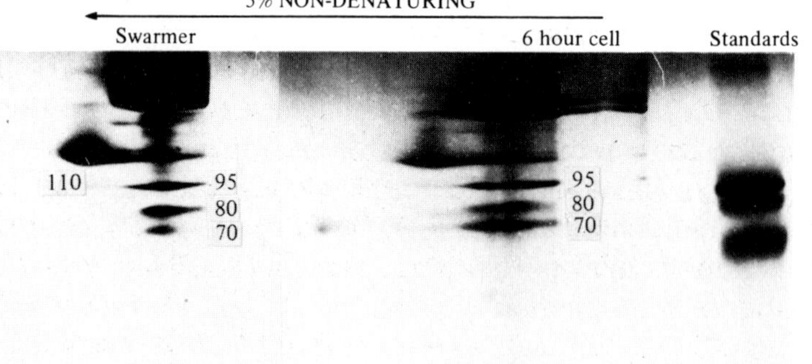

5% NON-DENATURING
Swarmer
6 hour cell
Standards
110
95
80
70
95
80
70
40
7.5–10% 4 M UREA

preparation of the cell lysates is that the nucleoid unfolds to give the characteristic whorls and loops (Fig. 11) argued to be a reflection of the hierarchal organization of the DNA (Worcel & Burgi, 1972; Pettijohn & Hecht, 1973). More detailed electron microscopic analysis of the lysates revealed cells with spheroplasts which had only ruptured on preparation for electron microscopy. These revealed that the nucleoid is organized into a beaded configuration (Fig. 13) somewhat analogous to the nucleosomal, and hierarchal organization of eukaryotic DNA (McGhee & Felsenfeld, 1980).

Pertinent questions with respect to nucleoid configuration relate to the role of the DNA binding proteins, both in the organization of the genomic DNA and their possible role(s) in the regulation of the differentiation cycle.

The presence of configurational DNA binding proteins in several prokaryotes is now well established (Rouvière-Yaniv, 1978; Guillen *et al.*, 1978; Rouviere-Yaniv & Yaniv, 1979). It is also evident that the best characterized DNA binding protein, HU, has the capacity, in the presence of nick-closing enzyme, to introduce superhelical turns in closed circular DNA, i.e. to generate a beaded configuration – a process which can be performed *in vitro* (Rouvière-Yaniv & Yaniv, 1979). Analogous proteins have been isolated from a wide range of prokaryotes and can readily be shown to be present in *R. vannielii* (Fig. 14).

A low molecular weight DNA binding protein, of molecular weight approximately 10 K, bears close analogy with the well documented HU protein. However, of particular interest is the observation that when a 0.4 M NaCl eluate from a DNA-cellulose column is analyzed by *non-denaturing* gel electrophoresis there is one major band of molecular weight approximately 110 K. When this is electrophoresed in the second dimension on a 10% SDS polyacrylamide gel it is apparent that the configurational protein, HU, is a constituent of this large complex (Fig. 14).

Fig. 10. RNA polymerase modification during the *R. vannielii* differentiation cycle. (*a*) 10% SDS polyacrylamide gel of samples from a DEAE sephadex column.

(*b*) 2-dimensional gel electrophoresis of RNA polymerase preparations.

Samples were run in 5% non-denaturing tube polyacrylamide gels in the first dimension, i.e. resolution of the holoenzyme. Subsequent electrophoresis in the second dimension was into a 7.5–10% 4 M urea polyacrylamide slab gel, i.e. denaturing.

Employing this procedure there is less chance of losing core-associated proteins, a major problem with respect to regulatory proteins associated with the RNA polymerase core, and of wrongly correlating others with the enzyme.

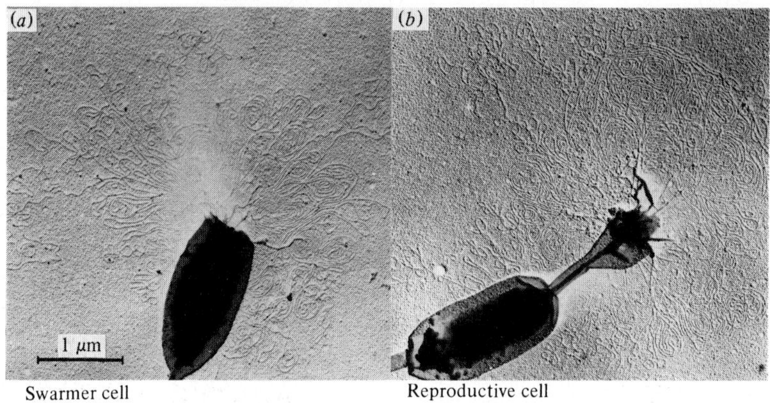

Fig. 11. Envelope associated nucleoids of *R. vannielii.* Cells from various stages in the differentiation cycle were gently lysed and centrifuged into 30–60% sucrose gradients. Samples from the various peaks were prepared for electron microscopy by the cytochrome C monolayer technique of Davis *et al.* (1971). Lysis was only apparent at the growth point of the cell – these having been identified previously by Whittenbury & Dow (1977). (*a*) Swarmer cell which has shed its flagella. (*b*) Reproductive cell with developing daughter.

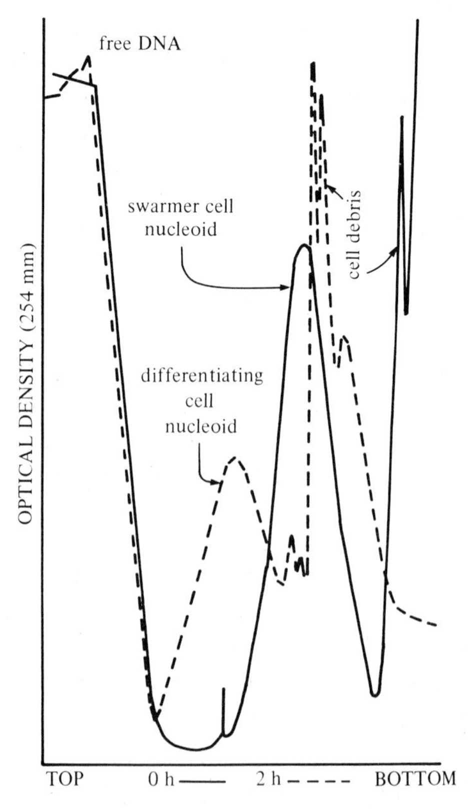

Fig. 12. Sucrose gradient (30–60%) profiles of *R. vannielii* cell lysates.

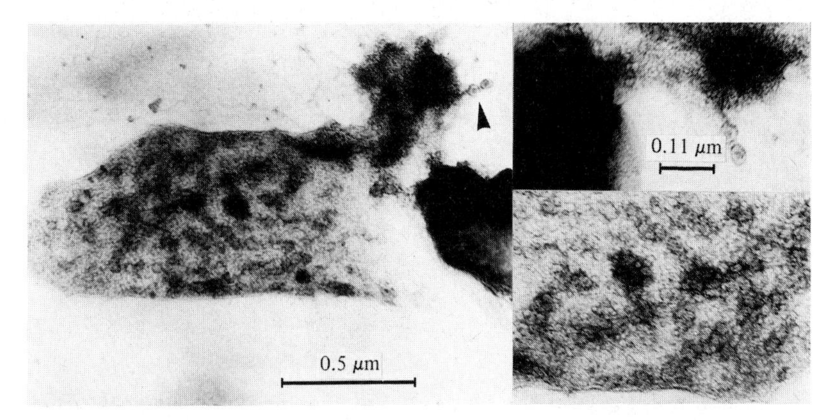

Fig. 13. Beaded configuration of the *R. vannielii* nucleoid. Samples from a 30–60% sucrose gradient which have only begun to lyse on cytochrome C spreading (Davis *et al.*, 1971) show the DNA to be organized ino a beaded configuration.

In vitro experiments using 0.4 M eluates (heat treated to remove residual RNA polymerase activity – HU is thermostable), λDNA and *E. coli* RNA polymerase, have shown that the 110 K complex has a marked stimulatory effect on transcription (Fig. 15). In the presence of a fixed amount of RNA polymerase, and for varying DNA template concentrations, the initial rates of transcription are considerably greater in the presence of the eluate. It is also clear that there is an optimal ratio of configuration protein to DNA template (Fig. 15*b*). These observations correlate closely with the data obtained previously for *E. coli* HU (Rouvière-Yaniv & Gros, 1975).

The questions of interest concerning condensation of the nucleoid and the concomitant involvement of DNA binding proteins, relate to the lack of DNA synthesis and the reduced transcriptional capacity in the swarmer cell, the release from this inactive state on the initiation of the differentiation cycle, and the organization and regulation of transcriptional capacity of the genome destined for the daughter cell. In the last instance there is evidence to indicate that daughter cell synthesis and the regulation of the differentiation cycle *per se* are orchestrated by the reproductive cell genome, and that the daughter cell genome is quiescent until close to physiological separation (Whittenbury & Dow, 1977).

Summary of the R. vannielii *swarmer cell differentiation cycle*

The first land-mark event of the differentiation cycle (Fig. 16) is the

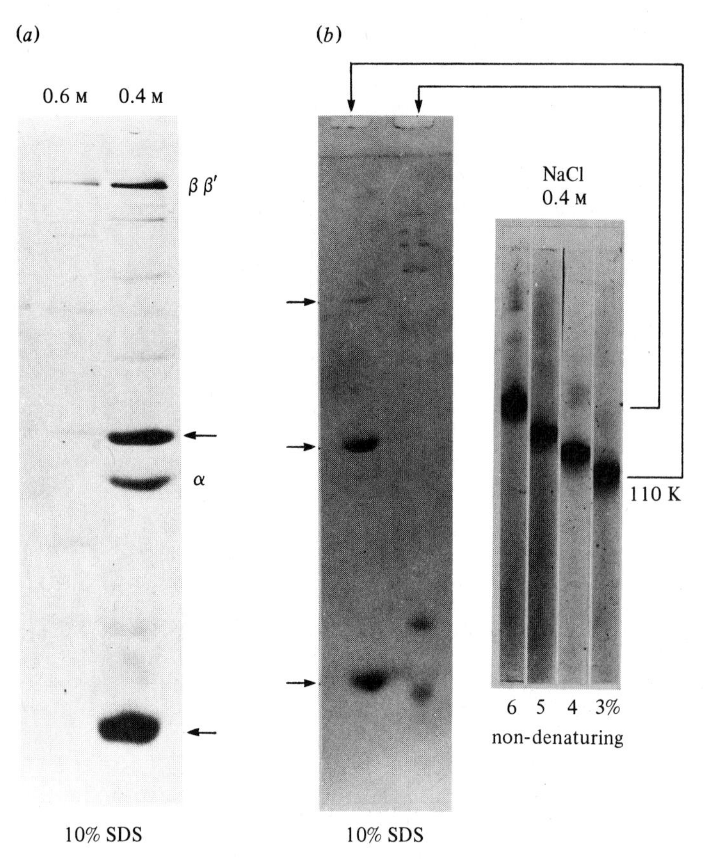

Fig. 14. DNA binding proteins from *R. vannielii.* The DNA binding proteins were purified from a cell lysate by the procedure of Rouvière-Yaniv & Gros (1975). (a) 0.4 M and 0.6 M NaCl eluates from a DNA-cellulose column analysed on a 10% SDS polyacrylamide gel. (b) Analysis of a 0.4 M NaCl eluate from a DNA-cellulose column by *non-denaturing* polyacrylamide gel electrophoresis indicates that the greater proportion of the DNA binding proteins eluted are in the form of a large (molecular weight approximately 110 K) complex. The second, higher molecular weight band, is the RNA polymerase. Subsequent analysis of the 110 K complex on a 10% SDS polyacrylamide gel shows it is composed of two major peptides, one of molecular weight approximately 10 K (presumptive HU), the other of molecular weight approximately 50 K and a minor constituent.

synchronous shedding of the flagella (filament and hook) followed by marked incorporation of diaminopimelic acid into the cell wall (Dow, unpublished). The latter is not accompanied by a change in cell volume and most probably reflects reorganization of the cell wall on shedding of the flagella. At this stage there is also a stimulation of photopigment synthesis, as measured by the incorporation of α aminolaevulinic acid (Dow, unpublished). The rate of protein and RNA synthesis begins to rise but the most interesting

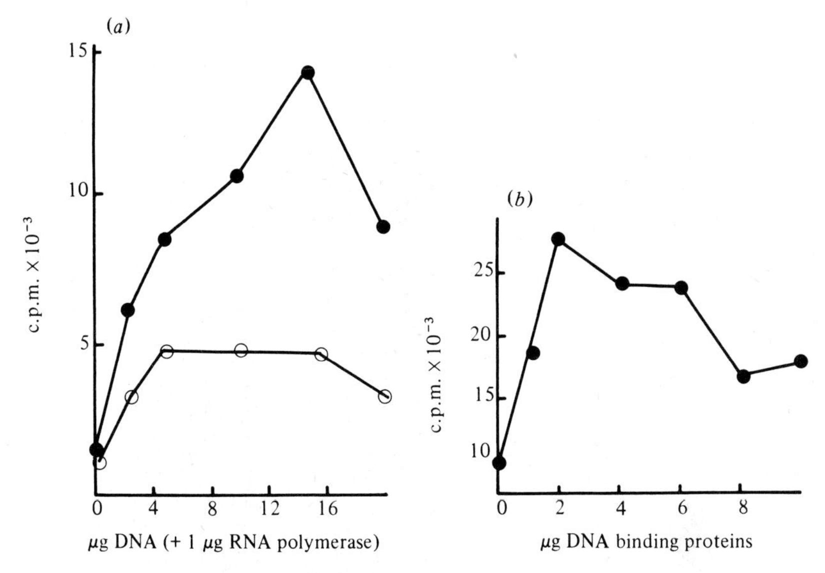

Fig. 15. Effect of the 110 K DNA binding complex on the *in vitro* transcription of DNA. (*a*) Increasing concentrations of λDNA with 1μg RNA polymerase were transcribed in the presence (●) and absence (○) of 4μg of 0.4 M NaCl eluate (heat treated to inactivate the residual RNA polymerase). The reactions were performed at 37 °C for 10 minutes. (*b*) The effect of increasing concentration of 0.4 M NaCl eluate (heat treated as above) on the transcription of 4μg of λDNA by 1μg RNA polymerase. The reaction conditions were as in (*a*).

switch is the initiation of rRNA synthesis at a point 15–20 minutes into differentiation (a cell cycle time of 0.05). With the synthesis of the prostheca, DNA synthesis is initiated and is followed shortly afterwards by a significant change in the rates of RNA and protein synthesis, the latter being concomitant with the initiation of daughter cell synthesis. Several sequential qualitative changes in protein synthesis are evident, particularly during the maturation phase and can be correlated with specific morphologies. There are no sudden changes in the holoenzyme structure of the RNA polymerase (as have yet been detected); the apparent changes are more gradual and the consequences of this with respect to transcriptional specificity have still to be elucidated. In the context of this discussion the most interesting event is the change in nucleoid configuration. The non-differentiating swarmer cell, i.e. the growth precursor cell, has a very condensed nucleoid; with the initiation of the differentiation cycle, prior to the initiation of DNA synthesis, the nucleoid assumes a more relaxed format. The subsequent partitioning of two transcriptionally distinct nucleoids, i.e. between reproductive mother

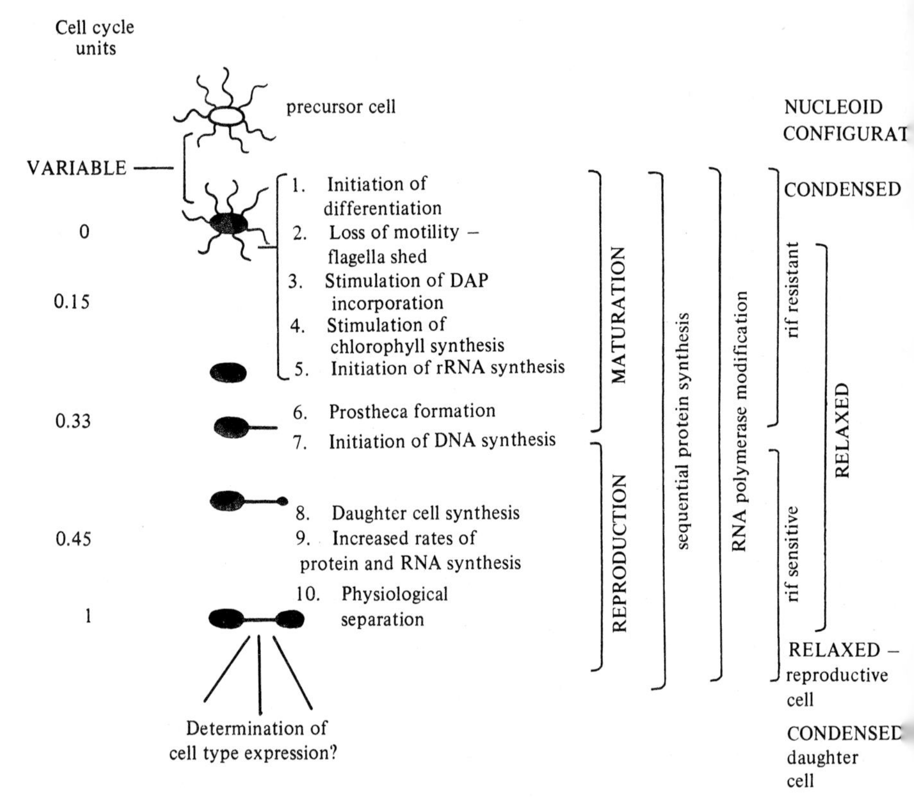

Fig. 16. *R. vannielii* swarmer cell differentiation cycle.

cell and daughter cell, raises many questions pertaining to how two nucleoids can be differentially controlled in what is essentially one cell.

Growth precursor cell

Inhibition of swarmer cell differentiation

Early observations on swarmer cell formation during phototrophic batch growth suggest that in the late exponential–early stationary phase the swarmer cell population reaches a peak *and does not decline* (Dow & France, 1980) i.e. they are not entering the differentiation cycle. By analysing sequential cell volume distribution profiles of swarmer cells during phototrophic batch growth, it has become apparent that as the optical density of the culture increases the number of swarmer cells embarking upon the differentiation sequence decreases – even though all nutrients are in

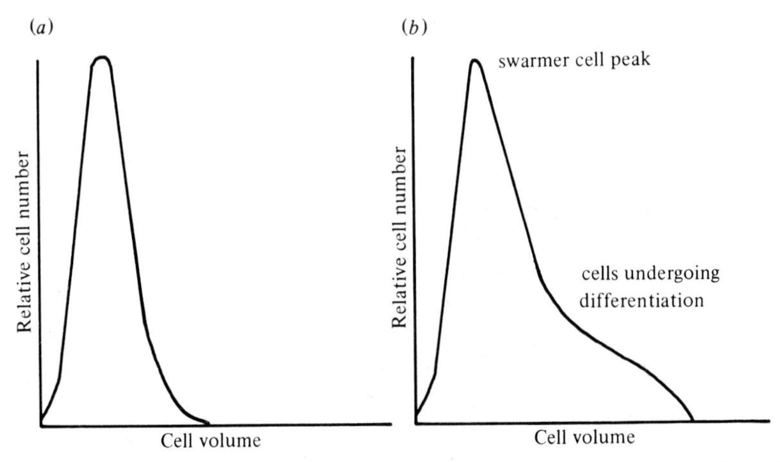

Fig. 17. Cell volume distribution profiles during batch culture. Using an electronic particle counter and cell volume distribution analyser it is possible to monitor the cell volume changes during phototrophic growth. (*a*) When the culture becomes light limited the swarmer cell numbers continue to increase but none initiate differentiation – hence the symmetry of the swarmer cell peak. (*b*) In the early-mid exponential growth phase swarmer cells are continuously produced and undergo differentiation. The right hand shoulder is a consequence of cells with prostheca, and of cells undergoing daughter cell formation.

These data have been corroborated by microscopic observation.

excess with the exception of the amount of light energy available to each cell. In essence, therefore, swarmer cells continue to be formed by the reproductive cells but are inhibited from initiating further development because of light limitation (Fig. 17). With the selective synchronization procedure (Whittenbury & Dow, 1977) the optical density of the culture is decreased (light available per cell is, therefore, increased) and the swarmer cell population synchronously enters the differentiation cycle. Such a regulatory phenomenon seems to be fully in keeping with the functional role being ascribed to this form (i.e. the gp cell). The extreme example of delayed development encountered was when phototrophically grown swarmer cells were kept anaerobically in the dark. Under such conditions they remained motile for the 16 hours of the experiment without initiating differentiation. On exposure to light at 16 hours differentiation was initiated although the degree of synchrony was far less than at two to three hours. This observation allows us to correlate the inhibition or holding step with the lack of light energy; anaerobically in the dark there is no alternative means of producing sufficient ATP to permit differentiation to run to its conclusion.

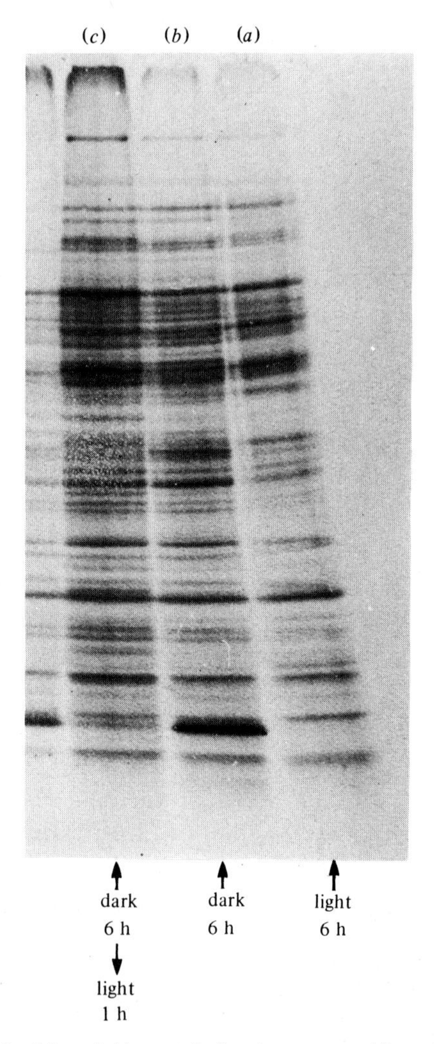

Fig. 18. Autoradiograph of the soluble protein fractions prepared from the following swarmer cell populations. A mid-late exponential phototrophic culture was selectively synchronized and the swarmer cell populations incubated in the presence of ^{35}S-methionine as follows: (*a*) Cultured phototrophically for 6 hours; (*b*) Maintained anaerobically in the dark for 6 hours; (*c*) Maintained anerobically in the dark for 6 hours followed by 1 hour in the light.

'Inhibition protein'

Variation in the cultural regimes of synchronous swarmer cells have revealed the presence of a soluble protein (molecular weight approximately 12 K), the synthesis of which correlates with the inhibition of differentiation (Fig. 18). Although readily visualized autoradiographically, this low molecular weight protein is present

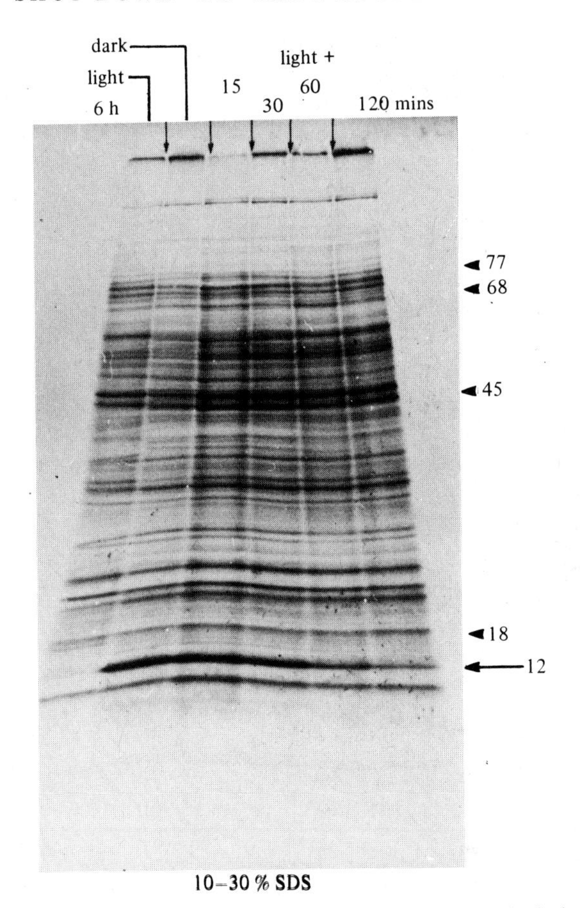

Fig. 19. Autoradiograph showing the switch off of the inhibition protein during the swarmer cell maturation period.

A selectively synchronized swarmer cell population was cultured in the presence of ^{35}S methionine, initially in the dark and then phototrophically. Samples were removed for analysis at regular intervals and the soluble protein fractions examined on a 10–30% SDS polyacrylamide gel which was subsequently dried and autoradiographed.

only in very low concentrations within the cell, i.e. it has a high rate of synthesis and turnover – two key criteria for a regulatory protein. When inhibited swarmer cells are allowed to initiate differentiation the synthesis of this protein stops and it is removed from the soluble protein fraction prior to prostheca formation and the initiation of DNA synthesis (Fig. 19). There is, therefore, a positive correlation between inhibition of differentiation and the presence of this protein; however, its functional role has yet to be elucidated. Nevertheless this is a possible regulatory protein, uniquely associ-

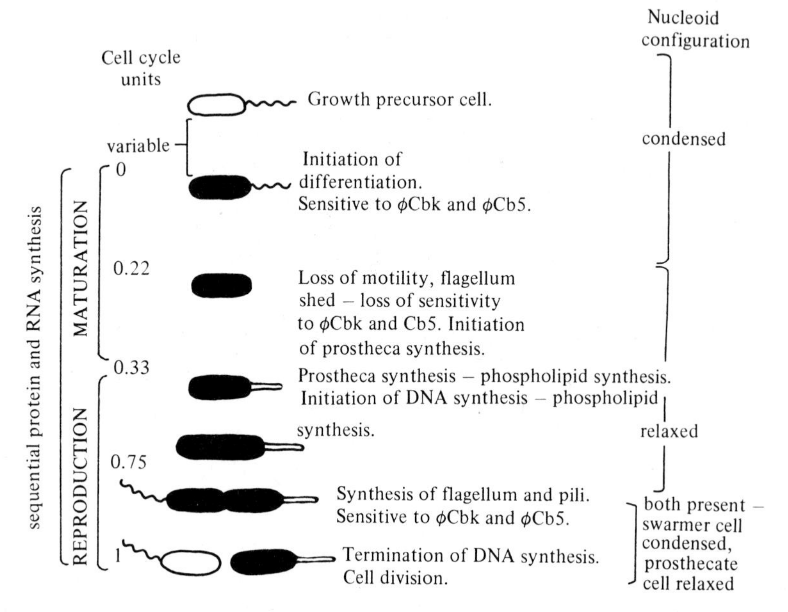

Fig. 20. Temporal events of the *Caulobacter* cell cycle.

ated with a particular stage of the cell cycle, the synthesis of which is switched off prior to the initiation of differentiation.

There is little doubt however that this gp cell is subject to a regulatory process which ensures that the commitment to differentiate can only be undertaken when the available 'energy levels' are above a threshold value.

CAULOBACTER CRESCENTUS

Although physiologically dissimilar, both *R. vannielii* and *C. crescentus* show a considerable degree of analogy with respect to their swarmer cell cycles. The precise sequence of morphological changes that characterize *Caulobacter* growth are well defined, the organism following an obligate differentiation cycle which leads to asymmetric division generating a stalked or prosthecate cell and a motile swarmer cell – two physiologically and functionally distinct cell types (Poindexter, 1964; Shapiro, 1976) (Fig. 20). In essence the swarmer cell cycle of *R. vannielii* and *C. crescentus* can be represented by a generalized cell cycle (Fig. 21).

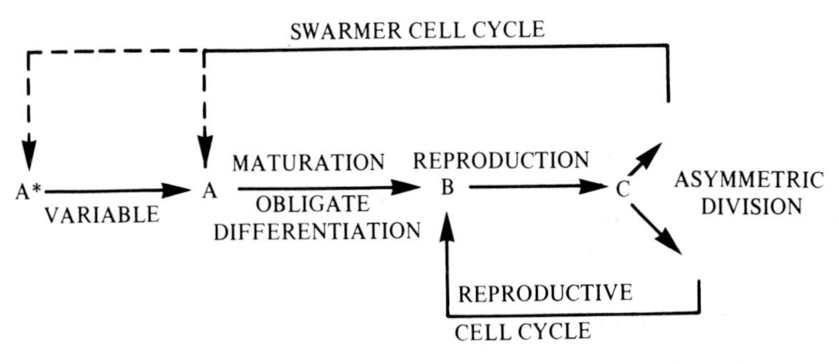

Fig. 21. Generalized cell cycle. A*, growth precursor cell – dispersal phase; A, swarmer cell committed to differentiation; B, prosthecate cell – reproductive unit; C, dividing cell.

Swarmer cell cycle of C. crescentus

The obligate dimorphic nature of the *Caulobacter* cell cycle has been extensively exploited as an experimental model for the study of the control and regulation of both temporal gene expression and the spatial organization of gene products within the cell (see, for example, Shapiro, 1976; Poindexter, 1981a,b). However we will only abstract evidence from the literature relating to those facets of the cycle which are relevant to the proposition that the swarmer cell is, in effect, a growth precursor cell.

Protein synthesis

During the *C. crescentus* cell cycle there are both qualitative and quantitative differences in the proteins expressed. The best characterized of these events is synthesis of the flagella subunits (Shapiro & Maizel, 1973; Osley, Sheffery & Newton, 1977; Sheffery & Newton, 1981), however, a considerable amount of data is also now available concerning the expression of cytoplasmic proteins (Cheung & Newton, 1977; Iba, Fakuda & Okada, 1978; Agabian, Evinger & Parker, 1979; Mulhausen & Agabian, 1981), membrane associated proteins (Evinger & Agabian, 1977, 1979; Agabian *et al.*, 1979) and DNA binding proteins (Evinger & Agabian, 1979). Although the majority of proteins are synthesized continuously, there are a few expressed only at defined stages in the cell cycle and for defined periods and which are found exclusively in one or other of the two cell types (Agabian *et al.*, 1979).

Two dimensional electrophoretic analysis of cell extracts of pulse-labelled, synchronized cells (Mulhausen & Agabian, 1981) led to the conclusion that the protein profile showed a discrete temporal pattern with specific periods of gene activity which could be correlated with the specific morphological stages of the two cell types. Three distinct periods of specific protein synthesis (gene activity) occur; (a) the swarmer to prosthecate cell stage, i.e. maturation phase, (b) the prosthecate cell growth stage, i.e. reproductive phase and (c) the cell division stage. The majority of the changes in protein synthesis are associated with the maturation phase. Existing data (Iba *et al.*, 1978) also suggests that the rate of protein synthesis in the swarmer cell is less than that in the prosthecate cell – an analogous situation to that found in *R. vannielii.*

RNA synthesis

The RNA inhibitor, rifampicin, blocks all stages of the differentiation cycle indicating that RNA synthesis is essential throughout the cycle for normal differentiation (Newton, 1972). In addition, incorporation of ^{3}H-deoxyguanosine into RNA during synchronous growth has shown that RNA synthesis is low during the maturation phase, i.e. in the swarmer cell, and rises to a maximal rate during reproduction (Degnen & Newton, 1972).

An unusual characteristic of *Caulobacter* RNA is the presence of adenylic acid-rich tracts associated with a metabolically unstable, non-ribosomal RNA (Ohta, Sanders & Newton, 1975; Ohta *et al.*, 1978). There is no direct evidence other than the rapid turnover, however, to suggest that this poly(A) RNA is a messenger species. As no significant differences in the rate of poly(A) synthesis have been observed in the morphologically distinct cell types it is difficult to align these observations in the context of the differentiation cycle. However, it has been postulated, for example, that these polyA sequences could be required for processing (Nakazato, Venkatesan & Edmonds, 1975), or alternatively they may play a role with respect to mRNA specificity and temporal phenotypic expression.

In vitro transcription studies (Swoboda, Dow & Vitkovic, 1982a) have shown that isolated envelope associated nucleoids (Swoboda, Dow & Vitkovic, 1982b) continue to synthesize RNA chains initiated *in vivo*, i.e. prior to lysis, but do not reinitiate. Under the assay regimes reported, three major classes of RNA were synthesized *in vitro*; the two larger species were heterodisperse and had electrophoretic mobilities equivalent to 16–20 S and 23–35 S, i.e. rRNA, the third class contained RNA in the region of 3–4 S (Fig.

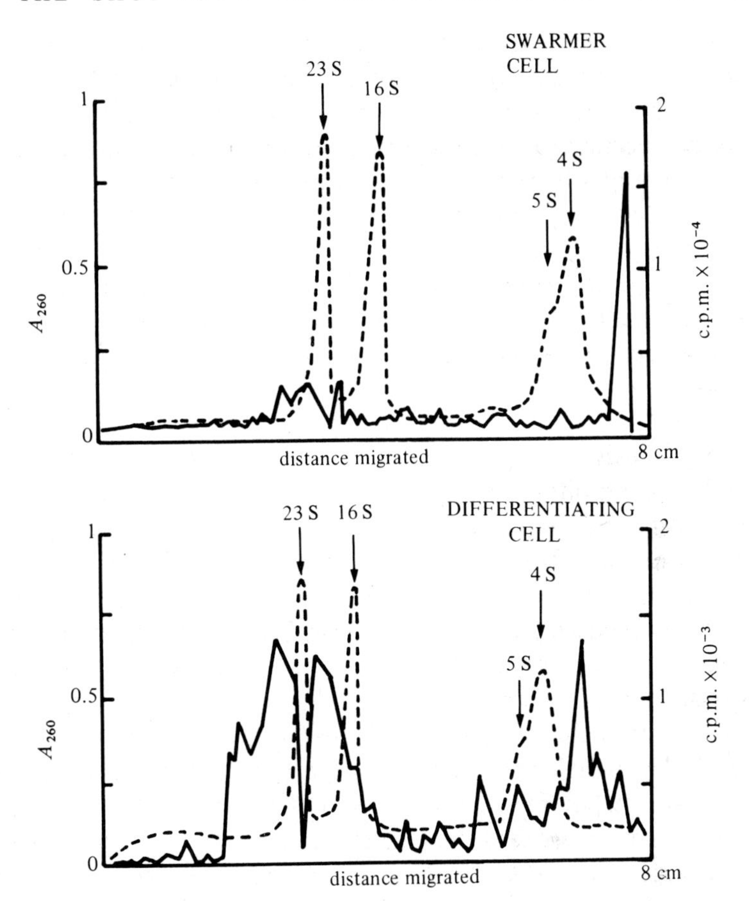

Fig. 22. Polyacrylamide gel electrophoresis of RNA synthesized *in vitro* on fast sedimenting, i.e. swarmer, envelope associated nucleoids and slow sedimenting, i.e. stalk or prosthecate cell, envelope associated nucleoids (Swoboda, Dow & Vitkovic, 1982a).

The unlabelled marker RNAs were extracted from *E. coli* by the method of Kirby (1965). Radioactivity in transcripts (——). A_{260} of *E coli* rRNA and tRNA (---).

22). *In vitro* coupled transcription and translation experiments indicated that biologically active mRNA was present. The point of importance from this work is that the fast-sedimenting (condensed) nucleoid (7100 S) primarily associated with the inhibited swarmer cells does not generate rRNA. In the inhibited swarmer cell the rRNA genes are either repressed and/or switched off.

RNA polymerase modification

The subunit structure of *C. crescentus* DNA dependent RNA polymerase consists of β' (165 K), β (155 K), α (44–48 K) and σ

(96–101 K) and is similar to other bacterial RNA polymerases (Bendis & Shapiro, 1973; Amemiya, Wu & Shapiro, 1977; Cheung & Newton, 1978). The RNA polymerases isolated at various times during the differentiation cycle were found to be identical, not only with respect to their subunit composition but also to their *in vitro* activity with bacteriophage T_2 DNA as a template (Bendis & Shapiro, 1973). However, these data should be interpreted with caution since it may be argued that differential transcriptional specificity *in vivo* could be mediated by a loosely bound minor protein(s) which was lost during purification.

Characterization of the purified enzyme (Cheung & Newton, 1978) has demonstrated that the holoenzyme and core polymerase catalyze polyadenylic acid synthesis in the absence of a template. In view of the previous reports (Ohta *et al.*, 1975; Ohta *et al.*, 1978) that a significant fraction of the mRNA of *C. crescentus* is polyadenylated, it is tempting to speculate that the enzyme may well produce a special subset of poly(A) mRNA species. However, the functional significance of poly(A) RNA in the context of the Caulobacter cell cycle is as yet unclear.

DNA synthesis

DNA synthesis during the *C. crescentus* cell cycle is periodic. There is no detectable DNA synthesis in the swarmer cell, this being initiated concomitant with prostheca formation and continuing until the predivisional stage, i.e. DNA synthesis is restricted to the prosthecate stage of the cycle (Degnen & Newton, 1972; Iba, Fukuda & Okada, 1977; Evinger & Agabian, 1977). At the time of completion of chromosome replication or at division an event occurs that delays chromosome replication in the progeny swarmer cell but not in the prosthecate cell. That is to say that after division the prosthecate cell reinitiates DNA synthesis whilst the progeny swarmer cell must undergo the obligate differentiation cycle, synthesize a prostheca and then initiate DNA synthesis. There are therefore two vegetative cycles being expressed, the swarmer cell cycle and the prosthecate or reproductive cell cycle. In terms of the prokaryotic cycle described by Helmstetter, Cooper, Pierucci & Revelas (1968), the swarmer cell undergoes an I phase, the period prior to the initiation of DNA synthesis when it is envisaged that the cell becomes competent to initiate chromosome replication, and then goes into the C phase, which is chromosome replication, and finally into the D phase, which is the time between completion of chromosome replication and cell division. The prosthecate cell has only a C+D cycle – there is no detectable I phase. This concept of the *C. crescentus* cycle will be enlarged upon later.

Closely correlated with the periodicity of DNA synthesis are changes in the sedimentation values of the envelope associated nucleoids (Evinger & Agabian, 1979; Swoboda *et al.*, 1982b). The nucleoid from the swarmer cell has a high sedimentation value (>6 000 S–7 100 S) whereas the nucleoid from the prosthecate cell has a much lower sedimentation value (3 000 S–5 600 S), and the pre-divisional cell appears to have both (Evinger & Ababian, 1979). The hierarchal organization of the *Caulobacter* nucleoid, as with *R. vannielii*, has been shown to be a beaded configuration (Swoboda *et al.*, 1982a). However it is data on the DNA binding proteins which is of particular interest. Evinger & Agabian (1979) have shown not only that there are distinct proteins associated with the nucleoids at specific stages of differentiation, but that a 26 K membrane protein is uniquely associated with the swarmer cell and, moreover, preferentially segregates with this nucleoid at division.

These variations in nucleoid structure, possibly mediated by differential association with proteins, may serve two roles, the maintenance of the tertiary structure of the DNA, and/or the regulation of cellular processes such as DNA replication and selective transcription at precise points in the differentiation programme. The latter possibility is highlighted in the pre-divisional cell where, as with *R. vannielii*, there are two nucleoids which are differentially expressed in the one cell cytoplasm. The apparent differential specificity of particular DNA binding proteins may hold the key to this question.

Inhibition of swarmer cell differentiation

Following cell division, the swarmer cell enters a period of reduced biosynthetic activity, even in high nutrient regimes, prior to the initiation of differentiation. Iba, Fukuda & Okada (1975) have reported that in the late exponential growth phase (nutrient broth, aerated at 30 °C) *C. crescentus* swarmer cells accumulated in the culture to 80% of the total cell population. Variation in aeration (O_2 tension) and temperature regimes in several culture media have yielded similar results (Swoboda, 1979). The detailed analysis of *Caulobacter* sp. C3, a strain which produces swarmer and prosthecate cells which differ significantly in cell volume (and hence can be easily resolved using an electronic cell volume distribution analyser

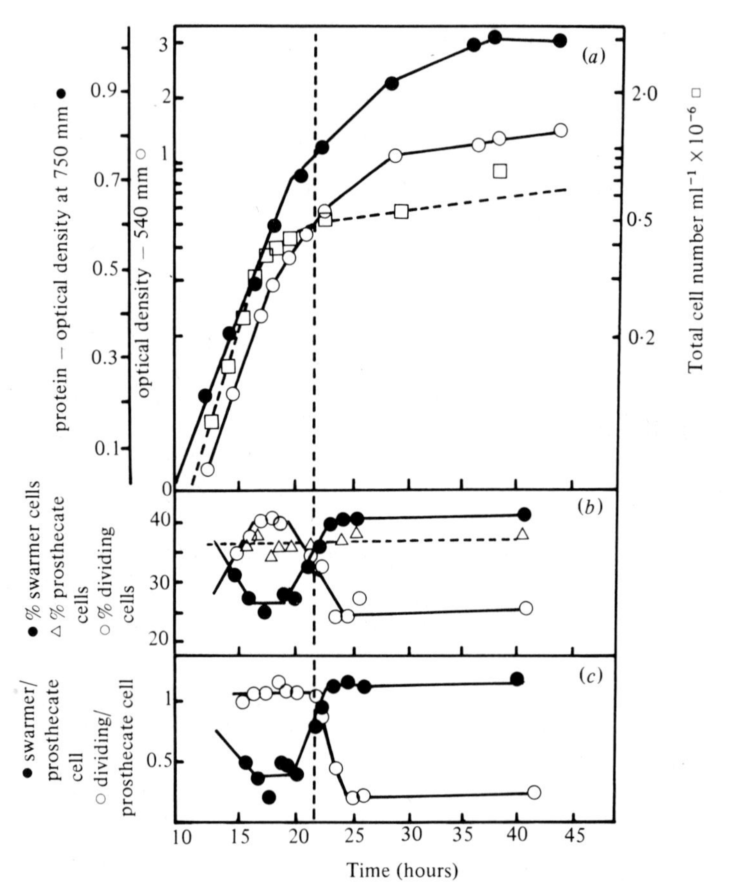

Fig. 23. Cell type expression during batch growth of *Caulobacter* sp. C3.

(*a*) Batch growth curve expressed in terms of cell mass (○), total cellular protein (●) and total cell number (□).

(*b*) Proportion of swarmer (●), prosthecate (△) and dividing cells (○).

(*c*) Ratio of swarmer (●) and dividing (○) cells as a fraction of the prosthecate cells. The latter remain relatively constant throughout growth.

Duplicate data were obtained by electronic cell volume distribution analysis, light and electron microscopy.

into the various cell types) has shown that as the late exponential growth phase is reached swarmer cells accumulate (Fig. 23). This data indicates, therefore, that *Caulobacter* possesses a regulatory mechanism which ensures that the swarmer cell does not initiate differentiation unless the environmental conditions are conducive to completion of the cell cycle with the generation of a reproductive cell, i.e. a situation very similar to that found with *R. vannielii* swarmer cells.

These observations contradict other findings where, under nutri-

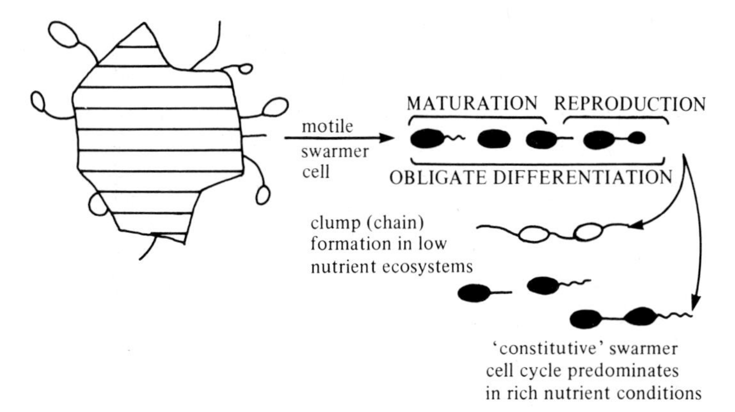

Fig. 24. *Hyphomicrobium* cell cycle. Many *Hyphomicrobium* species in the natural environment are in the form of tangled multicellular networks encased in metal hydroxide deposits (Tyler & Marshall, 1967; Tyler, 1970; Lawrence, 1978). Careful removal of these metal hydroxide deposits reveals chains of cells – similar to the multicellular chain complexes of *R. vannielii*, but which do not show any recognizable physiological separation between the cells. These findings suggest that in the natural environment *Hyphomicrobium* under nutrient stress colonises interfaces (Marshall, 1980) and can induce swarmer cell formation in response to specific stimulii. The recognized cell cycle, fundamentally very similar to the 'simplified' cell cycle of *R. vannielii*, may well be a laboratory artefact being equivalent to constitutive swarmer cell formation, i.e. high substrate concentrations repress the formation of multicellular complexes.

ent limitation when entering the stationary phase in batch culture, the swarmer cells develop into prosthecate cells. The latter do not reproduce but direct biosynthesis towards prostheca elongation (Poindexter, 1981a).

An explanation of this apparent contradictory observation may well be as follows. On entering the period of nutrient limitation the swarmer cells do indeed cease to differentiate and for a period of time accumulate. Subsequently a percentage of cells will become non-viable and may well lyse. The consequences of that might be cryptic growth, allowing the swarmer cells (non-prosthecate cells) to differentiate. No new swarmer cells would be produced as the reproductive cycle would have been halted by the nutrient limitation imposing a physiological stress.

HYPHOMICROBIUM

Since its isolation *Hyphomicrobium*, a chemoorganotroph which buds, has been considered to be a non-photosynthetic counterpart of *R. vannielii*, and morphologically, the analogy is very striking (Fig. 24) (Whittenbury & Dow, 1977). The *Hyphomicrobium* cell cycle (Zavarzin, 1960; Leifson, 1964; Harder & Attwood, 1978) is almost indistinguishable from the 'simplified' (see earlier) swarmer cell cycle of *R. vannielii.*

The Hyphomicrobium *swarmer cell cycle*

Hyphomicrobium cell division is asymmetric and yields a sessile, prosthecate, reproductive cell and a motile swarmer cell (Fig. 24), which undergoes an obligate differentiation cycle which is fundamentally indistinguishable from that of *R. vannielii* or *C. crescentus.* On initiation of differentiation the swarmer cell loses motility (shedding its flagellum), synthesizes a prostheca and initiates daughter cell formation at the prostheca tip. The daughter cell, as it attains maturity, synthesizes a flagellum and is released from the prostheca. Two distinct vegetative cycles are then initiated, the swarmer cell cycle and the reproductive cycle as is the case for *R. vannielii* and *C. crescentus.* Analysis of the differentiation cycle (Moore & Hirsch, 1973; Lawrence, 1978; Wali, Hudson, Danald & Weiner, 1980) has shown that the swarmer cell undergoes a period of maturation (referred to as the G_1 period by Wali *et al.* (1980) but which more correctly corresponds to the prokaryotic I phase) prior to the synthesis of a prostheca and the initiation of the reproductive cycle (equivalent to $C + D$ in cell cycle terms).

Protein, RNA and DNA synthesis

Incorporation of radiolabelled amino acids (Dow & Lawrence, unpublished) indicated that there was a significant level of protein turnover in the swarmer cell but, as with *R. vannielii*, there is no net increase in biomass. After initiation of differentiation there is an enchancement of the rate of protein synthesis during maturation with a very marked increase upon initiation of daughter cell formation (Wali *et al.*, 1980). In addition, subsequent studies have shown a pattern of sequential protein syntheses during the differentiation cycle (Matzen, Dow and Hirsch, unpublished).

Synchronized swarmer cells (obtained by the selective filtration system of Whittenbury & Dow (1977) were pulse labelled with ^{14}C-glutamic acid. Subsequent analysis of the soluble protein extracts by 2-D electrophoresis revealed 13 proteins unique to the swarmer cell, 3 unique to the prosthecate cell, and 4 attributable to the cells actively synthesizing daughter cells. The majority of the qualitative changes were associated with the maturation phase.

As with protein synthesis, RNA synthesis is lowest in the swarmer cell (Wali *et al.*, 1980) increasing during the differentiation cycle. No information is available concerning rRNA synthesis or RNA polymerase modification and/or specificity.

Data on DNA replication during the cycle (Zavarzin, 1960; Moore & Hirsch, 1973; Wali *et al.*, 1980) indicates that there is no DNA synthesis in the swarmer cell; this is initiated at a point during the maturation phase, possibly coinciding with the termination of prostheca synthesis (Wali *et al.*, 1980). The weight of evidence, therefore, is consistent with there being no DNA synthesis in the swarmer cell. Preliminary data also indicates that the swarmer cell nucleoid is in a more condensed format to that of the prosthecate cell (Dow & Swoboda, unpublished).

Physiologically, therefore, the swarmer cell of *Hyphomicrobium* is fundamentally indistinguishable from those of *R. vannielii* and *C. crescentus*.

Inhibition of swarmer cell differentiation

Analysis of cell type distribution during batch growth of *Hyphomicrobium* yields data which supports the contention of there being a regulatory switch, environmentally stimulated, governing the initiation of swarmer cell differentiation (Fig. 25). Determination of the relative numbers of each of the cell types, either by electron microscopy or cell volume distribution analysis (Dow & Lawrence, 1980), revealed the following: during exponential growth the population remains relatively stable with respect to cell type distribution. However, with the onset of nutrient limitation the ratio of swarmer cells to prosthecate cells dramatically increases. With the onset of nutrient stress reproduction continues, i.e. swarmer cells continue to be formed but they are inhibited from initiating the differentiation sequence, hence the inverse nature of the ratios of the various cell types.

Although we know less about the cell cycle of *Hyphomicrobium* than we do of the cell cycles of *R. vannielii* or *C. crescentus*, there can be little doubt that in all of these organisms the properties of the swarmer cells are fundamentally similar and that they serve a similar role – of dispersal and survival.

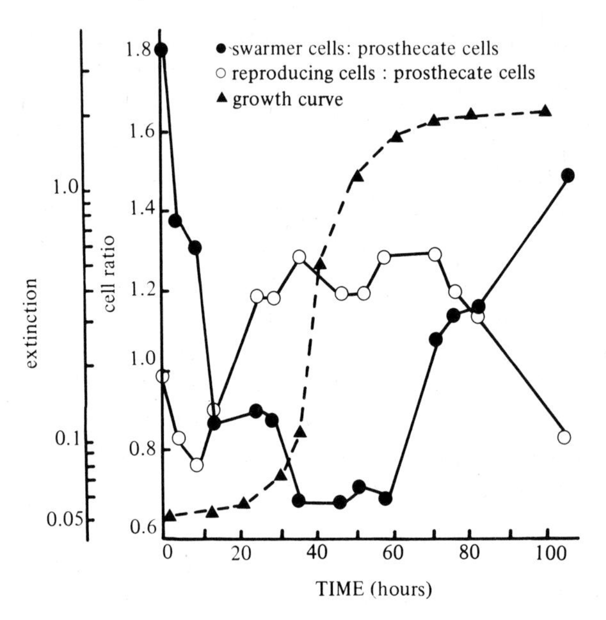

Fig. 25. Ratios of the relative numbers of *Hyphomicrobium* swarmer cells, prosthecate cells and reproducing cells plotted against time.

THE SWARMER CELL – THE GROWTH PRECURSOR CELL?

An analysis of the cell cycles of three distinct and unrelated species of bacteria, *Rhodomicrobium vannielii*, *Caulobacter crescentus* and a species of *Hyphomicrobium* has revealed a cell entity common to them all – the swarmer cell with the following properties.

(a) They are motile, the period of motility being unrelated to other cell cycle events which are temporarily related both in duration and initiation.
(b) They possess a low endogenous metabolic activity. Both protein and RNA synthesis occurs but at a very low rate in comparison to that observed in the subsequent part of the cell cycle.
(c) There is no rRNA synthesis, the genes being either switched off or repressed.
(d) There is no DNA synthesis.
(e) The nucleoid is in a highly condensed format with a high sedimentation value.
(f) The cell appears to have the capacity to monitor its environment, its differentiation being delayed in unfavourable circumstances. This would mean it has ‘environmental receptors’

which, on contact with key, unidentified as yet, factors or substances, trigger off differentiation – or, alternatively, the 'environmental receptors' may also be sensitive to inhibitors. In other words, a parallel situation to chemotaxis may exist.

The majority of studies on microbial adaptation have centred on the physiological and biochemical response induced in laboratory culture, be this batch or continuous, and are confined almost entirely to those exhibited when growth is unrestricted. Whilst such data has been of unquestionable value, Koch (1971) highlighted the inappropriateness of applying such studies to microbial events occurring in the natural environment by highlighting the 'feast and famine' existence of *E. coli* in the mammalian intestine. His observations were concerned with the capacity of *E. coli* to respond to high and sporadic nutrient inputs in terms of uptake (transport) and subsequent concomitant protein synthesis; this led to the conclusion that *E. coli* is adapted to respond quickly to unannounced nutrient excess and to withstand chronic starvation for significant time periods. Since Koch's article, our understanding of the fundamental physiology of *E. coli* at the biochemical and molecular level has increased immensely, but we are still faced with the problem of relating such data to situation(s) prevalent in the natural environment – by no means a simple matter.

When discussing adaptation to nutrient flux, one of the most extreme situations is that of oligotrophy where the adaptive responses are geared to continued nutrient limitation. Microbiological examination of such environments reveals the presence of microbes which have very distinct morphologies (Hirsch, 1979; Dow & Lawrence, 1980) the most prominent of which are the integral cellular appendages or prosthecae (Staley, 1968). In general these appendages are regarded as a morphological response to nutrient limitation, their primary function being to increase the surface/volume ratio so enhancing the uptake capacity of the cell (Pate & Ordal, 1965; Larson & Pate, 1976; Whittenbury & Dow, 1977). In non-prosthecate bacteria a similar effect can be produced by the cell becoming smaller, i.e. tending towards a sphere. The best known example of this is the sphere-rod transition of *Arthrobacter* (Ensign & Wolfe, 1964); readily utilizable substrates which support high exponential growth rates promote the rod morphology whilst nutrient limitation enhances the coccal form (Luscombe & Gray, 1974). In the latter case – the coccal form – at least a two-fold increase in the s/v ratio is established, i.e. a greater surface area is presented to

Table 1. *Oligotrophic characteristics exhibited by* Arthrobacter *spp.*

High growth rates	Low growth rates
(readily utilizable substrate)	(poor substrate)
$D>0.25\ h^{-1}$ cell volume =	$D<0.2\ h^{-1}$ cell volume =
0.45–0.6 μm^3	0.16–0.24 μm^3

(figures for *A. globiformis*, Luscombe & Gray, 1974)
Highest growth rate ($\sim\mu_{max}$) = 0.37 h^{-1}
Lowest growth rate ($\sim\mu_{min}$) = 0.01 h^{-1} (0.027 μ_{max}), 90% survival
With a 'normal' heterotroph loss of viability becomes significant as μ_{min} approaches 0.06 μ_{max} (Pirt, 1972).
Active transport of glucose is inducible, K_m = 0.8 mM (Krulwich & Ensign, 1969), which is considerably lower than for enterics (K_m 0.05–0.1 mM).
With the onset of starvation the endogenous metabolism is reduced to 10% of that observed in enterics or pseudomonads under similar conditions. *Arthrobacter* survive 10–100 times longer than the latter (Ensign, 1970; Poindexter, 1981b).
Vegetative cells very resistant to desiccation (Boylan, 1973).

Table 2. *Comparative data obtained for a* Spirillum *and a* Pseudomonas *isolated by chemostat enrichment from pond water (Matin & Veldkamp, 1978)*

	$D>0.05\ h^{-1}$ *Spirillum*	Pond water chemostat enrichment	$D<0.3\ h^{-1}$ *Pseudomonas*
μ_{max}	0.35 h^{-1}		0.64 h^{-1}
μ_{min}	0.01 h^{-1} (<3% μ_{max})		0.05–0.08 h^{-1} (6–12% μ_{max})
K_s (lactate)	23 μM		91 μM
K_m	5.8 μM		20 μM
Residual lactate at D 0.15 h^{-1}	17 μM		28 μM
Maintenance energy	0.016 glactate/g cell dry wt/h		0.066 g lactate/g cell dry wt/h

After Poindexter (1981b).

the environment, and this is the morphological form thought to predominate in nature (Mulder, 1963). Further physiological characteristics which enhance the capacity of *Arthrobacter* to cope with oligotrophic conditions are summarized in Table 1.

A second study which demonstrates many of the physiological characteristics desirable for existence in oligotrophic environments was the chemostat enrichments of Matin & Veldkamp (1978) in which it was shown that a *Spirillum* was selectively enriched when $D = 0.05\ h^{-1}$ and a *Pseudomonas* when $D = 0.3\ h^{-1}$. The comparative data from these experiments are given in Table 2.

Studies of growth kinetics in part explain why one organism may be better adapted for growth in a particular environment than another but they fail to provide us with an understanding of the mechanisms involved. Whilst such data is of general importance in this particular context it has been derived from heterogeneous populations, i.e. with cells in every stage of the cell cycle, and therefore raises the question as to whether the cells in such populations are physiologically and functionally identical (as is assumed in classical continuous culture theory); in essence the measured parameters are averaged for the population. Also, the norm is for the nutrient and cultural regimes to be far removed from those encountered in the natural environment. Although it is important to recognize the inherent limitations of routine laboratory culture and to treat data derived from such with a degree of caution, it is difficult to envisage satisfactory experimental protocols to circumvent these inadequacies.

In this article it is not discrete structural and physiological adaptations *per se* that we have considered but its purpose in a particular instance – the growth precursor cell.

DOES THE GROWTH PRECURSOR CELL CONCEPT APPLY TO OTHER BACTERIA, THE CYANOBACTERIA IN PARTICULAR?

Other species of bacteria which probably give rise to growth precursor cells as an obligate and transient stage in their cell cycles include all other known budding and prosthecate bacteria (e.g. *Rhodopseudomonas palustris* and *Nitrobacter* species), *Bdellovibrio* species (in the motile, non-growth phase prior to host invasion), *Sphaerotilus* species, and species of Cyanobacteria.

An interesting point to explore, of course, is whether or not such a cell cycle stage is expressed by the morphologically more 'conventional' forms of bacteria, the rods and cocci – and, in particular, *E. coli.* Two observations with regards to *E. coli* encourage further research in this respect: (a) when grown with a generation time in excess of 60 minutes *E. coli* grows polarly (Donachie & Begg, 1970) – as do the budding bacteria – leading to the possibility that the two cells formed at division differ from each other, in age and physiological state, the 'younger' cell possibly being a growth precursor cell, and (b) when grown with a generation time in excess of 3 hours (Dow, unpublished), *E. coli* lysates contain two nucleoids which differ in sedimentation values – one in a relaxed state, the other in a condensed format.

This evidence, slight as it is, is, nevertheless, *prima facie* evidence for the existence of a growth precursor cell in cell cycles of *E. coli* grown with long generation times (a situation more akin to their existence in their natural environment).

The last phase of this article is concerned with the possibility of establishing whether the concept of a growth precursor cell stretches to the cyanobacteria. Here the position is complex because of the bewildering variety of cell forms, cell cycles and differentiation patterns expressed by the cyanobacteria. In addition, it needs to be borne in mind that the morphologically more interesting of the cyanobacteria are *multicellular prokaryotes* – exhibiting cell–cell cooperation and dependence – and, consequently, the *growth precursor cell concept* may be of a more sophisticated nature than that previously discussed, i.e. the growth precursor cell may occur in more than one form in the one organism, depending upon the particular cell cycle being invoked.

The following discussion, therefore, highlights some of the cell cycles, cell types, forms of multicellular organization and differentiation patterns which may have stemmed from a growth precursor cell, either singly or as a multicellular entity (e.g. hormogonia).

CYANOBACTERIA

It is among the cyanobacteria that prokaryotes achieve their greatest degree of cellular specialization and these morphologically specialized cells sometimes occur in the context of complex, environmentally controlled cellular cycles. Some of the filamentous cyanobacteria exhibit such clear-cut specialization of function and physiology between their differentiated cell types that it is reasonable to consider species of *Anabaena*, for example, as multicellular structures, with all that implies, with regard to structural specialization, altered gene expression and interactive behaviour of the component cell types. In addition, species of *Anabaena* would seem to be prime examples of organisms likely to form growth precursor cells of one or more types as a base from which to differentiate in response to environmental dictates.

Cell cycles of cyanobacteria

There is extensive evidence that many species of cyanobacteria deviate from the simple division pattern of two equal daughter cells that is generally assumed to be the characteristic bacterial pattern. In examining the selected examples presented here it should be remembered that unicellular species (apart from the examples already mentioned) appear to divide equally and a vegetative cell population is uniform in growth and structure.

The unequal division rule of Anabaena

The growth of individual filaments of *Anabaena catenula* on agar plates was followed by photomicroscopy for periods of time equivalent to several cell divisions by Mitchison and Wilcox (1972) who noted that cell division was asymmetrical and yielded a 'small' and 'large' daughter cell. The asymmetrical division of *Anabaena* sp. is related to the development of a heterocyst pattern in that heterocysts only develop, providing the appropriate environmental trigger is present, from a small daughter cell. In fact, similar observations had been made in the nineteenth century but had been criticised and overlooked for many years until the relatively recent rekindling of interest in heterocyst formation (see Carr, 1979). The distribution of small and large daughter cells down a filament is not random. If any given cell in a filament has arisen as the left daughter of a division, then, when it divides, its left daughter will be the smaller of the two daughter cells (Mitchison & Wilcox, 1972); this may be seen in Fig. 26. Since all cells, at any defined growth rate, reach the same size before they divide, it follows that the small daughter cell takes a longer period of time to reach division than does the larger daughter (Fig. 26). This results in a degree of asynchrony being maintained down a filament of vegetative cells, not just with respect to cell size, but also most probably with respect to the status of replication. Thus during the induction of heterocysts the selection of candidate cells is made from a heterogeneous population of vegetative cells distributed periodically through a filament. The extent to which the asymmetric division rule of *Anabaena catenula* applies to all heterocyst producing cyanobacteria is not known, but certainly several other species of *Anabaena* divide in the unequal fashion described by Mitchison and Wilcox. The small cell referred to here could well be analogous to the growth precursor cell described earlier – in this context as a platform for differentiation, not a property of the large cell.

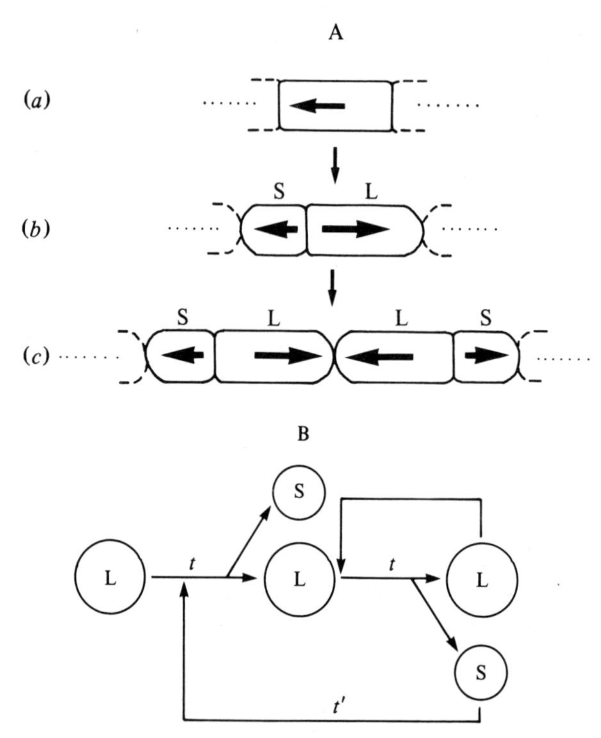

Fig. 26(A) The division rule. In (a) a cell of an *Anabaena* filament is represented with an arrow pointing away from the newly formed septum. The rest of the filament is indicated by dotted outlines. In (b) the two daughter cells are shown, with their relative sizes denoted by S = small, L = large. Arrows have been given to the daughters, pointing away from the new septum. In (c) the second generation is shown, using the same notation (from Mitchison & Wilcox, 1972).

(B) A diagrammatic representation of the altered lengths of the cell-cycle consequent upon asymmetric division in exponential culture. Whereas the large daughter divides after the same time period (t) of its (large) parent, the small daughter cell will require a longer length of time (t') to reach division depending upon its degree of 'smallness'. The difference between t and t' may vary with species and culture conditions.

Division patterns in some unicellular cyanobacteria

There are two orders of cyanobacteria that reproduce by the formation of spherical cells, smaller in size than that of the mother cell. The simpler of these two, the genera *Chamaesiphon*, reproduce exclusively by budding. The ovoid or elongated cells, attached to solid material in the environment, form and liberate small spherical cells from the apical pole and have been described in detail by Waterbury & Stanier (1977). The life cycle of these budding cyanobacteria is thus comparable to that of photosynthetic bacteria such as *Rhodopseudomonas palustris*. Nothing is known of environ-

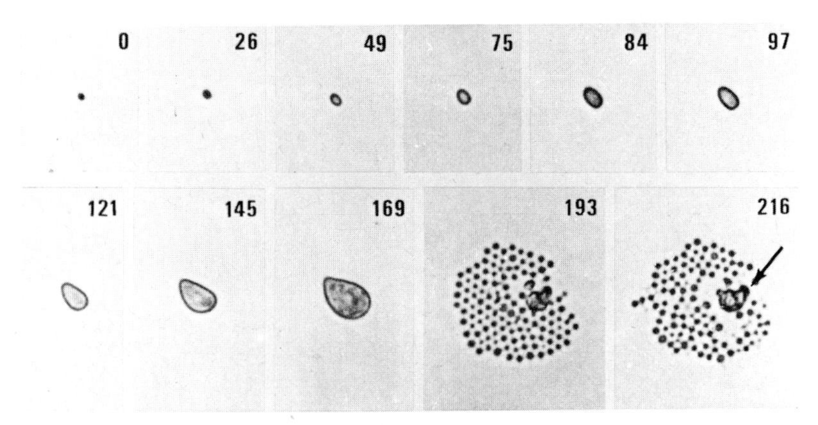

Fig. 27. The development of *Dermocarpella*. Growth of the large cell continues for 169 hours when multiple fission results in release of numerous baeocytes; the arrow indicating the remains of the basal cell after baeocyte release. ×500 (from Waterbury & Stanier, 1978).

mental alteration or control of this division pattern, but, again, this cell could be a growth precursor cell.

A more striking deviation from the cell division pattern of unicellular cyanobacteria is presented by species of pleurocapsalean cyanobacteria (Waterbury & Stanier, 1978; Waterbury, 1979). The group Pleurocapsa is defined as those cyanobacteria that are capable of reproducing by multiple fission, which in turn is defined as the division of the vegetative cell by a series of rapid, successive binary fissions into spherical daughter cells (Waterbury, 1979). The important feature of this is that there is no growth during the multiple division process, with the result that the volume of each daughter is half that of its immediately preceding parent, and progressively much less than that of the ancestral cell that initiated the multiple division process. The developmental sequence of a pleurocapsalean organism is shown in Fig. 27. The small cell released at the end of the multiple division period had been termed an 'endospore', but Stanier & Waterbury (1978) recognized the unsuitability of the name, indicating as it did comparability with the endospore of *Bacillus* to which it bore no meaningful similarity. They introduced the name baeocyte for the small daughter cells of cyanobacterial multiple fission. Organisms of the genera *Dermocarpella* (illustrated in Fig. 27) do not carry out normal binary fission; the baeocyte enlarges into an ovoid vegetative cell which then undergoes multiple fission. Other pleurocapsalean cyanobacteria, for example *Myxosarcina*, undergo repeated binary division of the developed baeocyte which leads to a large mass of vegetative cells –

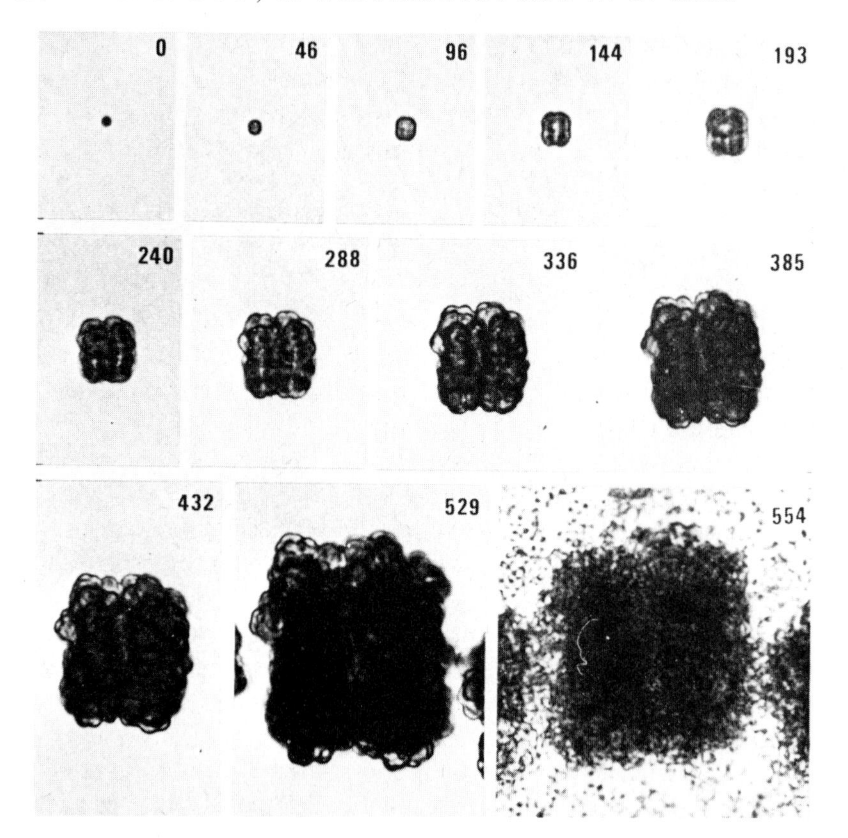

Fig. 28. Development of a single baeocyte of *Myxosarcina*, through 259 hours of vegetative binary fission leading to baeocyte release after multiple fission (554 hours) (from Waterbury & Stanier, 1978).

each one of which subsequently undergoes multiple fission (Fig. 28). The review of Waterbury & Stanier (1978) describes the modified cell division cycle of seven taxonomically meaningful groups of pleurocapsalean cyanobacteria and gives considerable fine-structural analysis of example species. The extent of environmental influence on these multiple fission cell cycles is largely unknown. Baeocytes of some species are phototactically motile. In at least one example, the inclusion of sucrose in the growth medium alters the form of vegetative growth from uniserate to multiserates. The number of binary fissions of vegetative cells that precede the onset of multiple fission to yield baeocytes is not fixed, favourable growth rates apparently inducing earlier multiple fission. In this respect the formation of new baeocytes from vegetative cells cannot be thought of as a reproductive strategy in response to an adverse environment.

It would be instructive to learn what are the constraints that limit vegetative binary division and cause the onset of multiple division, leading to such a radical reduction in cell size, and again, does the growth precursor cell concept embrace the baeocyte?

Light and morphology

All cyanobacteria are phototrophic and hence the various reported effects of light quality and intensity on their growth, structure and physiology are to be expected (see van Liere & Walsby, 1982). The adjustment of amount and identity of photosynthetic pigments to light variation is well studied, with the overall object of maximizing photosynthetic rate. In turn, the rate of photosynthesis will profoundly affect intermediary metabolism. Such generalizations hold for all cyanobacteria; here we will try to isolate distinct morphological effects consequent upon light variation, using only one or two selected species.

In a strain of *Nostoc muscorum* capable of heterotrophic growth, Lazaroff & Vishniac (1961) showed that light intensity markedly affected growth form. At saturating light intensity *N. muscorum* was entirely filamentous, whilst in the dark the organism adopted an aseriate morphology in which clumps of numerous vegetative cells were loosely associated together. Intermediate light intensities produced transitional morphological arrangements. Similar structural changes were observed (Evans, Foulds & Carr, 1976) in *Chlorogloeopsis fritschii* (Fig. 29); heterotrophic, or photoheterotrophic, growth suppressed the filamentous form of this organism. Photoautotrophic culture favoured the presence of filaments of cells, which as the culture aged, changed to the aseriate form. Thus in *C. fritschii* the presence of an organic substrate appeared to override the influence of light on morphological arrangement. *C. fritschii* produces heterocysts when deprived of adequate nitrogen. When the phenomenon is induced the first detectable change is in the division pattern of the vegetative cells, this changes from equal to a significant proportion of cells undergoing unequal divisions (Foulds & Carr, 1981).

An important feature of all *Nostoc* species is the presence of motile trichomes which are termed 'hormogonia', these are short (approximately 10 cells) undifferentiated filaments, sometimes with tapered end-cells. Hormogonia arise from aseriate or vegetative filaments in the light, and after a period of motility, in which there is

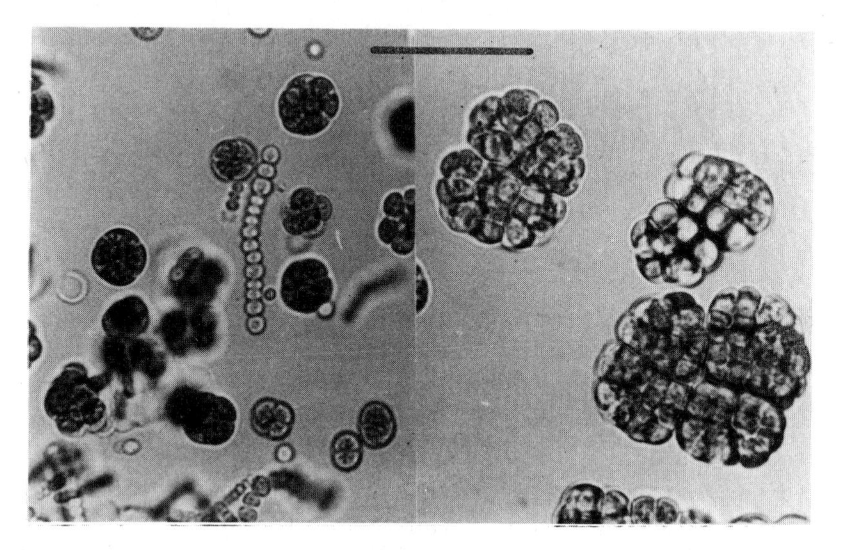

Fig. 29. Photoautrotrophic culture of *Chlorogloeopsis fritschii* (left) showing characteristic mixture of cell types. The single cell form is characteristic of older cutures whilst the filamentous form occurs only in photoautotrophic growth. The aseriate, clumped form of growth is the only type present after heterotrophic growth in the dark (right) (bar = 10 μm) (From Evans, Foulds & Carr, 1976).

no apparent growth as measured by cell division, they develop heterocysts, divide in a plane parallel to the axis of the filament, and thus turn into the aseriate form of structure. These in turn give rise to differentiated filaments containing vegetative cells and heterocysts; a simplified diagram of the sequence of cell cycle changes occurring in *Nostoc muscorum* is presented in Fig. 30 (Lazaroff & Vishniac, 1961). A detailed account of the experimental control of *N. muscorum* development by light of defined wavelength is given by Lazaroff (1972) who provides, in an appendix, a valuable historical account of early concepts regarding the developmental biology of heterocyst-containing cyanobacteria, with extensive references to the early literature. The role and physiology of hormogonia, short undifferentiated motile filaments, raise some interesting questions. There is some evidence that the hormogonia of Calothrix species (B. A. Whitton, personal communication) do not divide or grow, but have a defined lifetime with motility being their most important feature. They could, therefore, be dispersal agents – or growth precursor cells – with the role of environmental factors on their formation and outgrowth being similar to that applying to *Rhodomicrobium vannielii*. The aseriate stage of morphology in

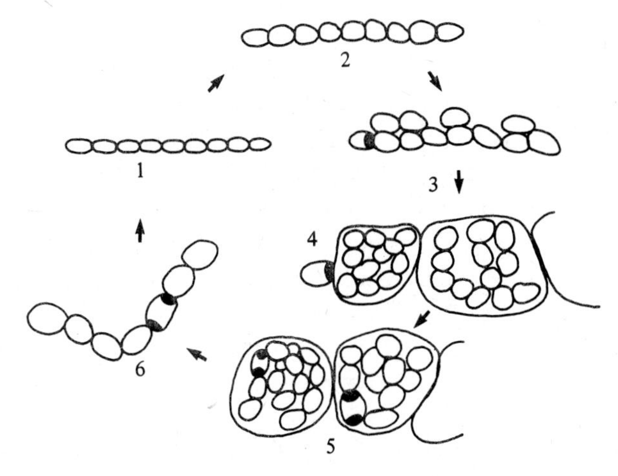

Fig. 30. Cyclic sequence of stages in the development of *Nostoc muscorum*. (1) Hormogonia with tapered terminal cells. (2) Terminal cells begin differentiation to heterocytes, intercalary cells have enlarged to vegetative size. (3) Division of vegetative cells in a plane parallel to that of the filament, mature heterocysts. (4) Cluster of vegetative cells resulting from division, enclosed in gelatinous material to form aseriate clumps. (5) Formation of a filament within each clump. (6) Release of a heterocyst containing filament of vegetative cells (After Lazaroff, 1972).

Nostoc species is maintained under dark, heterotrophic conditions. Exposure of aseriate cultures to red light (650 nm) induces the formation of motile hormogonia and this induction is reversed by simultaneous or subsequent exposure to green light (Lazaroff & Schiff, 1962; Lazaroff, 1972). A similar photomorphogenesis has been described in *Nostoc commune* (Robinson & Miller, 1970), with green light being, on a quantal basis, three times more effective than red. The action spectra and repeated photoreversibility of the process led the latter workers to suggest that a phytochrome-like photoreceptor system may be involved. In both *Nostoc* species the induction of hormogonia from aseriate cultures was achieved in darkness by the introduction into the culture of a heat-labile substance excreted into the growth media of cultures containing motile hormogonia (Robinson & Miller, 1970). The chemical identity of this material is not known, nor have we any precise knowledge of when it is produced, relative to the actual formation of hormogonia themselves. Clearly an investigation of the properties of hormogonia from the standpoint of our thesis – the growth precursor cell (or cells in this instance) – is now merited.

Heterocysts: structure, function and development

Heterocysts have been recognized as specialized cells in cyanobacteria since the nineteenth century. Their characteristic appearance in the light microscope, at least of those present in species of *Anabaena*, *Calothrix* and *Nostoc*, has encouraged wide-ranging speculation regarding their function, which has been resolved only by knowledge of their unequivocal role, in *Anabaena* species, as sites of nitrogen fixation. Structures resembling heterocysts in cyanobacterial filaments have been observed in Precambrian rock (Licari & Cloud, 1972), although whether these have any functional analogy (even assuming the structural similarity is real) to contemporary heterocysts is an open question. Nevertheless the appearance in electron microscopic pictures of Precambrian rocks of multi-cellular types of prokaryotic entities is a striking illustration of the antiquity of microbial differentiation.

Heterocyst biology has been reviewed from several standpoints during recent years and a full account of these specialized cells will not be presented here (see Adams & Carr, 1981; Wolk, 1982). The inverse relationship between heterocyst presence and available nitrogen was observed many years ago (Fogg, 1949) but the suggestion that heterocysts were the sites of nitrogen fixation (Fay *et al.*, 1968) awaited a much fuller understanding of the enzymic process. Because nitrogenase requires anaerobic conditions there was incompatibility therefore with O_2-evolving photosynthesis in the same cells and the idea of separation of the two activities was evolved. The heterocyst contains a functional PSI photosystem but not a PSII photosystem, and hence its relative anaerobic state, would permit nitrogen function.

Heterocysts are formed by several genera of cyanobacteria (*Anabaena, Nostoc, Cylindrospermum, Calothrix* etc.) and in all examples that have been examined, heterocyst frequency is increased by growth in the absence of available nitrogen in the form of nitrate or ammonium salts. In many species the adequate supply of such salts results in the complete absence of fully-formed heterocysts. The intracellular control of heterocyst development is much more complex than the external, environmental trigger of nitrogen availability. The assimilation and metabolism of nitrogen, via ammonium ions, to glutamine is probably necessary before it can exert a regulatory control. The intracellular level of glutamine, or of the enzyme glutamine synthetase, is likely to be crucial in the onset of nitrogenase synthesis. We know that the formation of a mature heterocyst involves a whole series of stages, the process of heterocyst formation having several sequential commitment times. A vegetative cell of *Anabaena cylindrica* requires several hours before the heterocyst developmental process becomes irreversible, and a further period of about 8 hours for the maturation process to be completed.

Further flexibility of the cyanobacterial response to the environmental signals causing heterocyst formation in *Anabaena cylindrica* is provided by the fact that light is required in the early stages of heterocyst formation. After the initiation of heterocyst formation by removal of ammonium ions an 8-hour period of illumination is necessary for the full heterocyst response to be expressed over the course of the next 10 hours, during which the light is not necessary (Bradley & Carr, 1977). The early light-dependent phase is that period during which serial commitment takes place and it is in the later 'dark' (light independent) period that the morphological changes of heterocyst formation are seen. *Anabaena cylindrica* is an obligate phototroph, therefore construction of heterocysts from vegetative cells in the dark emphasises the fact that the process is one of recycling of existing cell material, rather than one of *de novo* growth. The requirement for light in order that the 'no-nitrogen' environmental trigger be effective in producing heterocyst formation is understandable; an obligately-phototrophic organism in the dark needs to conserve materials for endogenous metabolism, rather than expend them on cell differentiation. A noteworthy feature of the dark–light heterocyst induction experiment is the 9-hour period of time that is necessary for the full (9%) heterocyst frequency to become expressed. Since the individual cells of *Anabaena cylindrica* filaments are normally unsynchronized the necessity for this time period can be explained by making the assumption that initiation of heterocyst development can occur only when a vegetative cell is at a particular point in its cell cycle. There is evidence from inhibitor studies that this is indeed the case (see Adams & Carr, 1981). Mitomycin, an inhibitor of DNA synthesis, prevented the onset of heterocyst formation when added to a culture of *Anabaena cylindrica* that had been deprived of available nitrogen. After a period of 10 hours of induction, the addition of mitomycin became progressively less effective in preventing heterocyst formation. It would, therefore, appear that DNA synthesis is necessary, for the initiation of heterocyst formation, but once that process had begun, mitomycin will not interfere with the maturation stages.

The primary environmental signal for heterocyst formation in *Anabaena cylindrica*, is lack of available nitrogen, constrained by a second environmental factor, light, and by an internal factor, the stage of DNA replication. These interact together so as to produce a coordinated filament response to the lack of nitrogen. Thus a short period of nitrogen deprivation, in the light, will elicit from *Anabaena cylindrica* only a small percentage of vegetative cells differentiating to become heterocysts. The maximum response, approximately 9% heterocysts, will be produced only by a sustained, approximately 10 hour, period of nitrogen deprivation in the light. Because heterocysts cannot propagate, there is a clear selective advantage to *Anabaena cylindrica* in a filament devoting only the minimum number of cells to heterocyst formation consistent with having a sufficient number of sites of nitrogen fixation necessary to maintain a growth rate appropriate to the other environmental conditions, such as light intensity, temperature and nutrient supply.

It may be calculated that producing 1% heterocysts extra to requirement, i.e. 10% instead of 9%, will produce an approximately 20% reduction in population size after thirty generations, making the assumption that the functional life of a heterocyst is two generations (Adams & Carr, 1981). A further environmental control over heterocyst formation has been observed in *Nostoc* sp. 6310 (Rippka & Stanier, 1978). In the absence of oxygen the heterocyst development was arrested at the preheterocyst stage when external wall structures had not yet been completed. Introduction of oxygen allowed the developmental process to complete, whereas maintenance of anaerobic conditions led to eventual de-differentiation back to a vegetative cell; *Nostoc* sp. 6310 will develop a functional nitrogenase anerobically. These experiments illustrate the distinction between nitrogenase synthesis and heterocyst formation and also the ability of the organism to complete heterocyst differentiation only when that differentiation is required; that is, under aerobic conditions when the specialized envelope of heterocysts presumably contributes to maintaining a relatively low oxygen tension within the specialized cell.

In addition to the presence of nitrogenase, heterocysts have several other enzymes that have markedly increased, or decreased specific activities (see Wolk, 1982). These changes in the enzymic complement of heterocysts can all be related to its specialization for nitrogen fixation, particularly with respect to the supply of ATP and reducing capacity for nitrogenase, and to the assimilation of the fixed nitrogen (see Adams & Carr, 1981). Heterocysts do not contain the two key enzymes of the reductive pentose phosphate pathway ribulose-1,5-bisphosphate carboxylase and ribulose 5 phosphate kinase, consequently they are largely dependent upon the adjacent vegetative cells for fixed carbon. Glutamine is the carrier molecule that exports nitrogen from the heterocyst into the adjacent support cells. Their specialization of metabolism, together with the characteristic structural features of the heterocyst envelope, indicates the considerable degree of selective gene expression that must be called for by heterocyst formation. Groups of polypeptides synthesized at different stages of heterocyst development by an *Anabaena* species were recognized by the incorporation of $^{35}SO_4$, gel electrophoresis and autoradiography (Fleming & Haselkorn, 1974). This procedure also pointed to individual polypeptides whose synthesis stopped during the developmental period. The most direct examination of altered gene expression during heterocyst formation has involved the recognition of transcriptional products made in the entire filament by probes of cloned genes known to be involved in heterocyst development, *nif* and the carboxylase gene. The expression of the genes associated with the *nif* operon have been shown to require anaerobic conditions, an important observation in view of the fact that heterocysts fix nitrogen under aerobic conditions. The induction of nitrogenase genes, by lack of available nitrogen, switches off the transcription of the carboxylase gene or, much less likely, induces the breakdown of its transcript (Haselkorn *et al.*, 1982). This experimental approach of Haselkorn and colleagues offers, through the use of molecular genetic techniques, a most productive way of analysing the heterocyst development in response to the causative environmental factors.

In the context of this story of differentiation, our point of focus is the small cell and its possible analogy to the swarmer cell described earlier. Insufficient evidence, clearly, is available to make a strong case now – but we would suggest that studies on the physiological state of the small cell, in contrast to that of the large cell, are worth pursuing to elicit whether the analogy has real substance in molecular biological terms.

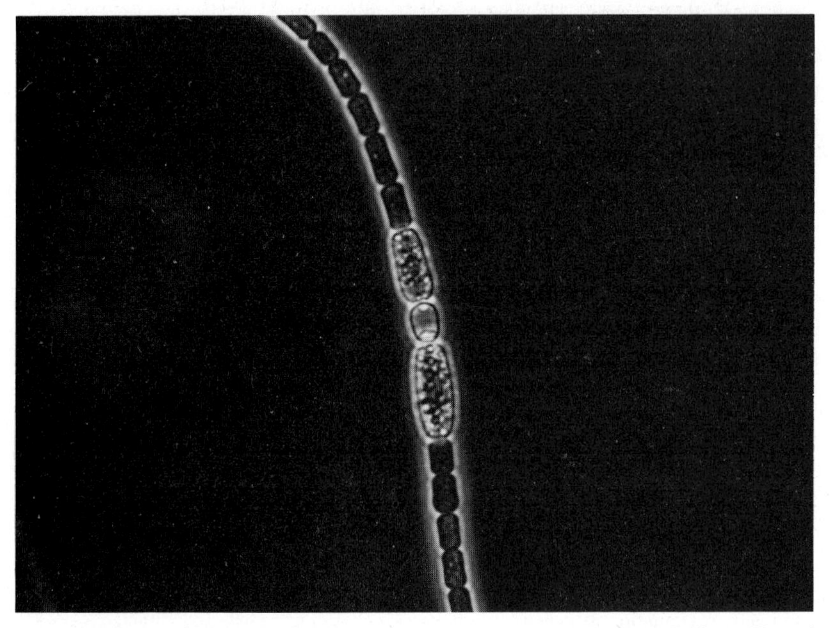

Fig. 31. Formation of akinetes adjacent to a heterocyst in *Anabaena cylindrica*. Note the granular appearance of the akinete and, characteristic for this species, the unequal rate of their development (×750).

THE FORMATION OF AKINETES

Occurrence, structure and function

Akinetes are specialized single-cells occurring in some members of Nostocaceae, Rivalariaceae and Stigonemataceae; they are readily recognized under the light microscope. The akinete is usually larger than a vegetative cell, with thicker wall structures and a more granulated cytoplasm, possibly due to accumulation of cyanophycin and glycogen bodies. Most heterocyst forming cyanobacteria also form akinetes, and these two differentiated cell types occur in a fixed, physical, species-variable relationship. Thus many *Anabaena* always develop akinetes immediately adjacent to existing heterocysts (Fig. 31), whilst some *Nostoc* species form akinetes midway between heterocysts. There is some evidence of greater resistance to cold and desiccation and considerable support for the view that they serve as perennation of dormancy structures (see Wolk, 1965; Nichols & Carr, 1978; Nichols & Adams, 1982).

Many of these features are illustrated in Fig. 32, which shows a culture of *Anabaena* sp. C4 that has developed one filament with a

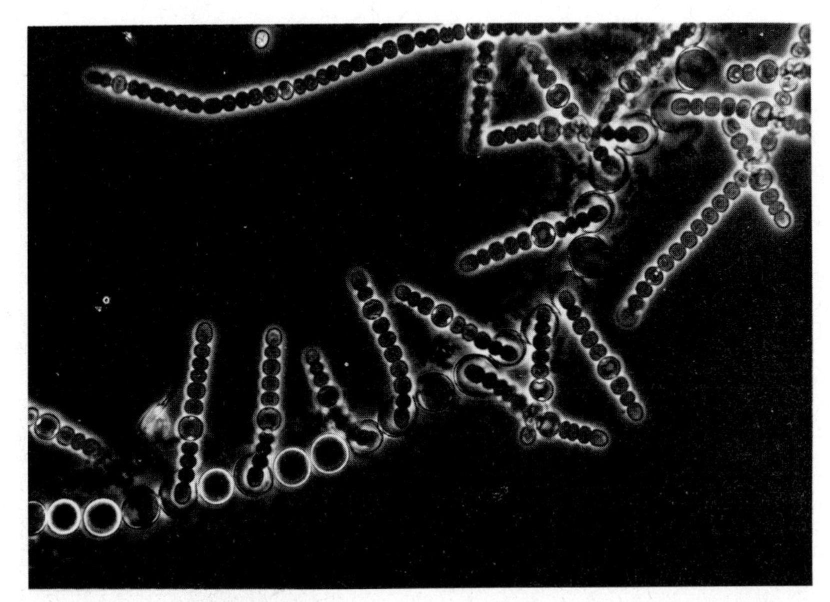

Fig. 32. Morphological differentiation in *Anabaena* sp. CA. Note the string of akinetes (lower left) germinating to produce short vegetative filaments with intercalary heterocysts. Photography by Dr D. G. Adams.

series of akinetes, some of which have germinated. Germination results in a short filament of vegetative cells containing intercalary heterocysts. An older emerged filament (upper left) shows the subsequent growth of such a filament, with the heterocysts already showing a regular pattern, the mid-interheterocyst cells beginning to differentiate into new heterocysts. The relatively large size, and spherical appearance, of akinetes of this species is evident.

Some attention has been directed to examining the extra wall layers that are clearly present in the akinetes of *Cylindrospermum* sp. and *Anabaena* species (Wildon & Mercer, 1963; Braune, 1980) but which appear to be absent in akinetes from *Aphanizomemon flos-aquae* (Wildman, Loescher & Winger, 1975). These differences exemplify the difficulty of deriving generalizations about cyanobacterial akinetes, not only with respect to structure, but also to function and to the extent that their metabolism is altered to facilitate that function. A problem common to the interpretation of most of the early studies was that the preparations of akinetes examined were heterogeneous with respect to the stage of development. Preparations contained akinetes in late stages of maturation, akinetes as such and akinetes in varying stages of germination.

Biochemical investigations on such mixed populations have considerably less value than can be obtained from electron microscopical studies on selected cells in them. In an examination of akinetes developed synchronously from *Nostoc* PCC7524 the cellular content of DNA, RNA, protein and phycocyanin were similar to those of vegetative cells, whilst chlorophyll and glycogen were greater (Sutherland, Herdman & Stewart, 1979). In contrast, Simon (1977a) estimated that whereas DNA content of akinetes from *Cylindrospermum* sp. was similar to that of vegetative cells, the RNA content was much greater, producing a surprisingly high RNA:DNA ratio of 54 compared with 13 for the vegetative cell.

Control of akinete development

In contrast to the heterocyst, the principle causative environmental factor producing akinetes is far from understood. The accounts in the literature (see Nichols & Adams, 1982) could indicate, in various species of cyanobacteria, that the deprivation of most major nutrients, and alteration in a range of environmental factors, leads to the induction of akinetes. It is quite possible that more than one of these apparently conflicting causative agents may be operative among the range of cyanobacteria that form akinetes. What does appear, at least as a generalization, is that akinete formation is environmentally produced, in that an environmental factor always limits the growth of a culture and akinete formation is associated with the ending of exponential growth. The potentially interesting experiments seeking to determine whether akinetes are produced from cyanobacteria in a chemostat culture have not yet been reported.

The effect of growth at different light intensities on the development of akinetes has been examined in *Anabaena cylindrica* (Nichols, Adams & Carr, 1980) and in *Nostoc* PCC7524 (Sutherland, Herdman & Stewart, 1979). In both these sets of experiments akinetes began to form shortly after the cultures entered the non-exponential phase of growth due to light limitation (Fig. 33). Since these cultures were light-limited, the onset of akinete formation occurred at a lower culture density when the incident light intensity was reduced. Indeed there was an approximately linear relationship between culture density at onset of akinete formation and light intensity. An interpretation of these results would be that light-limitation produces akinete formation, rather than any limitation of a nutrient component of the growth medium. However, it

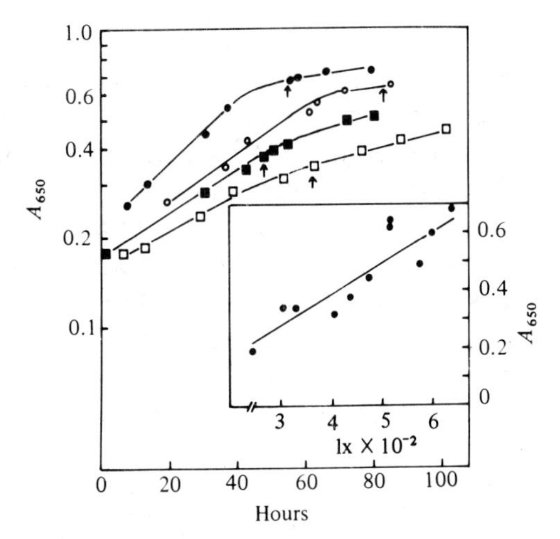

Fig. 33. Growth and time of onset of akinete formation (indicated by arrows) as a function of light intensity. Light intensities were 640 lx (●) 520 lx (○), 440 lx (■), and 305 lx (□). The inset figure shows the relationship between A_{650} at which akinetes were first evidence and light intensity at which the culture was grown.

can be seen that reduction of growth rate by light-limitation as such does not produce akinete formation – otherwise cultures grown as 305 lx (with a slower growth rate than those at 645 lx) would produce akinetes in the exponential phase of growth (Fig. 33). This point is particularly well illustrated in the experiments with *Nostoc* PCC7524 (Sutherland *et al.*, 1979). It may be that akinete formation is produced as a consequence of the reduction in division, which is itself a product of light limitation. It would be valuable to examine carefully the relationship of akinete formation to growth kinetics using a cyanobacterium that produces akinetes in response to a nutrient, say phosphate, limitation. It has been observed by many workers, and described in detail in *Anabaena cylindrica* by Simon (1977a) that akinete formation is preceded by a perturbation of the division pattern of the vegetative cell that is to become an akinete – leading to a significant increase in cell size.

The excretion into the culture filtrate of a molecule ($C_7H_5O_5N$, empirical formula) by an old, akinete-containing culture of *Cylindrospermum licheniforme* that induces akinetes in a young culture of the same organism has been reported (Fischer & Wolk, 1976). The tentative chemical structure is of two fused five-membered rings of a lactam and thioketone nature (Hirosawa & Wolk, 1979). Presumably the organism derives a selective advantage by being able to induce a larger proportion of its vegetative cells to form akinetes by virtue of this chemical messenger.

Germination of akinetes

The morphological form of the filament resulting from akinete germination varies considerably between species, principally with regard to how many cell divisions have occurred prior to emergence. An undivided cell is extracted from a germinating akinete of *Anabaena variabilis* by rupture of one pole, leaving an empty shell of akinete wall structure (Braune, 1980). In *Cylindrospermum* sp. one or two divisions were clearly visible before germination (Miller & Lang, 1968) and in *Anabaena* sp. CA several divisions were necessary to yield a filament with an intercalary heterocyst, leading to the release of an already differentiated filament of cells (See Nichols & Adams, 1982; Fig. 32). The environmental factor producing germination for which there is most evidence is light. *Nostoc* PCC 7524 akinetes germinated following an increase in light intensity and without any introduction of fresh culture media (Sutherland *et al.*, 1979). Yamamoto (1976) showed that the light-induced germination of akinetes from *Anabaena cylindrica* was inhibited by DCMU indicating an involvement of photosynthesis. However, Kaushik & Kumar (1970) observed germination of akinetes of *Anabaena* and *Fischerella mucicola* by green light of a spectral range that did not support photosynthetic growth. Braune (1979) constructed an action spectra of germination response, and concluded that phycocyanin was the most effective photoreceptor for akinete germination. These laboratory experiments are in accord with several observations on natural populations in which an increase in available light has been implicated as the factor producing akinete germination (see Nichols & Adams, 1982). There is a report of akinetes of a natural population of *Aphanizomenon flos-aquae* which, under laboratory conditions, still required the six month period to germinate that is operative in the over wintering of field populations (Wildman *et al.*, 1975). Confirmation of this interesting observation would indicate the possibility of specific lag period being built into stages of differentiation, and, obviously, the possibility of a precursor cell stage playing a holding role.

CONCLUSION

We conclude that the concept of the growth precursor cell is valid, that it probably applies to a range of bacterial species – indeed the growth precursor cell may be a part of all bacterial cell cycles – and that not only does it act as a survival cell but as a 'platform' from which differentiation into one of a variety of cell types may proceed depending upon as yet unspecified environmental triggers.

This development of the idea of the growth precursor cell also emphasizes the value of studying unusual microbes with accentuated morphological properties, e.g. the budding and the stalked bacteria and the cyanobacteria. Properties which at first sight appear to be characteristic of only a few bacteria, e.g. polar growth and condensed genomes, prove to have their counterparts in the more widely studied bacteria, e.g. *E. coli.* More importantly, such studies serve to highlight the importance of reviewing microbial behaviour under conditions more attuned to those of their existence in the wild, rather than under the conditions chosen to suit convenience in the laboratory.

Obviously it is too early to take a dogmatic stance about the nature and biological properties of growth precursor cells; much remains to be investigated concerning their molecular biology and their biochemistry under conditions which appear to restrict them to the precursor state. There may also be categories of growth precursor cells; some species of cyanobacteria, for instance, may give rise to two more, each serving as a platform for differentiation and development dictated by a particular set of environmental conditions.

Acknowledgement

We wish to thank U. K. Swoboda, A. D. France, N. W. Scott, D. G. Adams, A. Lawrence, L. Potts, D. Porter and C. Oakley for their contribution to the work and ideas incorporated in this article.

REFERENCES

AGABIAN, N., EVINGER, M. & PARKER, G. (1979) Generation of asymmetry during development: Segregation of type-specific proteins in Caulobacter. *Journal of Cell Biology*, **81,** 123–36.

ADAMS, D. G. & CARR, N. G. (1981). The developmental biology of heterocyst and akinete formation in cyanobacteria. *CRC Critical Reviews in Microbiology*, **9,** 45–100.

AMEMIYA, K., WU, C. W. & SHAPIRO, L. (1977). *Caulobacter crescentus* RNA polymerase: Purification and characterization of holoenzyme and core polymerase. *Journal of Biological Chemistry*, **252,** 4157–65.

BENDIS, I. K. & SHAPIRO, L. (1973). Deoxyribonucleic acid-dependent ribonucleic acid polymerase of *Caulobacter crescentus. Journal of Bacteriology*, **115,** 848–57.

BOHIN, J. P., BEN KHALIFA, K., GUILLEN, N., SCHAEFFER, P. & HIRSCHBEIN, L. (1982). Phenotypic expression *in vivo* and transforming activity *in vitro*: Two related functions of folded bacterial chromosomes. *Molecular and General Genetics*, **185,** 65–8.

BOYLEN, C. W. (1973). Survival of *Arthrobacter crystallopoietes* during prolonged periods of extreme desiccation. *Journal of Bacteriology*, **113,** 33–7.

BRADLEY, S. & CARR, N. G. (1977). Heterocyst development in *Anabaena cylindrica*. The necessity for light as an initial trigger and sequential stages of commitment. *Journal of General Microbiology*, **101,** 291–7.

BRAUNE, W. (1979). C-phycocyanin – the main photoreceptor in the light dependent germination process of *Anabaena* akinetes. *Archives of Microbiology*, **122,** 289–95.

BRAUNE, W. (1980). Structural aspects of akinete germination in the cyanobacterium *Anabaena variabilis*. *Archives of Microbiology*, **126,** 257–61.

BURGESS, R. R. (1969). Separation and characterization of the subunits of ribonucleic acid polymerase. *Journal of Biological Chemistry*, **244,** 6168–76.

CARR, N. G. (1979). Differentiation in filamentous cyanobacteria. In *Developmental Biology of Prokaryotes*, ed. J. H. Parish, pp. 167–201. Oxford: Blackwells Scientific Publications.

CHEUNG, K. K. & NEWTON, A. (1977). Patterns of protein synthesis during development in *Caulobacter crescentus*. *Developmental Biology*, **56,** 417–25.

CHEUNG, K. K. AND NEWTON, A. (1978). Polyadenylic acid synthesis activity of purified DNA-dependent RNA polymerase from Caulobacter. *Journal of Biological Chemistry*, **253,** 2254–61.

DAVIS, R. W., SIMON, M. & DAVIDSON, N. (1971). Electron microscope heteroduplex methods for mapping regions in base sequence homology in nucleic acids. *Methods in Enzymology*, **21,** 413–28.

DEGNEN, S. T. & NEWTON, A. (1972). Chromosome replication during development in *Caulobacter crescentus*. *Journal of Molecular Biology*, **64,** 671–80.

DOI, R. H. (1977). Role of ribonucleic acid polymerase in gene selection in procaryotes. *Bacteriological Reviews*, **41,** 568–94.

DONACHIE, W. D. & BEGG, K. J. (1970). Growth of the bacterial cell. *Nature*, **277,** 1220–4.

DOW, C. S. & FRANCE, A. D. (1980). Simplified vegetative cell cycle of *Rhodomicrobium vannielii*. *Journal of General Microbiology*, **117,** 47–55.

DOW, C. S. & LAWRENCE, A. (1980). Microbial growth and survival in oligotrophic freshwater environments. In *Microbial Growth and Survival in Extremes of Environment*, Society for Applied Bacteriology Technical Series 15, ed G. W. Gould & J. E. L. Carry, pp. 1–19. Academic Press.

DOW, C. S. & WHITTENBURY, R. (1980). Prokaryotic form and function. In *Contemporary Microbial Ecology*, ed. D. C. Ellwood, J. N. Hedger, M. J. Latham, J. M. Lynch & J. H. Slater, pp. 391–417. Academic Press.

DRLICA, K. & WORCEL, A. (1975). Conformational transitions in *Escherichia coli* chromosome: Analysis by viscometry and sedimentation. *Journal of Molecular Biology*, **98,** 393.

ENSIGN, J. C. (1970). Long-term starvation survival of rod and spherical cells of *Arthrobacter crystallopoietes*. *Journal of Bacteriology*, **103,** 569–77.

ENSIGN, J. C. & WOLFE, R. S. (1964). Nutritional control of morphogenesis in *Arthrobacter crystallopoietes*. *Journal of Bacteriology*, **87,** 924–32.

EVANS, E. H., FOULDS, I. & CARR, N. G. (1976). Environmental conditions and morphological variation in the blue-green alga *Chlorogloea fritschii*. *Journal of General Microbiology*, **92,** 147–55.

EVINGER, M. & AGABIAN, N. (1977). Envelope-associated nucleoid from *Caulobacter crescentus* stalked and swarmer cells. *Journal of Bacteriology*, **132,** 294–301.

EVINGER, M. & AGABIAN, N. (1979). *Caulobacter crescentus* nucleoid: Analysis of sedimentation behaviour and protein composition during the cell cycle. *Proceedings of the National Academy of Sciences, USA*, **76,** 175–8.

FAY, P., STEWARD, W. D. P., WALSBY, A. E. & FOGG, G. E. (1968). Is the heterocyst the site of nitrogen fixation in blue-green algae? *Nature, London*, **220,** 810–12.

FISCHER, R. W. & WOLK, C. P. (1976). Substance stimulating the differentiation of spores of the blue-green alga *Cylindrospermum lichenforme. Nature, London*, **259,** 394–5.

FLEMING, H. & HASELKORN, R. (1974). The program of protein synthesis during heterocyst differentiation in nitrogen-fixing blue-green algae. *Cell*, **3,** 159–70.

FOGG, G. E. (1949). Growth and heterocyst production in *Anabaena cylindrica* Lemm. 11. in relation to carbon and nitrogen metabolism. *Annals of Botany, NS*, **15,** 241–59.

FOULDS, I. J. & CARR, N. G. (1981). Unequal cell division preceding heterocyst development in *Chlorogloeopsis fritschii. FEMS Microbiology Letters*, **10,** 223–6.

GILLERT, M., O'DEA, M. H., ITOH, T. & TOMIZAWA, J. (1976). Novobiocin and coumermycin inhibit DNA supercoiling catalysed by DNA gyrase. *Proceedings of the National Academy of Sciences, USA*, **73,** 4474–8.

GROSS, G., ENGBAEK, F., FLAMMANG, T. & BURGESS, R. (1976). Rapid micromethod for the purification of *Escherichia coli* Ribonucleic acid polymerase and the preparation of bacterial extracts active in ribonucleic acid synthesis. *Journal of Bacteriology*, **128,** 382–9.

GUILLEN, N. (1979). Thèse de Doctorat d'Etat ès-Sciences Naturelles. Université de Paris Sud, France.

GUILLEN, N., LE HEGARAT, F., FLEURY, A. M. & HIRSCHBEIN, L. (1978). Folded chromosomes of vegetative *Bacillus subtilis:* Composition and properties. *Nucleic Acids Research*, **5,** 475–89.

HARDER, W. & ATTWOOD, M. M. (1978). Biology, physiology and biochemistry of Hyphomicrobia. *Advances in Microbial Physiology*, **17,** 303–56.

HASELKORN, R., CURTIS, S. E., FISHER, R. M., MAZUR, B. J., MEVARECH, M., RICE, D., NAGARAJU, R., ROBINSON, S. J. & TULI, R. (1982). Cloning and physical characterization of Anabaena genes that code for important functions in heterocyst differentiation: nitrogenase, gluamine synthetase and RuBP carboxylase. In *Cyanobacteria: Cell Differentiation and Function*, ed. G. G. Papageorigiou & L. Packer. Elsevier.

HELMSTETTER, C., COOPER, S., PIERUCCI, O. & REVELAS, E. (1968). On the bacterial life sequence. *Cold Spring Harbor Symposium Quantitative Biology*, **33,** 809–22.

HIROSAWA, T. & WOLK, C. P. (1979). Isolation and characterisation of a substance which stimulates the formation of akinetes in the cyanobacterium *Cylindrospermum licheniforme. Journal of General Microbiology,* **114,** 433–41.

HIRSCH, P. (1979). Life under conditions of low nutrient concentrations. In *Strategies of Microbial Life in Extreme Environments*, ed. M. Shilo, pp. 357–72. Dahlem Conference Life Sciences Research Report 13, Berlin.

HIRSCHBEIN, L. & GUILLEN, N. (1981). Characterization, assay and use of isolated bacterial nucleoids. In *Methods of Biochemical Analysis*, 28.

IBA, H., FUKUDA, A. & OKADA, Y. (1975). Synchronous cell differentiation in *Caulobacter crescentus. Japanese Journal of Microbiology*, **19,** 441–6.

IBA, H., FUKUDA, A. & OKADA, Y. (1977). Chromosome replication in *Caulobacter crescentus* growing in nutrient broth. *Journal of Bacteriology*, **129,** 1192–7.

IBA, H., FUKUDA, A. & OKADA, Y. (1978). Rate of major protein synthesis during the cell cycle of *Caulobacter crescentus. Journal of Bacteriology*, **135,** 647–55.

JEEJI-BAI, N. (1977). Morphological variation of some species of *Calothrix* and *Fortien. Archiv für Protistenkunde*, **119,** 367–87.
KAUSHIK, M. & KUMAR, H. D. (1970). The effect of light on growth and development of two nitrogen-fixing blue-green algae. *Archiv für Mikrobiologie*, **74,** 52–7.
KIRBY, K. S. (1965). Isolation and characterization of ribosomal ribonucleic acid. *Biochemical Journal*, **96,** 266–9.
KOCH, A. L. (1971). The adaptive response of *Escherichia coli* to a feast and famine existence. *Advances in Microbial Physiology*, **6,** 147–217.
KORNBERG, T., LOCKWOOD, A. & WORCEL, A. (1974). Replication of *Escherichia coli* chromosome with a soluble enzyme system. *Proceedings of the National Academy of Sciences, USA*, **71,** 3189–93.
KRULWICH, T. A. & ENSIGN, J. C. (1969). Alteration of glucose metabolism of *Arthrobacter crystallopoietes* by compounds which induce sphere to rod morphogenesis. *Journal of Bacteriology*, **97,** 526–34.
LAEMMLI, U. K. (1970). Cleavage of structural proteins during the assembly of the head of bacteriophage T_4. *Nature*, **227,** 680–5.
LARSON, R. J. & PATE, P. L. (1976). Glucose transport in isolated prosthecae of *Asticcacanlis biprosthecum. Journal of Bacteriology*, **126,** 282–93.
LAWRENCE, A. (1978). Microbial diversity – a consequence of the aquatic environment. Ph.D. thesis, University of Warwick.
LAZAROFF, N. (1972). Experimental control of nostocacean development. In *Taxonomy and Biology of Blue-Green Algae*, ed. T. V. Desikachary, pp. 521–44. Madras: University of Madras.
LAZAROFF, N. & SCHIFF, J. A. (1962). Action spectrum for development photo-induction of the blue-green alga *Nostoc muscorum. Science, NY*, **137,** 603–4.
LAZAROFF, N. & VISHNIAC, W. (1961). The effect of light on the developmental cycle of *Nostoc muscorum*, a filamentous blue-green alga. *Journal of General Microbiology*, **25,** 365–74.
LEIFSON, E. (1964). *Hyphomicrobium neptunium* sp. n. *Antonie van Leeuwenhoek Journal of Microbiology and Serology*, **30,** 249–56.
LICARI, G. R. & CLOUD, P. (1972). Prokaryotic algae associated with Australian proterozoic stromatolites. *Proceedings of the National Academy of Sciences, USA*, **69,** 2500–4.
LUSCOMBE, B. M. & GRAY, T. R. G. (1974). Characteristics of arthrobacter grown in continuous culture. *Journal of General Microbiology*, **82,** 213–22.
MCGHEE, J. D. & FELSENFELD, G. (1980). Nucleosome structure. *Annual Review of Biochemistry*, **49,** 1115–56.
MARSHALL, K. C. (1980). Bacterial adhesion in natural environments. In *Microbial Adhesion to Surfaces*, ed. R. C. W. Berkeley, J. M. Lynch, J. Melling, P. R. Rutter & B. Vincent, pp. 187–93. Ellis Harwood Ltd.
MATIN, A. & VELDKAMP, H. (1978). Physiological basis of the selective advantage of Spirillum sp. in carbon-limited environment. *Journal of General Microbiology*, **105,** 187–97.
MILLER, M. M. & LANG, N. J. (1968). The fine structure of akinete formation and germination in *Cylindrospermum. Archiv für Mikrobiologie*, **60,** 303–13.
MITCHISON, G. J. & WILCOX, M. (1972). Rule governing cell division in *Anabaena. Nature, London*, **239,** 110–11.
MIZUUCHI, K., GELLERT, M. & NASH, H. (1978). Involvement of supertwisted DNA in integrative recombination of bacteriophage lambda. *Journal of Molecular Biology*, **121,** 375–92.
MOORE, R. L. & HIRSCH, P. (1973). First generation synchrony of isolated Hyphomicrobium swarmer populations. *Journal of Bacteriology*, **116,** 418–23.

MULDER, E. G. (1963). Arthrobacter. In *Principles and Applications in Aquatic Microbiology*, ed. H. Heukelekian & N. C. Dondero, pp. 254–79. John Wiley and Son, New York.

MULHAUSEN, M. & AGABIAN, N. (1981). Regulation of polypeptide synthesis during Caulobacter development: two-dimensional gel analysus. *Journal of Bacteriology*, **148,** 163–73.

NAKAZATO, H., VENKATESAN, S. & EDMONDS, M. (1975). Polyadenylic acid sequences in *E. coli* messenger RNA. *Nature*, **256,** 144–6.

NEWTON, A. (1972). Role of transcription in the temporal control of development in *Caulobacter crescentus. Proceedings of the National Academy of Sciences, USA*, **69,** 447–51.

NICHOLS, J. M., ADAMS, D. G. & CARR, N. G. (1980). Effect of canavanine and other amino acid analogues on akinete formation in the cyanobacterium *Anabaena cylindrica. Archives of Microbiology*, **127,** 67–75.

NICHOLS, J. M. & ADAMS, D. G. (1982). Akinetes. In *The Biology of Cyanobacteria*, ed. N. G. Carr & B. A. Whitton, pp. 387–412. Oxford: Blackwells Scientific Publications.

NICHOLS, J. M. & CARR, N. G. (1978). Akinetes of Cyanobacteria. In *Spores VII*, ed. C. H. Chambliss & J. C. Vary, pp. 335–43. Washington, DC: American Society for Microbiology.

O'FARRELL, P. H. (1975). High resolution two-dimensional electrophoresis of proteins. *Journal of Biological Chemistry*, **250,** 4007–21.

OHTA, N., SANDERS, M. & NEWTON, A. (1975). Poly (Adenylic Acid) Sequences in the RNA of *Caulobacter crescentus. Proceedings of the National Academy of Sciences, USA*, **72,** 2343–6.

OHTA, N., SANDERS, M. & NEWTON, A. (1978). Characterization of unstable poly(A)-RNA in *Caulobacter crescentus. Biochimica et Biophysica Acta*, **517,** 65–75.

OSLEY, M. A., SHEFFERY, M. & NEWTON, A. (1977). Regulation of flagellum synthesis in the cell cycle of Caulobacter: Dependence on DNA replication. *Cell*, **12,** 393–400.

PATE, J. L. & ORDAL, E. J. (1965). The fine structure of two unusual stalked bacteria. *Journal of Cell Biology*, **27,** 133–50.

PETTIJOHN, D. E. (1976). Procaryotic DNA in nucleoid structure. *CRC Critical Reviews in Biochemistry*, **4,** 175–202.

PETTIJOHN, D. & HECHT, R. (1973). RNA molecules bounded to the folded bacterial genome stabilize DNA folds and segregate domains of supercoiling. *Cold Spring Harbour Symposium Quantitative Biology*, **38,** 31–35.

PIRT, S. J. (1972). Prospects and problems in continuous flow culture of micro-oganisms. *Journal of Applied Chemical Biotechnology*, **22,** 55–64.

POINDEXTER, J. S. (1964). Biological properties and classification of the Caulobacter group. *Bacterial Reviews*, **28,** 231–95.

POINDEXTER, J. S. (1981a). The Caulobacters: Ubiquitous unusual bacteria. *Microbiological Reviews*, **45,** 123–79.

POINDEXTER, J. S. (1981b). Oligotrophy. Feast and famine existence. *Advances in Microbial Ecology*, **5,** 63–89.

POTTS, L. & DOW, C. S. (1979). Nucleic acid synthesis during the developmental cycle of the *Rhodomicrobium vannielii* swarm cell. FEMS Microbiology Letters, **6,** 393–5.

RIPPKA, R. & STANIER, R. Y. (1978). The effects of anaerobiosis on nitrogenase synthesis and heterocyst development by nostocacean cyanobacteria. *Journal of General Microbiology*, **105,** 83–94.

ROBINSON, B. L. & MILLER, J. H. (1970). Photomorphogenesis in the blue-green alga *Nostoc commune* 584. *Physiologia Plantarum*, **23,** 461–72.

Rouvière-Yaniv, J. (1978). Localization of the HU protein on the *Escherichia coli* nucleoid. *Cold Spring Harbour Symposium on Quantitative Biology*, **42,** 439–47.

Rouvière-Yaniv, J. & Gros, F. (1975). Characterization of a novel, low-molecular weight DNA binding protein from *Escherichia coli. Proceedings of the National Academy of Sciences, USA*, **72,** 3428–32.

Rouvière-Yaniv, J. & Yaniv, M. (1979). *E. coli* binding protein HU forms nucleosome-like structure with circular double-stranded DNA. *Cell*, **17,** 265–74.

Shapiro, L. (1976). Differentiation in the Caulobacter cell cycle. *Annual Review of Microbiology*, **30,** 377–407.

Shapiro, L. & Maizel, J. V. (1973). Synthesis and structure of *Caulobacter crescentus* flagella. *Journal of Bacteriology*, **113,** 478–85.

Sheffery, H. & Newton, A. (1981). Regulation of periodic protein synthesis in the cell cycle: control of initiation and termination of flagellar gene expression. *Cell*, **24,** 49–57.

Simon, R. D. (1977a). Macromolecular composition of spores from the filamentous cyanobacterium *Anabaena cylindrica. Journal of Bacteriology*, **129,** 1154–5.

Simon, R. D. (1977b). Sporulation in the filamentous cyanobacterium *Anabaena cylindrica.* The course of spore formation. *Archives of Microbiology*, **111,** 283–8.

Sinclair, C. & Whitton, B. A. (1977). Influence of nutrient deficiency on hair formation in the Rivulariaceae. *British Phycological Journal*, **12,** 287–313.

Staley, J. T. (1968). Prosthecomicrobium and Ancalomicrobium: New prosthecate freshwater bacteria. *Journal of Bacteriology*, **95,** 1921–42.

Stonington, O. G. & Pettijohn, D. E. (1971). The folded genome of *Escherichia coli* isolated in a protein-DNA-RNA complex. *Proceedings of the National Academy of Sciences, USA*, **68,** 6–9.

Sutherland, J. M., Herdman, M. & Stewart, W. D. P. (1979). Akinetes of the cyanobacterium Nostoc PCC 7524: macromolecular composition, structure and control of differentiation. *Journal of General Microbiology*, **115,** 273–87.

Swoboda, U. K. (1979). Studies on the growth and development of Caulobacter. Ph.D. thesis, Open University.

Swoboda, U. K., Dow, C. S. & Vitokovic, L. (1982a). *In vitro* transcription and translation directed by *Caulobacter crescentus* CB15 nucleoids. *Journal of General Microbiology*, **128,** 291–301.

Swoboda, U. K., Dow, C. S. and Vitkovic, L. (1982b). Nucleoids of *Caulobacter crescentus* CB15. *Journal of General Microbiology*, **128,** 279–89.

Tyler, P. A. (1970). Hypomicrobium and the oxidation of manganese in aquatic ecosystems. *Antonie van Leeuwenhoek*, **36,** 567–78.

Tyler, P. A. & Marshall, K. C. (1967). Pleomorphy in stalked budding bacteria. *Journal of Bacteriology*, **93,** 1132–6.

Van Liere, L. & Walsby, A. E. (1982). Interactions of cyanobacteria with light. In *The Biology of Cyanobacteria*, ed N. G. Carr & B. A. Whitton. Oxford: Blackwells Scientific Publications.

Wali, T. M., Hudson, G. R., Danald, D. A. & Weiner, R. M. (1980). Timing of swarmer cell cycle morphogenesis and macromolecular synthesis by *Hyphomicrobium neptunium* in synchronous culture. *Journal of Bacteriology*, **144,** 406–12.

Waterbury, J. B. (1979). Developmental patterns of pleurocapsalean cyanobacteria. In *Developmental Biology of Prokaryotes*, ed. J. H. Parish, pp. 203–26. Oxford: Blackwells Scientific Publications.

Waterbury, J. B. & Stanier, R. Y. (1977). Two unicellular cyanobacteria which reproduce by budding. *Archives of Microbiology*, **115,** 249–57.

Waterbury, J. B. & Stanier, R. Y. (1978). Pattern of growth and development in pleurocapsalean cyanobacteria. *Microbiology Reviews*, **42,** 2–44.

WHITTENBURY, R. & DOW, C. S. (1977). Morphogenesis and differentiation in *Rhodomicrobium vannielii* and other budding and prosthecate bacteria. *Bacteriological Reviews*, **41,** 754–808.

WILDMAN, R. B., LOESCHER, J. H. & WINGER, C. L. (1975). Development and germination of akinetes of *Aphanizomenon flos-aquae. Journal of Phycology*, **11,** 96–104.

WILDON, D. C. & MERCER, F. V. (1963). The ultrastructure of the heterocyst and akinete of the blue-green algae. *Archives für Mikrobiologie*, **47,** 19–31.

WOLK, C. P. (1965). Control of sporulation in a blue-green algae. *Development Biology*, **12,** 15–35.

WOLK, C. P. (1982). Heterocysts. In *The Biology of Cyanobacteria*, ed. N. G. Carr & B. A. Whitton, pp. 359–86. Oxford: Scientific Publications.

WORCEL, A. & BURGI, E. (1972). On the structure of the folded chromosome of *Escherichia coli. Journal of Molecular Biology*, **71,** 127–47.

WRAY, W., DOULIKAS, T., WRAY, V. P. & HANCOCK, R. (1981). Silver staining of proteins in polyacrylamide gels. *Analytical Biochemistry*, **118,** 197–203.

YAMAMOTO, Y. (1976). Effect of some physical and chemical factors on the germination of akinetes of *Anabaena cylindrica. Journal of General and Applied Microbiology, Tokyo*, **22,** 311–23.

ZAVARZIN, G. A (1960). The life cycle and nuclear apparatus in *Hyphomicrobium vulgare Stutzer* and *Hartleb. Microbiology USSR*, **29,** 38–42.

EXTRACELLULAR ENZYME–SUBSTRATE INTERACTIONS IN SOIL

RICHARD G. BURNS

Biological Laboratory, University of Kent, Canterbury, Kent CT2 7NJ, UK

INTRODUCTION

Microbiologists with an interest in the ecology of terrestrial environments are broadly divided into two camps: that which believes that informative studies must be conducted *in situ*; and that which supports the notion that carefully controlled experiments in the laboratory are the most revealing. In this author's opinion allegiance to the first philosophy is akin to studying the ecology of (say) a tropical rain-forest from 5 000 m up in an aeroplane. Whilst it is reasonable to suppose that, from his remote perch, the observer may discover a great deal about such events as the diurnal and seasonal patterns of plant and animal life, he will learn little about the properties and inter-relationships of individual species and the factors which determine their role in the forest ecosystem. The experienced ecologist knows that to expand his knowledge of the forest community he must examine its components at close quarters. By analogy, an investigation into the microbial ecology of soil requires not merely microbial counts or an assessment of respiration rates but also the observation of individual cells and communities of cells and their metabolic characteristics together with measurements of the physical and chemical properties of the microenvironment in which they reside. Unfortunately few techniques exist to enable the detailed study of soil environments in the field, and thus collection and transference of a sample to the laboratory is a requirement rather than a mere convenience. In other words, the soil microbiologist's justifiable concern about the disruption caused to the environment he is interested in should be tempered by the sheer necessity of removing that environment to the laboratory in order to study it. This apparent paradox in experimental microbial ecology (i.e. destructive sampling of a habitat followed by dissection and study *in vitro*) should not be alarming because scientific discovery and the gradual development of an holistic view of a subject often proceeds from an analytical phase (in which the variables are carefully controlled), through a synthetic phase (in which individual observations are combined to provide a more realistic picture), to a study of complex systems.

Although it has been hailed as a great revelation by some, it should come as no surprise that microorganisms, like plants and animals, are components of highly integrated communities. Bacteria, fungi, protozoa, algae and viruses in soil do not carry out their degradative and synthetic functions in isolation but rely upon a plethora of inter-relationships with other microbial species (as well as with plant roots and the soil fauna). Moreover, the physical and chemical properties of soil have a profound influence on microbial activity; an influence that is not apparent from studies in the absence of soil. As a result of its community structure, the soil microbial population possesses a remarkable spectrum of metabolic properties and is resilient to all sorts of natural and anthropogenic perturbations.

In the long run the success or failure of individual species within the soil community will be determined by the availability of suitable substrates and the capacity of the microorganism to respond to those substrates. This chapter is concerned with the preparation of exogenous compounds for microbial utilization and the difficulties experienced by soil microorganisms in detecting and transforming these potential substrates.

COMPONENTS AND PROPERTIES OF THE SOIL ENVIRONMENT

Introduction

The subject of soil microbial ecology is in its analytical phase and thus the destructive investigation of the soil environment is necessary in order to discover the fundamental biological, physical and chemical properties of the mm^3 or μm^3 volume microsites in which microorganisms reside. The concept of the microbial microenvironment of soil and its experimental complexity have been discussed by many authors in recent years (e.g. Stotzky, 1974; Hattori & Hattori, 1976; Burns, 1980; Stotzky & Burns, 1982) and one general feature is apparent, namely that microbial activity predominates on, or in close proximity to, the surfaces of soil particulates. This phenomenon can be demonstrated by adsorption experiments with microorganisms, substrates and soil components, and confirmed by observations of microorganisms *in situ* using transmission and scanning electron microscopy.

The most influential soil particles, as far as microbial activity is concerned, are the colloidal-size clays and humic materials. This is due to three important physical properties possessed by soil colloids: (1) they have high surface to volume ratios and, when separated from whole soil, tend to have a significant proportion of the total microflora associated with them; (2) they have ionic properties such that, as a general rule, cations are attracted to surfaces and anions are repulsed; and (3) they have a high affinity for water molecules. Consequently, microorganisms, substrates, enzymes, metabolites and inorganic nutrients tend to be concentrated at clay/water and humic/water interfaces and, correspondingly, depleted in zones distant from those interfaces. The other principle soil particulates, sand and silt, have low unit surface areas, do not accumulate organic matter to any great extent and do not retain water and inorganic ions. The coarser soil fractions do, however, promote gas and solvent diffusion and therefore have an important role in the biology of soil even if they only rarely attract and support an abundant microbiota.

Thus it can be seen that the heterogeneous environment of the soil is very different from that to which the microorganism is exposed in traditional laboratory investigations. In soil a biochemically and physiologically diverse community of microorganisms is competing for fluctuating levels of unevenly distributed and frequently insoluble nutrients and is subject to fluctuations in pH, pO_2, ionic concentration and relative humidity. In contrast, the study of microorganisms *in vitro* has usually involved uniformly dispersed axenic cultures grown in conditions of excess soluble substrate and at optimum pH values, temperatures and levels of aeration.

Structure and properties of soil colloids

The principal colloid-size particles in soil are: clay minerals; oxides and hydroxides of aluminium, iron, manganese and titanium; humic substances; polysaccharides; viruses; and bacterial cells.

The clay minerals are primarily crystalline aluminosilicates composed of silicon oxide tetrahedra and aluminium hydroxide octahedra sheets which are associated either in a 1:1 ratio (i.e.

Si-Al, Si-Al etc.) or a 2:1 ratio (i.e. Si-Al-Si, Si-Al-Si etc.). These combinations are known as layers and are bound together firmly by hydrogen bonds (kaolinite-serpentine group) or are only weakly associated through van der Waals' forces (smectite-vermiculite group). Intermediate types of layer silicate minerals are found as well as amorphous forms. As a consequence of the two types of bonds, 2:1 clays, especially smectites (e.g. montmorillonite), expand upon wetting exposing an internal surface area between adjacent silicon tetrahedral sheets; 1:1 layer clays, with the exception of hallosite, do not expand upon wetting.

Expanding clays in particular are subject to isomorphous substitution during weathering: a process by which structural cations (i.e. Si^{4+}, Al^{3+}) are replaced by those of a lower valency (e.g. Al^{3+}, Mg^{2+}, Fe^{2+}) and resulting in the negative charge deficit typical of layer silicates. This electronegativity can be quantified as the cation exchange capacity (CEC) of the clay because it is compensated by the adsorption of cations (e.g. Ca^{2+}, Mg^{2+}, K^+, Na^+, NH_4^+, H^+) from the soil solution. Anions may be negatively adsorbed, that is repulsed, in the vicinity of the clay surface although some clays, such as allophane, have high anion exchange capacities. The charged surface and its associated ions are described as the diffuse double layer (DDL), a zone that will vary in thickness from 0.5 to 100 nm depending upon the species and concentration of ions in the ambient medium.

In addition to the concentration and dilution effects on inorganic nutrients, organic ions are also influenced by the presence of anionic soil colloids. Thus many of the cationic and basic carbon and nitrogen sources for heterotrophic microorganisms (and substrates for extracellular enzymes) are associated with surfaces whilst acidic and non-ionic forms may be repulsed.

A further important consequence of the soil's cation exchange properties is that hydrogen ions will compete very effectively for adsorption sites within the DDL. This means that during periods of active microbial metabolism excreted and adsorbed hydrogen ions may give rise to interfacial pH values (pH_s) much lower than those encountered in the bulk phase (pH_b). As a result microenvironment variations in pH values (i.e. $\Delta pH = pH_b - pH_s$) may be as high as three pH units (McLaren & Skujins, 1968). In contrast, *Streptomyces* species growing in association with fungal hyphae were favoured by a localized pH *increase* due to ammonia production from chitin degradation (Williams & Mayfield, 1971). In biologically active soils it is probable that the pH of the microenvironment is constantly fluctuating and therefore an organism partially immersed in the DDL of a clay or humic colloid may have a wide and variable pH gradient from one end of the cell to the other. Full descriptions of the structure of soil clays can be found elsewhere (Gieseking, 1975; Arnold, 1978; Brown, Newman, Rayner & Weir, 1978).

Soil organic matter can be considered as being composed of three physically and chemically discernible fractions: (1) A physically recognizable component, containing recently incorporated animal, plant and microbial debris, in the early stages of disintegration and decay. (2) A biochemically recognizable, largely soluble and labile component (e.g. carbohydrates, peptides, organic acids) arising from the degradation of the material contained in fraction 1. (3) A brown-coloured polymeric component, largely phenolic but also containing polysaccharides and complexed and stable peptides, amino acids etc. The aromatic core structure is formed from the products of lignin breakdown but is also synthesized *de novo* by the soil microflora (and by chemical means) from non-aromatic as well as aromatic precursors. This is the humic fraction of soil and is traditionally differentiated into humic acids, fulvic acids and humins on the basis of its solubility in aqueous acids and bases. Humic acid is the major component of humates in fertile soil.

The third organic matter fraction ranks with the expanding lattice clays in terms of importance to biological activities in soil. This is because humic polymers have a tertiary structure which permits them to expand upon hydration exposing an extensive internal surface area that may have a significant influence on microbe/enzyme/substrate interactions. Humates are predominantly anionic with the principal functional groups being aliphatic and aromatic carboxyl and phenolic hydroxyl. The sign and magnitude of charge will vary according to the pH value of the soil solution and the isoelectric point of the constituent molecules of the humic colloid. Soil organic colloids and their aromatic and aliphatic moieties are discussed by Schnitzer & Khan (1972), Flaig, Beutelspacher & Rietz (1975) and Hayes & Swift (1978).

Soil organic fractions 1 and 2 are readily degradable in comparison to the most recalcitrant components of fraction 3. The most stable humic materials may have half-lives measured in centuries (Sørensen, 1975; Jenkinson & Rayner, 1977) and make a long-term contribution to the nutrient status and physical properties of soil. The mineralization of humus carbon and nitrogen (*ca* 2–10% per

Table 1. *Some properties of model clays and humic materials*

Colloid type	Layering	Swelling	Surface area[a] ($m^2\ g^{-1}$)	Cation exchange capacity ($\mu eq\ g^{-1}$)	Basal spacing[b] (nm)
Kaolinites	1:1	Non-expanding	10–50	20–100	0.72
Vermiculites	2:1	Expanding	500–750	1200–2500	0.93–1.57[c]
Smectites	2:1	Expanding	700–800	600–1300	0.95–2.2[c]
Humates	–	Expanding	500–800	2000–7500 (fulvic acids 5000–7500) (humic acids 3000–5000) (humins <3000)	–

[a]Determined by *N*-acetyl pyrimidinium bromide sorption.
[b]Distance between repeating layers.
[c]Variation due to level of hydration and species of interlayer cation.

annum) is believed to be compensated by new humus formation. Thus an equilibrium is established, especially in undisturbed aerobic soils.

Some of the properties of model clays and organic colloids are listed in Table 1. However, the direct application of this type of data to microbial activity in soil is uncertain because clays and humic materials (as well as other organic molecules, such as polysaccharides) are often intimately associated with each other and form 'organo-mineral complexes' (Theng, 1979). For example, humic acids may be bound to clays through divalent or polyvalent cations and, at acid pH values, low molecular weight fulvic acids may be adsorbed in the interlayer spaces of expanding lattice clays (Burchill, Hayes & Greenland, 1981; Hayes & Swift, 1981). Humins are also considered to be intimately associated with clay colloids.

Not surprisingly, organic polymers associated with soil minerals have a profound influence on the surface properties of the resulting organo-mineral complex, for instance, humic coatings may mask potentially reactive sites on a clay surface or prevent access to interlamellar areas. In addition, the C/N/S ratios of the organic matter, the relative proportions of humic and fulvic acids, and the stability of the organic colloids all vary according to the size of the inorganic particulates with which they are associated (Turchenek & Oades, 1979; Anderson, Saggar, Bettany & Stewart, 1981). Oxides and hydroxides of iron, aluminium and manganese participate in the binding of humic matter to clays (Gessa, Melis, Bellu & Testini, 1978) as well as interacting with the two components to reduce the density of their surface charges.

Bacteria, fungi and viruses have an overall negative charge in many soil environments because the pH is above the dissociation constant of the cell's exposed functional groups. The cell walls of Gram-positive bacteria contain teichoic acids towards their outer surface and these are anionic molecules whose sign and magnitude of charge is attributed to such factors as: phosphate ester groups, alanine ester residues, configuration of the teichoic acid within the wall and, of course, pH. The proteins contained in the outer membrane of Gram-negative bacteria confer on these organisms the possibility of bearing a positive or a negative charge depending upon the ionization of carboxyl or amino groups. Marshall (1976) has illustrated the pH-determined variability of charge shown by *Rhizobium* species. The components of fungal cell walls suggest a similar spectrum of pH-dependent ionogenic properties.

It can be seen from the foregoing that microbial cells, like the clays and humates, are likely to be charged colloids surrounded by a diffuse double layer in which organic and inorganic ions are concentrated. It is, however, important to remember the dynamic nature of growing microbial cells for at any particular moment the walls will contain some old material and some that has been deposited more recently. Thus it is reasonable to envisage the cell wall as having a non-uniform distribution of charge that varies with the age of the microorganism (Weiss, 1974). The cells of filamentous organisms, especially fungi, may have extremely variable properties according to their proximity to the growing tip (i.e. their age). Growth conditions will also affect the composition of the cell envelope (Braun, 1978). Humic colloids have changeable ionic properties because they are continually undergoing degradation, synthesis and polymerization. Clays are somewhat more predictable.

Soil water and soil atmosphere

Clay minerals and organic matter have a high affinity for water molecules and will retain moisture even in air-dried soils. For example, water held within the interlamellar space of expanding lattice clays may be up to four molecular layers thick and is only completely removed by oven-drying, whilst water is also adsorbed, although less tenaciously, on the surfaces of non-expanding layer silicates as well as to hydrous aluminium and iron oxides.

All forms of soil organic matter, ranging from humic matter to peat, contain a number of polar sites where water adsorption may take place. Carboxylic groups are common and at low relative humidities amide and hydroxyl groups may assume importance. Dehydration, however, may render organic matter somewhat resistant to subsequent rehydration. Incidentally, cycles of wetting and drying will solubilize potential substrates immobilized in humic polymers (Sørensen, 1974; Terry, 1980). Soil organic matter also contains a number of hydrophobic regions due to the presence of lipids and waxes, and even proteins and peptides which are hydrophilic overall contain hydrophobic moieties. Organic molecules adsorbed on or within mineral particles will doubtless affect the response of that mineral to water, although the relationship between organo-mineral complexes and water is incompletely understood. The nature of soil water has been discussed by Farmer (1978).

Needless to say the presence of soil water is crucial to microbial activity but this relationship may not be revealed by such measurements as soil water holding capacity or soil moisture content. Indeed the soil water tension (pF) is a more useful term as it indicates the actual availability of water to the microorganisms or to a reaction requiring water. High pF values indicate that water molecules are tightly held at the soil surface; low pF values mean that water is more accessible to microbes, plant roots and for extracellular hydrolytic reactions. Microorganisms vary widely in their tolerance of water stress and, as a general rule, fungi are more resilient than bacteria. Within the nitrogen cycle, nitrifiers are more susceptible to conditions of high pF than are some of the many ammonifying species, and ammonia may accumulate in soil locales with low levels of available water (Dommergues, 1977). The production of secondary metabolites, such as antibiotics, may be influenced by water tension (Wong & Griffin, 1974) possibly as a result of a shift to the post-exponential growth phase when water becomes a limiting factor.

The volume of water in soil and its strength of retention at soil surfaces affects the amount of dissolved oxygen. Oxygen is poorly soluble in water and environments can become oxygen-deficient very rapidly. Perhaps the most common stimulus for anaerobiosis in soils is the addition of readily degradable materials, such as crop residues. The pulse of substrate will stimulate aerobic heterotrophic microorganisms (and therefore oxygen consumption) and, because diffusion and dissolution of oxygen is not rapid enough to compensate, a shift from aerobic through facultative anaerobic to obligate anaerobic metabolism occurs (Terry & Tate, 1980). At the microenvironment level the easily degradable substrate may be a soluble root exudate or a single dead microbial cell. Thus, even a soil assessed as having an aerobic character overall will contain a large number of anaerobic microsites (Fluhler, Ardakani, Szuszkiewicz & Stolzy, 1976). Anaerobic microsites may also arise following organic matter decay because the reduction in bulk density leads to compaction.

The type and concentration of gases (other than oxygen) in the soil aqueous phase will vary according to microbial activity and carbon dioxide, hydrogen, methane, ethylene and nitrogen oxides will vary with time and show gradients over very short distances.

EXOGENOUS SUBSTRATES IN SOIL

Introduction

Vast quantities of plant, animal and microbial organic residues are deposited on or released into the soil each year. Estimates of annual accretion of plant debris (the principal source of soil organic matter) are difficult to calculate and vary enormously according to species and age of vegetation, soil type and climate. However, figures of between 0.5 and 15 tonnes ha^{-1} are reported in the literature (Williams & Gray, 1974; Stout, Goh & Rafter, 1981) and estimates of global deposition are of the order of 1.5×10^{12} tonnes of plant-carbon (Schlesinger, 1977). The highest proportion of the plant detritus comes from litter and dead roots but exudates and sloughed cells from growing plant roots (3.5 tonnes ha^{-1} in a wheat field – Coleman, 1976) and foliar leaching make a significant contribution. Soil fauna together with other animals and their excreta make a comparatively small annual contribution (<0.1 tonne ha^{-1}) whilst figures for microbial biomass range from 0.17 to 2.24 tonnes ha^{-1} (Jenkinson & Ladd, 1981).

Of course total soil organic matter levels in soil are determined not only by input of substrate but also the rate of its decomposition; organic carbon alone may account for as much as 70 tonnes ha^{-1} (Jenkinson & Powlson, 1976). A large number of factors will affect the rate of organic matter decay but perhaps 40–60% of it will be mineralized in the first year. The remainder is either assimilated by successive generations of microorganisms, with the gradual release of carbon dioxide, nitrate, phosphate etc., or becomes part of the refractile humic component of soil. As much as 20% of the original organic matter may remain after 5 years. Important factors influencing the rate of degradation include the diversity of the indigenous microbial population, soil type, climate, and the nature of the substrate with which the decomposer community must deal. For example, young plants and photosynthetic tissues have a much higher proportion of water-soluble carbohydrates than do their more mature counterparts which, in contrast, contain a higher percentage of less readily available cellulose, hemicelluloses and lignin. The supply of mineral nutrients – especially nitrogen – will have a powerful influence on the rate of organic matter decay. Plant residues are predominantly carbon, hydrogen and oxygen and are relatively low in nitrogen, phosphorus and sulphur. Animal and microbial cells on the other hand have higher nitrogen, phosphorus and sulphur contents but generally make a minor contribution to soil organic matter. As a result microorganisms are often nitrogen-limited in decomposer habitats and the C/N ratio of a substrate is a useful indicator of its rate of degradation.

This review is primarily concerned with the extracellular enzymic events responsible for organic matter degradation and the physicochemical properties of soil which influence those events. However, it is important to remember that the process of microbial and extracellular enzyme degradation may demand a number of prerequisite steps before the organic matter is in a receptive form. For instance, the early stages of degradation, especially of plant and animal detritus, are dominated by the soil fauna. These organisms contribute to the supply of suitable microbial substrates and to overall organic matter decomposition by: fragmenting tissues thereby permitting easier access for the microflora; channelling of woody material and thus improving aeration and hydration; acting as vectors for microbial propagules; and breaking down ingested matter.

The nature of exogenous substrates

Organic substances that are capable of supporting microbial growth occur in a variety of forms and in different physical relationships with the soil fabric. These forms and relationships will determine the availability of substrates, either directly to microorganisms or to their extracellular enzymes, and help to explain the differences between substrate/enzyme-microbe responses in soil as compared to those *in vitro*.

Organic molecules that are freely diffusible in the soil aqueous phase represent the only accessible form of substrate as far as a high proportion of the microbial population is concerned. These substrates include a large number of commonly occurring low molecular weight organic solutes such as monosaccharides, amino acids, organic acids and some aromatic monomers, and are components of organic materials (e.g. fertilisers, plant root exudates, leaf litter, microbial tissues, agrochemicals) newly added to the pool of soil organic debris, are metabolites secreted by soil organisms, or arise from the extracellular degradative activities of specialist components of the microflora. Whatever their origins, however, soluble substrates represent the final extracellular form of the organic molecule prior to uptake and intracellular metabolism and their presence is quite clearly a prerequisite for microbial growth and mineralization. In other words, such events as the solubilization and degradation of exogenous materials, together with any tactic response to diffusible products, are ways in which a microorganism actively increases its chances of being located in the vicinity of a suitable substrate. Presumably the main thrust in the evolution of microbial species in nutrient-limiting environments such as soil is towards maximizing exposure to suitable substrates.

Whilst some organic compounds in soil are in a form that is immediately available to microorganisms, a high proportion only become accessible after one or more extracellular transformations whilst a few are utilized very slowly or, in rare instances, are never in a suitable state for microbial absorption. The lack of availability of a potential substrate may be an intrinsic characteristic but very often organic materials which are poor microbial substrates in soil (or fail to support growth altogether) are entirely suitable when used in media in the absence of soil.

Macromolecules

A large number of polymeric organic compounds enter the soil each year as essential structural components of living tissues. Most will represent excellent sources of carbon, nitrogen and energy but will require extracellular transformations before they are in a suitable

Table 2. *Major macromolecular substrates in soil*

Macromolecular substrate (origin[a])	Structure	Molecule entering cell[b] (known or suspected)
Cellulose (P, M)	β-(1-4)-*D* linked glucan	Glucose, cellobiose
Hemicelluloses (P)	β-(1-4)-*D* linked xylan	Xylose, xylobiose
	Glucuronans	Glucuronic acid
	Galacturonans	Galacturonic acid
Pectin (P, M)	Xyloglucan	Xylose
	Galacturonans	Galacturonic acid
Starch (P, M)	α-(1-4) and α-(1-6) linked glucans	Glucose, maltose
Lignin (P)	Polymers of *p*-hydroxycinnamyl alcohols	Mono-lignols (coniferyl, sinapyl and *p*-coumaryl alcohols), di- and tri-lignols
Chitin (A, M)	β-(1-4)-linked *N*-acetylglucosamine	*N*-acetylglucosamine, chitobiose
Proteins and peptides (A,M, P)	Polymers of amino acids	Amino acids, low number peptides
Lipids (A, M, P)	Triglycerides, phospholipids	Glycerols, fatty acids
Peptidoglycan (M)	Polymers of *N*-acetylglucosamine and *N*-acetylmuramic acid with peptides	*N*-acetylglucosamine, *N*-acetylmuramic acid, amino acids, low-number peptides
Teichoic acid (M)	Polymers of polyol phosphates with saccharides and *D*-alanine	Glycerol, ribitol, mono- and di-saccharides, alanine
Microbial exopolysaccharides (M)	Mannans, dextrans, levans xanthans, pullulan, alginate	Mono- and di-saccharides

[a]P, plant; A, animal; M, microbial.
[b]Differs for different microorganisms.

form for microbial uptake and assimilation. A partial list of the most abundant high molecular weight substrates is given in Table 2 and details of their structure and occurrence can be found in a number of recent reviews (e.g. Dekker & Richards, 1976; Muzzarelli, 1977; Fan, Lee & Beardmore, 1980; Rogers, Perkins & Ward, 1981; Burns, 1982a). A few of the more important polymers are described here.

Cellulose is far and above the most abundant source of carbon in plant material. It is an insoluble polysaccharide composed of as many as 10 000 β-(1-4) linked glucose residues although only monomeric glucose, the dimer cellobiose and low-number oligosaccharides are taken up by microorganisms. The cellulose chains are arranged in structures known as fibrils which contain highly ordered stable crystalline regions and more loosely bound amorphous zones. The proportion of crystalline to amorphous cellulose will affect microbial and enzyme penetration and therefore susceptibility to hydrolysis.

Hemicelluloses are plant cell wall heteropolymers composed of a mixture of hexose sugars, pentose sugars and uronic acids. A third polysaccharide, pectin, is a minor yet important constituent of plant tissues and has a highly variable structure based on a methylated polyuronide. Cellulose, hemicellulose and pectin in combination with lignin, form a physically and chemically integrated structure that presents a series of problems to microorganisms utilizing plant cells as growth substrates.

Lignins are the most complex of the abundant plant-derived substrates and are polymers of coniferyl, sinapyl and *p*-coumaryl alcohols joined by a variety of intermonomer linkages. Whilst the likely monomeric constituents of lignin (e.g., vanillin, syringaldehyde, guaiacylglycerol, coniferaldehyde) are absorbed and metabolized by a large number of microbial species (Cain, 1980), a complicated and incompletely understood series of extracellular events must occur in order to release the low molecular weight fragments from native lignin.

A number of common microbial constituents will require extracellular depolymerization prior to their assimilation by other microorganisms, and include the glucosamine-containing chitin found in fungal cell walls (and in arthropods), as well as the bacterial peptidoglycans, lipopolysaccharides, teichoic acids, and a host of gums and mucilages. Microbial polysaccharides in soil are recorded as containing at least ten major sugars including glucose, mannose, rhamnose, fucose, galactose, xylose, arabinose, galacturonic acid, glucuronic acid, glucosamine and galactosamine. A whole range of somewhat persistent and structurally complex extracellular polysaccharides are produced by bacteria: dextrans (*Leuconostoc*), levans (*Pseudomonas, Bacillus*), xanthan (*Xanthomonas*), curdlan (*Agrobacterium*) and alginate (*Azotobacter, Pseudomonas*). The detection of extracellular polysaccharides in soil suggests that, even in a nutrient-limiting environment, there are periods of nutrient excess when overflow metabolism occurs.

Macromolecular substrates of animal origin include glycogen and glycoproteins whilst proteins, peptides and nucleic acids may all require depolymerization prior to cell uptake.

Clay–substrate complexes

Much of the organic carbon and nitrogen in soil is intimately associated with the mineral fraction. According to Greenland (1971), who collected evidence from a number of different soils, between 52 and 98% of soil carbon was bound to clay minerals whilst McGill & Paul (1976) showed that 40–50% of the amino acid nitrogen was adsorbed to the inorganic colloids.

It is well known that the susceptibility of polysaccharides to degradation may be modified by their close association with clay mineral surfaces. Studies of adsorption have included a range of plant, animal and microbial polysaccharides (e.g. cellulose, starch, pectins, glycogen, alginates, dextrans) and a number of adsorptive mechanisms have been suggested including ion exchange, polyvalent ion bridging, van der Waals forces, hydrogen bonding and complex formation (Theng, 1979; Burchill *et al.*, 1981). Some polysaccharides become bound within the interlamellar spaces of smectite clays reducing their accessibility to extracellular catalysts and completely removing them from direct contact with microorganisms. The tenacity of adsorption (and therefore likelihood of desorption) is determined by clay type, electrolyte concentration, other adsorbed ions, pH, water and, of course, the properties of the individual polysaccharide.

In addition to (and possibly because of) the effect of adsorption on substrate decomposition, polysaccharides have an important influence on soil aggregate formation and stabilization (Griffiths & Burns, 1972; Hayes, 1980). Cell wall polysaccharides are also involved in the retention of microorganisms at potentially repulsive surfaces (Marshall, 1980).

In general, soluble sugars, such as glucose, are poorly adsorbed to clays although the presence of methyl and amino substituents will clearly have a pH-dependent affect on surface interactions. Pyrimidines, purines and nucleic acids are adsorbed to clay surfaces (Cortez & Schnitzer, 1981) whilst detailed studies of amino acid, peptide and protein binding to clays have been summarized recently (Stotzky, 1980; Stotzky & Burns, 1982). It is apparent that such factors as molecular size and configuration of the adsorbate together with its isoelectric point and the pH at the surface of the clay are important in determining both the amount of amino compound immobilized and the strength and type of adsorptive mechanism involved. At acid pH values amino acids act as cations and are strongly adsorbed through ion exchange, whilst at their isoelectric point amino acids exist as dipolar ions and ion-dipole and hydrogen bonding become more important. Van der Waals forces are more important in the adsorption of larger size molecules such as proteins and nucleic acids.

In addition to the intrinsic recalcitrance of some of the components of humic matter, the formation of humic-clay complexes is an important phenomenon in stabilizing humic and fulvic acids (Guckert, Valla & Jacquin, 1975). The dominant mechanism by which humic and fulvic materials are associated with clays appears to be through complexing with calcium, iron, aluminium and other cations at the mineral surface (Greenland, 1971). Fulvic acids, but probably not humic acids, have been shown to penetrate the interlayer spaces of clays at acid pH values (Schnitzer, 1968). Incidentally, hydrous iron and aluminium oxides are often strongly associated with humic and fulvic materials.

Organic phosphates, such as phytin and hexose phosphates also form complexes with clay minerals resulting in an increase in their resistance to microbial decomposition (Perrott, Langdon & Wilson, 1974).

Some of the most detailed studies of organic matter adsorption to clays and the subsequent persistence of the adsorbate have involved substrates which are pesticides (Burns, 1975; Khan, 1978; Hill & Wright, 1979). The extreme case of rapid and irreversible adsorption is represented by the cationic bipyridyl herbicide, paraquat which is strongly adsorbed within the interlayer spaces of smectites and rendered biologically inactive (Burns & Audus, 1970). This, despite the fact that paraquat is an excellent carbon and nitrogen source for microorganisms when grown in the absence of clay (Anderson & Drew, 1972). Basic pesticides (e.g. *s*-triazines) will dissociate at acid pH values (according to the pK_a of the pesticide in question), become protonated at clay surfaces where hydrogen ions are concentrated, and then be adsorbed through cation exchange. Adsorbed basic pesticides may display extended resistance to degradation although even interlamellar adsorption appears reversible (Weber, 1970). Acidic pesticides tend to remain in their molecular form in soil or dissociate to form anionic species and be repulsed from like-charged clay surfaces. Non-ionic pesticides have less predictable adsorptive properties determined by their water solubility, volatility, molecular size and configuration etc.

The brief statements here concerning the adsorption of pesticides and other potential substrates to clays must be seen as broad generalizations, especially when one realises the large number of adsorptive mechanisms that are available, for example, cation exchange, anion exchange, van der Waals forces, pi-bonding, covalent bonding, hydrogen bonding, coordination complexing, hemisalt formation and entropy effects. Thus, it is possible to envisage all sorts of associations, weak and strong, taking place between opposite-charged, like-charged and non-charged substrates on one hand, and clay surfaces on the other.

Humic-substrate complexes

Organic substrates may be associated with the soil humates in three principal ways: adsorption to the charged surfaces, entrapment within the three-dimensional structure, and chemical bonding to the humic polymer. All these associations may reduce the accessibility of potential substrates.

Polysaccharides are particularly resistant to biodegradation in soil when compared to their rate of hydrolysis *in vitro*. This resistance is partly due to the formation of polysaccharide clay complexes (see page 262) but is also affected by the presence of humic polymers. Polysaccharides become linked into humic colloids (especially humins) through nucleophilic addition to quinone moieties or by strong hydrogen bonding (Martin, Parsa & Haider, 1978). These processes are mediated by enzymic or autooxidative events. Much of the recalcitrant soil polysaccharide is believed to be of microbial rather than plant origin because of the preponderance of amino-sugar units within them. (Cheshire, 1977; Benzing-Purdie, 1981). The products of polysaccharide degradation, along with other intrinsically easily catabolized substrates such as peptides and proteins, may also be stabilized after becoming an integral part of the largely aromatic humic polymer (Verma, Martin & Haider, 1975).

Many simple phenolic substances released from plant material or formed through microbial synthesis are readily degraded in pure culture. However, in soil benzene ring structures become components of new and intricate humic polymers or are linked to existing ones and have considerable resistence to enzymic attack. The simple addition of phenols to soils is not enough to prevent their degradation (Martin, Haider & Linhares, 1979; Sparling, Ord & Vaughan, 1981). Aromatic metabolites may also be stabilized by linkage into the humic-like polymers (melanins) within or released from the cell walls of certain fungi (Martin & Haider, 1976). Napthalenes, phenols, anthracenes, aflatoxins and griseofulvin, although inherently susceptible to aerobic microbial decomposition, may also be stabilized in soil humus (Mathur, 1971; Linhares & Martin, 1979;

Burns & Martin, 1983). Complexing fungal, bacterial and algal cell wall and cytoplasmic fractions to model humic polymers will significantly reduce the rate of microbial tissue decomposition (Verma & Martin, 1976; Martin & Haider, 1980; Pflug & Ziechmann, 1982).

In recent years there has been much interest in the occurrence of humic-pesticide residues in soil and controversy over the environmental hazards that bound pesticides may or may not present (Bartha, 1980). As far as the subject of this chapter is concerned, bound pesticide residues represent another type of substrate (albeit an exotic one) to which there is restricted microbial and enzymic access. For example, many agricultural chemicals are synthesized as or degraded to phenolic or anilinic compounds which may then undergo enzymic or chemical polymerization reactions and become linked to soil humus. Examples of stable humic-pesticide complexes cited in the literature include many aniline-based herbicides such as phenylcarbamates, phenylureas and acylanilides (e.g. Helling & Krivonak, 1978). Monophenol monooxygenases and peroxidases, which will bring about the oxidative coupling of pesticide-derived phenols, have been extracted from soils (Bollag, Liu & Minard, 1980; Sjoblad & Bollag, 1981). Bartha & Pramer (1970) and Hsu & Bartha (1974) have described the bonding of chloroanilines, released during the degradation of acylanilide pesticides (e.g. propanil), to humus. Such complexing may extend the residual life of chloroanilines in soil to as much as ten years. Wolf & Martin (1976) have presented evidence that the benzene ring portions of 2,4-D and chlorpropham are incorporated by some fungi into cell wall melanins. Subsequently, these humic-like materials are relatively resistant to biodegradation whilst the cell constituents containing side-chain carbon are readily utilized by microorganisms.

The ability of humic materials to confer recalcitrance on microbial products is of particular importance in the stabilization of extracellular enzymes (see page 280).

Recalcitrant molecules

It is a dangerous policy to describe a molecule as recalcitrant without reference to the time scale involved and an acceptance of the restricted nature of the research that has taken place using organic substances that do not seem at first sight to be readily available carbon, nitrogen or energy substrates. There is no doubt that in time (and one has to remember that synthetic molecules are a comparatively new addition to the organic matter pool) individual microbial species or communities of microorganisms may evolve ways of utilizing recalcitrant molecules. For instance, a combination of gene mutations and the transference of plasmids could provide a microbial species with a useful means of developing a novel catabolic capacity (Reineke & Knackmuss, 1979; Slater & Godwin, 1980). Additionally, the co-metabolism (or possibly abiological degradation) of a resistant molecule by one species may release a suitable substrate for another (Daughton & Hsieh, 1977) or within an individual species may even represent an evolutionary step on the way to acquiring competence in utilizing unusual molecules. Therefore, the presence of a suitable co-substrate or the correct combination of microbial species might be all that is necessary to trigger the biodegradation of a previously refractile substrate. Finally, the study of resistant molecules in anaerobic culture has been neglected as have the special conditions of a fluctuating aerobic/anaerobic environment common *in situ*. Yet these physical adjustments have been shown to stimulate the breakdown of recalcitrant pesticides (see Williams, 1977; Bull, 1980).

Having made this apologia it is apparent that natural and synthetic molecules do exist that are long-lived in the environment (i.e. persistence measured in years, decades, centuries and even millennia) and, furthermore, have even resisted attempts to rapidly degrade them under a variety of conditions *in vitro*. These material include man-made compounds such as organochlorine insectides (e.g. aldrin, chlordane, DDT), synthetic polymers (e.g. polyethylene, polyvinyl alcohol, cellulose acetate), alkylbenzene sulphonate surfactants and polychlorinated biphenols. Naturally occurring persistent molecules include some humic matter constituents, lignin, tannin and melanins: polyaromatics with molecular weights up to 2×10^6 daltons which may have half-lives, estimated by radiocarbon dating, of thousands of years (Jenkinson & Rayner, 1977; O'Brien & Stout, 1978).

The properties of organic molecules which render them totally unsuitable as microbial substrates include: excessive molecular size (and presumably the absence of extracellular enzymes capable of solubilizing the macromolecule), the nature and location of substituents, the occurrence of unusual bonds or bond sequences, the existence of highly condensed aromatic rings, and the sheer complexity of a heteropolymer that may require the concerted action of many enzymes and microbial species before it is in a form suitable for microbial uptake. Some organic molecules, such as solvents and phenolics, are actually inhibitory towards microorganisms and their enzymes.

Microbial structures (e.g. sclerotia, endospores, cysts) whose sole function is to resist environmental pressures, and palaeobiochemicals contained in fossils, are obviously unresponsive to microbial attack.

EXTRACELLULAR ENZYMES

Introduction

The outer layers of the microbial cell serve as a physical and functional barrier, controlling the passage of solutes to the cytoplasm and the movement of enzymes and metabolites into the surrounding medium. In other words the outer membranes of Gram-negative bacteria, the cell walls of Gram-positive bacteria, and the fibrillar structures which comprise fungal walls, form an interface between the protoplast and the extracellular environment. A multiplicity of uptake and secretion mechanisms have been reported in bacteria and fungi and include active and passive as well as selective and non-selective processes. There is no doubt that a clear understanding of the factors influencing the passage of solutes through microbial envelopes would provide an insight into the likely constraints to substrate utilization. However, our knowledge of this topic is restricted to a few substrates and enzymes and an even smaller number of microbial species. Moreover, research effort has been concentrated on pure cultures grown in artificial media; we know little of the factors influencing substrate utilization under the conditions prevailing in soils.

Perhaps the most informative *in vitro* studies concerning substrate uptake have taken place using Gram-negative bacteria and these organisms represent a particularly interesting group because the cell envelope has two distinct and spatially separated barriers through which molecules must pass, the outer membrane and the cytoplasmic membrane. It is less easy to distinguish any individual functions of the closely associated layers of Gram-positive cell walls. The outer membrane of Gram-negative bacteria is a highly specialized structure which forms a permeability barrier restricting the access of antibiotics, detergents and nutrients from the external environment to the periplasm and the cytoplasmic membrane. Studies using

Escherichia coli, *Pseudomonas aeruginosa* and *Salmonella typhimurium* suggest that membrane permeability is a somewhat selective process involving at least three different classes of proteins, namely, those forming pores (and called porins) which allow non-specific, passive diffusion of hydrophilic solutes with molecular weights of below about 650 daltons (e.g. amino acids, small peptides, mono- and di-saccharides, inorganic salts), those associated with channels used for the more specific penetration of molecules such as maltose, maltotriose and nucleosides, and finally those which transport relatively high molecular weight substances such as vitamin B12 and iron-chelator complexes. The second and third mechanisms may be particularly important when the solutes are in low concentrations in the external medium. The uptake of macromolecules and hydrophobic substances not effected by these proteins may occur through temporarily damaged or altered outer membranes – a possibility suggested by studies of deep rough mutants. Details of outer membrane structure and the transmembrane proteins used by Gram-negative bacteria can be found in reviews by Braun (1978), Nikaido & Nakae (1979) and Osborn & Wu (1980). A discussion of periplasmic carrier proteins and cytoplasmic membrane transport is beyond the purview of this chapter.

The molecular exclusion limit for solute uptake by Gram-positive bacteria appears to be much higher than that demonstrated for the Gram-negative organisms and a figure of 100 000 has been proposed for *Bacillus megaterium* (Scherrer & Gerhardt, 1971). It is not certain, however, whether pores equivalent to those in Gram-negative organisms even exist in Gram-positives (Scherrer, Berlin & Gerhardt, 1977). Pores have not been reported for fungal cell walls (although research has been almost entirely restricted to *Neurospora*) and any restrictions on uptake may be due to the inability of transport mechanisms to accommodate high molecular weight substrates (see Wolfinbarger, 1980).

If we now turn our attention to microbial enzymes, it is apparent that a large number of *in vitro* studies of cell-free extracellular enzymes are unconvincing because of the inadequate experimental techniques employed. To warrant being described as extracellular, an enzyme must be actively secreted by viable microorganisms and not arise from damaged or lysed cells. However, extracellular enzyme production is often at its peak during the post-exponential phase of growth when a significant proportion of the cell population is undergoing lysis and thus the detection of an enzyme in the supernatant liquid is not conclusive evidence of its normal functional location. It is true that some extracellular enzymes can only function outside the cell because either they catalyze the degradation of insoluble substrates, or they are known to assume their active configuration only after export. However, unequivocal claims for truly extracellular enzymes usually require the simultaneous monitoring of a known cytoplasmic enzyme (e.g. β-glucosidase) and, in the case of Gram-negative bacteria, a periplasmic enzyme marker such as alkaline phosphatase. These rigorous and necessary controls are often omitted. The study of extracellular enzymes attached to the cell wall (or retained within the periplasmic space) demands a different spectrum of physical and chemical techniques.

Of course, as far as the success of the microbial species is concerned it may not matter where the enzyme originates, only that there is enough of it to produce soluble substrate to satisfy the requirements of the colony. This is an ecological viewpoint which is emphasized elsewhere in the chapter.

A semantic problem common to the literature concerned with the location of microbial enzymes in relation to their cells, is the use of the term extracellular (Pollock, 1962; Glenn, 1976). Obviously enzymes are cytoplasmic, periplasmic or secreted through the cell wall, yet within these broad categories a number of other locations are possible. For instance, enzymes may be associated with membranes or be freely diffusible within the periplasm, or they may be partially attached to the outermost layer of the cell wall and project into the ambient environment. Moreover enzymes may be contained within the exopolysaccharide coatings present on some microorganisms, or they may diffuse into the surrounding aqueous phase. Furthermore, the same enzyme may exist in more than one location in the same species, for example, a periplasmic enzyme may leak to become a mural enzyme whilst *en route* to becoming a cell-free enzyme, or vary in location according to the species, for example, cell-free enzymes of Gram-positive bacteria have counterparts that are restricted to the periplasmic space in Gram-negative bacteria. Glenn (1976) suggested that truly extracellular enzymes were only produced by Gram-positive bacteria (e.g. *Bacillus, Streptomyces, Arthrobacter*). However, we now know of many species of *Pseudomonas, Flavobacterium, Aeromonas, Vibrio* and others which liberate enzymes into the surrounding medium. In this review extracellular is used to describe enzymes which are external to the cell wall and in contact with surrounding medium. Thus extracellular enzymes may be cell-wall bound as well as being located at a distance from their parent microorganism.

Two groups of enzymes that are predominantly extracellular, because of the molecular size of their substrates, are polysaccharases and proteinases. These are discussed here to illustrate some of the constraints placed upon the activity of depolymerases in the absence and in the presence of soil.

Polysaccharases

Given the number and complexity of plant, animal and microbial polysaccharides in the environment it is not surprising that there is an enormous range of polysaccharases. For ease of description a broad definition of a polysaccharase is adopted here and includes any enzyme that is involved in the degradation of sugar or sugar-based compounds made up of more than one molecule. Thus enzymes that split cellobiose and maltose are polysaccharases (as well as disaccharases) as well as the more familiar examples of cellulolytic and amylolytic enzymes.

Most polysaccharases are hydrolases, although there are a few examples (and many suggestions) of other groups of enzymes involved in depolymerization. The hydrolytic polysaccharases (glycosidases) that are of greatest relevance to this review are classified among those hydrolyzing *o*-glycosyl compounds (EC 3.2.1). They include the starch and glycogen polysaccharases, α- and β-amylase (EC 3.2.1.1 and 3.2.1.2) and exo-1,4-α-D-glucosidase (EC 3.2.1.3), the enzymes commonly implicated in cellulolysis, cellulase (EC 3.2.1.4), β-D-glucosidase (EC 3.2.1.21), exo-1,4-β-D-glucosidase (EC 3.2.1.74) and exocellobiohydrolase (EC 3.2.1.91) and considered in detail in this section, xylanases (EC 3.2.1.8/33/37/72), chitinases (EC 3.2.1.14/30), laminarin depolymerases (EC 3.2.1.6/39/58) and, finally, enzymes involved in the saccharification of galactans (EC 3.2.1.22/23), mannans (EC 3.2.1.24/25/37/77/79) and galacturonases (EC 3.2.1.15/67/82).

Non-hydrolytic enzymes involved in polysaccharide degradation include the lyases of pectin breakdown (EC 4.2.2.2/9/10) and two celluloytic enzymes, 'cellobiose oxidase', which converts cellobiose to cellobionic acid, and 'cellobiose-quinone oxidoreductase', which has a dual function in cellulose and lignin decay (Eriksson & Johnstrud, 1982).

The function of structural polysaccharides is to protect and support the tissues with which they are associated. In so doing they have formed an intricate network which, not surprisingly, presents a number of intrinsic and extrinsic problems (listed below) to microorganisms using the macromolecules as substrates.

(1) The proportion of stable crystalline regions contained in fibrillar polysaccharides such as cellulose and chitin. Crystalline zones are much less accessible to microbes and enzymes than are the amorphous areas. The degree of crystallinity varies according to the nature of the substrate (plant, fungal, arthropod) and will be a limiting factor in the rate of depolymerization.

(2) The degree of polymerization and the conformation and steric rigidity of the primary units. For example, cotton fibres contain as many as 10 000 glucose residues (molecular weight of polymer $>1.5 \times 10^6$) and have a total length of around 5 μm. Cellulose in mature wood on the other hand may contain $<$2 000 glucose units. Of course, microorganisms must contend with additional problems when degrading cellulose in wood (see (4) and (6)) when compared with cotton fibres which are almost pure cellulose.

(3) The extent of hydration of the polymer. In addition to the obvious requirement for water in hydrolytic processes, wet and swollen polysaccharides present a much greater surface area for enzyme adsorption.

(4) The presence in the native substrate of a large number of different polymers such as pectic substances, hemicelluloses and lignins. This implies that the release of soluble low molecular weight sugars from their physical and chemical combination within a recently deceased organism or tissue involves a range of different enzymes acting in concert and, more often than not, secreted by a number of different microbial species. Specifically cellulolytic microorganisms must either have the capacity to produce a range of other hydrolases (e.g. pectinases, 'hemicellulases', 'lignases') or be able to cooperate within a community of microorganisms – some specializing in xylan, arabinan or glucuronan hydrolysis, others in the depolymerization of polyphenolics.

(5) The physical barriers imposed by the meshwork of cell wall polysaccharides as well as specific protectants such as cutin and suberin. Filamentous microorganisms (i.e. fungi, actinomycetes) capable of vigorous and sustained growth are believed to be vitally important in the early stages of organic matter breakdown on account of their penetrative powers and their capacity to grow through regions containing unsuitable substrates.

(6) The bacteriostatic and fungistatic properties of some plant products, notably lignin and lignin constituents. These may prevent microbial growth as well as inhibiting the synthesis or activity of certain hydrolytic enzymes (Varadi, 1972; Vohra, Shirkot, Dhawan & Gupta, 1980). Inhibitors of endopolygalacturonase are associated with plant cell walls (Fisher, Anderson & Albersheim, 1973).

There are further constraints on the degradation of polysaccharides when they are located in soil and these are considered elsewhere in this chapter. Native cellulose, the dominant plant polymer, provides a valuable illustration of the difficulties which microorganisms are faced with and must overcome before they are able to utilize a polysaccharide.

Cellulose biodegradation

The first suggested mechanism that was widely adopted as an aid to understanding cellulose breakdown was that presented by Reese, Siu & Levinson (1950). This is the so-called C_l-C_x hypothesis which orginally proposed a sequence involving an activating or disaggregating non-hydrolytic enzyme, C_1, followed by a depolymerization process effected by a number of hydrolases, C_x. This view of cellulolysis is not nowadays thought to be particularly helpful for two reasons. Firstly, a non-hydrolytic, activating enzyme has proven difficult to find and, secondly, a number of parallel catalyses rather than a sequential process is responsible for glucose production in most well-understood systems. Attempts by many workers to relate their observations to the C_l-C_x hypothesis have led to unnecessary confusion and it is more helpful to envisage the hydrolysis of native cellulose as the result of the combined and synergistic efforts of at least three different groups of hydrolytic enzymes.

A composite picture of cellulose hydrolysis would involve the cooperative activities of endoglucanase, exoglucanase and cellobiase. The increased effectiveness of mixtures of endo- and exoglucanases and β-glucosidase on native cellulose hydrolysis (Selby & Maitland, 1967) may be due to the exoglucanase preventing the reformation of glycosidic bonds following endoglucanase cleavage. The tendency of glycosidic bonds to reform may be due to the highly ordered and hydrogen-bonded structure of cellulose, but this can be avoided by the rapid removal of cellobiose units following the initial breaking of β-(1–4) bonds. To facilitate efficient depolymerization, exo- and endoglucanases may even combine in an enzyme complex (Wood & McCrae, 1978a). Endo- and exoglucanases are inhibited by cellobiose and thus the activity of β-glucosidase is essential to maintain the momentum of hydrolysis. The comparative merits of the various concepts of cellulolysis are discussed in most recent reviews of the topic (e.g. Goksøyr & Eriksen, 1980; Lee & Fan, 1980).

A small number of microbial species have dominated studies of cellulose degradation in the past thirty years and much of our knowledge is derived from experiments using the fungi *Trichoderma viride, T. koningii* and the ligno-cellulose degrader *Phanerochaete chrysosporium*, and the bacterial genera *Cytophaga* and *Sporocytophaga.* In recent years greater attention has been paid to other cellulolytic microorganisms, especially those with a promise of biotechnological application (e.g. thermophilic species, actinomycetes). The immense complexity of cellulose enzymology is apparent from even a cursory appraisal of these studies as it is obvious that microorganisms differ widely in the way that they deal with cellulosic substrates. For instance, the fungi have been shown to produce both exo- and endoglucanases as well as the β-glucosidase, cellobiase, and are able to grow independently by the extensive hydrolysis of native cellulose. Many cellulolytic bacteria, on the other hand, do not appear to synthesize exoglucanase and must, therefore, be part of a community of species in soil which is capable of solubilizing highly ordered cellulose. Another major difference between the various cellulase systems is the number and molecular weights of the enzymes collectively termed exoglucanases, endoglucanases or β-glucosidases. For example, Wood & McCrae (1978b) identified four endoglucanases in *T. koningii* with molecular weights ranging from 13 000 to 48 000 – a range that may be necessary to allow enzyme penetration of the various forms of crystalline and amorphous cellulose. As many as five endoglucanases, one exoglucanase and two β-glucosidases have been described for *Phanerochaete chrysosporium* (Eriksson, 1978) and *Sporocytophaga myxococcoides* produces several endoglucanases (Osmundsvag & Goksøyr, 1975). A further important distinction between cellulolytic microorganisms is the location of the enzymes of the cellulase complex. Even with the proviso that experimental evidence is sometimes inconclusive (see page 268) it is apparent that some cellulases are intracellular, some are retained in the periplasm, some are attached to the cell wall, whilst others are secreted into the medium. When *Trichoderma* species are grown in liquid media, endo- and exoglucanases (and possibly β-glucosidase) are found in the culture filtrate. On the other hand, the gliding bacteria, *Cytophaga* and *Sporocytophaga*, may retain their cellulases on the outer membrane; a location consistent with the ability of these organisms to grow in close contact with cellulose fibrils.

There have been a number of studies of the biodegradation of cellulose and other polysaccharides in the presence of soil and soil components (e.g. Filip, 1973; Olness & Clapp, 1975; Marshman & Marshall, 1981b; Zunino *et al.*, 1982). Interpretation of the data, especially with regard to the enzymology, is difficult because any change in growth rate, substrate utilization or carbon dioxide evolution may be due to the direct influence of clay and humic surfaces (e.g. adsorption and concentration of the reactants) or indirectly through the soil particles adsorbing toxic metabolites, behaving as a buffer or serving as a source of trace elements. Attempts to differentiate between the direct and the indirect effects of soil surfaces on substrate utilization have involved experiments wherein the adsorbant and adsorbate are separated by a semi-permeable membrane (e.g. Filip, Haider & Martin, 1972). Marshman & Marshall (1981b) have added a further dimension to the study of substrate (starch)-clay-microbe interactions by using defined co-cultures of bacteria.

As a general rule expanding lattice clays and humic matter stimulate the growth rate and final yield of microbes utilizing soluble and insoluble carbohydrates (e.g. Filip *et al.*, 1972) but will sometimes reduce the loss of carbon as carbon dioxide. For example, Sørensen (1975, 1981) added labelled cellulose to soils with varying clay contents and recorded that the rate of substrate disappearance was similar in all soils (probably because the adsorption of cellulose to clays is minimal) but that high clay soils immobilized amino acid metabolites and cellular material and thus reduced further oxidation. He proposed that the presence of clay also increased the efficiency of microbial assimilation and calculated that, in the soil with the highest clay content (34%), carbon conversion efficiency (55%) approached that of bacteria grown under optimum conditions *in vitro*.

Few studies have been made of the adsorption of cellulases to soil components (Drozdowicz, 1971) and of their residual activity and stability. However, we know that narrow bands of smectite clays, inserted into cellulose agar, retard or prevent altogether the diffusion of cellulases or the radial growth of cellulolytic microorganisms (C. F. A. Hope & R. G. Burns, unpublished observations).

Proteinases

There are a large number of proteins available for proteolytic attack ranging from the fairly easily degraded non-structural proteins (e.g. enzymes) to the somewhat more recalcitrant structural and connective proteins (e.g. keratin, elastin). This variety of substrates explains the enormous diversity of proteinases (>100 listed in the I.U.B. Enzyme Nomenclature Handbook) many of them produced by bacteria or fungi (see Payne, 1980a for details). Cohen (1980) has suggested that there are virtually no proteins that are entirely immune from fungal hydrolysis. Whilst this chapter is primarily concerned with saprophytic exo-enzymes it is worth remembering that the invasiveness of many pathogenic microorganisms is due to their ability to penetrate physical barriers in the host through extracellular proteolytic attack. The classification of peptide hydrolases (EC 3.4) is somewhat unsatisfactory mainly due to similarity of reaction and the lack of strict substrate specificity. However, two principal groups of enzymes acting on peptide bonds are recognized, peptidases (EC 3.4.11 to 3.4.17) and proteinases or proteases (EC 3.4.21 to 3.4.24). The peptidases are exohydrolases which remove terminal amino acids or dipeptide units from proteins and peptides and those isolated from growth media usually have alkaline pH optima. The proteinases are endohydrolases which solubilize proteins by cleaving internal peptide bonds. Proteinases are divided into four classes according to their mode of action and are described as serine proteinases (EC 3.4.21), thiol- or SH-proteinases (EC 3.4.22), carboxyl-proteinases (EC 3.4.23) or metallo-proteinases (EC 3.4.24). Serine proteinases are often referred to as alkaline proteinases, SH- and metallo-proteinases as neutral proteinases, and carboxyl-proteinases as acid proteinases. This means that proteolytic enzymes, as a group, have a broad pH range of from 2 to 10. A small number of proteolytic enzymes, such as peptidoglycan (murein) endopeptidase, have unknown catalytic mechanisms and have been allocated the number EC 3.4.99. Extracellular proteolytic enzymes, like other groups of extracellular enzymes, share few common chemical or physical features. They have no general pattern of amino acid composition although a low incidence of cysteine residues is reported. Only a few are glycoproteins. However, proteinases and peptidases in general have relatively low molecular weights (20 000–40 000 daltons), often require divalent metal ions such as Zn^{2+} or Mn^{2+} for activity, and may exhibit enhanced stability in the presence of Ca^{2+}.

The utilization of exogenous proteins by microorganisms requires the hydrolysis of these high molecular weight substrates to peptides and amino acids before cellular absorption can occur. The direct uptake of proteins by microbial cells is infrequent although Cohen (1980) discusses the likelihood of protein permeases in fungi together with alternative uptake mechanisms such as micropinocytosis. The maximum size for oligopeptide uptake in Gram-negative bacteria and filamentous fungi is probably about five amino acid residues (molecular weight 600–700 daltons), that of yeasts somewhat larger (Naider, Becker & Katzir-Katchalski, 1973; Payne & Gilvarg, 1978). These restrictions may reflect the limits of cell wall porosity as well as the presence of suitable peptide transport systems. Presumably those cells lacking peptide transport systems yet having a requirement for peptides must secrete peptidases (or, in mixed communities, be associated with species that do) and possess amino acid transport proteins. Peptidases in bacteria and fungi are predominantly, but not exclusively, either cytoplasmic or periplasmic (Deutch & Soffer, 1978; Wolfinbarger, 1980). However, some bacteria, such as *Clostridium histolyticum, Aerobacter proteolytica* and *Bacillus subtilis*, produce both extracellular proteinases and aminopeptidases. The path of subsequent intracellular hydrolysis will vary from species to species; some microorganisms may have an obligate requirement for pre-formed amino acids, others may use amino acids as carbon and nitrogen sources.

The binding of proteins, peptides and amino acids to clay surfaces and the subsequent availability of the complexed substrates to microorganisms has been summarized by Stotzky (1980). Factors influencing binding and rate of breakdown include type of clay mineral (i.e. expanding or non-expanding lattice), concentration and valency of ions already occupying the clay surface, pH, and the properties of the adsorbate (e.g. molecular size, shape, number of potential binding sites, pI, concentration). Most proteins will enter the lattices of smectite clays and this will reduce their rate of depolymerization in comparison with proteins in the absence of adsorptive surfaces. Catalase and pepsin are exceptions to this in that they do not intercalate, perhaps because of size and adsorption to external positively charged sites, respectively. Furthermore, these proteins are completely resistant to microbial attack when in the adsorbed state and it was assumed that the binding mechanism effectively masked the sites for endo- and exopeptidase attack. The peptidases themselves were not inactivated. Interestingly, catalase activity was stimulated upon adsorption, possibly due to the exposure of active centres.

Some, but not all, complexed peptides and amino acids were available as microbial carbon and/or nitrogen sources even though all those tested were readily degraded when not adsorbed. The intricacy of microbe/clay/substrate interactions is illustrated by one series of observations: proline bound to Ca^{2+}-saturated or H^+-saturated montmorillonite was utilized only as a nitrogen source; arginine bound to H^+-montmorillonite was available as both a carbon and a nitrogen source, but when adsorbed to Al^{3+}-montmorillonite served as a nitrogen source (in medium containing dextrose) or a carbon source (in medium containing NH_4NO_3) but not both; aspartic acid and cysteine were unavailable when complexed with clays.

A recent study by Marshman & Marshall (1981a) prompted them to suggest that proteins (e.g. gelatin) are bound to two different sites on clay minerals. One of these sites was available to attack by bacterial proteinases and the other was not. The difference between these two sites was determined by the properties of the bacterial proteinases (which were different for the different bacterial species investigated), the clay minerals and the proteins. They also reported that growth was diauxic, protein hydrolysis occurring during the second phase of growth, and that the rate of utilization of protein, in comparison with substrate-only controls, was influenced by the protein/clay ratio. At low protein/clay ratio none was available for hydrolysis; at intermediate ratios growth rate but not yield was affected; and at high protein/clay ratios the clays had no effect on utilization.

No doubt in soil, where clay surfaces are heteroionic, are coated with organic molecules, and are subject to fluctuations in pH, the factors influencing the availability of amino compounds to microorganisms and their exoenzymes are even more complicated.

Proteins and peptides linked to model phenolic polymers by oxidative polymerization are stabilized such that their rate of microbial degradation in soil and pure culture is reduced by some 80–90%. Even the physical mixing of substrates with humic acid-type polymers reduced microbial decomposition by approximately one-third (Verma *et al.*, 1975). The strong hydrogen-bonding of protein molecules with the phenolic constituents of the polymers may hinder the approach of proteinase enzymes to the peptide bonds. Soil humic compounds, in contrast, may not combine with amino acids and, furthermore, do not interfere with microbiological decomposition (Schnitzer, Sowden & Ivarson, 1974).

ENZYMES IN SOIL: PROBLEMS AND POSSIBLE SOLUTIONS

Components of extracellular enzyme activity

Extracellular enzymes can be separated into a number of categories according to their location within the soil fabric (Burns, 1982b). Indeed, the measured activity of a particular enzyme (e.g. urease, arylsulphatase) or a group of enzymes (e.g. dehydrogenase, glucosidases) is a composite of activities belonging to two or more of the following categories.

(1) Enzymes whose usual functional location is within the cytoplasm of viable cells yet which remain active in (a) dead cells, (b) cell debris and (c) the soil aqueous phase. Many important carbon and nitrogen cycle hydrolases, such as β-glucosidases, β-galactosidases and urease, fall into this category. In contrast enzymes responsible for some of the central aspects of metabolism (e.g. glycolysis, oxidative phosphorylation) cannot function beyond cell death and lysis because they depend upon various co-factors, on some physiological property of the entire cell, or on being located alongside other enzymes.

(2) Periplasmic enzymes which are either (a) extracellular *per se* (according to the definition of extracellular – Glenn, 1976) or (b) become extracellular following leakage through damaged cell membranes. The artificiality of any classification of enzymes founded solely upon their site of activity is illustrated when one recalls that some of the periplasmic enzymes of Gram-negative bacteria (e.g. phosphatases) have their counterparts in the truly extracellular enzymes of Gram-positive bacteria and fungi, and those excreted by plant roots.

(3) Enzymes attached to the outer surfaces of cell walls or associated with extracellular polymeric coats (i.e. slime layers, capsules) and in contact with substrate in the surrounding environment. Many bacterial polysaccharases and proteinases are retained by the cell in this manner.

(4) Enzymes which are truly extracellular and are secreted into the aqueous phase of the soil during normal cell growth.

(5) Enzymes associated temporarily with exogenous soluble or insoluble substrates as enzyme–substrate complexes.

(6) Enzymes adsorbed to the anionic clay constituents of soil. Depending upon the clay type and the physicochemical properties of the immediate environment, adsorption of enzymic protein may be minimal or extensive. Expanding lattice clays, belonging to the smectite-vermiculite group, have a high potential for enzyme adsorption in comparison to kaolinite clays. Adsorption of an enzyme to a clay surface is not generally compatible with the retention of the enzyme's activity (Stotzky & Burns, 1982), but examples do exist (e.g. catalase – Stotzky, 1974; bacteriolytic endopeptidase – Haska, 1975, 1981).

(7) Enzymes intimately associated with the soil colloidal organic matter. The mechanisms involved in this tenacious relationship are far from certain but adsorption, entrapment and co-polymerization have all been proposed (Burns, 1977). Whatever, extracellular enzymes immobilized in this manner retain a proportion of their original activity and represent extracellular biological catalysts that are long-lived and are unrelated to the extant microbial biomass.

Not surprisingly, there are major difficulties in measuring enzyme activities in heterogeneous environments such as soil and these involve methodological decisions as well as problems in interpreting the data. For example, should assays be performed according to the well-established principles of enzymology (i.e. excess substrate, optimum pH, shaken reaction mixtures) or in conditions mimicking those encountered in soil (i.e. limiting and unevenly distributed substrate, sub-optimal and fluctuating pH, stationary incubation)? How are the independently reproducible data arising from the former approach related to those recorded using the 'realistic' parameters of the latter? These and other experimental problems have been discussed fully elsewhere (Burns, 1978; Stotzky & Burns, 1982).

The greatest interpretative problem in soil enzymology is differentiating between the various categories of enzymes described above. In practice, some useful clues are afforded by the conditions under which the assay is carried out. For instance, incorporating a microbial inhibitor (e.g. toluene, sodium azide) into the reaction mixture eliminates activity due to new enzymes produced by proliferating cells, and enzymes free in the aqueous phase (categories 1c, 2b, 4) are destroyed by air-drying field moist soil or by subjecting it to low levels (*ca* 2.5–5.0 Mrad) of γ-irradiation (Burns, Gregory, Lethbridge & Pettit, 1978). Further clarification of the various enzyme locations can be achieved by the judicious use of controls but mostly by obtaining crude extracts from soil (Burns, El-Sayed & McLaren, 1972; Pettit, Gregory, Freedman & Burns, 1977). Depending upon the extraction techniques employed, the resulting fractions will be free of microbial cells (categories 1a, 2a and 3), cell debris (category 1b) and inorganic particulate material (category 6).

The location of the various enzyme components in soil may also be determined histochemically to within 10 nm and Foster & Martin (1981) have stated that it should be possible, by using this technique to quantify enzymes in living tissues, cell debris, faecal material and adsorbed to soil surfaces.

Immobilized soil enzymes

A large number of enzymes have been described that are active in 'sterile' soil and/or soil extracts (Ladd, 1978) and are presumed to belong to categories 6 and 7 described above. However, evaluation of the experimental techniques employed reveals that at least some of the reported activities were associated with microbial growth during the assay period or were in extracts which did not contain the clay or humic adsorbents. These activities were not then, by definition, due to the immobilized or 'accumulated' enzymes of categories 6 and 7. Convincing evidence of immobilized enzymes is provided by studies using whole soil, clay and humic extracts, and the characterization of enzyme properties according to classic principles. In other words it is necessary to measure: stability to temperature, storage, proteolysis, solvents, etc.; to assay pH- and temperature-activity profiles; as well as to determine the traditional Michaelis-Menten kinetics (Stotzky & Burns, 1982). Lacking this stringent approach some claims for immobilized soil enzymes may even be describing abiological catalysis!

Notwithstanding, some, if not all, of the enzymes listed in Table 3 are immobilized on or in soil clays and humates and constitute a persistent extracellular biological catalyst. It can be seen that enzymes immobilized in this manner are predominantly hydrolases and that many of them also function within the microbial cell. Thus, it is reasonable to suppose that enzymes such as urease, arylsulpha-

Table 3. *Enzymes immobilized by soil clays and/or humates*

Class	EC number	Recommended name	Substrate
Oxidoreductases	1.7.3.3	Urate oxidase	Uric acid
	1.10.3.1	Catechol oxidase	Catechol
	1.10.3.2	Laccase	Phenylenediamine
	1.11.1.7	Peroxidase	Pyrogallol, chloroanilines[a]
	1.14.18.1	Monophenol monooxygenase	Catechol, pyrogallol, hydroquinone
Transferases	2.4.1.5	Dextransucrase	Sucrose
	2.4.1.10	Levansucrase	Sucrose
	2.8.1.1	Thiosulphate sulphur-transferase	Thiosulphate + cyanide
Hydrolases	3.1.1.1	Carboxylesterase	Hydroxy-methylcoumarin butyrate, Malathion[a]
	3.1.1.2	Arylesterase	Phenyl acetate
	3.1.1.3	Triacylglycerol lipase	4-Methyl umbelliferone nonanoate
	3.1.3.1	Alkaline phosphatase	*p*-Nitrophenyl phosphate
	3.1.3.2	Acid phosphatase	*p*-Nitrophenyl phosphate
	3.1.6.1	Arylsulphatase	*p*-Nitrophenyl sulphate
	3.2.1.1	α-Amylase	Starch
	3.2.1.2	β-Amylase	Starch
	3.2.1.4	Cellulase	Cellulose, carboxymethyl-cellulose
	3.2.1.6	Endo-1,3(4)-β-D glucanase	Laminarin
	3.2.1.8	Xylanase	Xylan
	3.2.1.2	β-Glucosidase	*p*-Nitrophenyl β-D glucoside, cellobiose
		Peptidases	*N*-benzoyl L-arginine amide, benzyloxycarbonyl phenylalanyl leucine
		Proteinases	Casein, gelatine
	3.5.1.1	Asparaginase	Asparagine
	3.5.1.2	Glutaminase	Glutamine
	3.5.1.4	Amidase	Formamide, acetamide
	3.5.1.5	Urease	Urea
	3.5.1.13	Aryl acylamidase	Propanil[a]
Lyases	4.1.1.15	Glutamate decarboxylase	Aspartic acid
	4.1.1.25	Tyrosine decarboxylase	Tyrosine
	4.1.1.28	Aromatic-L-amino acid decarboxylase	Tryptophan, DOPA

[a]Xenbiotic substrates (pesticides).

tase and the phosphatases have been immobilized either after being released from leaking or dead cells or possibly by incorporation into resistant cell constituents (e.g. melanin) prior to release. Other hydrolases were possibly (e.g. β-glucosidase, amidase, xylanase, peptidases) or definitely (e.g. cellulase, endo- 1,3(4)-β-D glucanase, proteinases) extracellular (in terms of their usual functional location) prior to immobilization, although there are obvious difficulties in envisaging immobilized enzyme (e.g. cellulase)-insoluble substrate (e.g. cellulose) interactions in soil. Some immobilized enzymes of uncertain classification are able to degrade xenobiotic compounds such as pesticides (Gibson & Burns, 1977; Burns & Edwards, 1980; Burns & Gibson, 1980). Important non-hydrolytic enzymes that can be assayed in crude humic extracts include oxidoreductases involved in humic polymer synthesis (Mayaudon, El Halfawi & Chalvignac, 1973; Suflita & Bollag, 1980) as well as in the incorporation of pesticide residues into soil organic matter (see page 265).

The significance of immobilized enzymes in biogeochemistry and agriculture is difficult to assess, but work with urease and other enzymes has suggested that soil enzymes have an important role in the rate of substrate degradation as well as being a resistant catalyst that will survive or recover from treatments which inhibit or eliminate the indigenous microflora (Lethbridge & Burns, 1976; Burns & Lethbridge, 1981).

Strategies for exogenous substrate utilization in soil

Soil microorganisms which depend upon the activity of their extracellular enzymes to provide them with soluble nutrients are faced with a number of problems. It is apparent that soil is an unfavourable environment for extracellular enzymes which are inactivated following adsorption to clay minerals, are denatured by physical and chemical factors, and serve as substrates for proteolytic microorganisms (Fig. 1). Experiments in which enzymes have either been added to non-sterile soil, or those in which production has been stimulated, reveal that new enzyme is rapidly destroyed (Roberge, 1970; Zantua & Bremner, 1976). Free proteins and amino acids are also rapidly degraded in soil (Schmidt, Putnam & Paul, 1960; Verma *et al.*, 1975). However, for an enzyme to be of benefit to its producer cell it is self-evident that it must avoid destruction for long enough to locate its substrate. Next, the enzyme must be capable of transforming the substrate which may be insoluble, exist in intimate association with other organic molecules, or be bound to clay and humic particles (see page 261) and is not, therefore, in its most accessible form. Furthermore, even if the enzyme can overcome these obstacles and bind with its substrate, the physical conditions may be unsuitable for catalysis (e.g. sub-optimal pH, dehydrated: hydrolases, anoxic:oxygenases). Following a successful enzyme–

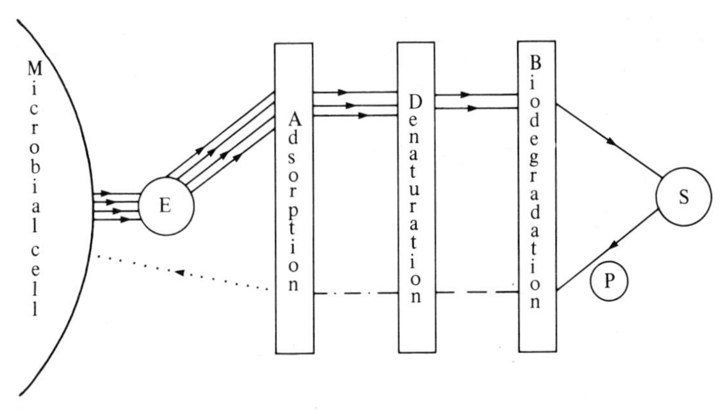

Fig. 1. The attenuation of enzyme–substrate interactions in soil by various biological and abiological forces. E, enzyme; S, substrate; P, product or trigger molecule (e.g. chemoattractant, inducer).

substrate interaction, the product must diffuse towards the source of the enzyme to serve directly as a growth substrate, as an enzyme inducer molecule, or as a signal molecule to enable the microorganism to make a tactic response to the substrate source. Of course, throughout its journey the product molecule, which is biochemically less complex than the original substrate, is also subject to adsorption, non-biological transformation and metabolism by a host of microbial species.

Having stated the hazards faced by extracellular enzymes in the presence of soil, it is difficult to conceive of these catalysts as serving any useful function *in situ*. Nevertheless, the fact remains that a large number of macromolecular substrates are presented to microorganisms and a large number of microbial species secrete extracellular enzymes and grow on the products of macromolecular biotransformations. Plant leaves, insect exoskeletons and microbial tissues *are* mineralized in soil! Therefore, we need to consider the various strategies which microorganisms may adopt when faced with exogenous substrates.

Four types of extracellular enzyme mechanisms can be recognized (Cohen, 1980) based on two forms of metabolic control, constitutive and inducible.

(1) Constitutive synthesis/induced or de-repressed secretion. If this type of control exists it suggests that microorganisms have an intracellular or cell bound store of unsecreted protein – possibly contained in the cytoplasm, restricted to the periplasmic space (of

Gram-negative bacteria), attached to the cell wall, or trapped within an extracellular polysaccharide coat.

(2) Constitutive synthesis/constitutive secretion. This mechanism implies that enzymes are being synthesized and secreted continuously, although possibly at a low level. It also means that the microorganism does not recognize the substrate or any other potential inducer.

(3) Induced synthesis/constitutive secretion. Induction requires the presence of an effector or inducer molecule, usually the substrate or some structurally related compound, which serves to initiate the synthesis of an enzyme *de novo*. Secretion of the enzyme follows automatically. Substrates demanding a number of catalytic steps prior to their utilization may stimulate sequential inductions, that is the induction of enzyme by the product of a previous enzyme reaction.

(4) De-repressed synthesis/constitutive secretion. De-repression involves the lifting of enzyme repression most often due to high levels of carbon, nitrogen and sulphur nutrient sources. Thus depletion of an easily assimilable substrate or of a catabolite gives rise to enzyme synthesis and secretion. De-repression and exoenzyme synthesis coincides with the post-exponential phase of growth in batch cultures and can therefore be a virtually continuous process in chemostat cultures grown under specific nutrient limitation. For example, proteinase synthesis may be stimulated and maintained under nitrogen-limited conditions and alkaline phosphatase under phosphorus limitation.

It is important to realize, when discussing the regulation of enzyme synthesis and secretion, that induction and repression are related to concentration, period of exposure and such factors as temperature, pH and oxygen level, as well as the chemical nature of the molecule. The same molecule that is an inducer under one set of circumstances may be a repressor under other environmental conditions, or if at a different concentration.

Let us now consider the relevance of these various types of regulation with regard to microbial activity in soil, remembering that the soil will generally contain sub-optimal concentrations of substrate unevenly distributed in both space and time. This means that at any particular moment (and probably for long periods of time) there may be no suitable substrate within the sphere of influence (say 50 μm) of a particular microorganism.

Although there is little evidence for mechanism (1) from investigations *in vitro* (and certainly no significant studies of regulation in soil), it could conceivably represent an economic method of utilizing exogenous polymeric substrates in soil. A polymeric substrate may require a number of enzymic transformations before it can be transported into the cytoplasm and it would be advantageous to compartmentalize the various depolymerases such that they are protected from dilution and inactivation. For example, in a sequence A→B→C→D, where A is an exogenous macromolecule and D is the monomeric constituent of A that is transported through the cytoplasmic membrane and metabolized, only the constitutive enzyme converting A to B need be truly extracellular. The other enzymes in the pathway could be maintained at constant levels within, for instance, the exopolymer coat, or attached to the outer membrane, or retained within the periplasmic space. Furthermore, the differential porosities of these compartments suggest that their outer boundaries could act as molecular sieves controlling the passage of decreasing molecular weight products from one enzyme-containing compartment to the next.

The idea that microorganisms in soil continuously synthesize and secrete enzymes (mechanism 2) does not appeal. This might prove to be a profitable strategy for microorganisms growing under the high nutrient, non-competitive conditions often encountered *in vitro*, or for obligate phytopathogens in the early stages of infection (e.g. pectate lyases of *Erwinia, Bacillus* and *Fusarium* species). However, such an approach would be energetically unfavourable in soil. Indeed in the prolonged absence of substrate, the nutritional and energy deficit brought about by continuous enzyme production would prove fatal to the microorganism.

The value of mechanism (3) to a microorganism in soil depends on the location and solubility of the substrate. For instance, if potential substrate is distantly located and non-diffusible then success (initially expressed as movement towards the substrate) depends upon the receipt of a reporter or trigger molecule, possibly released by the action of a 'scout' enzyme. The scout enzyme must therefore be constitutive. The notion of an enzyme being produced constitutively at low concentrations and then induced when substrate is detected is perfectly reasonable; however, a scout enzyme would have the same problems as the constitutive enzymes involved in mechanism (2). Closely located substrate may be recognized by the microorganism and induce efficient enzyme synthesis by contact.

De-repressed synthesis of enzymes (mechanism 4) is believed to be the strategy adopted by a large number of microbial species. Under the starvation conditions encountered in soil, de-repression due to substrate or product depletion would be the rule rather than the exception. This is all very well if the new enzyme is able to rapidly locate and interact with its substrate, but in the absence of the substrate then the same energetic problems arise as described for (2) and (3).

Unfortunately, the unsuitability of most of these mechanisms is evident when considering the hostility of the soil environment to extracellular enzymes (see page 282). Indeed a microorganism that adopts any strategy which is based upon a speculative role for extracellular enzymes has little chance of success. Instead the secretion of enzymes should be a response to a situation in which there is a high probability of finding a substrate or, at the very least, does not represent a serious drain on the energy resources of the organism. An example of the latter would be the periodic release of enzymes from damaged, dead and lysing cells within a population. These enzymes will only have an ephemeral existence in soil but during that period may locate and react with a substrate. Even if these enzymes fail (as, no doubt, they often will) their destruction does not represent an energy deficit to the remaining viable cells. Thus enzymes arising from moribund cells may represent the least risk to the species. Another possibility is for microorganisms to produce trigger molecules indirectly by creating an environment in which the spontaneous chemical cleavage of exogenous compounds occurs. Therefore, a different series of strategies could depend on the receipt by the microorganism of a trigger molecule which would inform the cell of the presence of a putative substrate. One such strategy involves chemoreception and taxis (Payne, 1980b). In this instance binding proteins located on the cell wall could act as chemoreceptors. Then, over a period of time, as more attractant is bound to the cell, the microorganism would respond to the gradient by moving towards the target. A choice between more than one attractant would be governed by their relative concentrations and binding affinities.

A method of actually attracting protein substrates without involving microbial movement was suggested by Cohen (1980). He reported that under starvation conditions non-motile *Neurospora crassa* and *Aspergillus nidulans* precipitated proteins onto their cell walls and thereby set up a concentration gradient that led to a

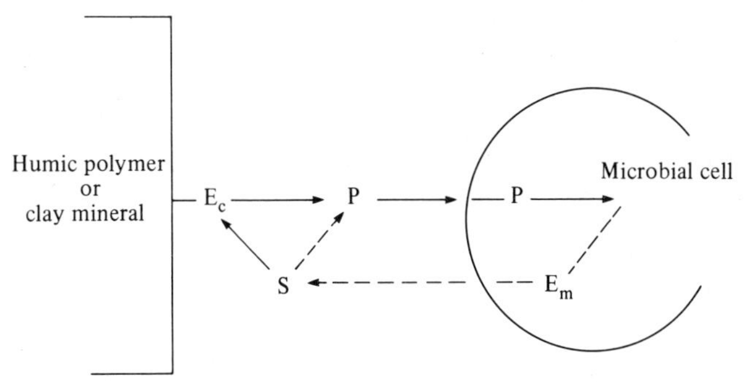

Fig. 2. Role of soil colloid–enzyme complex in substrate–microbe interaction. E_c, enzyme associated with humic polymer (or clay mineral); E_m, induced microbial enzyme; S, exogenous substrate; P, product.

diffusion-powered flow of soluble protein from a distant site. Of course, cell walls may gradually accumulate nutrients but only translocate them into the cell once a certain concentration has been reached. Enzyme induction would follow these preliminary steps.

Other responses to nutrient insufficiency are discussed by Tempest & Neijssell (1978 and Chapter 4) and may include a switch from low to high affinity catalytic and uptake systems, and the secretion of substances (e.g. chelating agents) which react with specific nutrients and thereby promote their uptake by the cell. Microorganisms must also regulate the rates of penetration of non-limiting nutrients so as to prevent the intracellular accumulation of intermediary metabolites.

Another strategy, involving immobilized soil enzymes in the microbial utilization of exogenous substrates, has been proposed (Burns 1979, 1980, 1982b). Its applicability depends directly on the widespread occurrence of immobilized enzymes and their stability in comparison with free extracellular enzymes, and indirectly on the general inadequacy of the other strategies for substrate detection. The major steps in the relationship are depicted in Fig. 2.

Molecule, S, on account of its size, insolubility, association with other substrates and soil constituents, or potential toxicity, is unsuitable in its present form as a growth substrate. Moreover, if S is insoluble or immobilized in some way the microorganism has no way of detecting its presence other than by direct contact following random movement. The immobilized enzyme (soil colloid–enzyme complex), E_c, is protected from inactivation and proteolysis and is certainly not subject to the control exerted on its microbial counterpart. In other words E_c is an indigenous constituent of soil capable

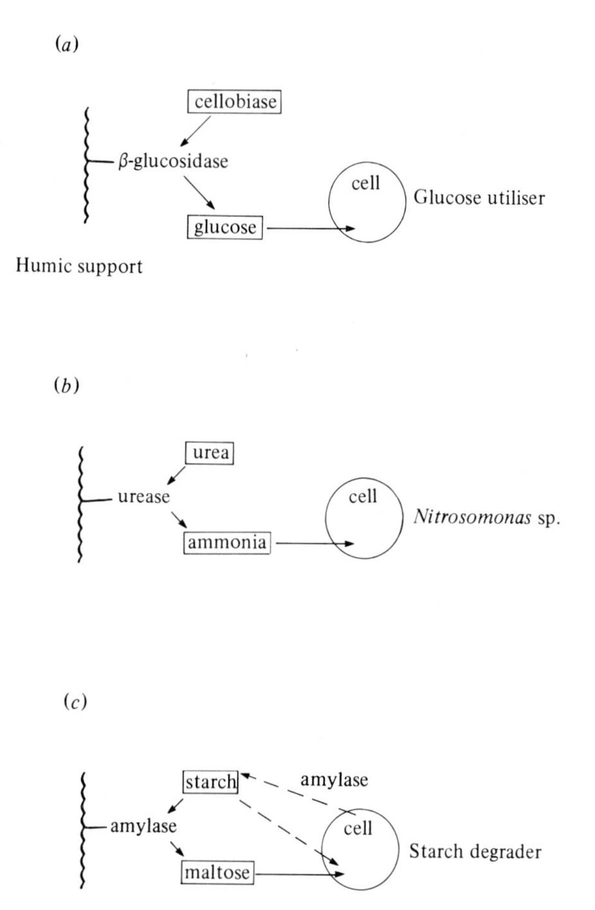

Fig. 3. Possible strategies involving immobilized soil enzymes in the utilization of exogenous substrates by microorganisms.

of responding rapidly to its specific substrate molecule, S. The product of this reaction, P, is metabolized by an adjacent cell or diffuses to a remote cell prior to uptake. In addition, if P is at an appropriate concentration (and therefore indicating the same for S) it may act as an enzyme inducer for the microorganism and/or as a stimulus for chemotaxis. The consequence of P being the trigger molecule referred to earlier is that the microorganism assumes the task of extracellular substrate decay, is brought in close contact with the substrate and is assured of exposure to the product (which will become the intracellular substrate).

Let us speculate as to how this strategy could work using specific examples of enzymes and substrates.

Firstly an example involving soil-β-glucosidase complexes (Fig. 3*a*). This immobilized enzyme is able to catalyse the cleavage of soluble cellobiose (and, in most assays, *p*-nitrophenyl β-D-glucoside) and has been described in soils and humic extracts (Batistic, Sarkar & Mayaudon, 1980; Hope, Alexander & Burns, 1980). The glucose produced is obviously an excellent carbon source and a chemoattractant for a great many microorganisms. In addition, more complicated relationships could involve regulation of microbial enzymes through catabolite repression, as well as the acceleration of cellulolysis by the removal of an end-product inhibitor (cellobiose) of the cellulases. β-Glucosidase will also release cellotriose and higher-oligomers from cellulose which may act as cellulase inducers in some microorganisms.

Urease is well known to form stable complexes with soil humic matter (Burns, Pukite & McLaren, 1972; Nannipieri, Ceccanti, Cervelli & Sequi, 1974; Pettit, Smith, Freedman & Burns, 1976) and can effect the rapid hydrolysis of urea to carbon dioxide and ammonia. Nitrifying chemoautotrophs, such as *Nitrosomonas* species, would obviously benefit from being in close proximity to a stable exogenous urease enzyme and would not need to possess a ureolytic capacity of their own (Fig. 3*b*). Carbon dioxide fixation by heterotrophs and its involvement in methanogenesis suggest further relationships between microorganisms and immobilized soil urease. Other immobilized ammonia-producing enzymes (e.g. asparaginase, amidase, D-glutaminase) may serve a similar function.

Although end-product inhibition is a common phenomenon in the regulation of extracellular enzyme synthesis, maltose acts as an inducer for amylase secretion in some microorganisms (Fogarty & Kelly, 1980). Therefore a soil–amylase complex (Pancholy & Rice, 1973) could serve as a stable detector of starch, produce the inducer disaccharide, and initiate the microbial degradation of the substrate (Fig. 3*c*). Amino acids and peptides may induce extracellular proteinase synthesis in some bacteria (Loriya, Bryukner & Egorov, 1977).

It is more difficult to envisage the catalysis of insoluble substrates, such as native cellulose or laminarin, by immobilized soil enzymes. However, this type of reaction may be possible (Lethbridge, Bull & Burns, 1978) if there is a high concentration of colloid-polysaccharase complexes in soil (thereby improving opportunities for substrate–enzyme contact) and if the colloidal and motile nature of the enzyme support is taken into account. Sugar dimers or low number oligomers released by immobilized enzymes could act as inducers.

It is attractive (even if somewhat fanciful) to think of soil colloid–enzyme complexes as having a crucial role in the cycling of carbon, nitrogen and other elements, particularly in conjunction with the notion of clays as contributing to protein formation and the beginning of life on Earth (Bernal, 1959). However, a far greater understanding of the soil environment along with the development of suitable experimental probes are needed before any of these possible strategies can be evaluated.

CONCLUSION

It is difficult to imagine a more complicated environment for microbial activity than that of soil. Moreover, the interplay of chemical, physical and biological influences and the presence of spatial and temporal gradients ensure that an understanding of microorganisms *in vitro* only provides clues to the microbiology of the soil.

Nowhere is the chasm between *in vitro* observations and events in soil more apparent than when comparing enzyme activities. This is because a high proportion of the potential substrate in soil is insoluble or otherwise unavailable to microorganisms and requires one or a series of transformations before it becomes suitable for cell uptake. Furthermore, and in contrast to most traditional studies *in vitro*, substrates are distributed in soil in an heterogeneous manner and there are even prolonged periods when substrates are absent altogether. After considering the characteristics of substrates, together with the physicochemical nature of the soil environment, it is obvious that the factors which regulate and control exoenzyme production and substrate transport *in vitro* can provide only a general guide to those processes in soil. Direct investigations of microbe–exoenzyme–substrate interactions in soil are few and far between. Indeed, our knowledge of the enzymology of polymer degradation by microorganisms in pure and mixed cultures is incomplete so it is not surprising that microbiologists have avoided introducing a further complication, the soil.

In this review I have highlighted the problems faced by those soil-dwelling microorganisms which depend upon the activities of their extracellular enzymes to provide them with substrates. A number of elaborate strategies allowing the detection of, response to, and utilization of exogenous substrates are presented and assessed. All suffer from a lack of experimental support and, in that sense, serve as signposts to future research in soil microbiology.

REFERENCES

Anderson, D. W., Saggar, S., Bettany, J. R. & Stewart, J. W. B. (1981). Particle size fractions and their use in studies of soil organic matter. I. The nature and distribution of forms of carbon, nitrogen, and sulphur. *Soil Science Society of America Journal,* **45,** 767–72.

Anderson, J. W. & Drew, E. A. (1972). Growth characteristics of a species of *Lipomyces* and its degradation of paraquat. *Journal of General Microbiology*, **70,** 43–58.

Arnold, P. W. (1978). Surface-electrolyte interactions. In *The Chemistry of Soil Constituents*, ed. D. J. Greenland & M. H. B. Hayes, pp. 355–404. Chichester: John Wiley & Sons.

Bartha, R. (1980). Pesticide residues in humus. *American Society for Microbiology News*, **46,** 356–60.

Bartha, R. & Pramer, D. (1970). Metabolism of acylanilide herbicides. *Advances in Applied Microbiology,* **13,** 317–41.

Batistic, L., Sarkar, J. M. & Mayaudon, J. (1980). Extraction, purification and properties of soil hydrolases. *Soil Biology and Biochemistry,* **12,** 59–63.

Benzing-Puride, L. (1981). Glucosamine and galactosamine distribution in soil as determined by gas liquid chromatography of soil hydrolysates: effect of acid strength and cations. *Soil Science Society of America Journal,* **45,** 66–70.

Bernal, J. D. (1959). *The Physical Basis of Life.* London: Routledge & Kegan Paul.

Bollag, J-M., Liu, S-Y. & Minard, R. D. (1980). Cross-coupling of phenolic humus constituents and 2,4-dichlorophenol. *Soil Science Society of America Journal*, **44,** 52–6.

Braun, V. (1978). Structure-function relationships of the Gram-negative bacterial cell envelope. In *Relations Between Structure and Function in the Prokaryotic Cell, Symposium of the Society for General Microbiology*, vol. 28, ed. R. Y. Stanier, H. J. Rogers & J. B. Ward, pp. 111–38. Cambridge University Press.

Brown, G., Newman, A. C. D., Rayner, J. H. & Weir, A. H. (1978). The structure and chemistry of soil clay minerals. In *The Chemistry of Soil Constituents*, ed. D. J. Greenland & M. H. B. Hayes, pp. 29–178. Chichester: John Wiley & Sons.

Bull, A. T. (1980). Biodegradation: some attitudes and strategies of microorganisms and microbiologists. In *Contemporary Microbial Ecology*, ed. D. C. Ellwood, J. N. Hedger, M. J. Latham, J. M. Lynch & J. H. Slater, pp. 107–36. London: Academic Press.

Burchill, S., Hayes, M. H. B. & Greenland, D. J. (1981). Adsorption. In *The Chemistry of Soil Processes*, ed. D. J. Greenland & M. H. B. Hayes, pp. 221–400. Chichester: John Wiley & Sons.

Burns, R. G. (1975). Factors affecting pesticide loss from soil. In *Soil Biochemis-*

try, vol. 4, ed. E. A. Paul & A. D. McLaren, pp. 103–41. New York: Marcel Dekker.

BURNS, R. G. (1977). Soil enzymology. *Science Progress (Oxford),* **64,** 275–85.

BURNS, R. G. (1978). Enzymes in soil: some theoretical and practical considerations. In *Soil Enzymes*, ed. R. G. Burns, pp. 295–339. London: Academic Press.

BURNS, R. G. (1979). Interaction of microorganisms, their substrates and their products with soil surfaces. In *Adhesion of Microorganisms to Surfaces*, ed. D. C. Ellwood, J. Melling & P. Rutter, pp. 109–38. London: Academic Press.

BURNS, R. B. (1980). Microbial adhesion to soil surfaces: consequences for growth and enzyme activities. In *Microbial Adhesion to Surfaces*, ed. R. C. W. Berkeley, J. M. Lynch, J. Melling, P. R. Rutter & B. Vincent, pp. 249–62. Chichester: Ellis Horwood.

BURNS, R. G. (1982a). Carbon mineralization by mixed cultures. In *Microbial Interactions and Communities*, ed. A. T. Bull & J. H. Slater, pp. 475–543. London: Academic Press.

BURNS, R. G. (1982b). Enzymes in soil: location and a possible role in microbial ecology. *Soil Biology and Biochemistry*, **14,** 421–5.

BURNS, R. G. & AUDUS, L. J. (1970). Distribution and breakdown of paraquat in soil. *Weed Research,* **10,** 49–58.

BURNS, R. G. & EDWARDS, J. A. (1980). Pesticide breakdown by soil enzymes. *Pesticide Science,* **11,** 506–12.

BURNS, R. G., EL-SAYED, M. H. & MCLAREN, A. D. (1972). Extraction of an urease-active organo-complex from soil. *Soil Biology and Biochemistry,* **4,** 107–8.

BURNS, R. G. & GIBSON, W. P. (1980). The disappearance of 2,4-D, diallate and malathion from soil and soil components. In *Agrochemicals in Soils*, ed. A. Banin & U. Kafkafi, pp. 149–59. Oxford: Pergamon Press.

BURNS, R. G., GREGORY, L. J., LETHBRIDGE, G. & PETTIT, N. M. (1978). The effect of γ-irradiation on soil enzyme stability. *Experientia,* **34,** 301–2.

BURNS, R. G. & LETHBRIDGE, G. (1981). The effect of pesticides and fertilisers on enzyme activities in soil. *Journal of the Science of Food and Agriculture,* **32,** 626–7.

BURNS, R. G. & MARTIN, J. P. (1983). Biodegradation of organic residues in soil. In *Microfloral and Microfaunal Interactions in Natural and Agro-Ecosystems*, ed. M. J. Mitchell. Martinus Nijhoff, The Hague (in press).

BURNS, R. G., PUKITE, A. H. & MCLAREN, A. D. (1972). Concerning the location and persistence of soil urease. *Soil Science Society of America Proceedings,* **36,** 308–11.

CAIN, R. B. (1980). The uptake and catabolism of lignin-related aromatic compounds and their regulation in microorganisms. In *Lignin Biodegradation: Microbiology, Chemistry and Potential Applications*, vol. 1, ed. T. K. Kirk, T. Higuchi & H.-M. Chang, pp. 21–60. Florida: CRC Press.

CHESHIRE, M. V. (1977). Origins and stability of soil polysaccharides. *Journal of Soil Science,* **28,** 1–10.

COHEN, B. L. (1980). Transport and utilization of proteins by fungi. In *Microorganisms and Nitrogen Sources*, ed. J. W. Payne, pp. 411–30. Chichester: John Wiley & Sons.

COLEMAN, D. C. (1976). A review of root production processes and their influence on soil biota in terrestrial ecosystems. In *The Role of Terrestrial and Aquatic Organisms in Decomposition Processes*, ed. J. M. Anderson & A. Macfadyen, pp. 417–34. Oxford: Blackwell Scientific Publications.

CORTEZ, J. & SCHNITZER, M. (1981). Reactions of nucleic acid bases with inorganic soil constituents. *Soil Biology and Biochemistry,* **13,** 173–8.

Daughton, C. G. & Hsieh, D. P. (1977). Parathion utilization by bacterial symbionts in a chemostat. *Applied and Environmental Microbiology,* **34,** 175–84.

Dekker, R. F. H. & Richards, G. N. (1976). Hemicellulases: their occurrence, purification, physicochemical properties and mode of action. *Advances in Carbohydrate Chemistry and Biochemistry,* **32,** 227–352.

Deutch, C. E. & Soffer, R. L. (1978). *Escherichia coli* mutants defective in dipeptidyl carboxypeptidases. *Proceedings of the National Academy of Sciences, USA,* **75,** 5998–6001.

Dommergues, Y. (1977). *Biologie du Sol.* Paris: Presses Universitaire de France.

Drozdowicz, A. (1971). The behaviour of cellulase in soil. *Revues of Microbiology*, **2,** 17–23.

Eriksson, K-E. (1978). Enzyme mechanisms involved in cellulose hydrolysis by the white rot fungus *Sporotrichum pulverulentum. Biotechnology and Bioengineering Symposium,* **10,** 317–32.

Eriksson, K-E. & Johnstrud, S. C. (1982). Mineralization of carbon. In *Experimental Microbial Ecology*, ed. R. G. Burns & J. H. Slater, pp. 134–53. Oxford: Blackwell Scientific Publications.

Estermann, E. F., Peterson, G. H. & McLaren, A. D. (1959). Digestion of clay-protein, lignin-protein, and silica-protein complexes by enzymes and bacteria. *Soil Science Society of America Proceedings,* **23,** 31–6.

Fan, L. T., Lee, Y-H. & Beardmore, D. H. (1980). Major chemical and physical features of cellulosic materials as substrates for enzymatic hydrolysis. *Advances in Biochemical Engineering,* **14,** 101–17.

Farmer, V. C. (1978). Water on particle surfaces. In *The Chemistry of Soil Constituents*, ed. D. J. Greenland & M. H. B. Hayes, pp. 405–48. Chichester: John Wiley & Sons.

Filip, Z. (1973). Clay minerals as a factor influencing the biochemical activity of soil microorganisms. *Folia Microbiologica,* **18,** 56–74.

Filip, Z., Haider, K. & Martin, J. P. (1972). Influence of clay minerals on growth and metabolic activity of *Epicoccum nigrum* and *Stachybotrys chartarum. Soil Biology and Biochemistry,* **4,** 134–45.

Fisher, M. L., Anderson, A. J. & Albersheim, P. (1973). Host-pathogen interactions. VI. Single plant protein efficiency inhibits endopolygalacturonases secreted by *Colletotrichum lindemuthianum* and *Aspergillus niger. Plant Physiology,* **51,** 489–91.

Flaig, W., Beutelspacher, H. & Rietz, E. (1975). Chemical composition and physical properties of humic substances. In *Soil Components*, vol. 1, ed. J. E. Gieseking, pp. 1–211. Berlin: Springer-Verlag.

Fluhler, H., Ardakani, M. S., Szuszkiewicz, T. E. & Stolzy, L. H. (1976). Field-measured nitrous oxide concentrations, redox potentials, oxygen diffusion rates, and oxygen partial pressures in relation to denitrification. *Soil Science,* **122,** 107–14.

Fogarty, W. M. & Kelly, C. T. (1980). Amylases, amyloglucosidases and related glucanases. In *Economic Microbiology*, vol. 5, ed. A. H. Rose, pp. 115–70. London: Academic Press.

Foster, R. C. & Martin, J. K. (1981). In situ analysis of soil components of biological origin. In *Soil Biochemistry*, vol. 5, ed. E. A. Paul & J. N. Ladd, pp. 75–111. New York: Marcel Dekker.

Gessa, C., Melis, P., Bellu, G. & Testini, C. (1978). Inactivation of clay pH-dependent charge in organo-mineral complexes. *Journal of Soil Science* **29,** 58–64.

Gibson, W. P. & Burns, R. G. (1977). The breakdown of malathion in soil and soil components. *Microbial Ecology,* **3,** 219–30.

Gieseking, J. E. (1975). *Soil Components. Volume 2 Inorganic Components.* New York: Springer Verlag.

Glenn, A. R. (1976). Production of extracellular proteins by bacteria. *Annual Review of Microbiology,* **30,** 41–62.

Goksøyr, J. & Eriksen, J. (1980). Cellulases. In *Economic Microbiology*, vol. 5, ed. A. H. Rose, pp. 283–330. London: Academic Press.

Greenland, D. J. (1971). Interactions between humic and fulvic acids and clays. *Soil Science,* **111,** 34–41.

Griffiths, E. & Burns, R. G. (1972). Interaction between phenolic substances and microbial polysaccharides in soil aggregation. *Plant and Soil,* **36,** 599–612.

Guckert, A., Valla, M. & Jacquin, F. (1975). Adsorption of humic acids and soil polysachharides on montmorillonite. *Pochvovedenye,* **2,** 41–7.

Haska, G. (1975). Influence of clay minerals on sorption of bacteriolytic enzymes. *Microbial Ecology,* **1,** 234–45.

Haska, G. (1981). Activity of bacteriolytic enzymes adsorbed to clays. *Microbial Ecology,* **7,** 331–4.

Hattori, T. & Hattori, R. (1976). The physical environment in soil microbiology: an attempt to extend principles of microbiology to soil microorganisms. *CRC Critical Reviews of Microbiology,* **4,** 423–61.

Hayes, M. H. B. (1980). The role of natural and synthetic polymers in stabilizing soil aggregates. In *Microbial Adhesion to Surfaces*, ed. R. C. W. Berkeley, J. M. Lynch, J. Melling, P. R. Rutter & B. Vincent, pp. 263–96. Chichester: Ellis Horwood.

Hayes, M. H. B. & Swift, R. S. (1978). The chemistry of soil organic colloids. In *The Chemistry of Soil Constituents*, ed. D. J. Greenland & M. H. B. Hayes, pp. 179–320. Chichester: John Wiley & Sons.

Hayes, M. H. B. & Swift, R. S. (1981). Organic colloids and organo-mineral associations. *Bulletin of the International Society of Soil Science,* **60,** 67–74.

Helling, C. S. & Krivonak, A. E. (1978). Physicochemical characteristics of bound dinitroaniline herbicides in soils. *Journal of Agriculture and Food Chemistry,* **26,** 1156–63.

Hill, I. R. & Wright, S. J. L. (1979). The behaviour and fate of pesticides in microbial environments. In *Pesticide Microbiology*, ed. I. R. Hill & S. J. L. Wright, pp. 79–136. London: Academic Press.

Hope, C. F. A., Alexander, J. M. & Burns, R. G. (1980). β-D-glucosidase activity in soil. *Society for General Microbiology Quarterly*, **8,** 41.

Hsu, T-S. & Bartha, R. (1974). Biodegradation of choloroaniline-humus complexes in soil and culture solution. *Soil Science,* **118,** 213–20.

Jenkinson, D. S. & Ladd, J. N. (1981). Microbial biomass in soil: measurement and turnover. In *Soil Biochemistry,* vol. 5, ed. E. A. Paul & J. N. Ladd, pp. 415–71. New York: Marcel Dekker.

Jenkinson, D. S. & Powlson, D. S. (1976). The effect of biocidal treatments on metabolism in soil V. *Soil Biology and Biochemistry,* **8,** 209–13.

Jenkinson, D. S. & Rayner, J. H. (1977). The turnover of soil organic matter in some of the Rothamsted classical experiments. *Soil Science,* **123,** 298–305.

Khan, S. V. (1978). The interaction of organic matter with pesticides. In *Soil Organic Matter*, ed. M. Schnitzer & S. V. Khan, pp. 137–71. Amsterdam: Elsevier.

Ladd, J. N. (1978). Origin and range of enzymes in soil. In *Soil Enzymes*, ed. R. G. Burns, pp. 51–96. London: Academic Press.

Lee, Y-H. & Fan, L. T. (1980). Properties and mode of action of cellulase. *Advances in Biochemical Engineering,* **17,** 101–29.

Lethbridge, G. & Burns, R. G. (1976). Inhibition of soil urease by organophosphorus insecticides. *Soil Biology and Biochemistry,* **8,** 99–102.

LETHBRIDGE, G., BULL, A. T. & BURNS, R. G. (1978). Assay and properties of 1,3 β-glucanase in soil. *Soil Biology and Biochemistry,* **10,** 389–91.
LINHARES, L. F. & MARTIN, J. P. (1979). Decomposition in soil of emodin, chrysophanic acid, and a mixture of anthroquinones synthesized by an *Aspergillus glaucus* isolate. *Soil Science Society of America Journal,* **43,** 940–5.
LORIYA, Z. K., BRYUKNER, B. & EGOROV, N. S. (1977). Effect of amino acids on the synthesis of exocellular protease in *Serratia marcescens. Microbiology USSR,* **6,** 31–5.
MCGILL, W. B. & PAUL, E. A. (1976). Fractionation of soil and ^{15}N nitrogen to separate the organic and clay interactions of immobilized N. *Canadian Journal of Soil Science,* **56,** 203–12.
MCLAREN, A. D. & SKUJINS, J. (1968). The physical environment of micro-organisms in soil. In *The Ecology of Soil Bacteria*, ed. T. R. G. Gray & D. Parkinson, pp. 3–24. Liverpool University Press.
MARSHALL, K. C. (1976). *Interfaces in Microbial Ecology.* Cambridge Massachusetts: Harvard University Press.
MARSHALL, K. C. (1980). Reactions of microorganisms, ions and macromolecules at interfaces. In *Contemporary Microbial Ecology*, ed. D. C. Ellwood, J. N. Hedger, M. J. Latham, J. M. Lynch & J. H. Slater, pp. 93–106. London: Academic Press.
MARSHMAN, N. A. & MARSHALL, K. C. (1981a). Bacterial growth on proteins in the presence of clay minerals. *Soil Biology and Biochemistry,* **13,** 127–34.
MARSHMAN, N. A. & MARSHALL, K. C. (1981b). Some effects of montmorillonite on the growth of mixed microbial cultures. *Soil Biology and Biochemistry,* **13,** 135–41.
MARTIN, J. P. & HAIDER, K. (1976). Decomposition of specifically carbon-14-labelled ferulic acid; free and linked into model humic acid-type polymers. *Soil Science Society of America Proceedings,* **40,** 377–80.
MARTIN, J. P. & HAIDER, K. (1980). Microbial degradation and stabilization of ^{14}C-labeled lignins, phenols, and phenolic polymers in relation to soil humus formation. In *Lignin Biodegradation: Microbiology, Chemistry and Potential Applications*, vol. 1, ed. T. K. Kirk, T. Higuchi & H–M. Chang, pp. 77–100. Boca Raton, Florida: CRC Press.
MARTIN, J. P., HAIDER, K. & LINHARES, L. (1979). Decomposition and stabilization of ring-^{14}C-labeled catechol in soil. *Soil Science Society of America Journal,* **43,** 100–4.
MARTIN, J. P., PARSA, A. A. & HAIDER, K. (1978). Influence of intimate association of humic polymers on biodegradation of {^{14}C} labeled organic substrates in soil. *Soil Biology and Biochemistry,* **10,** 483–6.
MATHUR, S. P. (1971). Characterization of soil humus through enzymatic degradation. *Soil Science,* **111,** 147–57.
MAYAUDON, J., EL HALFAWI, M. & CHALVIGNIC, M-A. (1973). Propriétés des diphénol oxydases extraites des sols. *Soil Biology and Biochemistry,* **5,** 369–83.
MUZZARELLI, R. A. A. (1977). *Chitin.* Oxford: Pergamon Press.
NAIDER, F., BECKER, J. M. & KATZIR-KATCHALSKI, E. (1973). Utilization of methionine-containing peptides and their derivatives by a methionine-requiring auxotroph of *Saccharomyces cerevisiae. Journal of Biological Chemistry,* **249,** 9–20.
NANNIPIERI, P., CECCANTI, B., CERVELLI, S. & SEQUI, P. (1974). Use of 0.1 M pyrophosphate to extract urease from a podzol. *Soil Biology and Biochemistry,* **6,** 359–62.
NIKAIDO, H. & NAKAE, T. (1979). The outer membrane of Gram-negative bacteria. *Advances in Microbial Physiology,* **20,** 163–250.

O'BRIEN, B. J. & STOUT, J. D. (1978). Movement and turnover of soil organic matter as indicated by carbon isotope measurements. *Soil Biology and Biochemistry,* **10,** 309–17.

OLNESS, A. & CLAPP, C. E. (1975). Influence of polysaccharide structure on dextran adsorption by montmorillonite. *Soil Biology and Biochemistry,* **7,** 113–18.

OSBORN, M. J. & WU, H. C. P. (1980). Proteins of the outer membrane of Gram-negative bacteria. *Annual Review of Microbiology,* **34,** 369–422.

OSMUNDSVAG, K. & GOKSØYR, J. (1975). Cellulases from *Sporocytophaga myxococcoides,* purification and properties. *European Journal of Biochemistry,* **57,** 405–9.

PANCHOLY, S. K. & RICE, E. L. (1973). Carbohydrases in soil as affected by successional stages of revegetation. *Soil Science Society of America Proceedings,* **37,** 227–9.

PAYNE, J. W. (ed.) (1980a). *Microorganisms and Nitrogen Sources.* Chichester: John Wiley & Sons.

PAYNE, J. W. (1980b). Transport and utilization of peptides by bacteria. In *Microorganisms and Nitrogen Sources*, ed. J. W. Payne, pp. 211–56. Chichester: John Wiley & Sons.

PAYNE, J. W. & GILVARG, C. (1978). Transport of peptides in bacteria. In *Bacterial Transport*, ed. B. P. Rosen, pp. 325–83. New York: Marcel Dekker.

PERROTT, K. W., LANGDON, A. G. & WILSON, A. T. (1974). Sorption of phosphate by aluminium and iron (III)-hydroxy species on mica surfaces. *Geoderma,* **12,** 223–31.

PETTIT, N. M., GREGORY, L. J., FREEDMAN, R. B. & BURNS, R. G. (1977). Differential stabilities of soil enzymes: assay and properties of phosphatase and arylsulphatase. *Biochimica et Biophysica Acta,* **485,** 357–66.

PETTIT, N. M., SMITH, A. R. J., FREEDMAN, R. B. & BURNS, R. G. (1976). Soil urease: activity, stability and kinetic properties. *Soil Biology and Biochemistry,* **8,** 479–84.

PFLUG, W. & ZIECHMANN, W. (1982). Humic acids and the disruption of bacterial cell walls by lysozyme. *Soil Biology and Biochemistry,* **14,** 165–6.

POLLOCK, M. R. (1962). Exoenzymes. In *The Bacteria*, vol. 4, ed. I.C. Gunsalus & R.Y. Stanier, pp. 121–78. New York: Academic Press.

REESE, E. T., SIU, R. G. H. & LEVINSON, H. S. (1950). The biological degradation of soluble cellulose derivatives and its relationship to the mechanism of cellulose hydrolysis. *Journal of Bacteriology,* **59,** 485–97.

REINEKE, W. & KNACKMUSS, H-J. (1979). Construction of haloaromatic utilizing bacteria. *Nature, London,* **277,** 385–6.

ROBERGE, M. R. (1970). Behavior of urease added to unsterilized, steam-sterilized, and gamma-radiation sterilized black spruce humus. *Canadian Journal of Microbiology,* **16,** 865–70.

ROGERS, H. T., PERKINS, H. R. & WARD, J. B. (1980). *Microbial Cell Walls and Membranes.* London: Chapman and Hall.

SCHERRER, R., BERLIN, E. & GERHARDT, P. (1977). Density, porosity and structure of dried cell walls isolated from *Bacillus megaterium* and *Saccharomyces cerevisiae. Journal of Bacteriology,* **129,** 1162–4.

SCHERRER, R. & GERHARDT, P. (1971). Molecular sieving by the *Bacillus megaterium* cell wall and protoplast. *Journal of Bacteriology,* **107,** 718–35.

SCHLESINGER, W. H. (1977). Carbon balance in terrestrial detritus. *Annual Reviews of Ecological Systems,* **8,** 51–81.

SCHMIDT, E. L., PUTNAM, H. D. & PAUL, E. A. (1960). Behavior of free amino acids in soil. *Soil Science Society of America Proceedings,* **24,** 107–9.

SCHNITZER, M. (1968). Reactions between organic matter and inorganic soil

constituents. *Transactions of the 9th International Congress of Soil Science, Adelaide*, Australia, **1,** 635–42.
SCHNITZER, M. & KHAN, S. U. (1972). *Humic Substances in the Environment.* New York: Marcel Dekker.
SCHNITZER, M., SOWDEN, F. J. & IVARSON, K. C. (1974). Humic acid reactions with amino acids. *Soil Biology and Biochemistry,* **6,** 401–7.
SELBY, K. & MAITLAND, C. C. (1967). The cellulase of *Trichoderma viride.* Separation of the components involved in the solubilization of cotton. *Biochemical Journal,* **104,** 716–24.
SJOBLAD, R. D. & BOLLAG, J-M. (1981). Oxidative coupling of aromatic compounds by enzymes from soil microorganisms. In *Soil Biochemistry*, vol. 5, ed. E. A. Paul & J. N. Ladd, pp. 113–52. New York: Marcel Dekker.
SLATER, J. H. & GODWIN, D. (1980). Microbial adaption and selection. In *Contemporary Microbial Ecology*, ed. D. C. Ellwood, J. N. Hedger, M. J. Latham, J. M. Lynch & J. H. Slater, pp. 137–60. London: Academic Press.
SØRENSEN, L. H. (1974). Rate of decomposition of organic matter in soil as influenced by repeated air-drying-rewetting and repeated additions of organic matter. *Soil Biology and Biochemistry,* **6,** 287–92.
SØRENSEN, L. H. (1975). The influence of clay on the rate of decay of amino acid metabolites synthesized in soils during the decomposition of cellulose. *Soil Biology and Biochemistry,* **7,** 171–7.
SØRENSEN, L. H. (1981). Carbon-nitrogen relationships during the humification of cellulose in soils containing different amounts of clay. *Soil Biology and Biochemistry,* **13,** 313–21.
SPARLING, G. P., ORD, B. G. & VAUGHAN, D. (1981). Changes in microbial biomass and activity in soils amended with phenolic acids. *Soil Biology and Biochemistry,* **13,** 455–60.
STOTZKY, G. (1974). Activity, ecology, and population dynamics of microorganisms in soil. In *Microbial Ecology*, ed. A. Laskin & H. Lechvalier, pp. 57–135. Cleveland, Ohio: CRC Press.
STOTZKY, G. (1980). Surface interactions between clay minerals and microbes, viruses and soluble organics, and the probable importance of these interactions to the ecology of microbes in soil. In *Microbial Adhesion to Surfaces*, ed. R. C. W. Berkeley, J. M. Lynch, J. Melling, P. R. Rutter & B. Vincent, pp. 231–47. Chichester: Ellis Horwood.
STOTZKY, G. & BURNS, R. G. (1982). The soil environment: clay-humus-microbe interactions. In *Experimental Microbial Ecology*, ed. R. G. Burns & J. H. Slater, pp. 105–33. Oxford: Blackwell Scientific Publications.
STOUT, J. D., GOH, K. M. & RAFTER, T. A. (1981). Chemistry and turnover of naturally occurring resistant organic compounds in soil. In *Soil Biochemistry*, vol. 5, ed. E. A. Paul & J. N. Ladd, pp. 1–73. New York: Marcel Dekker.
SUFLITA, J. M. & BOLLAG, J-M. (1980). Oxidative coupling activity in soil extracts. *Soil Biology and Biochemistry,* **12,** 177–83.
TEMPEST, D. W. & NEIJSSELL, O. M. (1978). Eco-physiological aspects of microbial growth in aerobic nutrient limited environments. *Advances in Microbial Ecology*, **2,** 105–53.
TERRY, R. E. (1980). Nitrogen mineralization in Florida histosols. *Soil Science Society of American Journal,* **44,** 747–50.
TERRY, R. E. & TATE, R. L. (1980). Denitrification as a pathway for nitrate removal from organic soils. *Soil Science,* **129,** 162–6.
THENG, B. K. G. (1979). *Formation and Properties of Clay-Polymer Complexes.* Amsterdam: Elsevier Scientific Publishing Company.

TURCHENEK, L. W. & OADES, J. M. (1979). Fractionation of organo-mineral complexes by sedimentation and density techniques. *Geoderma,* **21,** 311–43.

VARADI, J. (1972). The effect of aromatic compounds on cellulase and xylanase production of the fungi *Shizophyllum commune* and *Chaetomium globosum.* In *Biodeterioration of Materials*, vol. 2, ed. A. H. Walters & E. H. Hueck-van der Plas, pp. 129–35. London: Applied Science Publishers.

VERMA, L. & MARTIN, J. P. (1976). Decomposition of algal cells and components and their stabilization through complexing with model humic acid-type phenolic polymers. *Soil Biology and Biochemistry,* **8,** 85–90.

VERMA, L., MARTIN, J. P. & HAIDER, K. (1975) Decomposition of carbon-14-labeled proteins, peptides and amino acids; free and complexed with humic polymers. *Soil Science Society America Proceedings*, **39,** 279–84.

VOHRA, R. M., SHIRKOT, C. K. DHAWAN, S. & GUPTA, K. G. (1980). Effect of lignin and some of its components on the production and activity of cellulase(s) by *Trichoderma reesei. Biotechnology and Bioengineering,* **22,** 1497–1500.

WEBER, J. B. (1970). Mechanisms of adsorption of *s*-triazines by clay colloids and factors affecting plant availability. *Residue Reviews,* **32,** 93–128.

WEISS, L. (1974). Studies on cellular adhesion in tissue culture. XIV. Positively charged surface groups and the rate of cell adhesion. *Experimental Cell Research,* **83,** 311–18.

WILLIAMS, P. P. (1977). Metabolism of synthetic organic pesticides by anaerobic microorganisms. *Residue Reviews,* **66,** 63–135.

WILLIAMS, S. T. & GRAY, T. R. G. (1974). Decomposition of litter on the soil surface. In *Biology of Plant Litter Decomposition*, vol. 2, ed. C. H. Dickinson & G. J. F. Pugh, pp. 611–32. London: Academic Press.

WILLIAMS, S. T. & MAYFIELD, C. I. (1971). Studies on the ecology of actinomycetes in soil: III. The behaviour of neutrophilic streptomycetes in acid soil. *Soil Biology and Biochemistry,* **3,** 197–208.

WOLF, D. C. & MARTIN, J. P. (1976). Decomposition of fungal mycelia and humic-type polymers containing carbon-14 from ring and side-chain labeled 2,4-D and chlorpropham. *Soil Science Society of America Journal*, **40,** 700–4.

WOLFINBARGER, L. (1980). Transport and utilization of peptides by fungi. In *Microorganisms and Nitrogen Sources*, ed. J. W. Payne, pp. 281–300. Chichester: John Wiley & Sons.

WONG, P. T. W. & GRIFFIN, D. M. (1974). Effect of osmotic potential on streptomycete growth, antibiotic production and antagonism to fungi. *Soil Biology and Biochemistry,* **6,** 319–25.

WOOD, T. M. (1975). Properties and modes of action of cellulases. In *Cellulose as a Chemical and Energy Resource, Biotechnology and Bioengineering Symposium*, vol. 5, ed. C. R. Wilke, pp. 111–37. New York: John Wiley & Sons.

WOOD, T. M. & MCCRAE, S. I. (1978a). The mechanisms of cellulase action with particular reference to the C_1 component. In *Bioconversion of Cellulosic Substances into Energy, Chemicals and Microbial Protein*, ed. T. K. Ghose, pp. 111–41. New Delhi and Zurich: Indian Institute of Technology and Swiss Federal Institute.

WOOD, T. M. & MCCRAE, S. I. (1978b). The cellulases of *Trichoderma koningii. Biochemical Journal,* **171,** 61–72.

ZANTUA, M. I. & BREMNER, J. M. (1976). Production and persistence of urease activity in soils. *Soil Biology and Biochemistry,* **8,** 369–74.

ZUNINO, H., BORIE, F., AGUILERA, S., MARTIN, J. P. & HAIDER, K. (1982). Decomposition of ^{14}C-labeled glucose, plant and microbial products and phenols in volcanic ash-derived soils of Chile. *Soil Biology and Biochemistry,* **14,** 37–43.

MOVEMENT, TAXES AND CELLULAR INTERACTIONS IN THE RESPONSE OF MICROORGANISMS TO THE NATURAL ENVIRONMENT

ROBIN J. ROWBURY, JUDITH P. ARMITAGE AND CONRAD KING

Department of Botany and Microbiology and Department of Zoology, University College London, Gower Street, London WC1E 6BT, UK

INTRODUCTION

Motility and taxes are important in the natural environment, either to allow a microorganism to seek out a particular cellular site or cell type, for example in the homing of a gamete on its target or of a predator on its prey, to seek favourable physico-chemical conditions such as light, temperature, aeration or substrate, to avoid unfavourable conditions like the presence of toxic agents or to bring together organisms, under unfavourable conditions, to form resistant aggregates.

The aim of this article is to discuss the way in which motility and taxes of several distinct types allow prokaryotic and eukaryotic microorganisms to respond best to environmental factors. For each type of organism discussed, an account of its motility mechanism, as far as it is known, will be presented. This is important since advances in the understanding of taxis often depend on a sound knowledge of the underlying translocation mechanism. Thus knowledge of the way motile *Escherichia coli* cells respond to positive and negative chemotactic stimuli has followed closely from findings about the way rotation of the flagellar motor is governed.

The cell–cell interactions between motile microorganisms will also be considered because for many, some tactic behaviour depends not only on the independent movement of individual organisms but also on interplay between organisms involving either indirect interaction via exogenous components (e.g. chemical signalling between organisms as occurs in *Blepharisma* prior to conjugation) or direct cellular attachments leading to more efficient movement and tactic responses (e.g. as in swarming and aggregating myxobacteria) or both (as in the aggregating amoebae of *Dictyostelium*).

Although it is of paramount importance to establish how microorganisms move and how this movement is modulated to produce taxis, an attempt has been made here to emphasize how the behaviour and cellular interactions observed in the laboratory relate to the responses of the organism to its environmental conditions because motility is clearly significant in such conditions. Although motility and taxes in both liquid media and on solid surfaces will be considered, there will be some emphasis on the latter. The systems chosen for consideration are flagellar motility in bacteria (including a consideration of the swarming phenomenon which occurs for e.g. *Proteus* and *Vibrio* on solid surfaces), gliding motility particularly in myxobacteria and cytophagas and both amoeboid movement and ciliary/flagellar movement in protozoa.

FLAGELLATE BACTERIA

Most bacterial species move by rotating semi-rigid protein flagella, which are quite distinct from the flagella of eukaryotic organisms. The genetics and biochemistry of this bacterial behavioural system has been well reviewed by Macnab (1978), Koshland (1981) and Parkinson (1977).

Flagella structure

Individual flagella are too thin, 20 nm, to be distinguished *in vivo* by light microscopy, although groups of flagella working together as bundles, or sheathed flagella can just be seen. The cellular arrangement of flagella can vary from species to species. Thus *Escherichia coli* has several flagella apparently arranged at random over the bacterial surface whilst *Pseudomonas aeruginosa* has one or possibly two flagella located at the polar position. Other bacteria like *Rhodospirillum rubrum* have tufts of flagella at each pole, whilst in the case of the *Spirochaetes* these are arranged internally lying between the inner and outer membranes.

Bacterial flagella are made up of subunits of a single protein, flagellin, which appears to be arranged in 11 nearly longitudinal rows (Bode, Engel & Winklmair, 1972; Calladine, 1976). Isolated protein subunits are able to spontaneously polymerize to form flagella-like structures. There is evidence that flagellins may be similar throughout the bacterial kingdom, for example those from *Salmonella typhimurium* (Gm – ve) and *Bacillus subtilis* (Gm + ve) showing some sequence homology.

At the base of each flagellum is a hook, with a different protein composition to flagellin, functioning as a universal joint, allowing articulation of the flagellum. The flagellum passes through the outer membrane of Gram-negative bacteria via a pair of protein rings. It is thought that these rings (L and P) act as bushes. Flagellar rotation is accomplished by the rotation of two protein rings located close to the cytoplasmic membrane. One, the S-ring, is anchored in the cell wall, and the other, called the M-ring, rotates in the cytoplasmic membrane.

The genetics of flagellar formation and function

There are at least 29 genes involved in flagellar formation and function in *E. coli* and *S. typhimurium* (see Parkinson, 1977, and Iino, 1977, for reviews of genetics), their transcription being repressed in the presence of glucose, when motility is not required and derepressed with increase in cyclic AMP (Komeda, Suzuki, Ishidsu & Ino, 1975).

What determines the site of insertion of a new flagellum in the cell envelope is unknown, however, once inserted the flagellum grows by polymerization of the flagellin subunits at the distal tip of the flagellum. As no intracellular pool of flagellin can be detected it seems likely that flagellin is translated near the basal unit at the base of the flagellum and transported up the centre of the shaft to polymerize at the end until the maximum length is achieved.

The mechanism of flagella rotation and motility

The flagellum rotates as a semi-rigid helix driven by a rotary motor in the basal body (see reviews). It is the control of the direction of rotation of the rotary motor which is fundamental to the behaviour of bacteria. The flagella in most bacteria normally rotate in an anticlockwise direction. In polarly flagellate bacteria this results in a forward movement, the flagellum pushing like a ship's propeller, the body of the bacterium precessing slowly. In peritrichously flagellate bacteria the situation is similar, the flagella automatically form a bundle and operate as one flagellum pushing the cell forward. When the direction of flagella rotation reverses, by a mechanism as yet not well understood, the polarly flagellate bacterium is pulled along as efficiently in the new direction. However, reversal of the flagella rotation in peritrichously flagellate bacteria causes the flagella to deform and the bundle to fly apart, resulting in a tumble, directional movement only recommencing with the onset of anticlockwise rotation. In non-gradient environments peritrichously flagellate bacteria such as *E. coli* perform a three-dimensional random walk swimming in straight lines interspersed with a tumbling motion for about 0.1 s every 1.0 s. Polarly flagellate bacteria spend approximately equal times in clockwise and anticlockwise flagella rotation, relying on the Brownian collisions of the medium to cause a new swimming direction at the instant of the reversal in the direction of flagellar rotation.

Spirochaetes may move similarly but the flagella, whose basal bodies lie in the cytoplasmic membrane, are periplasmic. Flagella function may cause counter rotation between the protoplasmic cylinder. This would generate a 'corkscrew' rotation along the local body axis causing forward movement, especially in viscous environments (Berg, 1976). This model requires the spirochaetal body to be semi-rigid, rather than rigid like *Spirillum* and there is evidence from 'viscotaxis' studies that these bacteria can change the wavelength of the body helix.

It is interesting that polarly uniflagellate bacteria, which have shorter flagella (5 μm instead of 15 μm) than peritrichously flagellate bacteria move much faster. The speeds reached by a bacterium under full power range from over 70 μm/sec for *Pseudomonas* to about 30 μm/sec for *Salmonella*.

The power for flagella rotation comes not from ATP directly but from the proton motive force, leading to the hypothesis that the rotation of the S and M rings is driven directly by the movement of protons between these two rings (Khan & Macnab, 1980a,b; Manson, Tedesio & Berg, 1980). It is reasonable to assume (Berg, 1974; Manson, Tedesio & Berg, 1980) that a fixed number of protons move through the flagellar motor to perform one rotation. Calculations suggest this number to be at least 300. De Pamphilis & Adler (1971) observed a rotational symmetry in the flagellar motor, 16 fold for the M ring and probably 15 or 17 for the S ring, which may prove to be multiple channels for proton translocation. The mechanism by which the proton flux may be coupled to mechanical rotation has yet to be fully described. The speed of flagella rotation depends linearly on the pmf up to a maximum rate, although the threshold potential required for flagella rotation seems to vary between groups. Values of about 0 mV have been reported for photosynthetic bacteria, streptococci and pseudomonads (Glagolev & Skulachev, 1978; Manson, Tedesio & Berg, 1980; J. P. Armitage, unpublished observations) while −30 mV has been suggested for *Salmonella*, *E. coli* and *Bacillus* (Khan & Macnab, 1980a,b). The use of the pmf to drive flagella rotation directly allows flexibility. Either pH or $\Delta\psi$ components can be used, whether generated by the electron transport system, ATP hydrolysis, reaction centre or bacteriorhodopsin. There is some evidence that bacteria in particularly harsh, for example very alkaline, environments may use Na^+ instead of H^+ as the driving force (Hirota, Kitade & Imae, 1981).

Pseudomonas aeruginosa, a strict aerobe, can maintain a pmf using arginine in aerobic conditions and remains motile though metabolically inactive, increasing opportunities for escape. This may be related to the observation of Khan & Macnab that at low pmf values chemotaxis in *E. coli* is repressed and the bacteria swim smoothly, increasing the chance of escape from dangerous or energy-poor environments.

Mechanism of tactic response

A rapidly respiring bacterium having a single flagellum would use only 0.1% of its available energy for flagella rotation; however, even this amount would be wasteful if the result was a simple three-dimensional random walk.

In a well stirred, growth-limited chemostat, a non-motile mutant will outgrow its motile parent, suggesting that flagella synthesis and use can reduce growth rate. However, in a limiting environment not only does the motile parent outgrow the non-motile bacterium but a chemotactic strain will outgrow a motile but non-chemotactic mutant (Pilgram & Williams, 1976; Smith & Doetsch, 1969). In a competitive environment, therefore, the ability to move in a useful direction is an advantage.

The first recorded observation of bacteria moving to favourable environments was made by Engelmann in 1883 who observed photosynthetic bacteria moving towards light and CO_2 but away from oxygen. In recent times the best studied systems have been *E.*

coli, *S. typhimurium* and *B. subtilis*, whose chemotactic systems appear to show a common pattern.

About 20 different specific receptor proteins have been identified in *E. coli*. These are either periplasmic, where they have dual but separate functions in sugar transport and chemotactic sensing, or membrane bound, responding to amino acids. The receptors themselves interact with one of these different membrane bound methyl accepting chemotaxis proteins (MCPs), which may themselves serve as the specific amino acid receptors. A conformational change within these proteins then releases an unidentified cytoplasmic signal which controls the direction of rotation of the flagellum. If the bacterium is moving in a favourable direction the flagellum will rotate in an anticlockwise direction for longer. The MCP proteins focus the information from the receptors to the flagellum and are also responsible for adaptation to a new but stable environment. The MCP can accept from 1 to 4 methyl groups from a specific methyl transferase enzyme and this blocks signals from the MCP to the flagellum. This control system is sensitive only to a change in occupancy of MCP by receptor proteins. Under varying environmental conditions signals to the flagellum are reinitiated. For a review of this subject see Koshland (1981).

This is, however, not the only receptor system sending information to the flagellum; oxygen is sensed via the electron transport chain, light in photosynthetic bacteria via the reaction centre, weak acids and alkalis and components of the PTS system, directly through the internal cellular pH and the PTS system. All these environmental factors directly affect the membrane potential and there must therefore be a potential sensing system. Responses to membrane potential changes compete with signals from the MCP system, suggesting a common pathway to the flagellum.

The tactic response of bacteria can take several forms. The movement towards a favourable environment may have kinetic as well as regulated direction components. For example, a photosynthetic bacterium at very low light levels may swim towards the light source partly because its speed increases as the pmf rises in response to light-activated electron transport and partly due to a reduction in tumbling activity when moving in the same direction.

The environmental role of motility and taxis

The fact that many bacteria possess the means of moving around their environment and for biasing that movement in favourable directions suggests that it is important in their growth and survival. The most common environmental stimuli are chemical.

Chemotaxis

Nutrient poor environments

Marine bacteria are found in environments where nutrient concentrations in the nanomolar region are common. These organisms are generally motile when isolated and it therefore seems probable that chemotaxis plays some role in their survival. Some marine bacteria appear to have two different transport systems for nutrients, a high affinity one operating at nanomolar concentrations which is not associated with motility; the other lower affinity system comes into operation when substrate concentrations rise to micromolar levels and seems to be linked with motility and chemotaxis (Geesey & Morita, 1979). This dual system may allow the bacteria to conserve energy when nutrient concentrations are very low, but to use a search strategy in the vicinity of higher solute concentrations enabling them to locate or remain in the vicinity of a nutrient source.

Some marine pseudomonads have thresholds for chemotactic responses much lower than those of enteric bacteria (10^{-10} M rather than 10^{-6}) and will in addition respond to both L and D amino acid isomers (Chet & Mitchell, 1976; Torella & Morita, 1978), both useful adaptations for nutrient poor environments.

The role of motility in plant infection

Many bacteria associated with different plants show chemotaxis in the laboratory. However, the occurrence of chemotaxis in an artificial environment does not establish a role in plant infection or colonization *in vivo*. Recent work suggests that whilst chemotaxis may not be essential for host colonization, it certainly helps the organism locate it. For example, *Vibrio alginolyticus*, a colonizer of algae, was attracted to algal filtrates at levels exuded by the alga, apparently attracted by both amino acids and glycolic acid (Wood & Hayasaka, 1981). Rhizoplane bacteria were attracted to the nutrient-rich environment of the eel grass root system, the colonizing bacteria all proving to be highly motile (Sjoblad & Mitchell, 1979). However, the attraction process is probably non-specific. For example, *Pseudomonas lachrymans* was attracted to water droplets exuded by both resistant (cotton, lettuce) and susceptible plants (cucumber, melon). The attraction was directly correlated to the amino acid content of the water droplet (Chet, Zilberstein & Henis, 1973). A mixture of motile and non-motile organisms of *Pseudomonas phaseolicola* resulted in lesion formation on beans being caused predominantly by the motile bacteria, although systemic invasion did not involve motility. As no non-motile strains of this pathogen have been found in nature it suggests a survival advantage (Panopoulus & Schroth, 1974). This is supported by the discovery that *Erwinia amylovora* became motile on liberation from infected cells, but was non-motile inside the host (Raymundo & Ries, 1980). There is one report of *Xanthomonas oryzae* being attracted to water droplets only from susceptible rice plants and not from resistant ones, but the specific chemoattractant was not identified (Feng & Kno, 1975).

Rhizobium species are chemotactic to many compounds exuded from the roots of legumes; among them are sugars, amino acids and specific glycoproteins. These chemoattractants are released by many different plants, however, most of which the bacteria cannot nodulate (see Schmidt, 1979, for review). In many cases motility is also sensitive to heavy metals which are often found in inhibitory concentrations in soil (Bowra & Dilworth, 1981). Experiments carried out on motile and non-motile *Rhizobium* mutants suggest that motility and chemotaxis, though not necessary for nodulation, do provide a competitive advantage in the field (Ames, Schluederberg & Bergman, 1980; Currier & Strobel, 1977). Non-motile mutants could infect and produce nodules on legumes as efficiently as motile chemotactic *Rhizobium*, however; if motile and non-motile strains were mixed together the nodule formation on alfalfa plants was almost always achieved by the motile strain, even if mixed in ratios 1:1 000 (Ames & Bergman, 1981). Motile *Rhizobium* seem capable of migrating 2 cm/day through soil, so if chemotaxis were not involved motility would still increase the chances of the organism meeting a legume root. Once in its vicinity, an attractant gradient would increase the chances of nodule formation.

Chemotaxis may also be important in cyanobacterial nitrogen fixation. Paerl & Kellar (1978) used *Anabaena* isolated from lakes, which often have bacteria associated with the heterocysts. These bacteria have been shown to be chemotactic to the cyanobacterial excretion products which they metabolize. When *Anabaena* was grown in axenic culture, oxygen inhibition of nitrogenase activity occurred at oxygen levels which did not cause inhibition in *Anabaena* with bacterial associates. Possibly in a natural environment the bacteria remove oxygen in a microzone around the heterocyst, allowing the cyanobacteria concurrently to fix nitrogen and to maintain a high photosynthetic rate in the oxygen-rich surface waters.

Animal pathogens

There is evidence that *Aeromonas* fish pathogens were attracted to fish mucous extracts (Hazen, Raker & Esch, 1979), whereas the mammalian pathogens *Vibrio cholerae*, *E. coli* and *Salmonella typhimurium* were attracted to intestinal mucosa (Allweiss, Dostal, Carey, Edwards & Freter, 1977; Freter, 1981). High concentrations of motile and non-motile strains in contact with mucous membranes were all able to invade the tissues; however, when cell concentrations were low the motile chemotactic cells had a greater chance of adhesion and subsequent invasion than did the non-motile cells. Again chemotaxis appeared important in locating the invasion site rather than in the process of invasion. Experiments with HeLa and mucosal cells show that attachment by bacteria involved two stages. The first was reversible and was facilitated by chemotaxis and motility and presumably involved the attraction to the cell surface by the chemoattractants, whilst the second irreversible stage involved pili but not motility (Jones, Richardson & Uhlman, 1981; Uhlman & Jones, 1982). Studies on the relationship between motility, chemotaxis and the virulence of strains of *Pseudomonas aeruginosa* have shown that avirulent strains tend to have defective chemotactic responses. The most chemotactic strain was the most virulent with LD_{50} of 10 cfu whereas a motile but reduced chemotactic strain had an LD_{50} of 10^7 cfu (Moulton, Young & Montie, 1980; Craven & Montie, 1981).

Bdellovibrio

Bdellovibrio, a bacterial parasite of other bacteria, often lives in nutritionally poor waters. The organism has no endogenous energy storage compounds and yet it has a very high respiration rate. Its prey is small and almost certainly well separated from it in this type of environment so that motility and chemotaxis may therefore be important in locating bacteria. *Bdellovibrio* when isolated exhibited chemotaxis toward amino acids, organic acids and oxygen, though they did not respond to exudates of *E. coli* (Straley & Conti, 1977). Chemotaxis may be used to locate high concentrations of prey or other utilizable energy sources such as glutamate, and also in a negative sense to avoid open waters (La Marre, Straley & Conti, 1977; Straley, La Marre, Lawrence & Conti, 1979).

The role of chemotaxis in conjugation

There is evidence that in both *Salmonella* (Bezdek & Soska, 1972) and *Streptococci* (Dunny, Brown & Clewell, 1978) recipient strains produce a pheromone; that for *Streptococci* is a trypsin-sensitive, heat-resistant, nuclease-resistant agent which induces mating with donor cells. After transfer of the plasmid, production of this factor (clump-inducing factor CIA) by the recipient cell was inhibited but the bacteria became responsive to the pheromone in the medium.

Aerotaxis

Many obligately aerobic or facultatively anaerobic bacteria preferentially migrate to an oxygen-rich environment.

Positive aerotaxis requires a functioning cytochrome chain whose action is expressed through alteration in membrane pmf (Lazlo & Taylor, 1981; Niwano & Taylor, 1982). An increase in pmf seems to act by increasing the speed of the bacterium moving in the favourable direction and as a signal to the flagella, causing tumbling if the bacterium moved away and suppressing tumbling as it moved towards high oxygen. This signal seemed to compete with signals from the *Salmonella* MCPs and the PTS system, ensuring that a strong positive aerotactic signal did not lead the bacteria out of a nutrient rich environment.

A similar system appears to apply with the purple photosynthetic bacteria. *Rhodopseudomonas* and *Rhodospirillum* spp. can grow anaerobically using photosynthesis or aerobically as heterotrophs in the dark. They show positive phototaxis in the one case and positive aerotaxis in the other, but as high oxygen environment is harmful to photosynthetically growing bacteria, the two responses must be regulated. Induction of the terminal oxidase in response to oxygen and the resulting rise in pmf could be the stimulus for positive aerotaxis in photosynthetic bacteria (J. P. Armitage, unpublished), as seemed to be the case for *Salmonella*. Similarly, a membrane potential rise through operation of the reaction centre was apparently the phototactic stimulus (Armitage & Evans, 1981). Whilst a signal to the flagellum was probably an increase in pmf in both cases, there are almost certainly alternative negative signals. The presence of a negative aerotactic signal among obligate anaerobes is deduced by their rapid migration away from oxygen; however, the nature of this process is unknown.

Phototaxis

Motile purple photosynthetic bacteria move up a useful light gradient by an increase in speed due to increased pmf and as a result of specific signals to the flagellum. Both photokinetic and photophobic responses are observed; thus the photosensing system allows bacteria to congregate at optimum positions to best use their energy supply. *Halobacterium halobium* is a halophilic bacterium having a unique light-driven proton pump; it also responds to changes in light intensity. Light at 370 nm detected by photosystem 370 acts as a repellant and light detected by bacteriorhodopsin at 565 nm as an attractant (see Hildebrand, 1978, for review). An increase in light at 565 nm caused an increase in pmf whilst an increase in UV light led to an increase in passive membrane permeability to protons and reduced pmf. A second retinal protein has now also been implicated in positive phototactic responses.

When growing in conditions of high intensity light, low oxygen which induces the formation of the purple membrane, chemotaxis to amino acids and sugars was inhibited, the inhibition being relieved when bacteriorhodopsin was inhibited. These bacteria showed a positive chemotactic response to useful metabolites when growing as heterotrophic aerobes but not when light was being used as the energy source (Schimz & Hildebrand, 1979). Both the purple photosynthetic bacteria and *Halobacterium* adapt to alterations in light-induced energy levels. A methylation system similar to that described earlier in *E. coli* may be present in *Halobacterium* (Schimz, 1981), but has not been detected in photosynthetic bacteria.

Salmonella typhimurium and *E. coli* respond to intense flashes of blue light by tumbling. It seems probable that the blue light photooxidizes flavins in the electron transport chain causing a temporary fall in pmf (Taylor, Miller, Warrick & Koshland, 1979).

The purple bacteria growing photosynthetically in nature require the maximum light of a useful wavelength while still remaining anaerobic. One example is the environment below the thermocline in a stratified lake or in the muds of shallow water. Using the action spectrum they are able to move towards useful wavelengths by increasing speed, and by changing direction by tumbling if there is a reduction in light intensity. However, if the organism using this phototactic system moves into an oxygenated environment there will be no increase in pmf through the cytochrome chain since terminal oxidase production is suppressed in illuminated anaerobic environments. A negative response in oxidation of the anaerobic system will lead to a tumble and a bias in movement back towards the anaerobic, if less well illuminated, environment.

Thermotaxis

Bacteria move faster as temperatures rise due probably to an increase in metabolic rate. In addition, there appears to be a thermosensory system in *E. coli* similar to the chemosensory system. A decrease in temperature caused tumbling whereas an increase in temperature caused acceleration (Schneider & Doetsch, 1977), and possibly a fall in tumbling rate as well (Maeda, Imae, Shioi & Oogana, 1976). In *E. coli*, mutants unable to respond to L-serine were unaffected by temperature changes. It is possible that the L-serine receptor may undergo a temperature-dependent conformational change leading to altered signals to flagella (Maeda & Imae, 1979). The part played by a thermosensory response in nature is uncertain.

Mechanical taxes

External forces may also tend to orient a bacterium in a favourable direction, producing a positive or negative taxis but without obvious cellular receptors or signalling systems.

Magnetotaxis

In 1974 Blakemore isolated a bacterium from marine mud which apparently migrated towards the Earth's geomagnetic north pole. Since then similar geotactic bacteria have been isolated from fresh and marine muds in many areas of the world.

Magnetotactic bacteria isolated so far range from cocci to spirilla and deposit intracellular chains of single domain sized magnetite crystals (for review see Blakemore & Frankel, 1981). These crystals physically orientate living or dead bacteria in a north–south plane (Moench & Konetzka, 1978). Living bacteria are able to move at speeds of about 85 μm/sec in many cases up *or* down the magnetic lines of force. Another tactic response, for example negative aerotaxis, may determine the final direction. There is evidence that some species isolated in the Northern hemisphere do migrate towards the north, or towards the south if isolated from the Southern hemisphere, reversing the preferred orientations if the polarity of the intracellular dipole was reversed. These bacteria all live in aquatic sediments where currents and disturbances could move them long distances from their environment into open water; so that downward-directed motility, especially if linked with conventional taxis, would increase the chance of a cell returning to its natural habitat.

Viscotaxis

It is difficult to assess the effect of viscosity on an organism as small as a bacterium because of the possibility of internal regions of low viscosity within a viscous matrix (Berg & Turner, 1979). However, in experiments in which the physical mechanisms involved in increasing the viscosity of media were taken into account, it appeared that bacteria with external flagella, irrespective of whether the cell body was rod-shaped or spiral, became immobilized at lower viscosities than spiral bacteria with internalized flagella such as *Spirochaetes* (Greenberg & Canale-Parola, 1977). Thus the internalized flagellum may be relatively insensitive to viscous constraints. Petrino & Doetsch (1978) have shown that the spirochaete *Leptospira* responds positively to an increase in viscosity of the medium. It is likely here that the viscous forces cause a mechanical deformation of the semi-rigid helix of the leptospiral cell body, increasing its hydrodynamic efficiency and consequently its swimming speed. Leptospires and other spirochaetes tend to enjoy viscous environments. Thus free-living organisms favour sediments whilst parasitic species invade mucous membranes. The ability to swim fast in viscous environments may be important in the invasion of host tissue. Recent work has suggested that *Rhizobium* species with apparently complex rigid flagella are able to move efficiently in very viscous environments because of reduced flagellar distortion with increasing viscosity (Gotz, Limmer, Ober & Schmitt, 1982).

Swarming

Some bacteria exhibit an unusual pattern of motility on solid media known as swarming. There is a dramatic increase in the flagella number per unit cell wall after a brief period of growth on the surface of solid media, often accompanied by an inhibition of division and followed by the synchronous movement of these long, highly flagellate bacteria away from the parent colony which consists of short sparsely flagellate cells. Distances of up to 3 cm can be covered in about two hours. After swarming the long cells divide and shorten, lose their extra flagella and grow normally for several generations before bacteria at the edge of the now ring-shaped colony again move outwards, the cycle continuing until the solid surface is covered by several concentric rings of growth (see Williams & Schwarzhoff, 1978, for review).

Flagellate bacteria have difficulty in moving long distances over solid surfaces, since they are inhibited by surface tension. The increase in flagella numbers and the movement of the cells in groups or rafts creates an 'organism' big enough and motile enough to overcome this problem. All swarming bacteria have been isolated from environments where surfaces play an important role. *Proteus* species have all been intestinal or urinary tract strains and no soil isolates have been shown to swarm (Stahl & Williams, 1981). The vibrio-like *Beneckea* species which swarm are marine bacteria which tend to be found on algal surfaces (Baumann & Baumann, 1977).

Proteus mirabilis and *Beneckea alginolytica* are two quite different swarming species and have been investigated thoroughly. They both produce very long non-septate multinucleate filaments with large numbers of flagella at the edge of the swarm colony. The increase in flagella formation was the first swarming event in *Proteus* (Armitage & Smith, 1978), implying a derepression of flagellin synthesis and a relaxing of control over the number of flagella insertion sites. The flagellin from swarming or non-swarming cells appears to be identical. Both *Beneckea* and *Vibrio* normally have polar sheathed flagella. On the initiation of swarming, peritrichous antigenically different flagella are formed. Cell elongation follows the increase in flagella synthesis in *Proteus* and in *Beneckea*, however, the trigger for moving away from the parent colony may differ. The evidence from experiments with *Beneckea* suggests that negative chemotaxis is involved, guiding the bacteria away from the parent colony (Ulitzur, 1975); this explanation is questionable in the *Proteus* system. Thus, Williams, Anderson, Hoffman, Schwarzhoff & Leonard (1976) showed that non-chemotactic mutants of *Proteus* still swarm. The long swarming cells of *Proteus* appear to be metabolically inactive whilst swarming, suggesting that they developed solely in order to move to fresh environments (Armitage, 1981). As swarming ends, the normal metabolic activities return in parallel with cell division.

Conclusions

Flagellate bacteria have a complex mechanism for sampling and exploring their environment. They move by proton powered rotary motors, selectively sensing changes in physico-chemical conditions and biasing their movement in a favourable direction by controlling the rotation of the flagellar motor. They can transduce and integrate signals from many different receptors and balance their response to remain in an optimum environment as well as adapting to any longer-term changes which may occur.

Although in the laboratory an organism may lose its motility after many generations in rich broth culture, in the natural environment it is evident that the ability to move towards optimum energy sources to sites of adhesion and invasion or away from toxins gives an organism a major advantage over non-motile bacteria.

GLIDING MOVEMENT AND CHEMOTAXIS ON SOLID SURFACES

Swarming bacteria like *Proteus* may move over solid surfaces; however, another kind of motion called gliding motility has been developed by a number of organisms. These include the blue–green algae and their colourless relatives (e.g. *Beggiatoa*, *Vitreoscilla*, *Leucothrix* and *Thiothrix*) together with the flexible gliding bacteria of the myxobacter and flexibacter groups. The blue–green algae and their relatives will not be considered in any detail here.

The scavenging gliding bacteria

All these organisms are characteristic of solid surfaces in natural environments and are especially common in aerobic soils. The cytophagas, flexibacters, herpetosiphons and their allies are also widespread in marine and freshwater environments where fruiting myxobacteria may also be found possibly associated with the surfaces of seaweed or water plants (Burnham, Collart & Highison, 1981).

The organisms mentioned above are essentially scavengers. The fruiting myxobacteria and cytophagas, though unicellular, move as concerted swarms degrading living and dead bacteria, other microbes and decaying organic matter using a powerful battery of lytic enzymes. The flexibacters, herpetosiphons and others of the group occupying a similar niche consist of multicellular filaments.

The fruiting myxobacteria are aerobic Gram-negative prokaryotes with high G + C ratios close to 70%. They inhabit environments rich in decaying organic matter such as dung (Kaiser, Manoil & Dworkin, 1979). When well provided with nutrients they remain relatively static as vegetative cells. As substrate levels fall they can glide either individually or more efficiently as rafts with their long axes parallel (Henrichsen, 1972) to regions with higher substrate concentrations. It is said that operating as multicellular rafts can lead to the retention of high concentrations of extracellular hydrolytic enzymes near the cells that need them. Growth in homogeneous culture on the same substrates that need hydrolysis leads only to low levels of these enzymes and growth is nutrient limited at low initial cell concentrations (Rosenberg, Keller & Dworkin, 1977). It is also possible that hydrolytic enzymes are retained close to the cell in the slime excreted during gliding activity (Stanier, 1942; Dworkin, 1972).

Although most fruiting myxobacteria occur in soil, a few have adapted to aqueous environments where some organisms have modified their communal mode of life by forming cell aggregates in which susceptible microorganisms are trapped. The close packing of organisms now serves to ensure that the level of bacteriolysins and other degradative enzymes (and the level of substrates and nutrients released by lysis of entrapped organisms) is high.

The cytophagas are also Gram-negative, gliding bacteria but differ from fruiting myxobacteria in several ways, especially in their GC ratios (*ca* 33–43%) which are much lower (Mandel & Leadbetter, 1965; McCurdy & Wolf, 1967; Mandel & Lewin, 1969) and in their failure to form fruiting bodies. The cytophagas are aerobic or occasionally facultatively anaerobic (Warke & Dhala, 1968). The cytophagas, common in wet or marshy soils and in marine or freshwater habitats (Christensen, 1977) are also found in drier places alongside the myxobacteria. They are gliding scavengers with considerable nutritional versatility especially in breaking down structural and storage polymers from eukaryotic organisms. The latter include cellulose (Stanier, 1942; Charpentier, 1963, 1965), chitin (Stanier, 1942; Warke & Dhala, 1968) and seaweed polysaccharides (Stanier, 1941, 1942; Lewin & Lounsbery, 1969). Many can lyse other bacteria as well as fungi, algae and yeasts (Stewart & Brown, 1969; Ensign & Wolfe, 1965; Peterson, Gillespie & Cook, 1969).

The flexibacters are generally considerably longer than the cytophagas, sometimes forming sizeable multicellular filaments (Lewin & Lounsbery, 1969).

There are a number of related groups of flexible scavenging bacteria which are found particularly in marine locations, e.g. *Microscilla*, *Saprospira*, *Flexithrix* and *Herpetosiphon*. The latter group are fairly characteristic of these genera. The herpetosiphons are filamentous, multiseptate, multinucleate gliding Gram-negative bacteria closely related to the cytophagas and flexibacters. It has been suggested that they play a scavenging role particularly in lakes. The herpetosiphons are less versatile than the cytophagas in their ability to degrade macromolecules and resemble the flexibacters in this respect (Lewin & Lounsbery, 1969; Quinn & Skerman, 1980).

Swarming in cytophagas

Although individual cells can glide, cytophagas and flexibacters 'swarm' on solid surfaces by a process apparently similar to that described for fruiting myxobacteria. These organisms can cover a typical agar plate with spreading growth in 2 to 3 days. As with fruiting myxobacteria, 'rafts' or larger masses of organisms are involved in swarming (Henrichsen, 1972). As discussed above, degradative enzymes may be concentrated in the swarm and in addition activities may be associated with excreted slime polysaccharides in myxobacters and cytophagas (Duckworth & Turney, 1969; Christison & Martin, 1971).

The ecological importance of the flexible gliding bacteria

This group of gliding bacteria are of major importance in the environment as scavengers. In dry soil, mainly the fruiting myxobacters and, in damp marshy soils, mainly the cytophagas and flexibacters, bring about the breakdown of organic materials from bacterial, animal, insect, plant and fungal sources. Especially important is the role of cytophagas (and to a lesser extent fruiting myxobacters) in the breakdown of the structural and storage polymers of eukaryotic cells which are often resistant to other degradation processes (e.g. starch, cellulose, chitin and keratin). The cytophagas probably play a critical role in humification in damp soil. The importance of this group of organisms in stabilizing the soil structure may depend on their ability to produce slime. Soil stability is governed by the level of soil polysaccharides present and much of the latter is contributed by the flexible soil gliders as slime polysaccharides (Webley, Duff, Bacon & Farmer, 1965).

In freshwater and marine environments, the cytophagas and related species degrade bacterial, algal, animal and plant polymers. Thus, marine cytophagas such as *C. diffluens* will hydrolyse agar, cellulose and other polymers (Stanier, 1942) whilst marine and freshwater flexibacters isolated by Lewin & Lounsbery (1969) attacked agar, cellulose and gelatin. In aqueous environments, organisms of this group can digest other living microorganisms and may, therefore, reduce the numbers of bacteria or algae when they are present in large numbers. For example, herpetosiphon species can grow on bacteria present in weakly supplemented lake water, whilst on solid laboratory media their filaments may grow well by engulfing them (Quinn & Skerman, 1980). Degradation of eukaryotic and blue–green algae by enzymes from cytophagas has been demonstrated by Peterson, Gillespie & Cook (1969) and Shilo (1970). Interestingly, a strain of *Myxococcus xanthus* isolated from freshwater forms massive clumps around the lysing filaments of its cyanobacterial prey (Burnham, Collart & Higgison, 1981). As lysis proceeded, it appeared that more cyanobacteria in the medium were trapped and possibly transported to the centres of spherules by the gliding activity of the myxobacteria. Clearly myxobacterial cultures of this sort might be useful in the control of undesirable cyanobacterial blooms in lakes or in reservoirs.

Cytophagas and related organisms also play an important role in sewage treatment plants. They occur commonly in activated sludge and in trickling filter systems and are most prevalent in winter samples. Güde (1980) found that a high percentage of such sewage isolates attacked some or all of cellulose, chitin, dextrans, pectin xylans and gelatin, and they probably destroy other sewage microorganisms including bacteria and algae (Stewart & Brown, 1969). These organisms are also found in large numbers in the rumens of cattle and other ruminants where they play a part in the degradation of cellulose and other plant polymers.

Cytophagas are serious fish pathogens and fish spoilage organisms, thus *C. psychrophila* and *C. columnaris* are associated with disorders such as cold-water disease of salmon (Pacha, 1968). Non-pathogenic cytophagas isolated from the surfaces of freshwater fish by Pacha & Porter (1968) are probably associated with fish spoilage.

In almost all the above situations, the activities of these organisms will be aided by their ability to glide over solid surfaces, and their role as scavengers will be aided by their tendency to aggregate and hence concentrate the lytic enzymes that they produce.

Gliding motility

Gliding is a smooth continuous movement, not associated with the presence of flagella or other obvious external structures and occurs only when the organism is in contact with solid surfaces. These rod-shaped organisms move along their long axes and can do so individually or more commonly in small groups or 'rafts' with their long axes parallel (Henrichsen, 1972).

Twitching motility which occurs for example in *Moraxella* and *Acinetobacter* species may be related to gliding. Twitching is, as the name suggests, a less smooth mode of progression. The organisms, usually singly, move in discrete jerks upon which there is superimposed a very slow continuous movement. The jerking motion does not always occur along the long axis of the cell. The organisms which show twitching motility are very short rods, and the lack of flexibility which this suggests may not allow a smooth gliding movement. Alternatively, the mechanism of twitching may be quite distinct from that of gliding. This is supported by the finding (Henrichsen, 1972) that the 'twitching' of individual cells of *Acinetobacter* is apparently dependent on the presence of fimbriae; in contrast fimbriae are *not* needed for the gliding of *individual* cells.

Gliding motility needs to be considered from at least three points of view. Firstly, what is the gliding mechanism? Secondly, how do the cells associate during swarming and what environmental conditions induce this process? Thirdly, are the cellular associations which occur during fruiting aggregation in the myxobacteria the same as, or related to, those which take place during swarming and what mechanisms govern the process?

The mechanism of gliding

Four main theories have been proposed to explain gliding motility. It has been claimed that polar fimbriae (pili) are related to gliding (MacRae & McCurdy, 1976). Movement might stem from continual fimbrial retraction and extrusion to give a 'punting' motion. It appears, however, that fimbriae are not essential for gliding since strains which have lost them can still glide individually though their capacity for group motility is lost. Supporting this view, Dobson, McCurdy & MacRae (1979) showed that antiserum to myxobacterial fimbriae inhibited swarming though not the movement of individual organisms.

A second suggestion is that gliding is directly connected with slime secretion since gliding organisms appear to be associated with a ribbon or trail of slime. It is possible that if slime production were unidirectional it might *propel* the organism forward. The slow smooth progress of gliding organisms would fit such an idea. For such a mode of propulsion to operate, the organs of secretion would need to be orientated correctly as has been shown for cellulose-producing pores of *Acetobacter xylinum* (Brown, Willison & Richardson, 1976). Because the slime produced by gliding organisms is amorphous (in contrast to cellulose fibres) it is hard to be sure whether production is unidirectional from a defined group of structures or not. Holes have been reported in the outer membrane at the polar region of many gliding bacteria (MacRae, Dobson & McCurdy, 1977) but these are probably related to production of fimbriae rather than of slime, and anyway they are absent from many organisms. Ridgway (1977) showed that the outer surface of *Flexibacter polymorphus* is studded with goblet-shaped structures which frequently appear to be extruding a fibre, which may be the slime for this organism. The structures are not located in a spatially distinct group and accordingly are unlikely to secrete slime unidirectionally.

Thirdly, several workers have demonstrated fibrillar or tubular structures in gliding organisms and some have proposed that these have contractile properties which move the organisms by an inching mechanism. Pate & Ordal (1967) observed fibrils running longitudinally on the inner side of the outer membrane of *C. columnaris*. Presumably contraction of such fibrils could give movement by inching; however, Burchard & Brown (1973) suggested that they were an artefact of preparation formed from intermembrane particles and therefore not related to gliding. MacRae & McCurdy (1975) have observed filaments and tubules in the cytoplasm of *Chondromyces crocatus* but there is no evidence that these are involved in gliding. Lysis of many gliding bacteria produces the tubular rhapidosomes which can also be seen in sections of organisms if certain preparative methods are used. These structures appear to have a similar composition to that of the plasma membrane and, accordingly, they may be simply artefacts of preparation also (Pate, Johnson & Ordal, 1967; Kuhrt & Pate, 1973).

Although both the submembrane structures seen by Pate & Ordal (1967) and the rhapidosomes may be artefacts and not required for gliding, bundles of fibres or tubules may still be involved in this phenomenon.

Schmidt-Lorenz & Kuhlwein (1968) described bundles of filaments running just beneath the cytoplasmic membrane and longitudinally in cells of *M. xanthus*. They claimed (from finding ring structures of a similar diameter to these fibres in some sections) that the structures were tubular. Schmidt-Lorenz (1969) extended these studies, claiming first that the 13–15 nm diameter tubules were composed of spherical subunits arranged spirally, and second that the structures originated from basal granules and then ran the full length of the organism with a short stub (considered to be a flagellum) extruding from the pole.

Burchard, Burchard & Kloetzel (1977) also observed bundles of fibres running longitudinally in the cytoplasm of *M. xanthus*. These closely resembled some illustrated by Schmidt-Lorenz & Kuhlwein (1968) and showed regular striations 12 nm apart. The bundles appeared to terminate in the envelope near one pole and to run most of the length of the organism but not to end beyond the cell boundary. These workers found that one non-motile mutant (*mgl* – 1) had fibres with irregular striations and accordingly they proposed that the structures function in gliding.

Schmidt-Lorenz (1969) proposed that the structures he observed were modified flagella and functioned by rhythmic contractions which produced an inching movement. Bacterial flagella do not, however, contract and accordingly if the tubules observed are contractile, they are probably not related to flagella. Burchard, Burchard & Kloetzel (1977) agreed with the proposed mechanism but suggested that the fibres would need to be anchored at both ends to function in this way, in fact they were only able to see fibres fixed at one end. Schmidt-Lorenz proposed that contraction produced a beating of the fibres and that the inching motion produced in the flexible cell stemmed from this. If such a mechanism were possible, clearly the fibre bundle would need anchoring at one end only. Both suggested mechanisms postulate that gliding involves inching of the organism. Such movement has not, however, been observed, i.e. there are no apparent morphological changes during gliding consistent with such a mode of progression, and the movement of latex spheres on the surface of gliding organisms (Pate & Chang, 1979) could not occur by such a mechanism.

Finally, it has recently been suggested that the exterior surfaces of gliding organisms are covered completely or over large areas by hundreds of structures which by rotating at the surface of the organism cause it to glide over solid surfaces (Pate & Chang, 1979). The evidence for this view comes first from the behaviour of tiny latex spheres which are transported over the outer surface of *Cytophaga johnsonia* from both peripheral areas and polar regions as though organs of motility cover the entire organism. That such transport is one manifestation of a gliding mechanism is suggested by the finding that the rotary structures were related to the basal regions of flagella. Structures were isolated as 'rings' 20 nm in diameter and 10 nm thick, although they were frequently found in pairs 20 nm thick. They had a tendency to associate in long ribbons and appeared in treated preparations as a net-like structure covering large parts of the surface of *Cytophaga johnsonia* and *Flexibacter*, though absent from one non-motile strain. Additional clues to the identity of these structures came from the finding that gliding motility resembles flagellar motility in its sensitivity to inhibitors (Pate & Chang, 1979).

The claim that gliding motility is essentially flagellar motility on solid surfaces using 'sawn-off' flagella can be seriously questioned, as can the claim that the structures seen by Pate & Chang (1979) are indeed flagellar motors. Firstly, with the exception of swarming *Proteus*, flagella are relatively sparse on the surface of bacteria compared with the almost complete covering of large areas of *Flexibacter columnaris* by the ring-like structures seen by Pate & Chang (1979). These arrays resemble the net of a structural protein element more than an organ of motility. Secondly, gliding motility seems to differ from flagellar movement in that some gliding bacteria when feeding appear to move directly towards substrates when feeding or during the aggregation stages in fruiting body formation, whilst flagellate bacteria only show a *net* movement towards higher substrate concentrations through mechanisms involving tumbling activity and straight runs up a substrate concentration gradient. Thirdly, if the rings observed on gliding organisms are related to rotary flagellar motors then the total energy needed to rotate them continuously would be considerable because of their large numbers.

The precise location in the membrane of the rings seen by Pate & Chang (1979) has not been established. If there are structures analogous to flagellar basal units and hooks these would have to penetrate the entire envelope with structures corresponding to the M and S rings of the cytoplasmic membrane region (where the power for rotation is supplied) and further structures leading to the outside. The double rings isolated by Pate & Chang (1979) might correspond to the M and S rings of flagella; if so they ought to be located in the cytoplasmic membrane region. In fact the goblet-like structures of *F. polymorphus* (Ridgway, 1977) which Pate & Chang (1979) suggested may be related to their rings, appear to penetrate *outwards* from the cell with the innermost layer located just below the outer membrane (Ridgway, 1977).

Though the ideas of Pate & Chang (1979) suggest a fascinating relationship between gliding and flagellar motility, real evidence is still needed.

The regulation of gliding

Mechanism governing swarming and individual gliding

Many gliding bacteria move together in swarms over solid surfaces; this is particularly true for the predatory gliding bacteria considered above. On rich media, the colonies formed by these organisms are discrete with sharp edges. On poor media, however, the colonies are diffuse and spreading. If the edges of such colonies are observed, there is generally a marked association of the organisms and movement outwards appears to be by spearheads of organisms (Henrichsen, 1972). Individual bacteria can also move outwards but generally return to the swarm promptly. Henrichsen (1972) showed that spearheads move much more rapidly than do individual organisms.

To study regulation of the mechanism of gliding, several investigators have attempted to isolate non-motile mutants. Initially these were extremely rare but the discovery that, for some gliding organisms at least, there are two apparently independent types of gliding motility has made it relatively easy to isolate non-motile types.

Hodgkin & Kaiser (1979a,b) isolated non-motile mutants of *M. xanthus* from a strain (DK 101) subsequently found to be altered in social gliding. When non-motile strains were transduced with phage propagated on wild-type, two types of transductants appeared. One type were S-motile (S for social) and showed the ability to swarm though few individual cells could swim separately from the swarm. The S-motile transductants were said to be altered only in the motility of individual cells. The second type of transductant was A-motile (A for adventurous) since individual cells were able to move normally though they were unable to swarm. Hodgkin & Kaiser suggest that two gliding motility systems are present in *M. xanthus*. Genes in system A govern the motility of individual bacteria whilst genes in system B govern social gliding.

Hodgkin & Kaiser defined a number of mutants altered in the A system. Some of these are of the *agl* type (altered in *a*dventurous *gl*iding) and others of the *cgl* type, so-called because the presence of other strains can stimulate them to single cell or *c*onditional *gl*iding. Strains altered in the B system are defined as *sgl* mutants (mutants in *s*ocial *gl*iding). Only one class of mutant (*mgl* for *m*utual *gl*iding) is altered in both adventurous and social gliding.

Clearly if *agl* and *cgl* mutants show no individual gliding whereas rafts of cells can glide, there are likely to be two independent systems of gliding motility.

The presence of fimbriae is needed for social gliding because efimbriate mutants cannot swarm (and are classified as *sgl*) although individual organisms without fimbriae can move. It is possible, of course, that social gliding involves retraction and extrusion of fimbriae (see the earlier discussion of punting) whilst gliding in individual cells, which is still possible in efimbriate mutants, operates through a quite different mechanism. It is more likely, however, that the dependence of social gliding on fimbriae stems from their involvement in attachment. Thus EM studies of myxobacteria revealed that connections between swarming cells are established via the fimbriae (Dobson, McCurdy & MacRae, 1979).

Motility and cellular interactions during the formation of fruiting bodies

The formation of fruiting bodies in *M. xanthus* occurs as follows. Firstly vegetative swarms of myxobacteria move in a concerted fashion towards aggregation centres to give flat-topped colonies. With the continued inward movement, cells pile up to form translucent mounds. This stage is followed by autolysis of many of the organisms in the mound and the formation of myxospores. The aggregation stage and the subsequent piling-up clearly involve a concerted movement of the swarm towards the aggregation centres and increasingly firm associations between the aggregating cells.

Initiation of aggregation in *M. xanthus* is complex and there may be a series of diffusible triggering molecules involved. Two of these triggers are essentially nutritional. First, starvation is necessary for the induction of aggregation. Deficiencies in amino acids, carbon/energy sources, phosphate and so on (Manoil & Kaiser, 1980a) induce aggregation though there may be a single triggering metabolite or group of metabolites generated under a wide range of starvation conditions. The magic-spot compounds guanosine tetra and pentaphosphate accumulate during starvation in *M. xanthus* (Manoil & Kaiser, 1980a,b) and these compounds may trigger aggregation and fruiting body formation. Formation of guanosine polyphosphates in the presence of ADP, cAMP and related compounds may explain why the latter can induce fruiting body formation (Campos & Zusman, 1975).

M. xanthus will not begin to aggregate unless the cell density is relatively high, suggesting another level of control which is possibly due to the formation of adenosine in the starved cells. This nucleoside will induce aggregation and fruiting body formation in low-density starved cultures and will, in addition, reverse the inhibition of aggregation caused by hadacidin or the purine-binding agent Norit (Shimkets & Dworkin, 1981). It is suggested that adenosine formed by the starved cells is a trigger setting in motion the series of events which leads eventually to fruiting body formation.

It is not clear how adenosine induces aggregation and fruiting body formation; there is apparently no evidence so far for it having any direct chemotactic role. Adenosine may simply be acting as a signalling molecule, which if its concentration exceeds a certain level switches on aggregation.

Some mutants of *M. xanthus* defective both in aggregation and in fruiting body formation can complement one another so that one or both mutants of the pair can fruit together but not separately (Hagen, Bretscher & Kaiser, 1978). This may also indicate that diffusible signalling substances such as adenosine or diffusible chemotactic agents are produced during aggregation.

Stephens, Hegeman & White (1982) have claimed that aggregating organisms of the myxobacterium *Stigmatella aurantiaca* produce a diffusible signalling molecule which they refer to as a pheromone. This organism is particularly interesting from a developmental point of view because, when starved in the dark, it produces aggregations which eventually contain spores but no discrete fruiting bodies. In contrast, in the light, stalked fruiting bodies occur (Qualls, Stephens & White, 1978; Inouye, White & Inouye, 1980). Strikingly, more of the pheromone is produced by aggregating organisms in the light. Guanosine derivatives can substitute for light in inducing mature fruiting bodies in *Stigmatella* and the pheromone and GMP interact co-operatively. So far, there is no indication as to whether the pheromone or GMP have any direct chemotactic effect or whether they switch on the series of developmental responses related to aggregation.

There is now evidence that the cellular interactions taking place during aggregation of *M. xanthus* have some similarities to those which occur between cells in the vegetative swarm; however, there are also distinct differences. The finding that efimbriate mutants of *M. xanthus* which are defective in social gliding aggregate and fruit poorly implies that fimbriae are involved in aggregation (Dobson, McCurdy & MacRae, 1979) but it is possible that the mutation in such strains is a complex one affecting components other than fimbriae.

Even if fimbriae are involved in aggregation, there are other cellular components involved in the cell–cell associations occurring in this process because during aggregation, organisms of *M. xanthus* synthesize up to 2% of their cellular protein in the form of a lectin-like haemagglutinin (Cumsky & Zusman, 1979). This protein was not synthesized during glycerol-induced myxospore formation (no aggregation stage occurs in this process) and was formed in reduced amount in a mutant altered in both fruiting body formation and in social gliding. The haemagglutinin is not present in vegetative swarms but begins to be synthesized after about 8–10 hr of starvation and is then formed continuously during the rest of the aggregation and fruiting stages.

It is likely that the interactions between organisms in the early stages of aggregation involve fimbriae and that these may be the same interactions as those found in the vegetative swarm. In early stages it may simply be that the pattern of motility has become concerted with the organisms moving towards the aggregation centre, possibly under the influence of a chemotactic agent. Later on in the aggregation process there are evidently firmer associations between organisms mediated by newly synthesized haem-agglutinins.

There may be other components causing cellular associations in the later stages of fruiting body formation; for instance, the most important induced component S will self-assemble to S-deficient myxospores and this has been postulated to play a role in the interactions between myxospores in the fruiting body (Inouye, Inouye & Zusman, 1979).

Chemotaxis in swarming and aggregation

Swarms of feeding myxobacteria will move directly towards food sources as though moving towards a chemotactic attractant. Similarly, during aggregation, columns and waves of organisms move inwards towards the central aggregation point where the cell mound eventually forms again as if drawn by a positive chemotactic agent. One possible candidate for such an effector is cGMP because feeding myxococci are attracted both by this compound and by 5′AMP (Ho & McCurdy, 1979; Shimkets, Dworkin & Keller, 1979) and studies of its metabolism during aggregation are in accord with its being involved in the process. The level of extracellular cGMP was high during the aggregation period. Further studies of cGMP during aggregation and of the chemotactic effects of potential feeding attractants are obviously needed.

MOTILITY AND TAXES IN PROTOZOA

Chemotaxis plays an important role in the behaviour of many protozoa. Four main categories of behaviour can be recognized (van Houten, Hauser & Levendowsky, 1981):

(a) Accumulation – chemotactic signals have been shown to be involved in processes associated with location of food sources (particularly bacterial) and with mating processes. This latter case is very striking in some ciliates, where the normally autonomous cells indulge in a brief period of social behaviour under the influence of chemotactic signals leading to mating.

(b) Dispersal – these processes are generally associated with moving away from unfavourable conditions: this occurs with *Paramecium* exposed to extremes of pH.

(c) Host invasion – the process of chemotaxis could provide important possibilities for the correct host/tissue localization in response to chemical cues. An example is the preferential invasion of heart tissue by *Trypanosoma cruzi* – the protozoan responsible for Chagas disease.

(d) Settling processes – some protozoans have a sessile phase in their life cycle and it is thought that the motile forms seek out suitable substrates for development of the sessile stages using chemical cues. The settling behaviour of *Vorticella marina* (Langlois, 1975) on particular algal substrates is one example.

The existence of a large number of chemotactic processes in the protozoa is not in doubt, but their elucidation awaits further research. In this article only the better understood systems, particularly those concerned with reproductive behaviour and feeding, will be considered.

Protozoan locomotion; movement and chemotaxis in amoeboid protozoans

Two general methods of locomotion are found in the protozoa: (i) amoeboid movement (crawling); (ii) ciliary/flagellar movement.

Amoeboid movement requires both a force-generation system based on actin/myosin interactions (Taylor & Condeelis, 1979) and cell-substrate interactions producing the traction necessary for cell movement (King, Westwood, Cooper & Preston, 1979). Speeds achieved are generally low, ranging between 0.02 μm and 2.0 μm s^{-1} compared to values for ciliated protozoans from 50 to 3 000 μm s^{-1}.

Most amoeba have well developed powers of phagocytosis, a good example of which is the ingestion of large latex beads by *Acanthamoeba* (Korn & Weisman, 1967). It might therefore be expected that chemotaxis could play a major part in the detection of potential food sources (especially bacterial) which could then be phagocytosed. Experiments on *Acanthamoeba* by Linnemans, Blok & Horsten (1981) have shown that *E. coli* extracts can act as chemoattractants producing directed movement of the amoeba towards the extract source. The possible role of chemotaxis in the natural biology of, for example, soil amoebae such as *Acanthamoeba* and *Naegleria*, particularly in relation to the soil ecosystem, has so far not been studied. It is possible that amoebae discriminate,

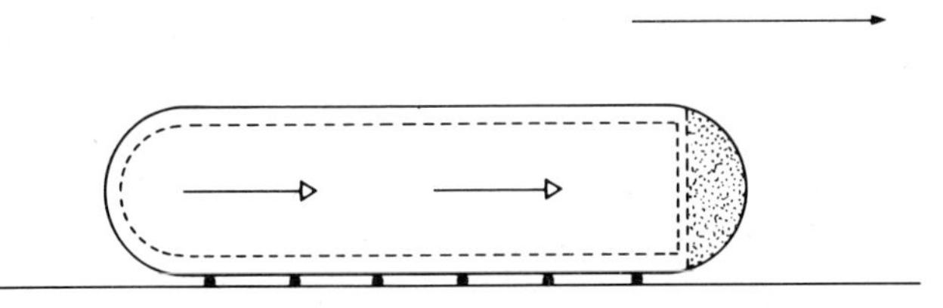

Fig. 1. Diagrammatic view of a moving amoeba on a substrate – viewed from the side. Arrows within cell denote cytoplasmic flow: ●, denotes adhesion forces between cell and substrate.

ingesting only those bacteria which produce the right chemotactic cues. The amount of experimental work carried out in this rewarding area has, however, been minimal. Factors from bacteria can act as chemotactic attractants for mammalian polymorphonuclear leucocytes (PMNs). Since protein synthesis in bacteria is initiated with *N*-formyl methionine, it was originally suggested that *N*-formyl methionyl peptides might be attractants for PMN cells. This has since been shown to be the case and some of these peptides can elicit a response at concentrations as low as 10^{-10} M. It would now be interesting to look at the effects of these peptides on a variety of amoebae, using the techniques and experimental approaches developed during studies on the chemotaxis of PMN cells (Zigmond, 1978, 1982).

Although the studies of Taylor and his associates (Taylor & Condeelis, 1979) have shown the potential role of actin, myosin and actin-binding proteins in the biochemical reactions associated with amoeboid movement, a general explanation governing such movement is lacking. In all theories of amoeboid locomotion, account must be taken of changes in the state of the cytoplasm, in particular to that between liquid colloid (sol) and more solid colloid states (gel). Recent work suggests that the gel state is produced by cross-linking of actin by actin-binding proteins and that the contraction event involving myosin occurs not in the gel but at the gel–sol interface (Hellewell & Taylor, 1979). Figure 1 shows a generalized view of an amoeba on a substrate, viewed from the side. At the anterior end there is normally a clear zone – the hyaline cap which has a watery consistency and is separated from the rest of the cytoplasm by the plasmagel sheet. In classical ideas of amoeboid locomotion, sol to gel conversions occur at the anterior whilst gel to sol conversions occur at the posterior ends of the cell and the effect of these changes generates cytoplasmic streaming from posterior to

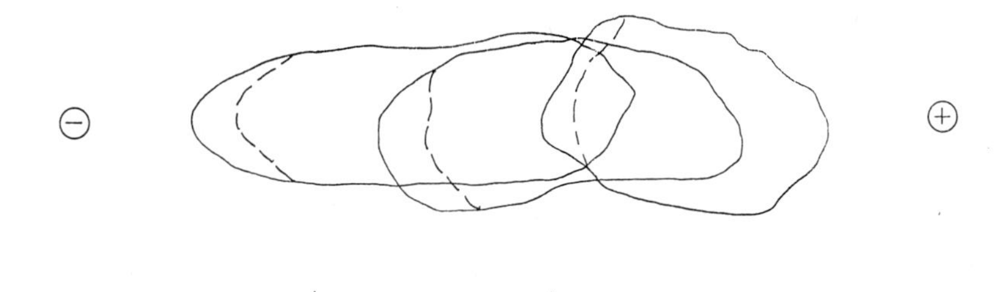

Fig. 2. A *Naegleria gruberi* amoeba moving under the influence of an electric field (Galvanotaxis). The cell profiles were drawn at 20, 30, and 40 seconds after the application of the current. The cell clearly shows polarized movement towards the cathode, the dotted line on each cell profile represents the posterior boundary of the hyaline cap.

anterior ends. However, it is also possible that gel to sol conversions could occur over the general gel layer of the amoeba, ectoplasm/cortical tube, and in the underlying (endoplasm) cytoplasm. The three main theories at present under consideration are set out below:

(a) Rear contraction model (Mast, 1926). Sometimes called the 'toothpaste tube hypothesis'. It is proposed that contraction of the ectoplasmic tube at the rear of the cell develops the force needed to move the wall contents forward. If relaxation at the front end of the cell takes place at the same time, then forward movement can occur. This theory requires that a pressure gradient is present along the posterior–anterior axis of the cell.

(b) Frontal contraction model (Allen, 1961). In this model it is proposed that forward thrust is achieved by restructuring the cytoplasm at the front end and 'relaxing' it at the rear end.

(c) Cortical tube contraction model (Grebecka & Grebecki, 1981). This is really a development of (a) in which it is suggested that the potential for contraction events can occur over the whole cortical tube. The hydrostatic pressure developed would bring about protrusion at the front end of the cell generating a pseudopodium. It is suggested that the front end region plays a major role in the steering process. This is conceptually important regarding the directionality of movement (polarity).

The development of polarity is in turn related to chemotactic responses. Figure 2 shows the cell outline of an amoeba moving in a polarized fashion due to the application of an electrical field. However, in many cases of amoeboid locomotion, a considerable amount of wandering in the lateral direction occurs, which is

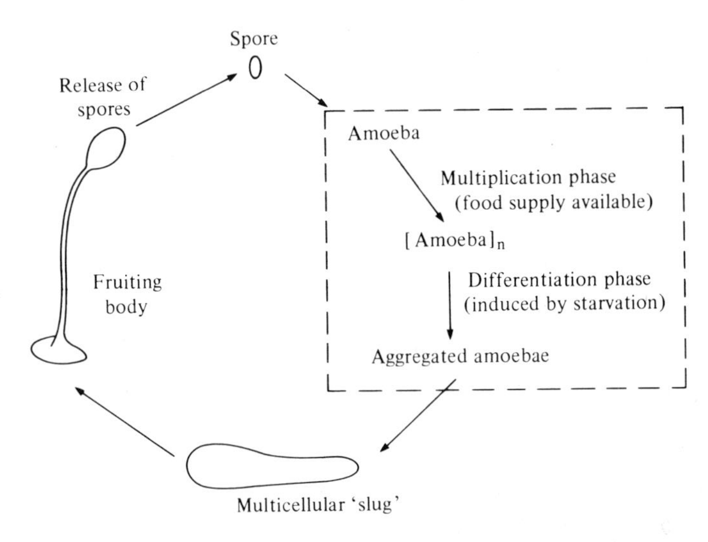

Fig. 3. Life cycle of *Dictyostelium discoideum*.

generally associated with changes in polarity. Thus, in the context of chemotaxis, changes resulting from contact with the chemoattractant could influence the force-generating system, the development of polarity and possibly cell-substrate interactions leading to changes in traction.

Chemotaxis in Dictyostelium discoideum

The organism studied in greatest detail in terms of chemotaxis amongst the protozoans has been the cellular slime mould *Dictyostelium discoideum* (Mato & Konijn, 1979). Its life cycle is shown in Fig. 3.

The part of the life cycle in which chemotaxis is important is shown within the box. During multiplication in the vegetative phase amoebae feed on locally available bacteria. In this process folic acid can act as a chemoattractant and it has been suggested that folic acid derived from bacterial sources can act as a food searching cue. However, major interest in the life cycle has focussed on the process by which the dispersed vegetative cells differentiate to become capable of aggregation to finally generate a multicellular 'slug'. The process of aggregation is induced by lack of nutrients. The chemoattractant for the aggregative amoebae has been identified as cyclic adenosine 3′,5′-monophosphate (cAMP). Propagation of the aggregation signal outwards from cell to cell is also by cAMP released into the environment by the amoebae themselves. The existence of an oscillating source of chemoattractant results in waves of amoeboid movement within the cell population. If a simple chemotactic gradient using cAMP is developed experimentally, however, the amoebae move continuously towards the chemotactic source. The aggregative amoebae can respond to artificial sources of cAMP by moving relatively quickly (up to $0.5\,\mu m\,s^{-1}$) (Alcantara & Monk, 1974) compared with rates in the absence of cAMP. The vegetative amoebae are about 100 times less sensitive to cAMP-induced chemotaxis than the aggregative amoebae, although the former are more sensitive to folic acid. The threshold cAMP concentration gradient for chemotactic responsiveness in aggregative amoebae was found to be about $10^{-9}\,\text{M}\,mm^{-1}$. The simplest model allowing detection of a chemotactic gradient by an amoeba would be one in which receptors for the chemoattractant were spread evenly over the cell surface but in the presence of a solute gradient receptor site occupancy would be greater at the anterior than at the posterior end, leading to intracellular signalling processes producing directed forward motion towards the chemotactic source. With reference to Fig. 1, assuming the chemoattractant source was to the right, this simple spatial model would require that the aggregative amoebae whose average length is $10\,\mu m$ would need to detect a very small gradient of the order of $10^{-11}\,\text{M}$ cAMP across this distance. This would require the ability to distinguish differences in cAMP levels of 1 part in 100. An alternative model has been suggested in which an assessment of spatial gradients is achieved by the amoeba pushing out small pseudopodia, about $0.5\,\mu m$ in length, in different directions. Using the same arguments, the chemosensory system would have to be able to detect differences in cAMP levels of 1 part in 2 000. Interestingly, the *N*-formyl peptides used in the studies of PMN leucocyte chemotaxis act at threshold values about the same order of magnitude as found in the cAMP/aggregative amoebae system described above. Diffusion of chemoattractant will be a major problem for detection of such low concentration gradients across short distances. Recent work on leucocytes has suggested that a temporal mechanism is involved in the chemosensory system in order to sense these very low levels of chemicals in the environment.

From the preceding discussion the importance of the cAMP receptor(s) present on the aggregative amoebae will be obvious.

cAMP has, of course, received a tremendous amount of attention over the last 25 years, particularly with regard to its role as 'the second messenger' in many hormone-regulated cellular activities (Pastan, 1972). More than 400 chemical analogues of cAMP have been synthesized since this work was published: certain of these have been labelled isotopically and have been important in elucidating the interaction of cAMP with *Dictyostelium* amoebae. The result of this work is that two distinct kinds of surface cAMP binding sites have been identified:

(i) A membrane-bound phosphodiesterase – this enzyme breaks down cAMP to AMP and destroys its chemoattractant properties.

(ii) A non-enzymic receptor which seems to be responsible for the chemotactic sensory input into the cell.

Both binding sites are most active at the onset of aggregation. It is important to realize that bound phosphodiesterase can act by altering the levels of cAMP locally, thus affecting the chemosensory input into the cell via non-enzyme cAMP receptors. The situation is further complicated by the existence of an extracellular phosphodiesterase which is secreted into the environment by the cells.

There are 10^5–10^6 non-enzymic cAMP receptors per aggregative cell and the cAMP-receptor association constant was found to be about 10^7–10^8 M, which fits well with the observation that the aggregative amoeba can respond down to 10^{-9} M cAMP.

Ligand–receptor interaction at the cell surface must in some way produce directed movement towards the chemoattractant source by induction of pseudopodium formation in a polarized fashion. Recent work has suggested that cyclic guanosine monophosphate (cGMP) is involved. Highest levels of cGMP were found 10 seconds after the addition of cAMP to aggregative cells, and fell again to their initial values within 30 seconds (Mato, Krens, Van Haastert & Konijn, 1977). The addition of cAMP to vegetative cells had no effect on cGMP levels though the addition of folic acid to these cells produced a rise in cGMP. The level of the enzyme producing cGMP (guanylcyclase) increased whilst vegetative cells were differentiating to become aggregative cells. There is therefore general support for the idea that cGMP has a 'second messenger' function. It is possible that cGMP plays a role in the phosphorylation of specific proteins required for directed movement, through the agency of specific protein kinases (Cohen, 1982).

Addition of cAMP under the right conditions can lead to desensitization of the cAMP receptors, particularly after repeated stimulation with cAMP when only the first stimulus may produce a response (as measured by cAMP-induced cGMP synthesis). Similar processes occur in the desensitization of hormone binding sites studied in mammalian systems.

In conclusion we can see that the studies on chemotaxis in *Dictyostelium discoideum* have yielded an enormous amount of information particularly concerning the chemosensory aspects of the process. Hopefully when more information becomes available on the general mechanism(s) of amoeboid locomotion the connection between directed movement and chemosensory inputs into the cell will become easier to study.

Chemotaxis and movement in flagellate and ciliate protozoa

The second general method of locomotion found in the protozoa employs flagella or cilia. The whip-like action of these appendages produces displacement of the surrounding aqueous medium leading to movement. Flagella are usually characterized by being relatively long (*ca* 10 μm) compared to cell length and only one or two will be present on a single cell, thus there is one present in *Trypanosoma* whilst *Chlamydomonas* possesses two. Cilia are generally shorter and much more numerous per cell. Estimates of the number of cilia on a single *Paramecium* vary from 3 000 to 14 000 (Wichterman, 1953). The basic structure of the flagellum and cilium are similar and this is shown in Fig. 4.

The importance of microtubules in the structure is clear, giving the classical 9 + 2 arrangement made up of 9 outer double microtubules and 2 central single microtubules. Each doublet consists of a complete microtubule (sub-fibre A) and an incomplete microtubule (sub-fibre B). Two arm-like structures are associated with sub-fibre A and these can form associations with the B sub-fibre of the adjacent doublet. These arms are composed of the protein dynein which has ATPase activity. The key event in ciliary or flagellar motility is the cyclical interaction between dynein and the tubulin protein present in the B sub-fibre. This mechanochemical cycle (Satir, Wais-Steider, Lebduska, Nasr & Avolio, 1981) requires the presence of ATP as an energy source, and this could be considered to be analogous to the myosin (ATPase)/actin interactions found in muscle systems. When a flagellum is treated with detergents to break the cell membrane, followed by proteolytic enzymes to break the restricting linkages between microtubules, the addition of ATP

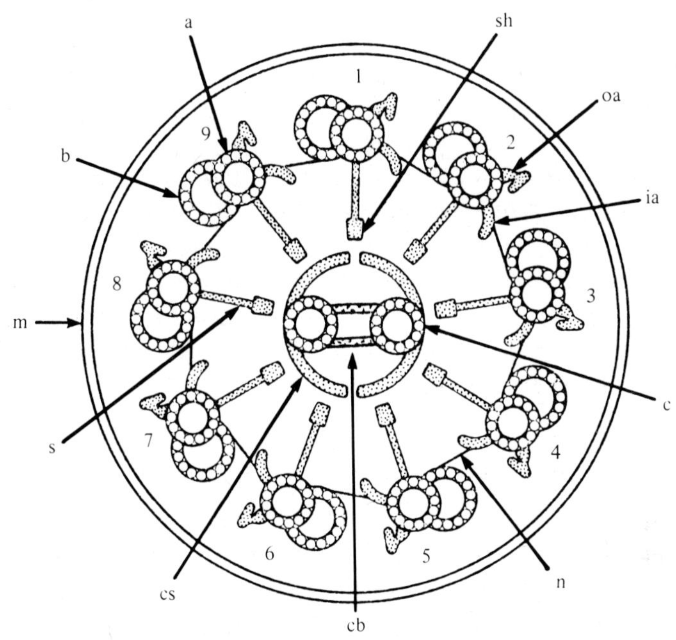

Fig. 4. Diagrammatic representation of a transverse section through a cilium or flagellum: (m), membrane; (a) and (b), A and B tubule of outer doublet; (oa) and (ia), outer and inner arm; (n), nexin link; (s), radial spoke; (sh), radial spoke head; (cs), projections from central pair; (c), central pair microtubules; (cb), central bridge. Reproduced by kind permission of Dr. H. Mohri.

brings about disintegration of the structure of the flagellum by active movements of the constituent doublets past one another. This elegant demonstration by Summers & Gibbons (1971) clearly shows that potential force is available although the mechanism for generation of the wide variety of wave forms found in cilia and flagella is still far from clear.

Subsequent discussion will be concerned with ciliary activity in protozoa. The simplest way to view one beat of an individual cilium is to consider it to consist of an effective stroke and a recovery stroke (Fig. 5). Recently, Sugino & Naitoh (1982) have carried out a computer based study on the beating of a single *Paramecium* cilium using simulations of the cross bridging dynein-tubulin interactions. Their computer based models of cilium bending agreed well with experimental observations in *Paramecium*. Essentially the model proposes that the outer doublets slide with respect to one another in a highly ordered way. If we consider a single pair of doublets, local activation starts at the cilium base and moves distally to the tip. The activation process is transmitted clockwise around the cilium and

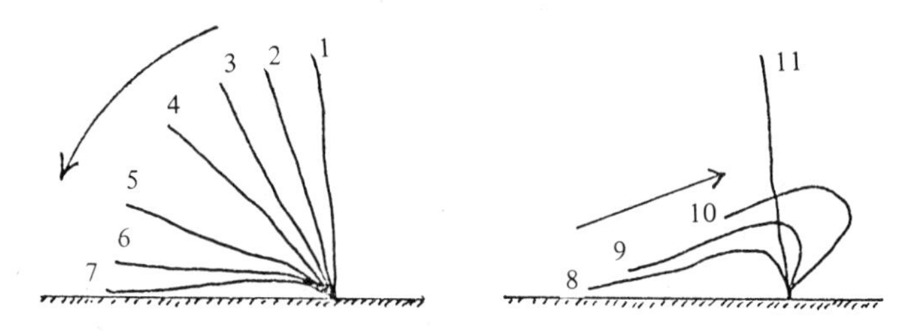

Fig. 5. Diagram to show ciliary beat cycle. Numbers denote sequential position of the cilium in the cycle.

incorporates a delay mechanism. In one complete cycle it appears that the first sliding would occur at the 5–6 pair (see Fig. 4), and subsequent interactions would be transmitted in the order: 7, 8, 9, 1, 2, 3, 4. Interestingly, these results suggest that the natural cycle of this cilium would start with a recovery stroke (associated with the activation of doublets 6 to 9) and proceed to the effective stroke (associated with the activation of doublets 1 to 4). It is important to remember that in *Paramecium* a three-dimensional movement is developed and not the simple planar model depicted in Fig. 5. The frequency of the beat of the cilium can be altered, potentially resulting in an increase or decrease in the speed of locomotion.

One distinctive feature of the movement of cilia when they are arranged in groups is that their pattern of beating is co-ordinated to form metachronal waves – classically compared to the effects of a breeze moving over a field of corn (Fig. 6).

Although it might be thought that co-ordination occurred by transmission of 'information' via the cilia bases (neuroidal theory), experimental evidence suggests that co-ordination is the result of viscous coupling between adjacent cilia through the aqueous medium.

Figure 7 considers the co-ordination of ciliary beating in a hypothetical ciliate in relation to the angle the metachronal waves make with the anterior–posterior axis of the cell. Figure 7(*a*) represents the simplest model in which the effective beat of the cilium is directly backwards, i.e. directly in line with the central axis of the cell and thus the angle = 0°. If the metachronal waves were to develop at an angle to the main axis, say 20°, as in Fig. 7(*b*), then the generated movement would have a spiral character. If the angle were to lie between 90° and 270°, e.g. 180° as illustrated in Fig. 7(*c*), then the result would be ciliary reversal and hence backward movement would occur. The way in which these various patterns of

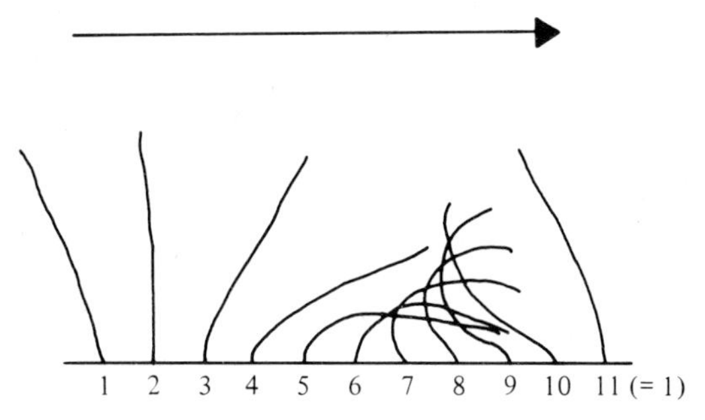

Fig. 6. Diagram of a metachronal wave. Cilia 1→4 are involved in the effective stroke and cilia 5→10 are involved in the recovery stroke. The arrow depicts the direction of the metachronal wave. In this example the direction of cell movement will be in the opposite direction to the direction of the metachronal waves (symplectic pattern). However, in some cases the direction of cell movement can be in the same direction as the metachronal wave movement (antiplectic pattern).

ciliary beat could be achieved in the context of the 9 + 2 structure of the cilium is unclear. One possibility is that rotation of the central pair of microtubules in the cilium (Omoto & Kung, 1979) might play some part in determining the angle of the metachronal wave.

Jennings (1906) found that ciliary beating in *Paramecium* could respond to a stimulus by a transient ciliary reversal – this response was termed 'the avoiding reaction' (Fig. 8). When a forward-swimming *Paramecium* collides with an object, the stimulus at the front end of the ciliate induces ciliary reversal and backward movement over a short distance results. Before resuming normal locomotion the cell rotates to some extent about its posterior end before resuming forward movement. Thus at the end of this avoiding reaction the ciliate is swimming in a new direction, hopefully avoiding the initial object. Jennings also found that the avoiding reaction could occur in response to contact with chemicals. Cations were more effective than anions and an increasing cation concentration could induce the avoiding reaction. Although mechanical stimulation of the anterior end induces the avoiding reaction, mechanical stimulation of the posterior end brings about an increased frequency of ciliary beating so that the paramecium moves forward faster.

Verworn had noted in 1889 that ciliary reversal could occur locally when *Paramecium* was placed in an electric field and electrophysiological approaches to ciliary movement (Kinosita & Murakami, 1967; Naitoh & Eckert, 1974; Eckert & Brehm, 1979) have provided substantial information on the modulation of ciliary

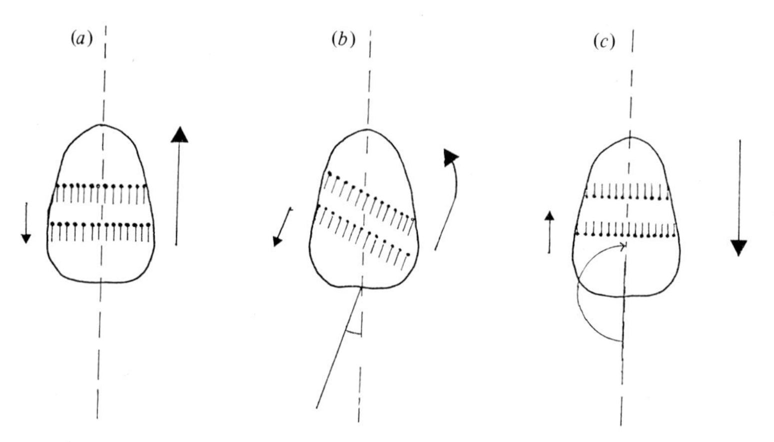

Fig. 7. Consideration of metachronal wave pattern relative to the anterior–posterior axis of a hypothetical ciliate. Only two rows of cilia have been considered and each cilium has been depicted as a solid dot (=cilium base) and the cilium shaft as at the end of the effective stroke. The small arrow represents the direction of the metachronal wave and the large arrow represents the direction of cell movement.

activity. The avoiding reaction has been shown to be due to Ca^{++} ion influx into the cell and depolarization of the cell membrane. This has been elegantly shown using microelectrode techniques to measure changes in membrane potential and using detergent-extracted *Paramecium* to study the effects of changes in ion concentrations. The selective permeability to ions characteristic of the cell membrane could be destroyed by mild treatment with detergent. The addition of ATP and Mg^{++} ions to these detergent extracted models led to ciliary beating in a direction which was dependent on the Ca^{++} ion level in the medium. When the Ca^{++} concentration was kept below 10^{-7} M, normal swimming occurred. However, when the Ca^{++} level was raised above 10^{-6} M ciliary reversal occurred. This result pointed to the existence of a Ca^{++} ion gating mechanism in the cell membrane which when open allowed extracellular Ca^{++} ions to enter the cell, raising the intracellular Ca^{++} ion levels to 10^{-6} M or more and inducing ciliary reversal. Mutants of *Paramecium* have been described which are not capable of ciliary reversal, i.e. can only move in a forward direction, termed ‘pawn’ mutants. Detergent extracted ‘pawn’ mutants are capable of backward movement in the presence of ATP, Mg^{++} ions and a high concentration of Ca^{++} ion and appear therefore to lack the normal Ca^{++} gating mechanism in the cell membrane.

Mechanical stimulation of the posterior end of *Paramecium*

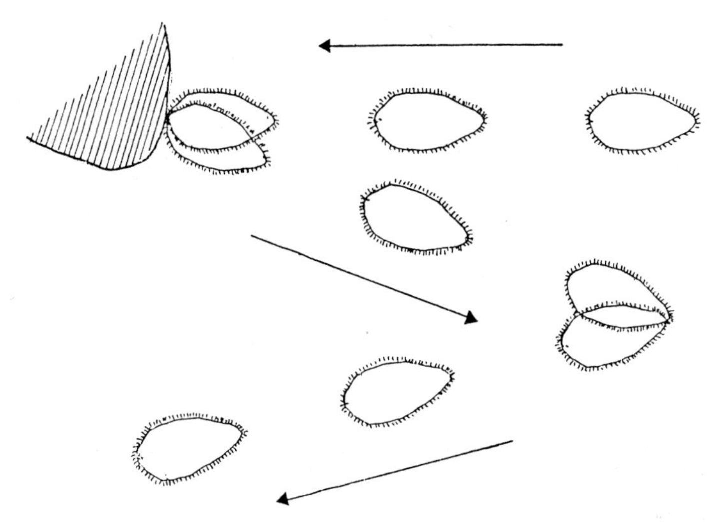

Fig. 8. The avoiding reaction of *Paramecium*. When the anterior end of a *Paramecium* collides with an obstacle, ciliary reversal occurs due to the transient influx of extracellular Ca^{++} ions. However, the presence of calcium pumps soon reduces the level of Ca^{++} ion in the cytoplasm to values below 10^{-6} M and forward swimming then resumes.

leading to increased frequency of ciliary beating is associated with an influx of K^+ ion and hyperpolarization of the cell membrane. This K^+ ion gating mechanism is not affected by deciliation whilst the calcium ion gating system is removed by this process. Thus topographical differences in sensory functions of the cell membrane must occur.

It can be seen that the presence of chemoreceptors and mechanoreceptors enables the ciliate to respond adaptively to various stimuli through changes in the bioelectric properties of the cell membrane.

Three properties of the ciliary motile process could be modulated:

(a) Frequency of the ciliary beat.

(b) Angle of the metachronal wave relative to the anterior–posterior axis of the cell.

(c) Frequency of the avoiding reaction.

(a) and (b) will determine the speed of movement and (c) will determine the rate of change of direction. Thus the motile pattern can consist of successive ‘runs’ and ‘turns’ in a manner similar to that found in motile bacteria. However, it should be stressed that many ciliates have the ability to alter the rate of movement during a ‘run’ in contrast to the constancy of the rate of movement observed in bacterial ‘runs’.

Changes in swimming behaviour of ciliates due to chemicals in the environment have been known for some time. Jennings (1906) at the beginning of this century showed aggregation of *Paramecium* around a bubble of CO_2 compared with the situation when a bubble of air was used. This accumulation reaction was found to occur in many cases using dilute solutions of acids, as was the case with carbonic acid just described. The experimental techniques have now become more sophisticated, for example, when using a T-shaped maze (Van Houten, Hauser & Levandowsky, 1981). In this method the entry arm containing the cells under test is allowed to connect with one arm (T) containing the solution under test and the other arm (C) containing the control solution. After a suitable time interval, say 30 minutes, the connection between arms T and C is closed and the number of cells in each arm is counted. If:

(i) No chemoattraction or repulsion occurs: number of cells in T = number of cells in C;

(ii) Chemoattraction occurs: number of cells in T is greater than number of cells in C;

(iii) Repulsion occurs: number of cells in T is less than the number of cells in C.

Using this technique it can be shown that *Paramecium* is attracted to fermentation and excretory products such as acetate, lactate, ammonium ion and folate. This suggests that these chemicals are acting as food-seeking cues for the ciliate.

Studies on the frequency of the avoiding reactions could be considered to be analogous to those on the tumbling behaviour of motile bacteria. Jennings (1906) showed that:

(1) Frequency of avoiding reaction (F_{AR}) decreases when swimming towards attractants;

(2) Frequency of avoiding reaction (F_{AR}) increases when swimming towards repellants. Since both the speed of movement and F_{AR} are functions of the membrane potential one would expect changes in F_{AR} to be associated with changes in speed of movement. Thus one would predict that under an attractant stimulus, the F_{AR} would fall and the speed of movement would rise due to hyperpolarization of the membrane. Experimentally it has been shown that *Paramecium caudatum* swims faster in attractants that decrease the F_{AR}. There is clearly a class of attractants and repellants that can act by modulating F_{AR} and thus affect the motile behaviour of *Paramecium* (Type I attractants/repellants). When these attractants are present there will be a larger mean free path of travel due to an

increased speed of locomotion (orthokinetic mechanism) and infrequent changes in direction (klinokinetic mechanism). These reactions will lead to accumulation since moving away from the source of attractant will lead to an increase in rate of change of direction (i.e. F_{AR} rises) and a decrease in motility. In the presence of repellants both orthokinetic and klinokinetic mechanisms will operate due to a depolarization of the membrane potential. Not surprisingly the 'pawn' mutants are unaffected by these type I attractants and repellants. However, some chemicals can act on these 'pawn' mutants (and on wild type cells) to produce attraction and repulsion (Type II attractants and repellants). Thus a type II repellant can depress the F_{AR} to almost zero and increase the speed, thereby making the mean free path very long. In this situation an orthokinetic mechanism alone seems to be operating.

Paramecium has very distinct responses to stimulation at the anterior end (due to Ca^{++} ion flux) and the posterior end (due to K^{+} ion flux). This distinct polarity coupled with a relatively great cell length (approx. 100–300 μm) could serve as a basis for detecting chemical gradients in space (Eckert, 1972) rather than utilizing a time based model as in bacteria.

Studies on the chemotactic behaviour of the well known ciliate genus *Tetrahymena* are a long way behind the detailed studies carried out on *Paramecium*. It is possible that *Tetrahymena* is less polarized physiologically in an anterior–posterior manner than *Paramecium*, and recently Almagor, Ron & Bar-Tana (1981) have suggested that a temporal sensing membrane is operating in *Tetrahymena* leading to alterations in the turning frequency during chemotaxis. Chemoreception can be triggered by metabolites and non-metabolites. This has been shown using D and L amino acids, including D and L leucine and D and L methionine. *Tetrahymena* is also responsive to the well characterised mammalian hormones – adrenaline (epinephrine) and nor-adrenaline (norepinephrine). The significance of this observation is unclear.

Tactic behaviour of ciliates related to conjugation

In general, ciliates only become competent to conjugate during nutritional deprivation. Conjugation is a process requiring the coming together of different mating types followed by the exchange of gametic nuclei between the two cells. This pairing process

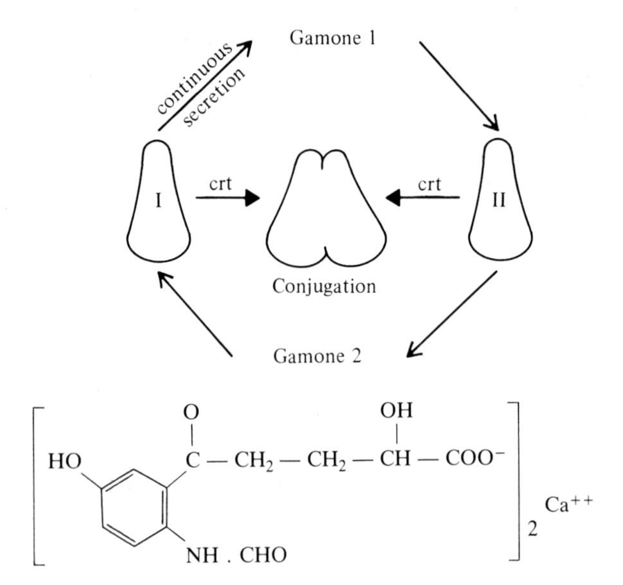

Fig. 9. Cell interactions during the early stages of conjugation in *Blepharisma japonicum*.
(i) Gamone 1 = Blepharmone (glycoprotein, MW ~ 20 000).
(ii) Gamone 2 = Blepharismone. This gamone can act as a chemoattractant for type I cells eliciting a positive response at concentrations as low as 10^{-15} g ml^{-1}.
(iii) Crt = conjugation receptive transformation.

requires mating specific signals, and two general types can be recognized (Miyake, 1981):

(i) The chemical signal specific for mating reaction remains bound to the cell surface and collision between the two mating types is required followed by specific interaction between the two individuals, e.g. as in *Paramecium* and *Tetrahymena*.

(ii) The chemical signals specific for the mating reaction are secreted into the medium and induce a change in ciliate behaviour resulting in pairing, e.g. *Blepharisma*.

Only type (ii) reactions will be considered here and this will be restricted to the elegant work carried out on *Blepharisma* by Miyake. Figure 9 shows the general interactions occurring between complementary mating types of *Blepharisma* (type I & II cells). It is important to stress that normal vegetative cells need to be subjected to mild starvation in order to differentiate to preconjugate cells.

Preconjugation type I cells secrete a chemical product – gamone I (Blepharmone). Generally gamone I is secreted from preconjugant type I cells autonomously. Under conditions of prolonged starvation, its secretion may be reduced and secretion can then be enhanced by the presence of gamone 2 (Blepharismone). Gamone 1 induces two properties in type II cells – gamone 2 production or secretion and transformation of the type II cells into a conjugation receptive form. It is assumed that gamone I reacts with a specific receptor on type II cells. This would produce an intracellular signal which activated specific enzyme systems associated with the conjugation process.

Interaction of gamone 2 with type I cells can elicit increased gamone 1 secretion (see above), production of a conjugation-receptive form of type I and acts as a chemoattractant cue for type I cells. The chemotactic signal acting on type I cells will enable type I and II cells to come together and allows temporary fusion of the cells to occur followed by nuclear exchange.

When different species of *Blepharisma* were compared it was found that gamone 1 was the species-specific gamone in contrast to gamone 2. Production of conjugation-receptive cells (in all four species tested) could be achieved by gamone 2 from any of the species.

CONCLUSIONS AND FURTHER DEVELOPMENTS

In relation to the systems considered here, there are clearly many areas where further concerted investigation is needed. The swarming phenomenon in *Proteus* still has a number of mysterious facets. In particular, the factors which trigger derepression of flagellar synthesis and inhibition of cell division are almost entirely unexplained; cyclic AMP can have a role in influencing both flagellar synthesis and division in some organisms, but changes in cAMP levels cannot be clearly implicated in the events which occur prior to swarming (Armitage, Rowbury & Smith, 1976), neither do any of the alterations in metabolism which precede swarming offer an obvious explanation for the changes in flagellation or division which follow. Of even more importance is the need to establish whether swarming involves negative chemotaxis. The claims of Williams, Anderson, Hoffman, Schwarzhoff & Leonard (1976) and Williams & Schwarzhoff (1978) that chemotaxis is not involved have surprisingly so far not been seriously challenged, although *Proteus mirabilis* shows negative chemotactic responses to a number of its metabolic end-products such as hydrogen sulphide, acetic acid and amines (J. P. Armitage, D. G. Smith & R. J. Rowbury, unpublished observations). The difficulties with this system result partly from the fact that neither flagellar derepression nor cell division inhibition take place in conventional media except on solid surfaces. Clearly it can be more difficult to study such a process but the role of taxis in swarming should be more readily amenable to study now that techniques for forming stable gradients of potentially chemotactic agents on solid media have been developed (Ho & McCurdy, 1979; Shimkets, Dworkin & Keller, 1979).

The mechanisms involved in the chemotactic responses of peritrichously flagellate bacteria are now well understood although the nature of the intracellular signal(s) which leads to a change in direction of rotation of the flagella has yet to be elucidated. There has been much less work on how chemotactic mechanisms aid the flagellate bacterium in its natural environment. Laboratory studies of such chemotaxis have thrown light on the responses of organisms in nutrient-rich environments but it is still intriguing to know how bacteria in the more usual nutrient-poor environments can detect and respond to the tiny concentration differences which presumably occur. It is now becoming clear that bacterial motility and chemotaxis not only aid in the search for usable energy sources and nutrients and in the avoiding of toxic agents but also facilitate infection of plant and animal hosts. This finding clearly opens a rewarding area of study. There is also the exciting possibility that as in the stages that precede mating in the protozoa described above (e.g. *Blepharisma*, Fig. 9), there may be chemotactic signalling involved in *bacterial* conjugation between donor and recipient strains of enterobacteria (Bezdek & Soska, 1972; J. P. Armitage & R. J. Rowbury, unpublished observations).

There is clearly need for further studies of gliding motility in prokaryotes. In particular, now that a range of non-motile mutants are available it may be possible to establish conclusively whether gliding is correlated with the presence of specific structural components. Further studies are also needed to establish how the energy for gliding is supplied; this area has been given scant attention so far. Although chemotaxis is clearly implicated in both food-seeking by swarms of gliders and aggregation in starved myxobacteria, few convincing conclusions can be made about the nature of the chemotactic agents involved. Experiments with mutants which are motile but non-chemotactic might throw light on these phenomena and on the possible role of cGMP.

Although it is clear that taxes are involved in many protozoan responses, very few (apart from the *Dictyostelium* system) have been studied in any depth. This is clearly an area where, using the techniques so far applied only to polymorphonuclear leucocytes for example, or to the amoebae of *Dictyostelium*, substantial advances can be expected.

REFERENCES

ALCANTARA, F. & MONK, M. (1974). Signal propagation during aggregation in the slime mould *Dictyostelium discoideum. Journal of General Microbiology*, **85,** 321–34.

ALLEN, R. D. (1961). A new theory of amoeboid movement and protoplasmic streaming. *Experimental Cell Research* (Suppl.), **8,** 17–31.

ALLWEISS, B., DOSTAL, J., CAREY, K. E., EDWARDS, T. F. & FRETER, R. (1977). The role of chemotaxis in the ecology of bacterial pathogens of mucosal surfaces. *Nature*, **260,** 448–50.

ALMAGOR, M., RON, A. & BAR-TANA, J. (1981). Chemotaxis in *Tetrahymena thermophila. Cell Motility*, **1,** 261–8.

AMES, P. & BERGMAN, K. (1981). Competitive advantage provided by bacterial motility in the formation of nodules by *Rhizobium melitoti. Journal of Bacteriology*, **148,** 728–9.

AMES, P., SCHLUEDERBERG, S. A. & BERGMAN, K. (1980). Behavioural mutants of *Rhizobium melitoti. Journal of Bacteriology*, **147,** 722–7.

ARMITAGE, J. P. (1981). Changes in metabolic activity of *Proteus mirabilis* during swarming. *Journal of General Microbiology*, **125,** 445–50.

ARMITAGE, J. P. & EVANS, M. C. W. (1981). The reaction centre in the phototactic and chemotactic response of photosynthetic bacteria. *FEMS Microbiological Letters*, **11,** 89–92.

ARMITAGE, J. P., ROWBURY, R. J. & SMITH, D. G. (1976). The role of cyclic adenosine monophosphate in the swarming phenomenon of *Proteus mirabilis. Experientia*, **32,** 1266–7.

ARMITAGE, J. P. & SMITH, D. G. (1978). Flagella development during swarmer differentiation in *Proteus mirabilis. FEMS Microbiological Letters*, **4,** 163–5.

BAUMANN, P. & BAUMANN, L. (1977). Biology of the marine enterobacteria: genera *Beneckea* and *Photobacterium. Annual Review of Microbiology*, **31,** 39–61.

BERG, H. C. (1974). Dynamic properties of bacterial flagellar motors. *Nature (Lond.)*, **249,** 77.

BERG, H. C. (1976). How Spirochaetes may swim. *Journal of Theoretical Biology*, **56,** 269–75.

BERG, H. C. & TURNER, L. (1979). Movement of microorganisms in viscous environments. *Nature (Lond.)*, **278,** 349–51.

BEZDEK, M. & SOSKA, J. (1972). Sex-determined chemotaxis in *Salmonella typhimurium* LT2. *Folia Microbiologica*, **17,** 366–9.

BLAKEMORE, R. P. & FRANKEL, R. B. (1981). Magnetic navigation in bacteria. *Scientific American*, **245,** 58–65.

BODE, W., ENGEL, J. & WINKLMAIR, D. (1972). A model of bacterial flagella based on small-angle X-ray scattering and hydrodynamic data which indicate an elongated shape of the flagellin protomer. *European Journal of Biochemistry*, **26,** 313–27.

Bowra, B. J. & Dilworth, M. J. (1981). Motility and chemotaxis towards sugars in *Rhizobium leguminosarum. Journal of General Microbiology*, **126,** 231–5.

Brown, R. M., Willison, J. H. M. & Richardson, C. L. (1976). Cellulose biosynthesis in *Acetobacter xylinum*: visualization of the site of synthesis and direct measurement of the *in vivo* process. *Proceedings of the National Academy of Sciences of the United States of America*, **73,** 4565–9.

Burchard, A. C., Burchard, R. P. & Kloetzel, J. A. (1977). Intracellular periodic structures in the gliding bacterium *Myxococcus xanthus. Journal of Bacteriology*, **132,** 666–72.

Burchard, R. P. & Brown, D. T. (1973). Surface structure of gliding bacteria after freeze-etching. *Journal of Bacteriology*, **114,** 1351–5.

Burnham, J. C., Collart, S. A. & Highison, B. W. (1981). Entrapment and lysis of the cyanobacterium of *Phormidium luridum* by aqueous colonies of *Myxococcus xanthus* PCO2. *Archives of Microbiology*, **129,** 285–94.

Calladine, C. R. (1976). Design requirements for the construction of bacterial flagella. *Journal of Theoretical Biology*, **57,** 469–89.

Campos, J. M. & Zusman, D. R. (1975). Regulation of development in *Myxococcus xanthus*: effect of 3′,5′-cyclic AMP, ADP and nutrition. *Proceedings of the National Academy of Sciences of the United States of America*, **72,** 518–22.

Charpentier, M. (1963). Recherches sur l'equipement enzymatique de *Sporocytophage myxococcoides. Annales Institut Pasteur (Paris)*, **105,** 174–8.

Charpentier, M. (1965). Etude de l'activite cellulalytique de *Sporocytophaga myxococcoides. Annales Institut Pasteur (Paris)*, **109,** 771–97.

Chet, I. & Mitchell, R. (1976). The relationship between chemical structure of attractants and chemotaxis by a marine bacterium. *Canadian Journal of Microbiology*, **22,** 1206–8.

Chet, I. Zilberstein, Y. & Henis, Y. (1973). Chemotaxis of *Pseudomonas lachrymans* to plant extracts and to water droplets collected from the leaf surfaces of resistant and susceptible plants. *Physiological Plant Pathology*, **3,** 473–9.

Christensen, P. J. (1977). The history, biology and taxonomy of the *Cytophaga* group. *Canadian Journal of Microbiology*, **23,** 1599–656.

Christison, J. & Martin, S. M. (1971). Isolation and preliminary characterisation of an extracellular protease of *Cytophaga* sp. *Canadian Journal of Microbiology*, **17,** 1207–17.

Cohen, P. (1982). The role of protein phosphorylation in neural and hormonal control of cellular activity. *Nature*, **296,** 613–20.

Craven, R. C. & Montie, T. C. (1981). Motility and chemotaxis of three strains of *Pseudomonas aeroginosa* used for virulence studies. *Canadian Journal of Microbiology*, **27,** 458–60.

Cumsky, M. & Zusman, D. R. (1979). Myxobacterial lectin of *Myxococcus xanthus. Proceedings of the National Academy of Sciences of the United States of America*, **76,** 5505–9.

Currier, W. W. & Strobel, G. A. (1977). Chemotaxis of *Rhizobium* spp. to a glycoprotein produced by birdsfoot trefoil roots. *Science*, **196,** 434–6.

De Pamphilis, M. L. & Adler, J. (1971). Fine structure and isolation of the hook-basal body complex of flagella from *Escherichia coli* and *Bacillus subtilis. Journal of Bacteriology*, **105,** 384–95.

Dobson, W. J., McCurdy, H. D. & Macrae, T. H. (1979). The function of fimbriae in *Myxococcus xanthus*. II. The role of fimbriae in cell–cell interactions. *Canadian Journal of Microbiology*, **25,** 1359–72.

Duckworth, M. & Turney, J. R. (1969). An extra cellular agarase from a *Cytophaga* species. *Biochemical Journal*, **113,** 139–42.

Dunny, G. M., Brown, B. L. & Clewell, D. B. (1978). Induced cell aggregation and mating in *Streptococcus faecalis* in evidence for a bacterial sex pheromone. *Proceedings of the National Academy of Sciences of the United States of America*, **75,** 3479–83.

Dworkin, M. (1972). The myxobacteria; a new direction in studies of procaryotic development. *Critical Reviews in Microbiology*, **2,** 435–52.

Eckert, R. (1972). Bioelectric control of ciliary activity. *Science*, **176,** 473–81.

Eckert, R. & Brehm, P. (1979). Ionic mechanisms of excitation in *Paramecium. Annual Reviews of Biophysics and Engineering*, **8,** 353–83.

Engelmann, Th.W. (1883). Bacterium photometricum: an article on the comparative physiology of the sense for light and color. *Pflugers Archiv fur der gersammte Physiologie*, **30,** 95–124.

Ensign, J. C. & Wolfe, R. S. (1965). Lysis of bacterial cell walls by an enzyme isolated from a myxobacter. *Journal of Bacteriology*, **90,** 395–402.

Feng, T-Y. & Kno, T-T. (1975). Bacterial leaf blight in rice plants. VI. Chemotactic responses of *Xanthomonas oryzae* to water droplets exudated from water pores on the leaf of rice plants. *Botanical Bulletin of the Academy of Sciences (Taipei)*, **16,** 126–36. Translated *Biological Abstracts* (1976), **61,** 4739.

Freter, R. (1981). Mechanisms of association of bacteria with mucosal surfaces in adhesion and microbial pathogenicity. *CIBA Foundation Symposium*, **80,** 36–55.

Geesey, G. G. & Morita, R. Y. (1979). Capture of arginine at low concentrations by a marine psychrophilic bacterium. *Applied & Environmental Microbiology*, **38,** 1092–7.

Glagolev, A. N. & Skulachev, V. P. (1978). The proton pump is a molecular engine of motile bacteria. *Nature*, **272,** 280–2.

Gotz, R., Limmer, N., Ober, K. & Schmitt, R. (1982). Motility and chemotaxis in two strains of *Rhizobium* with complex flagella. *Journal of General Microbiology*, **128,** 789–98.

Grebecka, L. & Grebecki, A. (1981). Testing motor functions of the frontal zone in the locomotion of *Amoeba proteus. Cell Biology International Reports*, **5,** 587–94.

Greenberg, E. P. & Canale-Parola, E. (1977). Motility of flagellate bacteria in viscous environments. *Journal of Bacteriology*, **132,** 356–8.

Gude, H. (1980). Occurrence of cytophagas in sewage plants. *Applied and Environmental Microbiology*, **39,** 756–63.

Hagen, D. C., Bretscher, A. P. & Kaiser, D. (1978). Synergism between morphological mutants of *Myxococcus xanthus. Developmental Biology*, **64,** 284–96.

Hazen, T. C., Raker, M. L. & Esch, G. W. (1979). Chemotaxis of *Aeromonas hydrophila. Abstracts of the Annual Meeting of the Society of American Microbiologists I*, **26.**

Hellewell, S. B. & Taylor, D. L. (1979). The contractile basis of amoeboid movement. IV. The solation-contraction coupling hypothesis. *Journal of Cell Biology*, **83,** 633–48.

Henrichsen, J. (1972). Bacterial surface translocation; a survey and classification. *Bacteriological Reviews*, **36,** 478–503.

Hildebrand, E. (1978). Bacterial phototaxis. In *Taxis and Behaviour*, ed. G. L. Hazelbauer, *Receptors and Recognition*, vol. 5, series B, pp. 37–68. Chapman & Hall.

Hirota, N., Kitade, M. & Imae, Y. (1981). Flagellar motors of alkalophilic *Bacillus* are powered by an electrochemical potential gradient of Na^+. *FEBS Letters*, **132,** 278–81.

Ho, J. & McCurdy, H. D. (1979). Demonstration of positive chemotaxis to cyclic

GMP and 5′-AMP in *Myxococcus xanthus* by means of a simple apparatus for generating practically stable concentration gradients. *Canadian Journal of Microbiology*, **25,** 1214–18.

HODGKIN, J. & KAISER, D. (1979a). Genetics of gliding motility in *Myxococcus xanthus* (Myxobacterales): Genes controlling movement of single cells. *Molecular and General Genetics*, **171,** 167–76.

HODGKIN, J. & KAISER, D. (1979b). Genetics of gliding motility in *Myxococcus xanthus*: Two gene systems control movement. *Molecular and General Genetics*, **171,** 177–91.

IINO, T. (1977). Genetics of structure and function of bacterial flagella. *Annual Review of Genetics*, **11,** 161–82.

INOUYE, M., INOUYE, S. & ZUSMAN, D. R. (1979). Biosynthesis and self-assembly of protein S, a development specific protein of *Myxococcus xanthus. Proceedings of the National Academy of Sciences of the United States of America*, **76,** 209–13.

INOUYE, S., WHITE, D. & INOUYE, M. (1980). Development of *Stigmatella aurantiaca*: effects of light and gene expression. *Journal of Bacteriology*, **141,** 1360–5.

JENNINGS, H. S. (1906). *Behaviour of the Lower Organisms*. New York: Columbia University Press.

JONES, G. W., RICHARDSON, C. A. & UHLMAN, D. (1981). The invasion of HeLa cells by *Salmonella typhimurium*: reversible and irreversible bacterial attachment and the role of bacterial motility. *Journal of General Microbiology*, **127,** 351–60.

KAISER, D., MANOIL, C. & DWORKIN, M. (1979). Myxobacteria: cell interactions, genetics and development. *Annual Review of Microbiology*, **33,** 595–639.

KHAN, S. & MACNAB, R. M. (1980a). The steady state counterclockwise/clockwise ratio of bacterial flagellar motors is regulated by proton motive force. *Journal of Molecular Biology*, **138,** 563–97.

KHAN, S. & MACNAB, R. M. (1980b). Proton chemical potential, proton electrical potential and bacterial motility. *Journal of Molecular Biology*, **138,** 599–614.

KING, C. A., WESTWOOD, R., COOPER, L. & PRESTON, T. M. (1979). Speed of locomotion of the soil amoeba *Naegleria gruberi* in media of different ionic compositions with special reference to interactions with the substratum. *Protoplasma*, **99,** 323–34.

KINOSITA, H. & MURAKAMI, A. (1967). Control of ciliary motion. *Physiological Reviews*, **47,** 53–82.

KOMEDA, Y., SUZUKI, H., ISHIDSU, J. & INO, T. (1975). Role of cyclic AMP in flagellatation of *Salmonella typhimurium. Molecular and General Genetics*, **142,** 289–98.

KORN, E. D. & WEISMAN, R. A. (1967). Phagocytosis of latex beads by *Acanthamoeba*: II. Electron microscopic study of the initial events. *Journal of Cell Biology*, **34,** 219–27.

KOSHLAND, D. E. JNR. (1981). Biochemistry of sensing and adaptation in a simple bacterial system. *Annual Review of Biochemistry*, **50,** 765–82.

KUHRT, M. & PATE, T. L. (1973). Isolation and characterisation of the tubules and plasma membrane of *Chondrococcus columnaris. Journal of Bacteriology*, **114,** 1309–18.

LA MARRE, A. G., STRALEY, S. G., CONTI, S. F. (1977). Chemotaxis toward amino acids by *Bdellovibrio bacteriovorus. Journal of Bacteriology*, **131,** 201–7.

LANGLOIS, G. A. (1975). Effect of algal exudates on substratum selection by motile telotrochs of the marine peritrich ciliate *Vorticella marina. Journal of Protozoology*, **20,** 115–23.

LAZLO, D. J. & TAYLOR, B. L. (1981). Aerotaxis in *Salmonella typhimurium*: role of electron transport. *Journal of Bacteriology*, **145,** 990–1001.

LEWIN, R. A. & LOUNSBERY, D. M. (1969). Isolation, cultivation and characterisation of flexibacteria. *Journal of General Microbiology*, **58**, 145–70.
LINNEMANS, W. A. M., BLOCK, F. & HORSTEN, T. M. (1981). Cell surface and motility in *Acanthamoeba*. In *2nd International Conference on the Biology and Pathogenicity of Small Free Living Amoebae, 1980*. US Department of Health & Human Sciences Center for Disease Control, Atlanta, Georgia, USA, **89**.
MACNAB, R. M. (1978). Bacterial motility and chemotaxis: the molecular biology of a behavioural system. *Chemical Rubber Company Critical Reviews of Biochemistry*, **5**, 291–341.
MACRAE, T. H., DOBSON, W. J. & MCCURDY, H. D. (1977). Fimbriation in gliding bacteria. *Canadian Journal of Microbiology*, **23**, 1096–108.
MACRAE, T. H. & MCCURDY, H. D. (1975). Ultrastructural studies of *Chondromyces crocatus* vegetative cells. *Canadian Journal of Microbiology*, **21**, 1815–26.
MACRAE, T. H. & MCCURDY, H. D. (1976). Evidence for motility-related fimbriae in the gliding micro-organism *Myxococcus xanthus*. *Canadian Journal of Microbiology*, **22**, 1589–93.
MAEDA, K. & IMAE, Y. (1979). Thermosensory transduction in *Escherichia coli* – inhibition of thermoresponse by L-serine. *Proceedings of the National Academy of Sciences of the United States of America*, **76**, 91–5.
MAEDA, K., IMAE, Y., SHIOI, J.-I & OOGANA, F. (1976). Effect of temperature on motility and chemotaxis of *Escherichia coli*. *Journal of Bacteriology*, **127**, 1039–46.
MANDEL, M. & LEADBETTER, E. R. (1965). Deoxyribonucleic acid base composition of myxobacteria. *Journal of Bacteriology*, **90**, 1795–6.
MANDEL, M. & LEWIN, R. A. (1969). Deoxyribonucleic acid base composition of flexibacteria. *Journal of General Microbiology*, **58**, 171–8.
MANOIL, C. & KAISER, D. (1980a). Guanosine pentaphosphate and guanosine tetraphosphate accumulation and induction of *Myxococcus xanthus* fruiting body development. *Journal of Bacteriology*, **141**, 305–15.
MANOIL, C. & KAISER, D. (1980b). Purine-containing compounds, including cyclicadenosine 3′,5′-menophosphate, induce fruiting of *Myxococcus xanthus* by nutritional imbalance. *Journal of Bacteriology*, **141**, 374–7.
MANSON, M. D., TEDESIO, P. M. & BERG, H. C. (1980). Energetics of flagella rotation in bacteria. *Journal of Molecular Biology*, **138**, 541–61.
MAST, S. O. (1926). Structure, movement, locomotion and stimulation in amoeba. *Journal of Morphology & Physiology*, **41**, 347–425.
MATO, J. M., KRENS, F. A., VAN HAASTERT, P. J. M. & KONIJN, T. M. (1977). cAMP-dependent-cyclic GMP accumulation in *Dictyostelium discoideum*. *Proceedings of the National Academy of Sciences of the United States of America*, **74**, 2348–51.
MATO, J. M. & KONIJN, T. M. (1979). Chemosensory transduction in *Dictyostelium discoideum*. In *Biochemistry and Physiology of Protozoa*, vol. 2, pp. 181–219, ed. M. Levandowsky & S. H. Hutner. New York: Academic Press.
MCCURDY, H. D. & WOLF, S. (1967). Deoxyribonucleic acid base composition of fruiting Myxobacterales. *Canadian Journal of Microbiology*, **13**, 1707–8.
MIYAKE, A. (1981). Physiology and biochemistry of conjugation in ciliates. In *Biochemistry and Physiology of Protozoa*, vol. 4, pp. 125–98, ed. M. Levandowsky & S. H. Hutner. New York: Academic Press.
MOENCH, T. T. & KONETZKA, W. A. (1978). A novel method for the isolation and study of a magnetotactic bacterium. *Archives of Microbiology*, **119**, 203–12.
MOULTON, R. C., YOUNG, L. & MONTIE, T. C. (1980). Motility and chemotaxis in relation to virulence of *Pseudomonas aeruginosa*. *Abstracts of the Annual Meeting of the American Society of Microbiology*, **B102**.

NAITOH, Y. & ECKERT, R. (1974). The control of ciliary activity in protozoa. In *Cilia and Flagella*, ed. M. Sleigh, pp. 305–52. London: Academic Press.
NIWANO, M. & TAYLOR, B. L. (1982). Novel sensory adaptation mechanism in bacterial chemotaxis to oxygen and phosphotransferase substrates. *Proceedings of the National Academy of Sciences of the United States of America,* **79,** 11–15.
OMOTO, C. K. & KUNG, C. (1979). The pair of central tubules rotates during ciliary beat in *Paramecium. Nature*, **279,** 532–4.
PACHA, R. E. (1968). Characteristics of *Cytophaga psychrophila* (Borg) isolated during outbreaks of bacterial cold water disease. *Applied Microbiology*, **16,** 97–101.
PACHA, R. E. & PORTER, S. (1968). Characteristics of myxobacteria isolated from the surface of fresh water fish. *Applied Microbiology*, **16,** 1901–6.
PAERL, H. W. & KELLAR, P. E. (1978). Significance of bacterial *Anabaena* (cyanophyceae) associations with respect of N_2 fixation in fresh water. *Journal of Phycology*, **14,** 254–60.
PANOPOULUS, N. J. & SCHROTH, M. N. (1974). Role of flagellar motility in the invasion of bean leaves by *Pseudomonas phaseolicola. Phytopathology*, **64,** 1389–97.
PARKINSON, J. S. (1977). Behavioural genetics in bacteria. *Annual Reviews of Genetics*, **11,** 397–414.
PASTAN, I. (1972). Cyclic AMP. *Scientific American*, **227,** 97–105.
PATE, J. L., JOHNSON, J. L. & ORDAL, E. J. (1967). The fine structure of *Chondrococcus columnaris.* II. Structure and formation of rhapidosomes. *Journal of Cell Biology*, **35,** 15–35.
PATE, J. L. & ORDAL, E. J. (1967). The fine structure of *Chondrococcus columnaris*. III. The surface layers of *C. columnaris. Journal of Cell Biology*, **35,** 37–51.
PATE, J. L. & CHANG, E. L. (1979). Evidence that gliding motility in procaryotic cells is driven by rotary assemblies in the cell envelope. *Current Microbiology*, **2,** 59–64.
PETERSON, E. A., GILLESPIE, D. C. & COOK, F. D. (1969). A wide spectrum antibiotic produced by a species of *Sorangium. Canadian Journal of Microbiology*, **12,** 221–30.
PETRINO, M. G. & DOETSCH, R. N. (1978). 'Viscotaxis', a new behavioural response of *Leptospira interrogans (biflexa)* Strain B16. *Journal of General Microbiology*, **109,** 113–17.
PILGRAM, W. K. & WILLIAMS, F. D. (1976). Survival value of chemotaxis in mixed cultures. *Canadian Journal of Microbiology*, **22,** 1771–3.
QUALLS, G. T., STEPHENS, K. & WHITE, D. (1978). Light stimulated morphogenesis in the fruiting myxobacteria *Stigmatella auratiaca. Science*, **201,** 444–5.
QUINN, G. R. & SKERMAN, V. B. D. (1980). Herpetosiphon – Nature's scavenger? *Current Microbiology*, **4,** 57–62.
RAYMUNDO, A. K. & RIES, S. M. (1980). Motility of *Erwinia amylovora. Phytopathology*, **70,** 1062–5.
RIDGWAY, H. F. (1977). Ultrastructural characterisation of goblet-shaped particles from the cell wall of *Flexibacter polymorphus. Canadian Journal of Microbiology*, **23,** 1201–13.
ROSENBERG, E., KELLER, K. H. & DWORKIN, M. (1977). Cell density dependent growth of *Myxococcus xanthus* on casein. *Journal of Bacteriology*, **129,** 770–7.
SATIR, P., WAIS-STEIDER, J., LEBDUSKA, S., NASR, A. & AVOLIO, J. (1981). The mechanochemical cycle of the dynein arm. *Cell Motility*, **1,** 303–28.
SCHIMZ, A. (1981). Methylation of membrane proteins is involved in chemosensory

and photosensory behaviour of *Halobacterium halobium*. *FEBS Letters*, **125,** 205–7.

SCHIMZ, A. & HILDEBRAND, E. (1979). Chemosensory responses of *Halobacterium halobium*. *Journal of Bacteriology*, **140,** 749–53.

SCHMIDT, E. L. (1979). Initiation of plant root–microbe interactions. *Annual Reviews of Microbiology*, **33,** 355–76.

SCHMIDT-LORENZ, W. (1969). The fine structure of the swarm cells of Myxobacteria. *Journal of Applied Bacteriology*, **32,** 22–3.

SCHMIDT-LORENZ, W. & KUHLWEIN, H. (1968). Intracellulare bewegungs – organellen der Myxobacterien. *Archiv für Mikrobiologie*, **60,** 95–8.

SCHNEIDER, W. R. JR. & DOETSCH, R. N. (1977). Temperature effects on bacterial movement. *Applied & Environmental Microbiology*, **34,** 695–700.

SHILO, M. (1970). Lysis of blue–green algae by myxobacter. *Journal of bacteriology*, **104,** 453–61.

SHIMKETS, L. J., DWORKIN, M. & KELLER, K. H. (1979). A method for establishing stable concentration gradients in agar suitable for studying chemotaxis on a solid surface. *Canadian Journal of Microbiology*, **25,** 1460–7.

SHIMKETS, L. & DWORKIN, M. (1981). Excreted adenosine is a cell density signal for the initiation of fruiting body formation in *Myxococcus xanthus*. *Developmental Biology*, **84,** 51–60.

SJOBLAD, R. D. & MITCHELL, R. (1979). Chemotactic responses of *Vibrio alginolyticus* to algal extracellular products. *Canadian Journal of Microbiology*, **25,** 964–7.

SMITH, J. C. & DOETSCH, R. N. (1968). Motility in *Pseudomonas fluoescens* with special reference to survival advantage and negative chemotaxis. *Life Science*, **7,** 875–86.

SMITH, J. L. & DOETSCH, R. N. (1969). Studies in negative chemotaxis and the survival value of motility in *Pseudomonas fluorescens*. *Journal of General Microbiology*, **55,** 379–91.

STAHL, M. L. & WILLIAMS, F. D. (1981). Immunofluorescent evidence of *Proteus mirabilis* swarm cell formation on sterilised rat faeces. *Applied & Environmental Microbiology*, **41,** 801–6.

STANIER, R. Y. (1941). Studies on marine agar-digesting bacteria. *Journal of Bacteriology*, **42,** 527–59.

STANIER, R. Y. (1942). The cytophaga group: a contribution to the biology of the myxobacteria. *Bacteriological Reviews*, **6,** 143–96.

STEPHENS, K., HEGEMAN, G. D. & WHITE, D. (1982). Pheromone produced by the myxobacterium *Stigmatella auratiaca*. *Journal of Bacteriology*, **149,** 739–47.

STEWART, J. R. & BROWN, R. M. (1969). Cytophaga that kills or lyzes algae. *Science*, **164,** 1523–4.

STRALEY, S. C. & CONTI, S. F. (1977). Chemotaxis by *Bdellovibrio bacteriovorus* towards prey. *Journal of Bacteriology*, **132,** 628–40.

STRALEY, S. C., LA MARRE, A. G., LAWRENCE, L. J. & CONTI, S. F. (1979). Chemotaxis of *Bdellovibrio bacteriovorus* towards pure compounds. *Journal of Bacteriology*, **135,** 634–42.

SUGINO, K. & NAITOH, Y. (1982). Simulated cross-bridge patterns corresponding to ciliary beating in *Paramecium*. *Nature*, **295,** 609–11.

SUMMERS, K. E. & GIBBONS, I. R. (1971). ATP induced sliding of tubules in trypsin treated flagella of sea urchin sperm. *Proceedings of the National Academy of Sciences of the United States of America*, **68,** 3092–6.

TAYLOR, B. L., MILLER, J. B., WARRICK, H. M. & KOSHLAND, D. E. JR. (1979). Electron acceptor taxis and blue light effect on bacterial chemotaxis. *Journal of Bacteriology*, **140,** 567–73.

TAYLOR, D. L. & CONDEELIS, J. S. (1979). Cytoplasmic structure and contractility in amoeboid cells. *International Review of Cytology*, **56,** 57–144.

TORELLA, F. & MORITA, R. Y. (1978). Chemotaxis towards amino acids by marine psychrotrophic and a marine psychrophilic bacterium: adaptation to low nutrient levels. *Abstracts of the Annual Meeting of the American Society for Microbiology*, **N15**.

UHLMAN, D. C. & JONES, G. W. (1982). Chemotaxis as a factor in interactions between HeLa cells and *Salmonella typhimirium. Journal of General Microbiology*, **128,** 415–18.

ULITZUR, S. (1975). The mechanism of swarming of *Vibrio alginolyticus. Archives of Microbiology*, **104,** 67–71.

VAN HOUTEN, J., HAUSER, D. C. R. & LEVANDOWSKY, M. (1981). Chemosensory behaviour in Protozoa. In *Biochemistry and Physiology of Protozoa*, vol. 4, pp. 67–124, ed. M. Levandowsky & S. H. Hutner. New York: Academic Press.

WARKE, D. M. & DHALA, S. A. (1968). Use of inhibitors for selective isolation and enumeration of cytophagas from natural habitats. *Journal of General Microbiology*, **51,** 43–8.

WEBLEY, D. M., DUFF, R. B., BACON, J. S. D. & FARMER, V. C. (1965). A study of polysaccharide-producing organisms occurring in the root region of certain pasture grasses. *Journal of Soil Science*, **16,** 149–57.

WICHTERMAN, R. (1953). *The biology of Paramecium.* New York: Blakiston Company Inc.

WILLIAMS, F. D. & SCHWARZHOFF, R. H. (1978). Nature of the swarming phenomenon in *Proteus. Annual Review of Microbiology*, **32,** 101–22.

WILLIAMS, F. D., ANDERSON, D. M., HOFFMAN, P. S., SCHWARZHOFF, R. H. & LEONARD, S. (1976). Evidence against the involvement of chemotaxis in swarming in *Proteus mirabilis. Journal of Bacteriology*, **127,** 237–8.

WOOD, D. C. & HAYASAKA, S. S. (1981). Chemotaxis of rhizoplane bacteria to amino acids comprising eel grass (*Zostera marina* L.) root exudate. *Journal of Experimental Marine Biology and Ecology*, **50,** 153–61.

ZIGMOND, S. (1978). Chemotaxis by polymorphonuclear leukocytes. *Journal of Cell Biology*, **77,** 269–87.

ZIGMOND, S. (1982). Polymorphonuclear leucocyte response to chemotactic gradients. In *Cell Behaviour*, ed. R. Bellairs, A. Curtis & G. Dunn, pp. 183–202. Cambridge University Press.

MICROBES AND SURFACES

JOHN N. WARDELL, CHARLES M. BROWN & BRIAN FLANNIGAN

Department of Brewing and Biological Sciences, Heriot Watt University, Chambers Street, Edinburgh EH1 1HX, UK

INTRODUCTION

Any survey of recent literature suggests that the significance of the relationships between microorganisms and surfaces may not be generally recognized by many microbiologists. One possible reason for this is that interests in microbial adhesion and colonization tend to be in terms of specific problems of which the phenomenon may apparently play only a minor role. For instance one might consider an infection process, corrosion of marine structures, some aspect of microbial physiology or many other areas of microbiology. The reality is that in the natural state all surfaces will become coated with a thin film of microorganisms of which bacteria will be the most numerous. The term 'surface' in this context is generally used to describe a solid-gas or solid-liquid interface and this will be the convention used here. Other interfaces, gas-liquid and liquid-liquid may be equally important when considering factors which affect the accumulation of bacteria. It is likely that in natural environments all bacteria spend some time associated with an interface; moreover, as will be discussed later, it may be argued that it is the microorganisms associated with interfaces, particularly solid surfaces, which show the greatest metabolic activity and therefore are the most important feature of the environment. While this article is concerned mainly with bacteria, a concluding section includes some current information on filamentous fungi.

THE EARLY WORK

Although by no means the earliest, it is the report of Zobell & Anderson (1936) which for many stimulated an interest in the relationship between bacteria and solid surfaces. This report recorded that the number of bacteria present in stored sea water and the rate of carbon mineralization was proportional to the surface area to volume ratio of the container used, and the authors proposed that this was due to nutrients being concentrated at the container surfaces. In experiments specifically designed to investigate Zobell's thesis, Heukelakian & Heller (1940) grew *Escherichia coli* in a phosphate buffer at a range of concentrations of glucose and peptone. At concentrations of 0.5–2.5 p.p.m. growth was restricted to flasks containing glass beads. The 'stimulatory' effect of the glass surfaces was observed at concentrations up to 25 p.p.m. glucose and peptone after which *Escherichia coli* grew equally well in the absence of the beads. It was also noted that at low concentrations of nutrients most of the bacterial cells were associated with the surfaces. In similar experiments growth of the natural population of stream water was stimulated by the addition of clean sand. However, Heukelakian and Heller did note that the addition of slime-covered stones to cultures of polluted and unpolluted waters produced lower increases in the population than might be expected. They suggested that this was due to the oxidation of adsorbed organic substrates by the biologically active film. The overall conclusions reached were that by concentrating available nutrients, inert surfaces enable bacteria to develop where the nutrient levels are too low for normal growth and that the development occurs either as bacterial slime or as colonial growth attached to the surfaces. However, once bioactive slime is established the stimulatory effect is greatly reduced.

Later experiments by Jannasch (1958), using *Bacillus subtilis* grown in peptone tap water medium and in the presence of chitin particles, confirmed the work of Heukelakin & Heller and suggested a critical or threshold concentration of nutrients above which cells grew distributed throughout the medium, and below which cells grew mainly associated with the chitin. Parallel observations on planktonic bacteria showed that the differences between plate counts and direct counts was due in part to the clumping of cells around small (in this case organic) particles. Results rather similar to those of Jannasch (1958) have been obtained more recently by Corpe (1970a) investigating the colonization of glass slides by a *Pseudomonas* sp.

The observations made during the early investigations are important because many natural environments are oligotrophic, that is the concentrations of available nutrients (compared with *inert* organic and inorganic particulate matter) are very low. This is especially true of aquatic environments where dilution and dispersion of available nutrients occurs through diffusion and water currents (Jannasch, 1958, 1979; Marshall, 1976). However, it is generally held that ions and various macromolecules are adsorbed at inert surfaces (see Zobell, 1943; Marshall, 1979, 1980a). As a result it is thought that the physiological basis of the stimulatory effect of surfaces on bacterial activity observed by, among others, Zobell & Anderson (1936) and Heukelakian & Heller (1940) is due to the presence of adsorbed nutrients (Marshall, 1976, 1980b; Meadows & Anderson, 1979; Wardell *et al.*, 1980). This seems to be supported by the evidence of Corpe (1970a) that bacteria attach to polymer coated surfaces in larger numbers than to uncoated surfaces, but other factors such as cross feeding, co-metabolism, interspecies hydrogen transfer and interspecies organism proton transfer may well stimulate the growth of microcolonies of single organisms and mixed communities (for a review of community interaction of this type see Slater, 1978). However, even this is regarded by many to be an over simplified view of events as in physico-chemical terms bacterial cells can be regarded as charged particles having an overall net negative charge at pH values encountered in natural environments (Burns, 1979; Harden & Harris, 1953; Ward & Berkeley 1980). Moreover, in attempts to explain the initial events of the adhesion process, bacteria are often considered to be colloidal systems (Marshall, 1976).

ATTACHMENT TO SURFACES

The whole subject area of microbial attachment to surfaces is confused. This is not due to any lack of information (Daniels (1972) with 482 reference lists over 1000 citations of microorganisms absorbing onto surfaces) but rather to the wide range of interests of those physical chemists, engineers, dentists, plant pathologists and microbiologists involved and their different approaches to the problem (see recent edited volumes by Bitton & Marshall, 1980; Ellwood *et al.*, 1979, and Berkeley *et al.*, 1981).

In practice the subject of bacterial adhesion can be discussed at three levels: in physico-chemical terms; at the primary film level; or at the macrofouling stage. In simplified terms that is, (a) the approach of single cells to a surface, (b) the adhesion of single cells and the development of a monolayer and (c) the development of a thick film of bacteria leading perhaps to colonization by other organisms (macrofouling). These are not in any way strict divisions or definitions and there must inevitably be considerable overlapping of the different stages (see also Fletcher *et al.*, 1980; Floodgate, 1972).

PHYSICO-CHEMICAL CONSIDERATIONS

Although several different forces have roles in the adhesion process (Daniels, 1980), it seems likely that for bacteria in aquatic systems those most closely involved are the London Van der Waal's attractive forces and the electrostatic (usually repulsive) forces (Daniels, 1980; Marshall, 1976; Marshall *et al.*, 1971a). The resultant potential energy of interaction determines the behaviour of the particle/cell at the surface (Marshall, 1976). Bacteria may be regarded as negatively charged particles due in large part to the structure of the cell envelope and this is similar for both Gram-positive and Gram-negative cells, despite their having outer walls and outer membranes of differing composition (Ward & Berkeley, 1980). The charge surrounding a particle is not of uniform density.

The interaction between a particle/cell and a solid surface is therefore the result of the interaction between the attractive forces and the repulsive forces arising from the overlapping of the electrical double layers. However, it must be remembered that the surface may be modified by the adsorption of ions and macromolecules so that its charge may be altered. For example, if the net surface charge was positive then the forces of electrostatic attraction would be overriding (see Marshall, 1976, 1980a). Moreover, whereas the thickness of the electrical double layer (given the symbol 1/K) varies with electrolyte concentration (the double layer becomes compressed at high electrolyte concentrations), the attractive forces vary only with separation. It follows that the resultant potential energy of interaction also varies with the electrolyte concentration (Marshall, 1976).

From Fig. 1 it can be seen that under both conditions there is a point at a very small distance from the solid surface at which the resultant force would favour attraction to the surface, the primary minimum. Under conditions of low electrolyte concentration there is, however, a large potential barrier to be overcome before that is

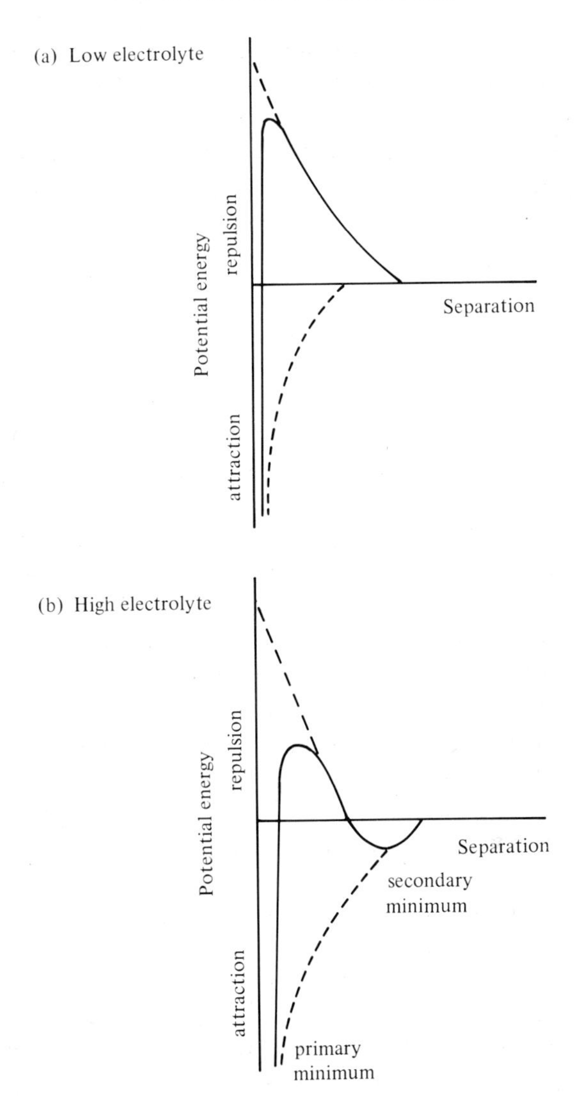

Fig. 1. Variation of the energies of repulsion and attraction with separation and the resultant potential energy of interaction (after Marshall, 1976).

reached. On the other hand, at high electrolyte concentrations there is a secondary minimum at some further distance away (5–10 nm) and this would enable a cell or particle to approach very close to the surface. Marshall *et al.* (1971a) described a reversible stage in the adsorption of an *Achromobacter* sp. where the number of cells adsorbing increased with increasing electrolyte concentration. That this was an effect due to the compression of the diffuse double layer was demonstrated by the fact that bacteria adsorbed at lower concentrations of divalent electrolyte than of monovalent electrolyte (Marshall *et al.*, 1971a).

ADHESION OF SINGLE CELLS AND THE FORMATION OF A MONOLAYER

It is generally accepted that the adhesion of single cells to a solid surface occurs in two stages, a reversible and an irreversible or time-dependent phase (Fletcher, 1977a; Floodgate, 1972; Gibbons 1977; Marshall & Bitton, 1980). Evidence for the reversible phase comes from the work of Marshall *et al.* (1971a) who described the reversible sorption to glass of two marine bacteria, a *Pseudomonas* sp. and an *Achromobacter* sp., while similar observations were made by Meadows (1971) and Corpe (1974).

The reversible phase, where cells are able to exhibit Brownian motion and flagellar movement, can be explained in terms of physico-chemical forces as discussed above (see also Floodgate, 1966). It would therefore be reasonable to expect that increasing the electrolyte concentration would increase the numbers of bacteria able to enter this phase since such action would reduce the repulsion forces by compressing the electrical double layer. Fletcher (1979) takes a different view and presents evidence that an increase in cation concentration, in this case Al^{3+} and La^{3+}, has a detrimental effect on the adhesive properties of a marine pseudomonad. Moreover, she proposed that the mechanism involved may be an interaction between the cations and an adhesive polymer. This suggests that it was not the *reversible* phase of attachment which was considered but the time-dependent phase.

The time-dependent or irreversible phase of attachment was recognized quite early (Zobell, 1943) but it did not receive much attention until 'rediscovered' in the early 1970s. This phase is characterized by the appearance of extracellular polymeric adhesives and although the nature and amounts of polymer produced vary considerably, for the majority of aquatic bacteria only small amounts of what are thought to be specific adhesive polymers are involved (Fletcher & Floodgate, 1973; Marshall, 1976; Marshall *et al.*, 1971a; Zobell, 1943). Indeed there is evidence that large amounts of polymer coating the surface may inhibit rather than aid attachment (Brown *et al.*, 1977). This is just a part of the complex effect which adsorbed substances have on bacterial adhesion. Evidence can be found that presorbed macromolecules both promote and inhibit adhesion (Corpe, 1970a; Meadows, 1971; Fletcher, 1976, 1977a; Fletcher & Leob, 1976). Small quantities of adsorbed materials may promote the accumulation of bacteria at surfaces due to the increased nutritional status of the interface, whereas completely coated surfaces will have a modified surface charge which, depending on the nature of the charge, could aid or inhibit the attachment process. There is also the possibility to consider of specific receptor sites on cells and surfaces being saturated and unavailable depending on the nature of the nutrient limitation (Brown *et al.*, 1977).

The chemical nature of many of the polymer adhesives has not been determined but staining with ruthenium red and alcian blue have shown several to be polysaccharides or glycoproteins (Corpe, 1970a, 1974; Fletcher & Floodgate, 1973) and these have subsequently been implicated on many occasions (Costerton *et al.*, 1978; Corpe, 1964; 1974; Gibbons, 1977; Harris & Mitchell, 1973; Humphrey, *et al.*, 1979). Whether such polymers are preformed or the result of *de novo* synthesis is uncertain. If bacteria are able to attach firmly within minutes (see Fletcher, 1980) then it is unlikely that synthesis of new polymer is required. This view would seem to be supported by reports that heat-killed and formaldehyde-treated cells do not attach whereas UV-killed cells do, suggesting that it is the integrity of the cell envelope which is important (Meadows, 1971; Corpe, 1974). In contrast, there is a wealth of evidence that some sort of physiological activity may be required. Original observations (discussed above) show evidence for time-dependence for firm adhesion; Hamada (1977) has shown that the adhesion of *Streptococcus mutans* to glass was due to the active synthesis of new glucan, irrespective of whether preformed glucan was available; *Veillonella* sp. cells with no innate adhesive ability are able to adhere to smooth surfaces when complexed with an extracellular polymer-producing enzyme (glucosyl transferase) originating from *S. salivarius* (Wittenberger *et al.*, 1977). The equivocal results obtained from experiments to investigate the effects of metabolic inhibitors on bacterial adhesion (Fletcher, 1980) suggest that time-dependent metabolic activity, in this latter case energy production and protein synthesis, may also be required. However, this requirement was shown to vary with the nature of the surface (Fletcher, 1980). Further evidence for the involvement of metabolic activity comes from reports that bacteria in the exponential phase of growth are able to attach faster than either stationary or death phase cells. Zobell (1943) and Marshall *et al.* (1971a) demonstrated this for irreversible sorption and Fletcher (1977b) for sorption over two hours. However, two points must be borne in mind: (a) it is possible that the exponential growth phase cells in both cases contained a higher proportion of motile (flagellate) organisms and thereby the chances of cells making contact with the surfaces must be increased; and (b) that in Fletcher's (1977b) experiments death phase cells also attached in appreciable numbers but at a lower rate than exponential growth phase cells.

The adhesion of single cells may also be facilitated by specialized appendages such as holdfasts, stalks, fimbriae and flagella (Zobell, 1943; Corpe, 1974; Meadows, 1971). The role of these structures was reviewed by Corpe in 1970 and again more recently in 1980. While these structures are undoubtedly involved in the process of attachment they can be regarded as special cases. Bacteria with stalks and holdfasts only appear late in the development of bacterial films, and fimbriae are known to have a range of specific receptors (lectins) which enable them to adhere to a range of substrates. The properties of fimbrae have been discussed by Ottow (1975) and recently Duguid & Old (1980) reviewed the adhesive properties of the fimbriae and other fibril antigens of the enterobacteriaceae. Special culture techniques are often required to produce these structures and this must cast doubts on their relevance to adhesion in natural environments. One aspect of these structures is, however, important. The problems bacteria have overcoming electrostatic repulsion could be reduced if the bacterial surface produced fine hair-like probes since this would enable the cell to make contact with the surface whilst held at a distance at the secondary minimum. This is possible because the net charge around a fine probe would be considerably less than that of the main part of the cell envelope.

GROWTH AND DEVELOPMENT OF THE MICROBIAL FILM

Once single cells have firmly attached, growth and division rapidly lead to the development of microcolonies (Wardell *et al.*, 1980) which, as they enlarge, coalesce to form a layer of cells covering the surface. In many cases this will be a monolayer which subsequently builds outwards away from the surface to become a complex film. In natural environments this would be composed of a variety of microorganisms. Often the initial distribution of cells across the surface is not uniform, this is particularly true of glass surfaces where the edges of the surface are colonized first (Zobell, 1943). The explanation is likely to be due to local variations in the surface charge density, since materials with a uniform crystal structure such as mica tend to have a uniform covering (Floodgate, 1972).

Film development has received little attention, however, some inferences may be drawn from the available data. As microcolonies enlarge they appear to do so along the plane of the surface until they have reached a certain size, whereupon development away from the surface also takes place. It seems that in many instances this is due to the presence of polymers within the developing film.

We have seen earlier that in the adhesion of single cells, small amounts of specific polymer are often present. There is now evidence that for the developing film more substantial amounts of polymer are involved. Such polymers are often described as present in reticular form or as fibrils (Corpe, 1964; Fletcher & Floodgate, 1973; Geesey *et al.*, 1977; Jones *et al.*, 1969; Marshall, 1976; Marshall *et al.*, 1971a, b). Moreover, from a number of reports it is clear that organisms growing on surfaces in a range of natural environments are surrounded by a reticulum of polymeric fibrils facilitating cell-cell and cell-surface bridging when seen in electron micrographs (Bauchop *et al.*, 1975; Costerton, 1980; Dempsey, 1981a, b; Geesey *et al.*, 1978; Jannasch & Wirsen, 1977; McMeekin *et al.*, 1979; Pearl, 1980; Costerton *et al.*, 1981; Ellwood *et al.*, 1982). It is likely therefore that the large amounts of polymer, noted amongst others by Corpe (1970a) and Fletcher & Floodgate (1973) are concerned not with the initial adhesion process but with the development of the microbial film; that is anchoring cells to each other and to the surface.

Evidence for this also comes from other workers who have demonstrated that bacteria which form flocs or clumps do so by the production of fibrous exopolymers which facilitate cell-cell bridging (Deinema & Zevenhuizen, 1971; Friedman *et al.*, 1968, 1969; Napoli *et al.*, 1975; Stanley & Rose, 1967). Similar speculations have been made by Corpe (1980).

The term 'glycocalyx' has been used by Costerton *et al.* (1978) to describe these networks of fibrils and it was further suggested that in natural environments such structures may be the functional surface of the cell, involved not only with adhesion *per se* but also providing an increased surface area for ion exchange and the adsorption of nutrient molecules, behaving in the same manner as an ion exchange resin.

At the present time it remains unclear whether such fibrils/reticula are in fact modified fimbriae or other filamentous structures (such as those found in the Enterobacteriaceae and increasingly in other bacterial groups) or, whether they are indeed a function of surface-associated growth. This latter must be questioned as there is evidence from Friedman *et al.* (1968, 1969) and Pearl (1980) that fibril formation also occurs in free-living bacteria.

THE METABOLIC ACTIVITY OF ATTACHED BACTERIA

From the evidence presented so far it can be inferred that attached bacteria may exhibit an elevated growth, or metabolic rate compared with that of comparable suspended populations. There are problems associated with direct measurements of growth rates at surfaces since it is difficult to distinguish between growth, the deposition, and attachment of new cells at the surface. Metabolic activity on the other hand can be determined by a number of methods some of which have been applied to attached microbial populations. Evidence in this area has shown that surface associated bacterial populations do have a higher metabolic rate than the suspended bacteria.

Jannasch & Pritchard (1972) showed that the presence of chitin particles could increase the respiratory activity of an *Achromobacter* sp., particularly at low concentrations of organic substrate. Similarly, an elevated respiration rate for attached bacteria was reported by Hendricks (1974). This latter result was also supported by evidence for an increased alkaline phosphatase activity although other enzymes examined showed no significant differences in activities between attached and free-living populations. Ellwood *et al.* (1982) have shown recently that growth of a *Pseudomonas* sp. on glass surfaces suspended in chemostat cultures may be several times higher than the imposed growth rate of the liquid cultures (see below).

Mineralization processes of surface-associated heterotrophic populations have also been investigated. The available evidence from ^{14}C glucose (Goulder, 1977) and ^{14}C glutamate (Ladd *et al.*, 1979) studies support the view that much of the heterotrophic activity in aquatic environments may be due to the surface-associated microbial population.

From the very earliest studies of surface associated growth it has been suggested that periphytic organisms are able to exploit nutrients adsorbed onto the surfaces. However, whilst this must take place and probably accounts for the initial 'attraction' of the surface it may not account for the complex microbial films which subsequently develop.

It has been postulated (see Costerton *et al.*, 1978, 1981) that the role of the microfibril net (or glycocalyx) may be to act in a manner similar to an ion exchange resin, trapping and releasing molecules, many of which may be utilized by the microorganisms. Thus organisms apparently 'trapped' on the surface would not be deprived of nutrients once those in the immediate vicinity had become depleted. Fluid flowing over the surface would bring further nutrients which would be adsorbed onto the microfibrils and would be subsequently utilized by the organisms. In this way an organism would not be disadvantaged by being confined to one site and may well be at an advantage over free-living, motile forms which must expend energy in order to move into favourable environments.

Pearl (1975) has noted that the production of microfibrils will require an expenditure of energy in the form of reduced organic structural components and has further suggested that this may be compensated for by the ability to increase growth and replication on favourable physical and chemical substrates. This does not necessarily contradict the discussion so far since the amount of energy required for the production of microfibrils may well be less than that utilized in movement. Additionally, it must be borne in mind that the production of fibrils takes place over a period of time and since the fibrils can be seen to link and are produced by many cells within a microcolony, the expenditure of energy and therefore, by inference, the benefits will be shared by the community.

Recently the fate of energy, in the form of protons, within microbial films has been discussed by Ellwood *et al.* (1982) in terms of the chemiosmotic theory.

Protons translocated out of the cell during respiration or the expenditure of ATP (adenosine triphosphate) generate an electrical potential across the cell membrane which can be viewed as a source of potential energy used to drive, in particular, substrate uptake and excretion of metabolic by-products against a concentration gradient. However, ATP can be generated by the inward flow of protons across the cell membrane; it would therefore be to the cells' advantage for translocated protons to be retained near the surface of the membrane. In the free living cell this would seem unlikely; protons would diffuse away from the cell. However, if cells were concentrated at a surface or held in close proximity to other cells then such diffusion would be inhibited. It has been speculated by Ellwood *et al.* (1982) that such a situation may arise within microbial films and that the cells in a microcolony would benefit from the resultant 'pool' of available protons. If, in addition to merely anchoring and entrapping cells, microfibrils were to act as ion exchange resin they may well provide a reservoir of protons available to drive chemiosmotic uptake and ATP-generating systems; this in itself may be sufficient to account for the increased rate of growth and metabolism seen in surface associated films.

SOME ECOLOGICAL AND INDUSTRIAL CONSIDERATIONS

An association between bacteria and surfaces has been observed in many environments. In marine (Corpe, 1970b; Marshall *et al.*, 1971b; Meadows & Anderson, 1968; Wiebe & Pomeroy, 1972) and freshwater (Brown *et al.*, 1977; Geesey *et al.*, 1977, 1978; Pearl, 1980) systems; in association with plant (Cundell *et al.*, 1977; Dazzo, 1980a, b) and animal surfaces (Bauchop *et al.*, 1975; McMeekin *et al.*, 1979; Walker & Nagy, 1980) as well as the more obvious examples of marine fouling and dental plaque formation. The last are perhaps examples of the detrimental effect of surface growth but there is a more positive aspect which is the exploitation of surface associated growth and flocculant behaviour in biological filters, sewage treatment and fermentation processes. It follows that for a variety of reasons an understanding of the factors influencing surface growth is of considerable importance.

We have discussed the importance of adhesion to surfaces in aquatic environments in some detail and it should be noted that for most of the other examples the same basic principles may be applied since a liquid-surface interface will be involved. There are of course exceptions, such as the phylosphere (leaf surface) and dental surfaces where an air-liquid-surface interface may be involved.

ADHESION TO PLANT AND ANIMAL SURFACES

Reports of interactions between microorganisms and plant or animal surfaces are numerous and are themselves a specialized field of study. For plant interactions the two examples given may serve to illustrate the variety of associations, for others, attention is drawn to the two recent reviews by Dazzo (1980a, b).

The association between *Rhizobium* spp. and legumes is well known. The bacterial cells adsorb onto root hairs with a specificity which is the basis of the taxonomy of these bacteria (Baldwin & Fred, 1929). Subsequent to adsorption the bacteria enter into a symbiotic relationship with the host plant by infecting the root, resulting in the appearance of the characteristic nodules. The plant is able to utilize nitrogen (as NH_4^+) which is fixed by the rhizobia within the nodules. There is evidence that the firm attachment of the bacteria to the root hairs is mediated by a complex linkage between a lectin (protein agglutinin, see Goldstein *et al.*, 1980) produced by the root hair tip and the exopolysaccharides of the rhizobium (Dazzo & Hubbell, 1975; Dazzo & Brill, 1979). Moreover, it has been reported that the ability to flocculate and produce cellulose microfibrils may aid the initial attachment, allowing the more complex linkage or bond formation to occur (Napoli *et al.*, 1975).

A different association exists between *Erwinia amylovora* and its hosts. This organism is a pathogen causing a dry wilt disease (fire blight) and the likely role of the bacterial exopolysaccharide is to promote the disease process. Ayers *et al.* (1979) were able to demonstrate that the production of extracellular polysaccharides by the bacteria correlated with its ability to cause the disease in host inoculation tests.

Although bacterial adhesion to animal surfaces is a large subject area much work has been done to elucidate the mechanisms involved in the attachment/adhesion process. Much of the available information refers to pathogenic organisms or to easily manipulated organisms like *Escherichia coli.* There is therefore a wealth of information available from work with members of the Enterobacteriaceae (see Arbuthnot & Smyth, 1979; Duguid & Old, 1980; Sugarman & Donta, 1979) and from oral microbiology (see reviews by Gibbons, 1977, 1980; Gibbons & Van Houte, 1975; Newman, 1980; Rutter, 1979). From these reviews and other work it is clear that the mechanisms of attachment to surfaces vary greatly. However, in a great many cases either polysaccharides or fibril structures (such as the fimbriae and other fibril antigens) have been implicated and whilst in many cases quite detailed investigations have been carried out to determine the molecular basis of adhesion, for others the mechanism remains uncertain. In spite of this it seems likely in the light of available evidence that the role of fibril antigens may be similar to that described for rhizobia. They enable the cell to approach and to be held at a small distance from the surface (the secondary minimum?) and this in turn facilitates the formation of more permanent bonds, perhaps between adhesins on the fibrils or on the cell coat and the surface.

THE USE OF CONTINUOUS CULTURES TO STUDY MICROBIAL ADHESION

The study of bacterial adhesion within continuous cultures is not new. Brown *et al.* (1977) reported on the effect of varying the nature of the limiting substrate on the colonization of aluminium foil by bacteria enriched from river water. Similarly, the sorption of both heterotrophic and enteric bacteria to glass surfaces during the continuous culture of presterilized river water was investigated by Hendricks (1974). Jannasch & Pritchard (1972) demonstrated that the presence of particulate matter in a chemostat could determine the outcome of an enrichment or competition experiment. The subject of sorption of bacteria to surfaces within continuous cultures has been further discussed Brown *et al.* (1978), Bull & Brown (1979), Wardell *et al.* (1980) and Ellwood *et al.* (1982). In addition, the effects of wall growth and the colonization of surfaces within continuous cultures has been discussed by Larsen & Dimmick (1964), Topiwala & Hamer (1971), and Wilkinson & Hamer (1974). In reality however, the potential offered by steady state continuous cultures to investigate microbial growth at a surface has not been fully exploited.

A simple mathematical model has been proposed to show the effect of wall growth in a single stage chemostat (Topiwala &

Hamer, 1971). For this model three assumptions were made: (1) bacteria adhere to the walls in contact with the culture as a monolayer; (2) after an initial build up, the adherent bacterial mass remains constant and progeny are released into the culture; (3) the growth rate of the attached bacteria is the same as that of the suspended population. Given these conditions, the relationship between the steady state biomass $\bar{x}$ and the limiting substrate $\bar{s}$ was shown to be dependent on a constant K, determined by the total mass of the adhered cells and therefore proportional to the surface area to volume ratio and the density of the bacterial film. The effect of K on $\bar{x}$ was found to be more pronounced when the medium substrate concentration was low (as one might expect), and that wall growth has the effect of extending the effective operation of the chemostat to values of the dilution rate exceeding the maximum specific growth rate. It was also demonstrated that high values of K would increase the value of the productivity term ($D\bar{x}$); therefore for biomass production increasing the available internal surface area to volume ratio would be beneficial. This correlates with effects noted by Larsen & Dimmick (1964) when adding glass wool to cultures of *Serratia marcescens*.

The model however, relates to an ideal system and although shown to be generally valid by later experimental evidence (Wilkinson & Hamer, 1974), in practice, the experimental results often differ from those predicted by the theory. The reasons for these anomalies lie with the assumptions made. Bacteria do not always grow as a monolayer, the thickness of a bacterial film varies markedly from the thin 'almost invisible' translucent films observed by Munsen & Bridges (1964) and Larsen & Dimmick (1964) to the much thicker films often observed in long term continuous cultures (Munsen & Bridges, 1964; Wilkinson & Hamer, 1974; Wardell & Brown, unpublished observations).

The second of the assumptions made, that the film biomass remains constant and progeny are released into the culture, is more difficult to assess. Since some films of bacteria do develop beyond a monolayer the biomass of attached films cannot always remain constant. However, it may be that for given environmental conditions there is an 'optimal thickness' for the film. Given that, or indeed if the situation were such that development of the film occurred outwards away from the surface, then the outermost layer of the film might act as a monolayer discharging progeny into the culture (see also Wilkinson & Hamer, 1974).

Assuming the growth rate of the liquid and surface populations to be the same is arguably the weakest point of the model. Larsen & Dimmick (1964) made the point that if nutrients, especially the limiting nutrient, are concentrated at the surfaces of a continuous culture vessel then the attached population would have access to a higher concentration of limiting nutrient than would the cells in the liquid phase. There would, in effect, be two populations growing at different rates. The evidence of Munsen & Bridges (1964), that the growth rates of prototrophs originating from wall growth and auxotrophs from the liquid culture were the same when measured, does not necessarily contradict this thesis, since we have previously discussed the possibility that it is only when bacteria are in contact with a surface that they exhibit a higher growth rate. One is therefore left with a situation where wall-surface growth does occur within continuous cultures to a degree determined by the organism, surface characteristics and culture conditions. The effect on the kinetics of continuous growth will therefore vary considerably. For large volume fermenters where the surface area to volume ratio is small the effect may be negligible but for small fermenters and cultural conditions where the initial growth-limiting substrate concentration (S_R) is low, the effect might be considerable; more than 90% of the cells in the culture may originate from wall growth and the experimental values of the washout dilution rate (D_c) and $D\bar{x}$ will not be as predicted by theory.

A potential problem of using continuous-flow systems for producing single cell protein is the possibility that the culture will become contaminated with pathogenic, or potentially pathogenic, organisms. Rokem *et al.* (1980) found that surface growth could be used to advantage in such systems. Using a *Pseudomonas* sp. growing on methanol to produce single cell protein they reported that pathogens, such as *Staphylococcus aureus* and *Salmonella typhimurium*, could become established on the internal surfaces of the culture vessel, growing on the excretion products of the pseudomonad, and providing a continuous inoculum into the culture. However, the deliberate introduction of soil bacteria led to the displacement of the pathogenic organisms by the preferential growth of the soil bacteria on the walls of the culture vessel. The pathogens, unable to compete successfully on the surfaces, or against the pseudomonad in the liquid phase, were gradually displaced from the culture.

A particular advantage of the use of continuous-flow cultures is that the attachment and growth of growing cultures may be studied over an extended time period, an advantage not possible in batch culture experiments. This allows the kinetics of attachment and growth to be followed using surfaces exposed for different time intervals in the on growing liquid population. Such studies have indicated that under both carbon and nitrogen limited conditions the growth of the surface associated organisms was similar to that of the liquid culture at low liquid population densities (about 10^5/ml) while at higher liquid population densities (10^6/ml and higher), the surface growth rate was up to six times higher than that of the liquid culture. This suface growth was accompanied by the production of microfibrils (visualized in the SEM) which were associated both with cell-surface and cell-cell adhesion. These microfibrils were not apparent in organisms from the liquid culture suggesting that their synthesis was stimulated by attachment to the surfaces used (Ellwood *et al.*, 1982).

ATTACHMENT OF FILAMENTOUS FUNGI TO SURFACES

The attachment of filamentous fungi to surfaces has not attracted the same attention as that of bacteria, but the role of filamentous fungi in binding soil particles into aggregates by adhesion as well as physical enmeshment is recognized (Aspiras *et al.*, 1971; Burns, 1980; Fletcher *et al.*, 1980; Stotzky, 1980; Sutton & Sheppard, 1976). An extracellular polysaccharide, PP floc, produced by *Pullularia (Aureobasidium) pullulans* and very closely related chemically to pullulan has been shown to flocculate clay. It has a possible use for removal of clay slimes from aqueous solutions in hydrometallurgical processes such as extraction of uranium potash (Zajic & Le Duy, 1973).

In addition to initial impaction and entrapment, the production of mucilaginous appendages by the ascospores appears to have an important role in the establishment of marine fungi on surfaces. The spores may, for example, be attached loosely by hair-like appendages (Jones & Moss, 1978), by drops of mucilage exuded from hollow spines or the end compartments of the spores (Schmidt, 1974; Kirk, 1976), or by mucilaginous fibrillar pads (Jones, Moss & Koch, 1980). Both germinating spores and hyphae developing from them are invested by a polysaccharide sheath which appears to act in adhesion to the surface. Sheathing polysaccharide has been observed by Hutching & Jones (1982) in freshwater hyphomycetes, some species of which had, additionally, simple or more elaborate appressoria; features which are more usually associated with plant pathogens (Staples & Macko, 1980). Generally, the formation of appressoria on leaf surfaces prior to invasion of the leaf tissue is held to involve host recognition, i.e. it requires a particular set of physical and chemical factors provided only by the host. Macko *et al.* (1978) however, showed that, as well as heat shock, 2-propenal could induce their formation without surface contact in the wheat stem rust fungus, *Puccinia graminis tritici*. Macko *et al.* (1976) found that five rust fungi could produce appressoria on artificial membranes in response to an entirely thigmotropic stimulus. Staples & Macko (1980) have reviewed formation of appressoria in plant pathogens and noted examples where purely physical and purely chemical stimuli are involved. Interestingly, Hill *et al.* (1980) have noted that appressorium formation in the grey mould *Botrytis cinerea* increases with the degree of hydrophobicity of the surface.

Mucilaginous sheaths thought to attach hyphae to leaf surfaces have been observed in three facultative parasites in the genus *Helminthosporium* (Hau & Rush, 1978; Potter *et al.*, 1980; Wheeler & Gantz, 1979). In *H. maydis* (Potter *et al.*, 1980) the sheath was observed to spread out from round the germ tube to form a thin film over the immediate leaf surface. This film was partly fibrillar and was most obvious at the tips of germ tubes and where appressoria were forming. These phenomena were observed on membrane filters or on paper in contact with leaves, but on membrane filters in contact with moist glass only the mucilaginous layer formed and no mucilage was formed by hyphae on agar.

Although it is often difficult to distinguish between purely chemical corrosion of rock and stone and corrosion caused by microbial activity (particularly in areas of urban pollution), Krumbein (1972, 1973) asserts that there is a clear relationship between chemolithotropic bacteria and superficial degradation of stone. Evidence reviewed by Strzelczyk (1981) suggests that the activity of sulphur-oxidizing and nitrifying bacteria is promoted by the supply of energy substrates by heterotrophic communities in microniches on the surface and within stone where organic substrates accumulate. In producing organic acids, heterotrophs may also be directly involved in the breakdown of stone. Calcareous rocks, e.g. limestone, chalk and marble, are naturally susceptible to solubilization by carbonic acid, but Duff *et al.* (1963) found that the Gram-negative rods most active in decomposition of a variety of siliceous rocks produced 2-

ketogluconic acid, which released silicates. Among other authors, Henderson & Duff (1963) and Krumbein (1972, 1973) have noted the apparent importance of filamentous fungi in weathering of rock and stone. Under laboratory conditions, a diverse range of fungi (Henderson & Duff, 1963) produce organic acids, such as acetic, citric, formic, fumaric and oxalic acid, and dissolve silicate rocks in broth culture. The levels of Mg, Fe, Al and Si solubilized from basalt by *Penicillium simplicissimum* show a high correlation with the amounts of citric acid produced (Silverman & Munoz, 1970), but weakly complexing acids, e.g. acetic acid, do not remove as many cations from silicates as citric acid, a stronger chelating agent (Huang & Keller, 1972).

In addition to removal of cations into water soluble complexes by chelation with organic acids, breakdown releases free silica, which may facilitate solution of phosphatic rocks. Not all complexes with organic acids are readily water soluble; using filamentous fungi from weathered rocks, Eckhardt (1980) demonstrated under laboratory conditions the formation of Fe gluconate crystals during solubilization of biotite. Availability of other ions, such as Ca^{2+} which increases gluconic acid production (Elnaghy & Megalla, 1975), and PO_4^{3-} which increases production of both gluconic and 2-ketogluconic acid, but suppresses oxalic acid production (Martin & Steel, 1954), may have a profound effect on chelation of cations in rock.

The nature of the rock will clearly affect its resistance to corrosion by different microorganisms. Although fungi generally released more silica from siliceous rocks than did bacteria, solution of calcium-containing rocks such as augite and talc was achieved more slowly by fungi secreting oxalic acid than by bacteria producing 2-ketogluconic acid (Henderson & Duff, 1963). However, Bassi & Chiatante (1976) concluded that fungi isolated from deteriorating marble were more active than bacteria in dissolving this calcareous rock, causing roughening of polished surfaces and the formation of pits and other depressions. Mehta, Torma & Muir (1979) found that the Al and Fe containing phases (labradorite, olivine, and pyrite) in basalt were preferentially eroded by acids of *Penicillium simplicissimum* and that the Ti-containing phases were most resistant. Silverman & Munoz (1971) found that although very little Ti was released from basalts, up to 80% of Ti in granitic rocks was leached out by this organism. Eckhardt (1980) found that less than 4% of the Al in amphibolite and orthoclase was released by *Aspergillus niger* in culture, but 59% was leached from the micaceous mineral biotite, although the three minerals contained approximately equal amounts of Al. He concluded that differences in crystalline structure were of importance in degradation.

The solubilization of insoluble phosphates and silicates by organic acids may also affect the permeability of rock (Strzelczyk, 1981). The widening of capillaries will allow the entry of water leading to erosion, partly as a result of increased absorption of acids from the atmosphere and partly as a result of freezing and thawing. In summarizing the effects of lichens on stone, Seal & Eggins (1981) have noted that in addition to the effects just mentioned the expansion of embedded hyphae of the mycobiont of lichens will contribute to the flaking of stone particles. The hyphae of free-living fungi will have a similar contribution to make to weathering, but it has also been shown that fungi can bring about the conversion of micas to partially expanded minerals akin to vermiculite (Weed, Davey & Cook, 1969). This conversion is brought about by the exchange of Na in solution for K in mica particles and the capacity of the fungus to mediate the conversion appears to depend on its effectiveness as a K sink. Using *Rhizoctonia* the authors found the greatest amount of conversion to occur when mica and fungus were kept separate, and concluded that encasement of mica particles by mycelium reduced diffusion of Na from the liquid medium into the mica.

Since much of the information on the breakdown of rock has been derived from purely laboratory experiments, it would be unwise to stress the role of filamentous fungi, and other microorganisms, in the weathering of rock to form soil and release ions for higher plant growth. It is also often difficult to elucidate the part played by microorganisms in the deterioration of stonework in buildings and other artefacts. However, Eckhardt (1978) found the same organic acids in the decaying stonework of a sandstone monument as were produced in culture by the fungi he isolated from the monument, and subsequently showed that the fungi caused leaching of cations from aluminosilicate minerals commonly present in building stone. Bassi & Chiatante (1976) concluded that deterioration of the external marble of Milan Cathedral was caused by filamentous fungi which derived nutriment from pigeon droppings contaminating the marble. As Strzelczyk (1981) has pointed out, heterotrophic bacteria, acidic components of the excrement, and atmospheric pollution may also be important factors in the deterioration of the marble.

Although not normally a problem in temperate regions, organic acids produced by filamentous fungi can cause extensive etching of optical glass in the tropics. Hutchinson (1947) observed deep etching immediately below mycelium radiating from detritus or from cork supports, which it was presumed were a source of nutrients. None of the fungi isolated from optical instruments by Richards (1949) was able to grow on clean glass lacking a nutrient supply, but Nagamuttu (1967) found that spores of *Aspergillus fumigatus, A. glaucus, Penicillium* sp. and *Scopulariopsis* sp. germinated and produced extensive mycelium around organic particles on otherwise clean glass. Recently, Smith & Nadim (1982) demonstrated that condensation of respiratory exhalation, cigarette smoke and either volatile vegetable or mineral oil would, like finger prints on the inert surface of glass, promote varying degrees of germination and vegetative growth.

The unavoidable contamination of aviation kerosene with water and inorganic nutrients provides conditions favourable for the growth of *Cladosporium resinae* which, in addition to causing blockages in fuel lines and gauges, causes corrosion of alloy fuel tanks. It has been demonstrated that acetic, dodecanoic, glycolic and glyoxylic acids produced by the fungus corrode fuel tank alloys (McKenzie, Akbar & Miller, 1976) and it has been proposed that a combination of localized concentrations of acid and differential oxygen concentration in dieso/water systems accounts for pitting in aluminium/4% copper alloys (Hansen, Tighe Ford & George, 1982). Pitting up to 0.6 mm deep, is restricted to the outer edge of the mycelium, where acid secretion and a steep oxygen gradient create 'hyphal valleys' in the metal surface.

REFERENCES

Arbuthnott, J. P. & Smyth, C. J. (1979). Bacterial adhesion in host/pathogen interactions in animals. In *Adhesion of Microorganisms to Surfaces*, ed. D. C. Ellwood, J. Melling & P. Rutter, pp. 165–98. Society for General Microbiology Publication. London: Academic Press.

Aspiras, R. B., Allen, O. N., Harris, R. F. & Chesters, G. (1971). Aggregate stabilization by filamentous micro-organisms. *Soil Science*, **112,** 282–4.

Ayers, A. R., Ayers, S. B. & Goodman, R. N. (1979). Extracellular polysaccharide of *Erwinia amylovora*: a correlation with virulence. *Applied and Environmental Microbiology*, **38,** 659–66.

Baldwin, I. L. & Fred, E. B. (1929). Nomenclature of the root-nodule bacteria of the leguminosae. *Journal of Bacteriology*, **17,** 141–50.

Bassi, M. & Chiatante, D. (1976). The role of pigeon excrement in stone biodeterioration. *International Biodeterioration Bulletin*, **12,** 73–9.

BAUCHOP, T., CLARKE, R. T. & NEWHOOK, J. C. (1975). Scanning electron microscope study of bacteria associated with the rumen epithelium of sheep. *Applied Microbiology*, **30,** 668–75.

BERKELEY, R. C. W., LYNCH, J. M., MELLING, J., RUTTER, P. & VINCENT, B. (1980). *Microbial Adhesion to Surfaces*. Chichester, England: Ellis-Horwood.

BITTON, G. & MARSHALL, K. C. (1980). *Adsorption of Micro-organisms to Surfaces*. New York: John Wiley and Sons.

BROWN, C. M., ELLWOOD, D. C. & HUNTER, J. R. (1977). Growth of bacteria at surfaces: influence of nutrient limitation. *FEMS Microbiology Letters*, **1,** 163–6.

BROWN, C. M., ELLWOOD, D. C. & HUNTER, J. R. (1978). Enrichments in a chemostat. In *Techniques for the Study of Mixed Populations*, ed. D. W. Lovelock, & K. Davies, pp. 213–22. London: Academic Press.

BULL, A. T. & BROWN, C. M. (1979). Continuous culture applications to microbial biochemistry. In *Microbial Biochemistry: International Review of Biochemistry*, vol. 21, ed. J. R. Quayle, pp. 177–226. Baltimore, USA: University Park Press.

BURNS, R. G. (1979). Interaction of microorganisms, their substrates and their products with soil surfaces. In *Adhesion of Microorganisms to Surfaces*, ed. D. C. Ellwood, J. Melling & P. Rutter, pp. 109–38. Society for General Microbiology Publication. London: Academic Press.

BURNS, R. G. (1980). Microbial adhesion to soil surfaces: consequences for growth and enzyme activities. In *Microbial Adhesion to Surfaces*, ed. R. C. W. Berkeley, J. M. Lynch, J. Melling, P. R. Rutter & B. Vincent, pp. 249–62. Chichester: Ellis Horwood.

CORPE, W. A. (1964). Factors influencing growth and polysaccharide formation by strains of *Chromobacterium violaceum*. *Journal of Bacteriology*, **88,** 1433–41.

CORPE, W. A. (1970a). An acidic polysaccharide produced by a primary film forming marine bacterium. *Developments in Industrial Microbiology*, **11,** 402–12.

CORPE, W. A. (1970b). Attachment of marine bacteria to solid surfaces. In *Adhesion in Biological Systems*, ed. R. S. Manley, pp. 73–87. New York: Academic Press.

CORPE, W. A. (1974). Periphytic marine bacteria and the formation of microbial films on solid surfaces. In *Effect of Ocean Environment on Microbial Activity*, ed. R. R. Coldwall & R. Y. Morita, pp. 397–417. Baltimore, USA: University Park Press.

CORPE, W. A. (1980). Microbial surface components involved in adsorption of microorganisms onto surfaces. In *Adsorption of Microorganisms to Surfaces*, ed. G. Bitton and K. C. Marshall, pp. 105–44. New York: John Wiley and Sons.

COSTERTON, J. W. (1980). Some techniques involved in the study of adsorption of microorganisms to surfaces. In *Adsorption of Microorganisms to Surfaces*, ed. G. Bitton & K. C. Marshall, pp. 403–23. New York: John Wiley & Sons.

COSTERTON, J. W., GEESEY, G. G. & CHENG, K. J. (1978). How bacteria stick. *Scientific American*, **238,** 76–85.

COSTERTON, J. W., IRVIN, R. T. & Cheng, K. J. (1981). The bacterial glycocalyx in nature and disease. *Annual Review of Microbiology*, **35,** 299–324.

CUNDELL, A. M., SLEETER, T. D. & MITCHELL, R. (1977). Microbial populations associated with the surface of the Brown alga *Ascophyllum nodosum*. *Microbial Ecology*, **4,** 81–91.

DANIELS, S. L. (1972). The adsorption of microorganisms onto solid surfaces — A review. *Developments in Industrial Microbiology*, **13,** 211–53.

DANIELS, S. L. (1980). Mechanisms involved in sorption of microorganisms to solid surfaces. In *Adsorption of Microorganisms to Surfaces*, ed. G. Bitton & K. C. Marshall, pp. 7–58. New York: John Wiley and Sons.

Dazzo, F. B. (1980a). Adsorption of microorganisms to roots and other plant surfaces. In *Adsorption of Microorganisms to Surfaces*, ed. G. Bitton & K. C. Marshall, pp. 253–316. New York: John Wiley and Sons.

Dazzo, F. B. (1980b). Microbial adhesion to plant surfaces. In *Microbial Adhesion to Surfaces*, ed. R. C. W. Berkeley, J. M. Lynch, J. Melling, P. R. Rutter & B. Vincent, pp. 311–28. Chichester, England: Ellis Horwood.

Dazzo, F. B. & Brill, W. J. (1979). Bacterial polysaccharide which binds *Rhizobium trifolii* to Clover root hairs. *Journal of Bacteriology*, **137,** 1326–73.

Dazzo, F. B. & Hubbell, D. H. (1975). Cross reactive antigens and lectin as determinants of symbiotic specificity in the *Rhizobium-Clover* association. *Applied Microbiology*, **30,** 1017–33.

Deinema, M. H. & Zevenhuizen, L. P. T. M. (1971). Formation of cellulose fibrils by Gram-negative bacteria and their role in bacterial flocculation. *Archives of Microbiology*, **78,** 42–57.

Dempsey, M. J. (1981a). Marine bacterial fouling: a scanning electron microscope study. *Marine Biology*, **61,** 305–15.

Dempsey, M. J. (1981b). Colonization of antifouling paints by marine bacteria. *Botanica Marina*, **24,** 185–91.

Duff, R. B., Webley, D. M. & Scott, R. O. (1963). Solubilization of minerals and related materials by 2 ketogluconic acid producing bacteria. *Soil Science*, **95,** 105–14.

Duguid, J. P. & Old, D. C. (1980). Adhesive properties of Enterobacteriaceae. In *Bacterial Adherence (Receptors and Recognition Series B 6)*, ed. E. H. Beachey, pp. 187–217. London: Chapman and Hall.

Eckhardt, F. E. W. (1978). Microorganisms and weathering of a sandstone monument. In *Environmental Biogeochemistry and Geomicrobiology*, vol. 2, ed. W. E. Krumbein, pp. 657–686. Ann Arbor, Michigan: Ann Arbor Science.

Eckhardt, F. E. W. (1980). Microbial degradation of silicates. Release of cations from aluminosilicate minerals by yeasts and filamentous fungi. In *Biodeterioration, Proceedings of the Fourth International Symposium, Berlin*, ed. T. A. Oxley, D. Allsopp & G. Becker, pp. 107–16. London: Pitman.

Ellwood, D. C., Keevil, C. W., Marsh, P. D., Brown, C. M. & Wardell, J. N. (1982). Surface associated growth. *Philosophical Transactions of the Royal Society, London*, **B297,** 517–32.

Ellwood, D. C., Melling, J. & Rutter, P. (eds.) (1979). *Adhesion of Microorganisms to Surfaces.* Society for General Microbiology Publication. London: Academic Press.

Elnaghy, M. A. & Megalla, S. E. (1975). Gluconic acid production by *Penicillium puberulum. Folia Microbiologica*, **20,** 504–8.

Fletcher, M. (1976). The effects of proteins on bacterial attachment to polystyrene. *Journal of General Microbiology*, **94,** 400–4.

Fletcher, M. (1977a). Attachment of marine bacteria to surfaces. In *Microbiology 1977*, ed. D. Schlessinger, pp. 407–10. Washington, DC: American Society for Microbiology.

Fletcher, M. (1977b). The effects of culture concentration, age time and temperature on bacterial attachment to polystyrene. *Canadian Journal of Microbiology*, **23,** 1–6.

Fletcher, M. (1979). The attachment of bacteria to surfaces in aquatic environments. In *Adhesion of Microorganisms to Surfaces*, ed. D. C. Ellwood, J. Melling & P. Rutter, pp. 88–108. Society for General Microbiology Publication. London: Academic Press.

Fletcher, M. (1980). The question of passive versus active attachment mechanism in non specific bacterial adhesion. In *Microbial Adhesion to Surfaces*, ed. R. C. W. Berkeley, J. M. Lynch, J. Melling, P. R. Rutter & B. Vincent, pp. 197–210. Chichester, England: Ellis Horwood.

FLETCHER, M. & FLOODGATE, G. D. (1973). An electron microscope demonstration of an acidic polysaccharide involved in the adhesion of a marine bacterium to solid surfaces. *Journal of General Microbiology*, **74,** 325–34.

FLETCHER, M., LATHAM, M. J., LYNCH, J. M. & RUTTER, P. R. (1980). The characteristics of interfaces and their role in microbial attachment. In *Microbial Adhesion to Surfaces*, ed. R. C. W. Berkeley, J. M Lynch, J. Melling, P R. Rutter & B. Vincent, pp. 67–78. Chichester, England: Ellis Horwood.

FLETCHER, M. & LEOB, G. I. (1976). The influence of substratum surface properties on the attachment of a marine bacterium. In *Colloid and Interface Science*, vol. 3, ed. M. Kerker, pp. 459–69. London: Academic Press.

FLOODGATE, G. D. (1966). Factors affecting the settlement of a marine bacterium. *Veröffentilchungen des Institutes fur Meeresforschung in Bremerhaven*, **2,** 265–70.

FLOODGATE, G. D. (1972). The mechanism of bacterial attachment to detritus in aquatic systems. *Memorie dell'Istituto italiano di idrobiologia dott*, **29** (suppl.), 309–23.

FRIEDMAN, B. A., DUGAN, P. R., PFISTER, R. M. & REMSEN, J. (1968). Fine structure and composition of the zoological matrix surrounding *Zoogloea Ramigera. Journal of Bacteriology*, **96,** 2144–53.

FRIEDMAN, B. A., DUGAN P. R., PFISTER, R. M. & REMSEN, J. (1969). Structure of exocellular polymers and their relationship to bacterial flocculation. *Journal of Bacteriology*, **98,** 1328–34.

GEESEY, G. G., RICHARDSON, W. T., YEOMANS H. G., IRVIN, R. T. & COSTERTON, J. W. (1977). Microscopic examination of natural sessile bacterial populations from an alpine stream. *Canadian Journal of Microbiology*, **23,** 1733–6.

GEESEY, G. G., MUTCH, R., COSTERTON, J. W. & GREEN, R. B. (1978). Sessile bacteria: an important component of the microbial population in small mountain streams. *Limnology and Oceanography*, **23,** 1214–23.

GIBBONS, R. J. (1977). Adherence of bacteria to host tissue. In *Microbiology 1977*, ed. D. Schlessinger, pp. 395–406. Washington, DC: American Society for Microbiology.

GIBBONS, R. J. (1980). The adhesion of bacteria to the surfaces of the mouth in *Microbial Adhesion to Surfaces*, ed. R. C. W. Berkeley, J. M Lynch, J. Melling, P. R. Rutter & B. Vincent, pp. 351–88. Chichester, England: Ellis Horwood.

GIBBONS, R. J. & VAN HOUTE, J. (1975). Bacterial adherence in oral microbial ecology. *Annual Review of Microbiology*, **29,** 19–44.

GOLDSTEIN, I. J., HUGHES, R. C., MONSIGNY, M., OSAWA, T. & SHARON, N. (1980). What should be called a lectin? *Nature, London*, **285,** 66.

GOULDER, R. (1977). Attached and free bacteria in an estuary with abundant suspended solids. *Journal of Applied Bacteriology*, **43,** 399–405.

HAMADA, S. (1977). New glucan synthesis as a pre requisite for adherence of *Strep. mutans* to smooth glass surfaces. *Microbios Letters*, **5,** 141–6.

HANSEN, D. J., TIGHE FORD, D. J. & GEORGE, G. C. (1982). Role of the mycelium in the corrosive activity of *Cladosporium resinal* in a dieso/water system. *International Biodeterioration Bulletin*, **17,** 103–12.

HARDEN, V. P. & HARRIS, J. O. (1953). The isoelectric point of bacterial cells. *Journal of Bacteriology*, **65,** 198–202.

HARRIS, R. H. & MITCHELL, R. (1973). The role of polymers in microbial aggregation. *Annual Review of Microbiology*, **27,** 27–50.

HAU, F. C. & RUSH, M. C. (1978). Interaction of *Helminthosporium oryzae* with susceptible and resistant rice cultivars during the prepenetration and penetration periods. *American Phytopathological Society 70th Annual Meeting*, Abstract No 61.

HENDERSON, M. E. K. & DUFF, R. B. (1963). The release of metallic and silicate ions from minerals, rocks, and soils by fungal activity. *Journal of Soil Science*, **14,** 236–46.

HENDRICKS, C. W. (1974). Sorption of heterotrophic and enteric bacteria to glass surfaces in the continuous culture of river water. *Applied Microbiology*, **28,** 572–8.

HEUKELAKIAN, H. & HELLER, A. (1940). Relation between food concentration and surface for bacterial growth. *Journal of Bacteriology*, **40,** 547–58.

HILL, G., STELLWAAG KITTLER, F. & SCHLOESSER, E. (1980). Substrate surface and appressoria formation by *Botrytis cinerea. Phytopathologische Zeitschrift*, **99,** 186–91.

HUANG, W. H. & KELLER, W. D. (1972). Organic acids as agents of chemical weathering of silicate minerals. *Nature, London*, **239,** 149–51.

HUMPHREY, B. A., DICKSON, M. R. & MARSHALL, K. C. (1979). Physicochemical and in situ observations on the adhesion of gliding bacteria to surfaces. *Archives of Microbiology*, **120,** 231–8.

HUTCHING, L. Z. & Jones, E. B. G. (1982). Spore adhesion and germination of freshwater hyphomycetes. *Bulletin of the British Mycological Society*, **16,** supplement **1**,4 (abstract only).

HUTCHINSON, W. G. (1947). The fouling of optical glass by micro organisms. *Journal of Bacteriology*, **54,** 45–6.

JANNASCH, H. W. (1958). Studies on planktonic bacteria by means of a direct membrane filter method. *Journal of General Microbiology*, **18,** 609–20.

JANNASCH, H. W. (1979). Microbial ecology of aquatic low nutrient habitats. In *Strategies of Microbial Life in Extreme Environments*, ed. M. Shilo, pp. 243–60. Berlin: Dahlem Konferenzen.

JANNASCH, H. W. & PRITCHARD, P. H. (1972). The role of inert particulate matter in the activity of aquatic microorganisms. *Memorie dell'Istituto italiano di idrobiologia dott*, **29**(Suppl.), 289–308.

JANNASCH, H. W. & WIRSEN, C. O. (1977). Microbial life in the deep sea. *Scientific American*, **236,** 42–52.

JONES, H. C., ROTH, I. L. & SANDERS, W. M. (1969). Electron microscope study of a slime layer. *Journal of Bacteriology*, **99,** 316–25.

JONES, E. B. G. & MOSS, S. T. (1978). Ascospore appendages of marine ascomycetes: an evaluation of appendages as taxonomic criteria. *Marine Biology*, **49,** 11–26.

JONES, E. B. G., MOSS, S. T. & J. KOCH (1980). Light and scanning electron microscope observations of the marine ascomycete *Crinigera maritima. Transactions of the British Mycological Society*, **74,** 625–31.

KIRK, P. W. (1976). Cytochemistry of marine fungal spores. In *Recent Advances in Aquatic Mycology*, ed. E. B. G. Jones, pp. 177–92. London: Elek Science.

KRUMBEIN, W. E. (1972). Role des microorganismes dans la genese, la diagenes et la degradation des roches en place. *Revue d'Ecologie et de Biologie du Sol*, **9,** 283–319.

KRUMBEIN, W. E. (1973). ¨Uber den Einflusz von Mikroorganismen auf die Bausteinver Witterung eine¨ Okologische Studie. *Deutsche Kunst und Kenkmalpflege*, pp. 54–71.

LADD, T. I., COSTERTON, J. W. & GEESEY, G. G. (1979). Determination of the heterotrophic activity of epilithic microbial populations. *Native Aquatic Bacteria: Enumeration, Activity and Ecology*. American Society for Testing and Materials STP 695 180–95.

Larsen, D. H. & Dimmick, R. L. (1964). Attachment and growth of bacteria on surfaces of continuous culture vessels. *Journal of Bacteriology*, **88,** 1380–7.

McKenzie, P., Akbar, S. A. & Miller, J. D. A. (1976). Fungal corrosion of aircraft fuel tank alloys. In *Microbial Corrosion Affecting the Petroleum Industry*, Technical Paper IP 77 001, pp 37–50. London: Institute of Petroleum.

Macko, V., Renwick, J. A. A. & Rissler, J. F. (1978). Acrolein induces differentiation of infection structures in the wheat stem rust fungus. *Science*, **199,** 442–3.

Macko, V., Staples, R. C., Yaniv, Z. & Granados, R. R. (1976). Self inhibitors of fungal spore germination. In *The Fungal Spore*, ed. D. J. Weber & W. M. Hess, pp. 73–98. New York: Wiley.

Marshall, K. C. (1976). *Interfaces in Microbial Ecology*. Harvard University Press, Cambridge, USA.

Marshall, K. C. (1979). Growth at interfaces. In *Strategies of Microbial Life in Extreme Environments*, ed. M. Shilo, pp. 281–90. Berlin: Dahlem Konferenzen.

Marshall, K. C. (1980a). Reactions of microorganisms, ions and macromolecules at interfaces. In *Contemporary Microbial Ecology*, ed. D. C. Ellwood, J. N. Hedges, M. J. Latham, J. M. Lynch & J. H. Slater, pp. 93–106. London: Academic Press.

Marshall, K. C. (1980b). Adsorption of microorganisms to soils and sediments. In *Adsorption of Microorganisms to Surfaces,* ed. G. Bitton & K. C. Marshall, pp. 317–29. New York: John Wiley and Sons.

Marshall, K. C. & Bitton, G. (1980). Microbial adhesion in perspective. In *Adsorption of Microorganisms to Surfaces*, ed. G. Bitton & K. C. Marshall, pp. 1–5. New York: John Wiley and Sons.

Marshall, K. C., Stout, R. & Mitchell, R. (1971a). Mechanisms of the initial events in the sorption of marine bacteria to surfaces. *Journal of General Microbiology*, **68,** 337–48.

Marshall, K. C., Stout, R. & Mitchell, R. (1971b). Selective sorption of bacteria from sea water. *Canadian Journal of Microbiology*, **17,** 1413–16.

Martin, S. M. & Steel, R. (1954). Effect of phosphate on production of organic acids by *Aspergillus niger. Canadian Journal of Microbiology*, **1,** 470–2.

McMeekin, T. A., Thomas C. J. & McCall, D. (1979). Scanning electron microscopy of microorganisms on chicken skin. *Journal of Applied Bacteriology*, **46,** 195–200.

Meadows, P. S. (1971). The attachment of bacteria to solid surfaces. *Archives of Microbiology*, **75,** 374–81.

Meadows, P. S. & Anderson, J. G. (1968). Microorganisms attached to marine sand grains. *Journal of Marine Biological Association (UK)*, **48,** 161–75.

Meadows, P. S. & Anderson, J. G. (1979). The microbiology of interfaces in the marine environment. *Progress in Industrial Microbiology*, **15,** 207–65.

Mehta, P. A., Torma, A. E. & Muir, L. E. (1979). Effect of environmental parameters on the efficiency of biodegradation of basalt rock by fungi. *Biotechnology and Bioengineering*, **21,** 875–85.

Munsen, R. J. & Bridges, B. A. (1964). "Take over" an unusual selection process in steady state cultures of *Escherichia coli. Journal of General Microbiology*, **37,** 411–18.

Nagamuttu, S. (1967). Moulds on optical glass and control measures. *International Biodeterioration Bulletin*, **3,** 25–7.

Napoli, C., Dazzo, F. B. & Hubbell, D. (1975). Production of cellulose microfibrils by *Rhizobium* sp. *Applied Microbiology*, **30,** 123–31.

NEWMAN, H. N. (1980). Retention of bacteria on oral surfaces. In *Adsorption of Microorganisms to Surfaces*, ed. G. G. Bitton & K. C. Marshall, pp. 207–51. New York: John Wiley and Sons.

OTTOW, J. C. G. (1975). Ecology, physiology and genetics of fimbriae and pili. *Annual Review of Microbiology*, **29,** 79–108.

PEARL, H. W. (1975). Microbial attachment to particles in marine and freshwater ecosystems. *Microbial Ecology*, **2,** 73–83.

PEARL, H. W. (1980). Attachment of microorganisms to living and detrital surfaces in freshwater systems. In *Adsorption of Microorganisms to Surfaces*, ed. G. Bitton & K. C. Marshall, pp. 375–402. New York: John Wiley and Sons.

POTTER, J. M., TIFFANY, L. H. & MARTINSON, Ç. M. (1980). Substrate effects on *Helminthosporium maydis* Race T conidium and germ tube morphology. *Phytopathology*, **70,** 715–19.

RICHARDS, C. W. (1949). Some fungous contaminants of optical instruments. *Journal of Bacteriology*, **58,** 453–5.

ROKEM, J. S., GOLDBERG, I. & MATELES, R. I. (1980). Growth of mixed cultures of bacteria on methanol. *Journal of General Microbiology*, **116,** 225–32.

RUTTER, P. (1979). The accumulation of organisms on the teeth. In *Adhesion of Microorganisms to Surfaces*, ed. D. C. Ellwood, J. Melling & P. Rutter, pp. 139–64. Society for General Microbiology Publication, London: Academic Press.

SCHMIDT, I. (1974). Höhere meerespilze der Ostsee Biologische Rundschau, **12,** 96–112.

SEAL, K. J. & EGGINS, H. O. W. (1981). The biodeterioration of materials. In *Essays in Applied Microbiology*, ed. J. R. Norris & M. H. Richmond. Chichester: Wiley.

SILVERMAN, M. P. & MUNOZ, E. F. (1970). Fungal attack on rock: solubilization and altered infrared spectra. *Science*, **169,** 985–7.

SILVERMAN, M. P. & MUNOZ, E. F. (1971). Fungal leaching of titanium from rocks. *Applied Microbiology*, **22,** 923–4.

SLATER, J. H. (1978). The role of microbial communities in the natural environment. In *The Oil industry and Microbial Ecosystems,* ed. K. W. A. Chater & H. J. Somerville, pp. 138–52. London: Heyden and Sons.

SMITH, R. N. & NADIM, L. M. (1982). Fungal growth on inert surfaces. In *Biodeterioration 5, Proceedings of the Fifth International Biodeterioration Symposium, Aberdeen.* Chichester: Wiley (in press).

STANLEY, S. O. & ROSE, A. H. (1967). On the clumping of *Corynebacterium xerosis* as affected by temperature. *Journal of General Microbiology*, **48,** 9–23.

STAPLES, R. C. & MACKO, V. (1980). Formation of infection structures as a recognition response in fungi. *Experimental Mycology*, **4,** 2–16.

STOTZKY, G. (1980). Surface interactions between clay minerals and microbes viruses and soluble organics, and the probable importance of these interactions to the ecology of microbes in soil. In *Microbial Adhesion to Surfaces*, ed. R. C. W. Berkeley, J. M. Lynch, J. Melling, P. R. Rutter & B. Vincent, pp. 231–47. Chichester: Ellis Horwood.

STRZELCZYK, A. B. (1981). Stone. In *Economic Microbiology*, vol 6, *Microbial Biodeterioration*, ed A. H. Rose, pp. 61–80. London: Academic Press.

SUGARMAN, B. & DONTA, S. T. (1979). Specificity of attachment of certain Enterobacteriaceae to mammalian cells. *Journal of General Microbiology*, **115,** 509–12.

SUTTON, J. C. & SHEPPARD, R. B. (1976). Aggregation of sand-dune soil by endomycorrhizal fungi. *Canadian Journal of Botany*, **54,** 326–33.

TOPIWALA, H. H. & HAMER, G. (1971). Effect of wall growth in steady state continuous cultures. *Biotechnology and Bioengineering*, **13,** 919–22.

WALKER, P. D. & NAGY, L. K. (1980). Adhesion of organisms to animal tissues. In

Microbial Adhesion to Surfaces, ed. R. C. W. Berkeley, J. M. Lynch, J. Melling, P. R. Rutter & B. Vincent, pp. 473–94. Chichester, England: Ellis Horwood.

Ward, J. B. & Berkeley, R. C. W. (1980). The microbial cell surface. In *Microbial Adhesion to Surfaces*, ed. R. C. W. Berkeley, J. M. Lynch, J. Melling, P. R. Rutter & B. Vincent, pp. 47–66. Chichester, England: Ellis Horwood.

Wardell, J. N., Brown, C. M. & Ellwood, D. C. (1980). A continuous culture study of the attachment of bacteria to surfaces. In *Microbial Adhesion to Surfaces*, ed. R. C. W. Berkeley, J. M. Lynch, J. Melling, P. R. Rutter & B. Vincent, pp. 211–30. Chichester, England: Ellis Horwood.

Weed, S. B., Davey, C. B. & Cook, M. G. (1969). Weathering of mica by Fungi. *Soil Science Society of America Proceedings*, **33,** 702–6.

Wheeler, H. & Gantz, D. (1979). Extracellular sheaths on hyphae of two species of *Helminthosporium*. *Mycologia*, **71,** 1127–35.

Wiebe, W. J. & Pomeroy, L. R. (1972). Microorganisms and their association with aggregates and detritus in the sea: a microscopic study. *Memorie dell'Istituto italiano di idrobiologia dott*, **29**(Suppl.), 325–52.

Wilkinson, T. G. & Hamer, G. (1974). Wall growth in mixed bacterial cultures growing on methane. *Biotechnology and Bioengineering*, **16,** 251–60.

Wittenberger, C. L., Beaman, A. J., Lee, L. N., McCabe, R. M. & Donkersloot, J. A. (1977). Possible role of *Strep. salivarius* glucosyl transferase in adherence of Veillonella to smooth surfaces. In *Microbiology 1977* ed. D. Schlessinger, pp. 417–26. Washington, DC: American Society for Microbiology.

Zajic, J. E. & LeDuy, A. (1973). Flocculant and chemical properties of a polysaccharide from *Pullularia pullulans*. *Applied Microbiology*, **25,** 628–35.

ZoBell, C. E. (1943). The effect of solid surfaces upon bacterial activity. *Journal of Bacteriology*, **46,** 39–56.

ZoBell, C. E. & Anderson, D. Q. (1936). Observations on the multiplication of bacteria in different volumes of stored sea water and the influence of oxygen tension and solid surfaces. *Biological Bulletin*, **71,** 324–42.

GENETIC INTERACTIONS AMONG MICROBIAL COMMUNITIES

DARRYL C. REANNEY*,
PETER C. GOWLAND† AND J. HOWARD SLATER†

* *Department of Microbiology, La Trobe University, Bundoora, Victoria 3083, Australia*

† *Department of Environmental Sciences, University of Warwick, Coventry CV4 7AL, West Midlands, UK*

INTRODUCTION

The complex structure of the biosphere results in the co-existence of rich and diverse microbial communities. Whilst diversity is certainly reduced in those habitats which from an anthropocentric viewpoint are extreme, few if any natural habitats contain single microbial species. Interactions, such as competition (Gause, 1934) or succession (Swift, 1982), have interested microbial ecologists for a considerable time. Only comparatively recently, however, have the intricate nature of the interactions occurring between different populations growing in the same habitat, and the consequences of these interactions, become apparent (Bull & Slater, 1982). Community structure will not be considered in detail here and the interested reader is referred elsewhere for such discussions (Bull & Slater, 1982; Slater, 1981; Slater & Lovatt, 1982). It will be assumed that interactions between two or more species are facts of microbial life in natural habitats whether these produce tenuous, non-obligatory associations or highly integrated, obligatory relationships (Slater, 1981).

Studies in microbial genetics have uncovered a wealth of mechanisms that allow defined DNA sequences to migrate among replicons inside and between cells (Cohen, 1976; Calos & Miller, 1980; Campbell, 1981; Kleckner, 1981). An extensive understanding of the mechanisms by which DNA may be mobilized, transferred and relocated within the same species or across diverse genetic boundaries is now available (Reanney, 1976, 1977; Reanney *et al.*, 1982). A naive extrapolation of these data from the precise, highly manipulated test tube environments to natural habitats, could easily generate the impression that prokaryotic genes are in a perpetual state of flux and that the identity of a given bacterial species is often compromised by the uptake of non-homologous DNA (Sonea & Panisset, 1976). Superficially the scene may appear to be set for a jamboree of unfettered movement of genes, particularly as the microbial flora responds to changing or novel environmental conditions. On the other hand the repeated isolation of virtually identical biotypes from specific ecological niches, such as the mammalian gut (Bauchop, 1980) or the rhizosphere (Bowen, 1980), suggests that bacterial genomes remain stable in spite of the potentially high rates of gene transfer which can undoubtedly occur or be made to occur under appropriate conditions.

This chapter explores the issue of genetic exchange on the one hand and genetic conservation on the other. In particular we suggest that the *in vitro* processes may be bad models of the *in vivo* realities. In a general sense this is not at all surprising since extrapolation in other areas has often proved to be disconcertingly wide of the mark. For example, enzymes in soil often fail to retain the same Michaelis-Menten kinetics that they display in a homogeneous solution (McLaren & Skujins, 1968). Chitinase adsorbed to kaolinite, for example, had a pH optimum approximately two units higher than that for the unadsorbed enzyme (McLaren & Estermann, 1957). Similarly the average generation times of bacteria in specific habitats may be over a hundred times longer than those observed in laboratory studies with the same organism (Gray & Williams, 1971). As we will document in this chapter DNA transfer processes in nature take place in a complex matrix of variable environmental factors which include structured microbial communities. Some of these factors aid gene transfer while others hinder or preclude them.

MICROBIAL COMMUNITY STRUCTURE

Given that complex, and frequently multiple, nutritional and physical interactions are established amongst natural microbial populations, the major questions are when, how and at what frequency do genetic transfer events occur. The most compelling evidence for the actual occurrence of these events in nature (as opposed to demonstrations of the existence of suitable mechanisms and the possibility that transfer events can occur in artificial laboratory systems) is seen under conditions when natural populations are significantly stressed and demand an adaptive response to a novel challenge. What then is the hard evidence for widespread DNA transfer in nature? There are two major chemical challenges which are peculiar consequences of man's 20th-century activities, and have demonstrated the facility for genetic exchange in natural populations of microorganisms. These will now be discussed.

Transfer of drug-resistance mechanisms

The earliest, and still perhaps the most convincing, documented evidence comes from the adaptation of bacteria to drugs whose concentrations have in particular niches been elevated by man and substantially altered from assumed natural concentrations (Falkow, 1975). The enzyme β-lactamase which confers resistance to penicillins and cephalosporins by a mechanism which cleaves the β-lactam ring, is widely distributed in both Gram-negative and Gram-positive bacteria. Matthew & Hedges (1976) compared by isoelectric focussing, 100 naturally-occurring drug-resistant plasmids encoding β-lactamase genes and found that 70 were indistinguishable. These data strongly suggested that the gene(s) responsible had diffused into multiple species by the transfer of common DNA elements, probably involving transposon Tn1 at various stages (Kleckner, 1977). Ambler (1980) compared the amino acid sequences of β-lactamase from *Staphylococcus aureus, Bacillus licheniformis, Bacillus cereus* and *Escherichia coli* and concluded that 'the amount of similarity is so great that divergence from a single ancestral gene is the most reasonable explanation'. Unless one can accept that the common ancestor of all Gram-negative and Gram-positive bacteria contained a β-lactamase which has essentially remained unaltered ever since, then one can only deduce that DNA transfers among various evolving strains of bacteria must have taken place. Such transfers probably occurred a long time ago since similarity matrices between the amino acid sequences showed that the β-lactamases from *S. aureus* and *B. licheniformis* had only a 56% match-up (Ambler, 1980). This provides evidence that drug-resistant mechanisms were advantageous to natural populations long before a massive drug challenge was introduced by man. It therefore seems likely that the need for resistance to natural, low levels of drugs produced by certain members of natural microflora has been, and still is, a necessary adaptive mechanism and also implies that the low levels of antibiotics produced by natural populations exert the desired negative influence on non-resistant organisms.

Much epidemiological evidence has accumulated which leaves little choice but to conclude that drug-resistance plasmids move readily amongst members of natural microflora to confer resistance to novel challenges. For example, Anderson (1975) demonstrated that resistance to chloramphenicol by many novel isolates of *Salmonella typhi* found throughout Central America was due to the spread of a single plasmid coding for chloramphenicol resistance. Although differences in the plasmids' structures were detected, they were remarkably similar (Anderson & Smith, 1972; Grindley *et al.*, 1972). Similarly, but in considerably more detail, the resistance to carbenicillin in hospital-isolated strains of *Pseudomonas aeruginosa* demonstrated without doubt the common nature of the responsible plasmid and also identified the primary strain responsible for the introduction of the resistance mechanism into this highly monitored environment (Lowbury *et al.*, 1969; Sykes & Richmond, 1970).

The ease with which drug-resistance plasmids may be transferred between different members of a mixed microbial community may easily be shown in the laboratory. Gowland & Slater (1982) demonstrated the transfer of the compatible plasmids R1 and TP120 between two different strains of *E. coli* K–12 in a continuous flow culture system using a water jacketed fixed bed column fermenter. The provision of surfaces in the column on which the members of the community might form more stable mating aggregates was thought to increase the frequency of plasmid transfer in the population. Thus a similar mating in a homogeneous chemostat produced significantly fewer recombinant bacteria (Gowland & Slater unpublished results). No antibiotic selection was required with the strains and plasmids employed either for maintenance or transfer of the plasmids. Recombinant isolates from the fermenter were shown by agarose gel electrophoresis to contain two plasmids, whose electrophoretic mobilities corresponded to those of plasmids R1 and TP120. Tests for auxotrophy of the recombinant isolates revealed that it was plasmid R1 which had transferred into the *E. coli* strain containing plasmid TP120. Furthermore, plate matings of a recombinant isolate with a nalidixic acid-resistant *E. coli* strain showed that unless both plasmids were specifically selected for, the R1 plasmid would preferentially transfer into the nalidixic acid-resistant *E. coli* strain.

Metabolism of unusual natural compounds and xenobiotic compounds

It will be argued later that the conservation of genetic material is a reality in spite of the availability of potentially active gene-transfer mechanisms (see page 394). It follows, however, that the realization of the potential for gene transfer will only be seen where the survival of a given microorganism or a complete community is challenged by novel, usually deleterious, conditions. An excellent example of this concerns the way in which microorganisms adapt to new conditions generated by the dispersal of large quantities of compounds which may either be toxic or consist of major sources of unusable carbon, energy and other nutrients. Similarly though less damaging to the environment, is the transfer of unusual, natural compounds, such as petrochemicals and their derivatives, into novel environmental situations. This is considered to be less of a problem conceptually since relevant metabolic capabilities have evolved elsewhere in the biosphere, as a result of the prolonged exposure of microorganisms to these natural products. The question in this case is whether the capability is easily transferred to organisms in the new habitat or whether the facility has to re-evolve. Moreover even if it can be introduced naturally or artificially, its existence as a functional activity integrated into the established communities is in doubt. This is particularly true concerning attempts made to engineer microorganisms in the laboratory to have desirable characteristics that might be disseminated to affected natural habitats. Thus, novel metabolic developments in response to major perturbations in the biosphere constitute 'a genuinely new dimension to microbiology . . . with xenobiotic compounds representing a major challenge to the metabolic versatility of microorganisms' (Slater & Bull, 1982). The magnitude of the problem can be demonstrated by calculations which suggest that enough herbicides were used in Great Britain in 1973 to generate an average concentration of 30 mg (kg soil)$^{-1}$ over the top 0.5 cm of agricultural land (Greaves *et al.*, 1976). More recently Munnecke (1978, 1981) has noted that over 500 different active pesticides are available in more than 5000 commercial formulations for agricultural use. In 1979 in the Western Hemisphere over 2.0×10^9 kg of pesticides were produced (Storck, 1980; Munnecke, 1981). It is not surprising to find that it is in these areas that some of the most interesting evidence of gene transfer events is becoming available.

The presence of a novel compound affects the microflora in one or other of two ways: either the material represents an additional growth substrate which can be exploited, or it is a toxic challenge which needs to be transformed to relieve growth inhibitory or potentially lethal effects. In both cases catabolic enzyme sequences need to be evolved to catalyse the necessary transformations. A traditional view would be that individual organisms would evolve the necessary mechanisms by an appropriate series of point mutations. It is certainly true that the modification of existing pathways to deal with a related group of compounds could take place in this manner. In view of the dissimilarity of many of the materials in question, it seems more probable however, that the route to the evolution of a novel capability depends on the reassortment of a pool of enzymes (and their genes) needed to catalyse the required transformations. Such a genetic pool will naturally be larger in the mixed population present in a natural habitat than that available in any single species (Slater & Somerville, 1979).

This model probably depends on the absence of absolute enzyme specificities and the exploitation of gratuitous or fortuitous activities of enzymes whose primary task is the metabolism of other normal substrates. This is clearly the case with many enzyme systems (Clarke, 1974; Hartley, 1974) whilst only a few show exclusive or single substrate specificities (Dagley, 1975). Flexibility accounts for the phenomenon of cometabolism (Dalton & Stirling, 1982). The variability of substrates attacked by particular cultures or enzymes is well illustrated in examples given by Knackmuss (1981) (Tables 1 and 2). The difficulty so far as single organisms are concerned is that the product of one reaction might not serve as the substrate for the next enzyme elaborated by that organism. In many instances these partial transformations occur without the causative organism deriving any benefit; that is, it is a cometabolic step. Nevertheless there can be a stoichiometric conversion to another compound which is exploited by the enzyme system of another organism. Accordingly the sequential metabolism of a particular compound may be dependent on a series of individual transformations, or partial sequences of the latter, which individually do little to benefit participating organisms. a number of illustrations are considered below.

There are two consequences of such a model for the evolution of a novel catabolic sequence. The primary event is the establishment of an interacting assemblage of microorganisms either in nature or in the laboratory during appropriate enrichment procedures. Careful

Table 1. *Relative oxygen uptake rates caused by the oxidation of various substituted aromatic compounds needed by populations enriched on benzoate and a number of alkylbenzoate analyses*

Assay substrate	Relative oxidation rates for cultures enriched on substrates indicated:				
	Benzoate	3-Methyl-benzoate	4-Methyl-benzoate	4-Ethyl-benzoate	4-Isopropyl-benzoate
benzoate	100	100	100	100	100
3-methylbenzoate	50	100	100	250	250
4-methylbenzoate	0	100	100	900	950
4-ethylbenzoate	0	70	70	900	950
3-trifluoromethyl benzoate	0	30	20	100	100
4-trifluoromethyl benzoate	0	10	10	150	100

After K. H. Engesses quoted in Knackmuss (1981).

Table 2. *Relative dioxygenation rates of benzoic acid and substituted benzoic acids by whole cells of an* Alcaligenes *species and a* Pseudomonas *species*

Assay substrate	Relative dioxygenation rates		
	Alcaligenes eutrophus Strain B9 (induced with benzoate)	*Pseudomonas* sp. Strain B13 (grown on 3-dichloro-benzoate)	*Pseudomonas putida* Strain mt-2 (grown on 3-methyl-benzoate)
benzoate	100	100	100
3-methylbenzoate	25	13	78
3-chlorobenzoate	15	31	48
3-bromobenzoate	8	9	34
4-methylbenzoate	1	1	63
4-chlorobenzoate	1	1	32
4-bromobenzoate	n.t.	1	22
3,5-dimethylbenzoate	1	1	10
3,5-dichlorobenzoate	1	1	27

After Reineke & Knackmuss (1978a). n.t., not tested.

analysis of many environments for synergistic communities capable of degrading unusual compounds show that they are readily selected (Table 3). The stability of the enriched community determines how rapidly the various elements of the new pathway might come together to produce the complete sequence in a single organism, if indeed it occurs at all. It is likely that in many instances apparent enrichments for pure cultures obscure the intermediate stage of community formation, especially if gene transfer events occur at high frequencies. A second consequence of these events ought to be high levels of gene transfer activity leading to a high incidence of partially complete or complete pathways encoded on appropriate transfer systems, especially plasmids. The rapidly expanding incidence of different catabolic plasmids seems to bear this prediction out.

Microbial community catabolism of halogenated aromatic compounds

The principles outlined in the preceding section have been demonstrated with great clarity by Knackmuss and his colleagues (Reineke & Knackmuss, 1978a, b, 1979; Hartmann *et al.*, 1979; Knackmuss, 1981). They have demonstrated that it is possible to recruit two complementary sequences of pathways involved in the catabolism of halogenated aromatic compounds; to rearrange the genetic information within a single organism; and, without the need for mutational events, to use pre-existing enzymes to degrade novel substrates. The catabolism of aromatic compounds and their halogenated analogues depends primarily on two dioxygenases and it is the stereospecificity of these enzymes which provide the key to understanding the routes by which novel strains might evolve through mixed cultures. The initial dioxygenase catalysed the hydroxylation of the primary substrates leading to catechol or substituted catechols. For many organisms such as *Pseudomonas* sp. strain B13 and *Alcaligenes eutrophus* strain B9, the specificity of this enzyme was sufficiently stringent to preclude the catabolism of certain halogenated aromatic compounds, whereas other strains, such as *Pseudomonas putida* strain mt-2, had much broader dioxygenase specificities (Table 2, Fig. 1). However, *Pseudomonas putida* strain mt-2 did not grow on many halogenated benzoates and attempts to evolve toluate-utilizing populations of *P. putida* mt-2 to use 3-chlorobenzoate in continuous-flow culture selection experiments failed (Knackmuss, 1981). The inability of this pseudomonad to use such compounds as

Table 3. *Synergistic microbial communities involved in the catabolism of unusual or xenobiotic compounds*

Compound	Reference	Compound	Reference
trichloroacetic acid	Jensen (1957)	isopropylphenylcarbamate	McClure (1970
2,2-dichloropropionic acid (Dalapon)	Senior *et al.* (1976)	cyclohexane	Stirling *et al.* (1976)
3,6-dichloropicolinic acid (Lontrel)	Lovatt *et al.* (1978)	cycloalkanes	Beam & Perry (1973, 1974)
	Slater & Lovatt (1982)	dodecylcyclohexane	Feinberg *et al.* (1980)
3,4-dichloropropionanilide	Bordeleau & Bartha (1968)	n-hexadecane	Schwarz *et al.* (1975)
		n-alkanes and other hydrocarbons	Miller & Johnson (1966)
2-(2-methyl 4-chloro) phenoxypropionicacid	Kilpi (1980)	crude oil	Horowitz *et al.* (1975)
chlorobenzoate	Hartmann *et al.* (1979) Reineke & Knackmuss (1979) Knackmuss (1981)	3-methyl heptane	Wodzinski & Johnson (1968)
4-chlorobiphenyl	Neu & Ballschmitter (1977)	benzoate	Ferry & Wolfe (1976)
			Cossar *et al.* (1981)
4,4′-dichlorobiphenyl	Tulp *et al.* (1978)	phenol	Pawlowsky & Howell (1973a,b,c)
polychlorinated biphenyls	Clark *et al.* (1979) Carey & Harvey (1978)	3,5-dihydroxyphenol	Osman *et al.* (1976)
0,0-diethyl *0*-2-isopropyl 4-methyl 6-pyrimidinylthiophosphate (Diazinon)	Gunner & Zuckerman (1968)	styrene	Seilicki *et al.* (1978)
		ligno-aromatic compounds	Balba & Evans (1977) Healey & Young (1978, 1979)
0,0-diethyl *0*-*p*-nitrophenyl phosporothioate (Parathion)	Daughton & Hsieh (1977)	linear alkylbenzene sulphonates	Johanides & Hršak (1976) Phillips & Hollis, cited in Slater & Lovatt (1982)
1,5-di-(2,4-dimethylphenyl) 3-methyl 1,3,5-triazapenta 1,4-diene (Amitraz)	Baker & Woods (1977)	alkylphenol ethoxylates	Baggi *et al.* (1978)
		2-ethyldecyl-1-decaethoxylate	Watson & Jones (1979)
		nitrosamines	Pickaver (1976)

After Slater & Bull (1982).

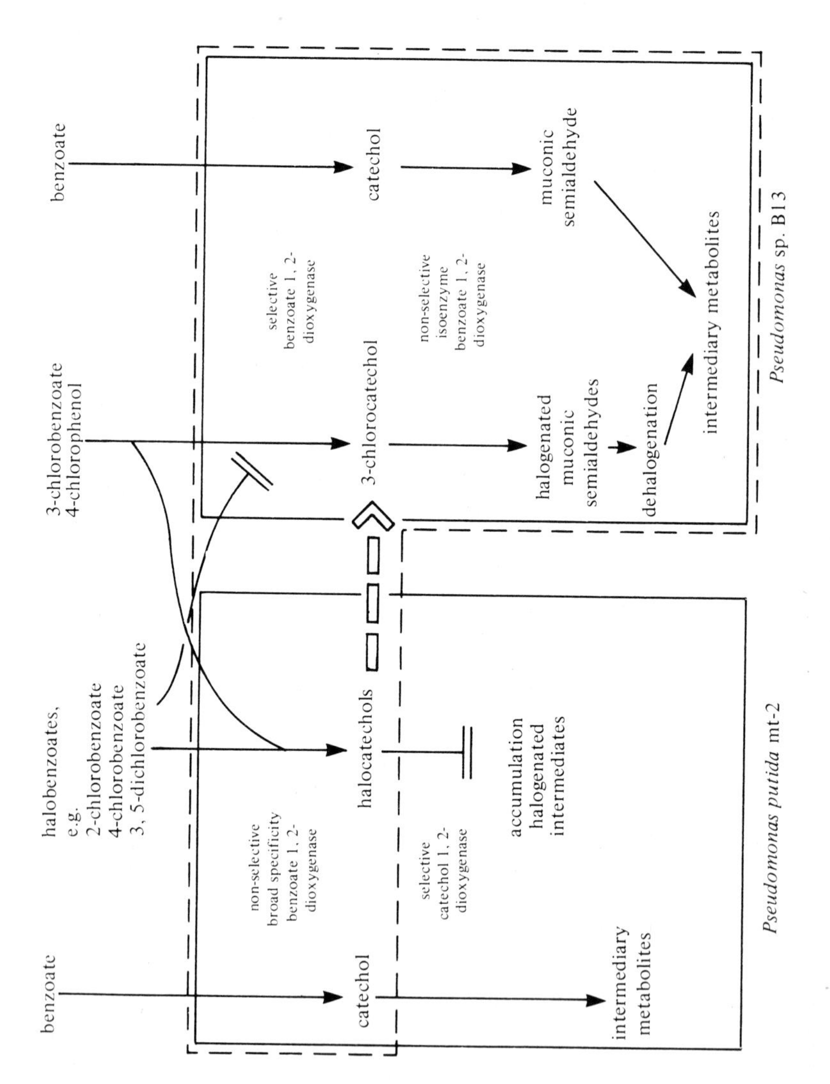

Fig. 1. The growth of *Pseudomonas putida* strain mt-2 and *Pseudomonas* sp. strain B13 on various halogenated benzoic acids. The hollow arrow indicates metabolite transfer if the two organisms grow as a synergistic community when growing on 2-chlorobenzoate, 4-chlorobenzoate or 3,5-dichlorobenzoate; the solid line indicates the initial distribution of enzymes, and hence genetic, potential between the two pseudomonads; the broken line indicates the necessary genetic rearrangements required to evolve a single organism capable of using the haloaromatic as the sources of carbon and energy in pure culture; = indicates a block in metabolism (after Slater & Bull, 1982).

growth substrates is due to the stringent substrate specificity of its catechol 1,2-dioxygenase, responsible for the oxygen-dependent ring cleavage stage, an enzyme which cannot metabolize substituted catechols. Conversely one of the catechol 1,2-dioxygenase isoenzymes (and the cycloisomerase catalysing the next step) synthesized by *Pseudomonas* sp. strain B13 was able to transform 4-chlorocatechol. Thus the combined activities of the two strains of pseudomonad brought together two enzymes with relaxed substrate specificity which in tandem operation enabled the conversion of previously non-growth compounds, such as 2-chlorobenzoate or 3,5-dichlorobenzoate, into usable intermediary metabolites (Fig. 1).

Mixed continuous culture selection experiments starting with these two organisms led to the formation of the anticipated strain having the total catabolic capability (Hartmann *et al.*, 1979). This evolutionary process appeared to be an easy one since the relaxed specificity benzoate 1,2-dioxygenase possessed by *Pseudomonas putida* strain mt-2 was encoded on plasmid pWWO (previously known as the TOL plasmid – Williams & Murray, 1974) and its transfer was responsible for the appearance of the new strain. Subsequently these experiments have been reproduced under more precise conditions by mating *P. putida* mt-2 with *Pseudomonas* sp. B13 and selecting for appropriate transconjugants (such as *Pseudomonas* sp. strain WR211) with the capacity to grow on *m*-toluate and subsequently 4-chlorobenzoate (Reineke & Knackmuss, 1979). More recently it has been shown that these events are more complex than the simple transfer of a plasmid with the desired characteristics (Williams, 1981, 1982). For example, the initial transconjugant could still not use 4-chlorobenzoate probably because the derived halocatechol was channelled down a competing *meta*-ring fission pathway which lead in *Pseudomonas* sp. strain B13 to non-productive metabolites (Williams, 1981). The additional mutational event required either reduction or elimination of the *meta*-cleavage pathway coded for by *Pseudomonas* sp. strain B13. In addition significant modifications of plasmid pWWO such as the elimination of a 38 kb fragment and its possible relocation within the chromosome occurred (Williams, 1981). These events demonstrate the high degree of DNA mobility associated with events which superficially appear to be simple.

Similar deletions and rearrangements have also been reported by Chatterjee & Chakrabarty (1981) for the same selection of 4-

chlorobenzoate utilizing strains of *Pseudomonas putida*. These experiments therefore reinforce the view that under appropriate conditions and with the correct selection pressures, gene flow and rearrangement events occur readily and at high enough frequencies to be significant factors in the evolution of novel strains.

Subsequently Chakrabarty and his colleagues have followed up the pioneering work of Knackmuss, using essentially the same strategies, to evolve strains of bacteria able to degrade 2,4,5-trichlorophenoxyacetic acid (2,4,5-T) (Kellogg *et al.*, 1981). Samples from a number of waste dump sites, together with known strains carrying plasmids coding for camphor, toluene, salicylic acid, *p*-chlorobiphenyl, 3-chlorobenzoate and 4-chlorobenzoate degradation, were inoculated into a chemostat fed with toluate, salicylate and chlorobenzoate (at 0.25 g l^{-1}) and 2,4,5-T (at 0.05 g l^{-1}). After prolonged cultivation (8 to 10 months) and under conditions of gradually increasing 2,4,5-T concentration, a culture evolved which could grow on 2,4,5-T as the sole carbon and energy source, eventually producing a culture which could degrade more than 70% of the available 2,4,5-T (starting at 1.50 g l^{-1}) in approximately 7 days. This approach is very similar to a number of previous continuous-flow culture enrichment procedures, particularly that of Daughton & Hsieh (1977) which led to the evolution of a synergistic community capable of degrading Parathion. From such an extraordinarily heterogeneous set of starting conditions, it is perhaps not surprising that, to date, the details of the selection events involved and the putative plasmid-mediated transfer events are obscure. Furthermore great caution must be exercised in ruling out simple selection of an existing 2,4,5-T-utilizing population from the various inocula samples, even though 'prolonged incubation of the enrichment flasks' apparently failed to detect such strains. It is now well established that mixed culture systems provide permissive environments within which previously quiescent metabolic capabilities may subsequently be expressed which would not normally be detected (Senior *et al.*, 1976). Moreover, as the following section demonstrates, selection of a desired characteristic can occur without invoking gene transfer mechanisms.

Microbial community catabolising of 2-chloropropionamide

We have demonstrated similar selection events occurring within a stable microbial community isolated for growth on 2-

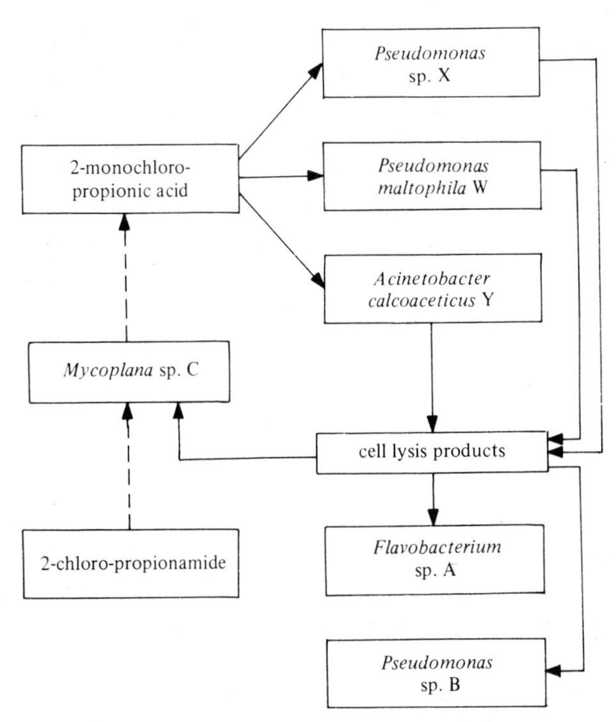

Fig. 2. A microbial community growing on 2-chloropropionamide as the sole carbon and energy source but dependent on the naturalistic interactions between four of the six organisms of the community (after Slater & Bull, 1982).

chloropropionamide (2CPA) as the growth-limiting substrate in a chemostat culture. The enrichment produced a consortium of at least six microorganisms (Fig. 2) from which it was not possible to isolate a single species able to use 2CPA as a carbon and energy source. It was however possible to isolate an organism, identified as *Mycoplana* sp. strain C, which grew readily on 2CPA as a nitrogen source but requiring carbon sources other than 2CPA or 2-monochloropropionic acid (2MCPA). The latter compound is the putative product of 2CPA deamination by an amidase which is specific for chlorinated amides and certainly did not lead to the release of chloride when 2CPA was used as a nitrogen source. Moreover with the realization that with respect to carbon metabolism the initial step was a cometabolic event stoichiometrically producing 2MCPA, the presence of 2MCPA-utilizing members of the community was readily established (Fig. 2). Unlike the Parathion community (Daughton & Hsieh, 1977) continuous growth, in this case for approximately 1 000 h at a dilution rate of 0.01 h^{-1} led to the isolation of *Pseudomonas* sp. strain 4 which was capable of slow growth on 2CPA as a carbon source ($\mu_{max} = 0.042\ h^{-1}$).

It was originally considered possible that *Pseudomonas* sp. strain 4 was originated by the transfer of the gene coding for the 2CPA amidase from *Mycoplana* strain C to a 2MCPA-utilizing strain of *Pseudomonas* present in the community. Recent evidence, however suggests that this was not the case since the amidase synthesized by *Pseudomonas* sp. strain 4 is markedly different, in terms of its substrate specificities to the *Mycoplana* amidase (R. C. Wyndham and J. H. Slater, unpublished observations). Great care has to be exercised in ascribing all community evolutionary events to gene transfer mechanisms.

Other interactions between loosely defined microbial communities

The preceding discussion on the specificity of tightly associated microbial communities catabolizing unusual substrates is widely believed to be atypical and unlike the situation found in more normal habitats. There is a tendency to consider a given microbial community, in the looser sense of those species which happen to occupy a particular habitat, as being composed of essentially independent, prototrophic populations demanding certain fundamental elemental requirements and appropriate carbon and energy sources. Basically the component populations exist as independent entities. In many environments, however, especially those which are microbially active, conditions are often such that a broad range of cellular components are available at concentrations sufficient to be used as exogenous sources without impairing the growth rate of the populations exploiting those resources. This is particularly important where genetic interactions between amino acids and vitamins are concerned, as we shall detail below.

In marine environments as much as 50% of photosynthetically fixed organic carbon may be released by members of the phytoplankton community (Fogg, 1966; Helleburst, 1974). Although primary photosynthetic products, such as glycollic acid, are the major constituents, these extracellular products include carbohydrates, lipids, peptides, organic phosphates, volatile materials, vitamins, toxins, enzymes and so on (Jones, 1982). Vitamins have been studied more thoroughly than other organic components and it has been concluded that their availability exceeds the normal requirements for many organisms in this habitat (Berland *et al.*, 1978; Jones, 1982). From one viewpoint, therefore, it is not surprising to find that vitamin auxotrophs can be readily isolated from these environments. However, what is more interesting is that it is also equally easy to isolate vitamin-producing phytoplankton and bacteria. This situation leads to the formation of interacting microbial communities dependent on the provision of and requirement for specific nutrients. Stirling *et al.* (1976) provided an excellent example of this type of community which is dependent on a biotin-requiring nocardia. Jolley & Jones (1977) demonstrated that a *Flavobacterium* species was closely associated with *Navicula muralis* needing biotin and, moreover, was auxotrophic for three amino acids, also produced and excreted by *N. muralis.*

In the rhizosphere the composition and characteristics of the microflora are principally influenced by the plant's root system (Bowen, 1980), in particular the root exudates. The released materials include many low molecular weight compounds, secretions, mucigels and autolysates (Rovira *et al.*, 1979) which may constitute up to 25% of the total root carbon available (Barber & Martin, 1976; Newman, 1978). Again these are conditions which favour the evolution of auxotrophic strains, but it also appears that in some cases the auxotrophic requirements are met by other microbial populations. It is a relatively simple matter to isolate such communities: for example, Jensen (1957) isolated from soil a bacterium which was dependent upon an exogenous supply of vitamin B_{12} produced by association with two *Streptomyces* species. Other examples are discussed by Slater (1981) and the frequency with which these assemblages are isolated would increase if more specific enrichment and isolation procedures, geared to their detection, were employed.

The question of interest at this point is why do such interactions occur? Intuitively the premise that prototrophic populations ought to be more competitive and therefore selected in natural habitats seems logical, particularly as many of these associations depend on the apparently wasteful synthesis of excreted metabolites by one organism to meet the needs of unrelated populations. In the light of the ease with which novel strains can be evolved for catabolic purposes, it seems unlikely that similar genetic exchange mechanisms could not operate in the case of the anabolic sequences required to produce prototrophic and therefore more competitive strains.

DNA EXCHANGE PROCESSES IN NATURE

With selected examples, the preceding section has demonstrated the need for genetic exchange mechanisms operating within the framework of the community structure. In the following sections the operations of the basic methods of DNA exchange between cells is discussed in relation to the factors likely to be encountered in nature.

Transforming DNA

Considered in terms of mass, the quantity of DNA in individual cells is small: a *Bacillus subtilis* cell for example contains only 0.0048 pg DNA; meristematic diploid cells of the bean, *Vicia faba*, in the apical root tip, 38.7 pg (Fasman, 1981); and the nuclei of animal cells, approximately 4 to 8 pg DNA per nucleus (Mahler & Cordes, 1966). However, these tiny quantities are counterbalanced by the large populations of cells found in specific niches or which make up multicellular organisms. Together 10^{13} cells of an average human body contain about 8 g of chemically pure DNA. The microbial flora of the human body consists of about 10^{14} microbial cells (Savage, 1977) whose combined DNA content adds about another 0.025 g to the total. It is difficult to estimate the physical mass of DNA in a complex ecosystem like an acre of garden with its mixed communities of microorganisms, worms, insects, plants and protozoa, but an overall figure of 30 g would not be unrealistic.

These calculations suggest that moribund or dead biomass may release very large quantities of DNA into the environment. Since soil is the sink for much of this DNA, the potential availability of extracellular DNA in nature can be gauged very roughly from its concentration in soil. Unfortunately the few available estimates are extrapolations. Using a value of 8.4 fg DNA $(\text{cell})^{-1}$ and a bacterial count of 1.1×10^{10} $(\text{g dry weight soil})^{-1}$ obtained by fluorescence microscopy, Torsvik & Goksøyr (1978) estimated that dry soil contained about 90 μg DNA $(\text{g soil})^{-1}$. Roughly comparable mean concentrations were obtained by Anderson (1961) who found a spread of values equivalent to 50 to 207 μg DNA $(\text{g soil})^{-1}$ in nine dry soil samples. Like all such estimates these values blur ecological reality by averaging DNA concentration across the whole of an arbitrary unit of the soil under investigation. Microbial populations in soils are heavily concentrated in specific microniches (Marshall,

1976; Weiss & Rheinheimer, 1978); the amount of DNA in these microniches may exceed the average value for that soil by an order of magnitude or more. It is also intuitively obvious that the amount of extracellular DNA in soil will be greatly affected by a complex range of variables that include soil type, pH, nature and number of resident microorganisms, ion exchange capacity and so on, so that the 'indicator' value of the DNA content of any soil sample is weak.

In what state is this soil DNA? While a good deal of evidence (Durand, 1966; Greaves & Wilson, 1970) suggests that DNA is quickly degraded in some soils, other studies indicate that soils can offer a significant degree of protection. Nucleic acids adsorb to clay minerals (Goring & Bartholomew, 1952) and this colloidal adsorption may protect the nucleic acids by either allowing more condensed molecular configurations and/or imposing spatial restrictions that delay or obviate inactivating reactions (Duboise *et al.*, 1979). Using X-ray diffraction techniques, Greaves & Wilson (1970) found that some nucleic acid internally adsorbed by montmorillonite was relatively resistant to degradation. Also the hydrolysis of organic substances, such as carbohydrates (Lynch & Cotnoir, 1956) or proteins (Ensminger & Gieseking, 1942), was restricted when the molecules were adsorbed to clays. The infectivity of viral RNA was increased about tenfold in the presence of bentonite and this enhanced infectivity was attributed to the inhibition of plant nucleases (Singer & Fraenkel-Conrat, 1961). Intriguingly, there is some arguable evidence that the addition of DNA to soil suppressed DNAase-producing bacteria (Greaves & Webley, 1965; Greaves & Wilson, 1970), although elsewhere DNAase activity has been shown to be important in structuring various microbial communities (Linton & Buckee, 1977).

The issue of DNA survival in soil is a complex one which does not permit simple generalizations. Nonetheless the existence of protective effects is consistent with the report that the amount of detectable bacterial-type DNA in soil was greater than could be accounted for by the number of bacterial cells (Anderson, 1958). It is tempting to speculate that the half-life of some extracellular DNA sequences in soil exceeds the half-life of those genes in their original cells. This speculation is consistent with Drake's (1974) observation that mutations in prokaryotic DNA occur as functions of absolute time not of numbers of generations. This means that more mutations may accumulate in quiescent DNA than are introduced by 'copying errors' during cell division (Drake, 1974).

An additional factor is that some bacterial species excrete or release DNA during growth (Catlin, 1960; Ephrati, 1968), and that this release seems pegged to specific phases in the life cycle. Thus, biologically active extracellular DNA accumulates in aging cultures of *Neisseria meningitides* (Catlin, 1960). Transformation by DNA appears to be an inducible characteristic which occurs at programmed metabolic intervals (Smith *et al.*, 1981); for example, transformable *Acinetobacter* strains became competent when entering stationary phase (Juni, 1972). Competence for transformation was raised to virtually 100% by transferring exponentially-growing *Haemophilus influenzae* cells to a medium which did not support growth: if these cells were then returned to a rich medium competence was quickly lost (Smith *et al.*, 1981). Transforming activity in *Bacillus subtilis* followed a cyclical pattern, peaking at the beginning of exponential growth and again at the start of the stationary growth phase (Ephrati, 1968). This may be a better representation of *in vivo* transformability than most. Whilst interpretation of the data is difficult and is confounded by differences in transformation mechanisms between Gram-positive and Gram-negative cells (Smith *et al.*, 1981) it is possible to infer – though cautiously – that the ability to release or accept extracellular DNA is favoured when the cells begin to face survival problems. These phenomena plus the ability of DNA to persist in soil in an extracellular state, suggest a coupling between DNA release and uptake that has novel evolutionary implications. Such a mechanism could allow bacterial genes to persist in the environment in a recoverable form long after the demise of their host cells. We believe this possibility deserves further attention.

If DNA from moribund or dead biomass is to have any biological future it must be introduced into and reproduced by viable recipient cells. Assuming that 'free' DNA can pass through barriers such as capsular materials, it must reach cell surfaces where its uptake is regulated by surface receptors. *Neisseriae* and *Haemophilus* species possess efficient mechanisms that discriminate against heterospecific DNA. Competent cells of *H. influenzae*, for example, contain a membrane protein which evidently recognizes a specific 11 base pair uptake sequence represented about 600 times in the genome (Sisco & Smith, 1979; Smith *et al.*, 1981). While Gram-negative species discriminate in favour of their own DNA, Gram-positive cells sometimes take up heterospecific DNA (Saunders, 1979; Smith *et al.*, 1981). Once such heterologous DNA has been integrated (intergenote formation), subsequent transformation of the original recipient strain by intergenotic DNA occurs with a frequency approaching that of homologous DNA (Wilson & Young, 1972; Biswas & Ravin, 1976). These studies suggest it is the lack of sequence homologies rather than the presence of antagonistic restriction systems (Arber & Linn, 1969) that is responsible for the relative inefficiency of heterologous transformation.

On balance it is evident that the chances of any free DNA fragment colonizing a living bacterial cell, especially a genetically foreign one, are vanishingly small. However, as discussed, this low probability factor is counterbalanced by the enormous physical quantitities of extracellular DNA released in many soils. It is tempting to speculate that transformation occurs with rather high frequencies in heavily colonized soils and that, in consequence, DNA from 'dead' biomass may still contribute significantly to future generations.

The apparent abundance of microbial DNA in soil allows multiple opportunities for genetic exchange across wide evolutionary gaps. Plant-bacterium interchanges seem especially likely. Bowen & Theodorou (1973) reported that cover by bacteria of root segments varied with root age: four-day old segments had from 1.2 to 15.7% cover while a 90-day old segment had 36.6% cover. This study suggested that the concentration of microbial cells (i.e. microbial DNA) increased at the very times moribund plant tissue is most likely to release its own DNA. Memories of past transfers may be found, for example, in the Ti DNA of the large plasmid responsible for the genesis of Crown Galls in *Agrobacterium* (Dellaporta & Pesano, 1981). The Ti DNA may be of plant origin (Roberts, 1982). It seems quite possible that this DNA was released from dead or dying plant cells and taken up by members of the *Rhizobiaceae* which characteristically inhabit the root zone. Since *A. tumefaciens* requires unusual amino acids called opines which it directs host tissue to produce after infection, the Ti plasmid creates a unique evolutionary niche for itself (Roberts, 1982). This strong selective advantage ensured that the foreign DNA in the Ti plasmid would eventually stabilize in suitable bacterial cells.

DNA in viral capsids

Most bacteriophage genomes probably exist in nature integrated into the DNA of their host cells. There is, for example, a quite remarkable coincidence between the average GC contents of bacteria and their phages for every taxon studied in sufficient detail. For example, the range of GC contents of *Bacillus* cells is 32 to 56% and that of *Bacillus* phages is 31 to 60%. This suggests that selection has treated phages DNA as a normal constituent of the bacterial DNA (Reanney & Ackermann, 1982).

While lysogeny seems rare in a few species of soil bacteria, such as *Pseudomonas putida*, in others, such as *Pseudomonas aeruginosa*, it can approach 100% (Patterson, 1965). Many bacteria are polylysogenic, carrying several phage genomes per cell (Ackermann & Smirnoff, 1975) and claims for eight prophages in one cell have been made (Holloway & Krishnapillai, 1975). The concentration of bacteriophage DNAs in soil may well mirror – within limits – the size of the bacterial cell population. The number of extracellular phage particles in soils is highly variable and is difficult to quantify accurately. Untreated soils yield few, if any, p.f.u.'s in many cases (Reanney & Marsh, 1973), although phage titres can be raised by treating soils with various additives. Robinson & Corke (1959) perfused various soils with distilled water and obtained titres of approximately 1.0×10^4 p.f.u. (ml perfusate)$^{-1}$. Reanney & Tan (1976) obtained p.f.u. counts of about 1.0×10^7 ml^{-1} by enriching soil with nutrients before sampling. However, these data gave no reliable indication of the concentration of 'extracellular' phages in the soils. All that can be deduced is that the number of potential phages per unit of soil could be very high (Reanney, 1976).

The behaviour of bacteriophages *in vitro* offers a misleading model of phage behaviour in soil (Reanney & Marsh, 1973). Viruses in soil are generally believed to be adsorbed to clays (Bitton, 1974; Duboise *et al.*, 1979) and, as in the case of extracellular DNA, this adsorption appears to afford the phages some protection against inactivation. When adsorbed to montmorillonite, phage T7 was noticeably resistant to inactivation, the addition of only 75 μg montmorillonite ml^{-1} significantly reducing the rate of inactivation (Bitton & Mitchell, 1974). Similar results were obtained by Gerba & Schaiberger (1975) with phage T2. Protective effects have also been reported for viruses other than phages: Miyamoto (1959) believed that clays act as 'carriers' of soil-borne plant viruses. Clays contaminated with wheat yellow mosaic virus and barley yellow mosaic virus evidently remained infective for 2 years (Miyamoto, 1958, 1959; Thung & Dijkstra, 1958). The granulosis virus of *Pieris brassicae* has been reported to survive in soil for years (David &

Gardiner, 1967) whilst the amount of nuclear polyhedrosis virus of the cabbage looper in surface loam has been reported to remain fairly constant for over 4 years (Jacques, 1964, 1967). These results must be viewed with caution however since many other reports speak of rapid viral die-off in soil (Duboise *et al.*, 1979).

The ability of bacteriophages to retain their infectivity in soil is the subject of some debate. Some reports (Bystricky *et al.*, 1975) indicated that adsorption of phages to materials, such as bentonite, did not affect the infectivity of the phages whereas other studies indicated the converse (Puck & Sagik, 1953). The issue has been investigated for *Arthrobacter* phages by Ostle & Holt (1979) who found that most phages added to solid phases having a high cation exchange capacity could not be eluted in an infective state. Since 75 to 90% of the bacteriophage antigen was recovered, it was concluded that viruses were being inactivated either by the adsorption process or by elution. Electron microscopy showed that 90% of eluted phages were damaged in their tail structures. It may be that the long non-contractile phage tails, such as those of the widespread B1 category (Reanney & Ackermann, 1982), are unusually sensitive to the shear forces likely to be encountered by phages in soil because of groundwater flow or run-off through the upper layers of soil during rainstorms.

In the light of the preceding discussion on transformation we question whether loss of infectivity should be equated with total elimination of phage DNA from evolutionary processes. It has still to be shown that the integrity of phage tails, for example, is always necessary for the transfer of phage-encapsulated genes in nature. *Streptococcus pneumoniae* contains a phage, PG24, which encapsidates host DNA in DNAase-resistant particles morphologically and physically similar to infectious phage. However, such phage particles did not carry the transfer of host genes process to completion, but rather the transfer required the development of competence for transformation and occurred via a DNAase-sensitive step (Porter *et al.*, 1979). Whether such pseudotransduction is a quirk of this system or whether it has a wider applicability remains to be determined. What we do suggest, with some reservations, is the possibility that the packaging of DNA in phage heads constitutes another means of prolonging the life expectancy of extracellular DNA in natural environments. Encapsidated genes will survive in situations where cells would perish, such as long periods without nutrients; periods of intense cold; or the presence of cytocidal chemicals, such as antibiotics. Also Robinson *et al.* (1970) found that some *Bacillus* phages were not eliminated from soil by doses of gamma radiation sufficient to inactivate viable microorganisms. Extrapolating from all these observations it is possible to speculate that the ability to alternate between a cellular and an extracellular state is part of the long-term survival strategy of microbial genes (Reanney, 1977; Kelly & Reanney, 1978; Davey & Reanney, 1980; Reanney, 1982). Transduction may have been selected, less for its ability to disperse variation, than for its ability to preserve nucleic acids from the adverse effects of environmental variables. This view acknowledges the selfish nature of viral genes (Doolittle & Sapienza, 1980) but sees the encapsidation of nucleic acids by viruses as part of a wider survival strategy imposed on all prokaryotic genes by the frequent need to withstand non-physiological conditions in many habitats.

The genus *Bacillus* harbours a variety of defective temperate phages of which phage PBSX is the prototype (Seaman *et al.*, 1964). These phages randomly encapsidate host DNA (Okamoto *et al.*, 1968). Since these particles do not contain their own coding genes, the DNA they enclose cannot be considered selfish DNA. The widespread abundance of genes encoding these particles in the genomes of many *Bacillus* species (Reanney & Ackermann, 1982) is consistent with the view that the indiscriminate packaging of genes *per se* has a strong survival advantage for soil-adapted species.

Bacteriophages have the potential to mediate extensive gene movement between different bacteria, but whether this potential is realised in nature is open to debate. Jones & Sneath (1970) found that phages crossed generic boundaries in only 0.3% of 26 681 attempted lytic reactions between taxa at various levels. Only the temperate phage P1 of known phages, has a genuinely wide host range. Phage P1 can infect 9 different bacterial genera (Murooka & Harada, 1979) and in addition can transfer genes into such diverse species as the *Myxobacteria* (Kaiser & Dworkin, 1975) and *Agrobacterium* (Murooka & Harada, 1979). However, if phage P1 which exists in host cells as a closed circular DNA, is placed in a plasmid incompatibility group and behaves essentially as a plasmid which has adopted a phage mode of dissemination (Reanney & Ackermann, 1982). In an important sense phage P1 may be the exception that proves the rule that bacteriophages are normally very restricted in their host ranges.

A limited host range does not mean that phages are ineffective agents of gene transfer among their own host species. The accumulation of drug resistance in *Staphylococcus aureus* is known to be due to the cell-to-cell transfer of staphylococcal R plasmids carried in phage heads (Lacey, 1975). Significantly, however, transfer was enhanced as cell density increased and appeared actually to require cell-to-cell contact (Lacey, 1975). On balance, transduction is probably effective in disseminating genetic information where populations of cells are large and concentrated, but where populations are sparse and dispersed, as in resting soil, the potential for DNA transfer by phages is probably rather low.

DNA transfer by conjugation

Whilst transformation and transduction are capable of creating new genetic structures by DNA transfers, the burden of evidence summarized to date suggests that these processes act conservatively in most cases. The situation is different for conjugation. Conjugative plasmids (Novick *et al.*, 1976) promote cell-to-cell transfer of DNA by special transfer mechanisms encoded in specific transfer (*tra*) genes (Clark & Warren, 1979). Conjugative plasmids are classified into a variety of incompatibility (Inc) groups on the basis of their ability to co-exist stably in one cell (Datta, 1979) and, in general, identical or related plasmids are incompatible while heterologous plasmids are compatible (Falkow, 1975).

The essential differences between transformation, transduction and conjugation are not difficult to rationalize. Transformation normally requires extensive nucleotide homology between entering and resident DNAs (Notani & Setlow, 1974; Smith *et al.*, 1981) and so transformation tends to be most efficient when the donor and recipient cells are genetically similar or identical. Temperate phages normally require pre-existing regions of homology to integrate into the DNA of their host lysogens (Nash, 1981); hence the host ranges of temperate phages tend to cluster tightly round specific taxonomic groupings (Reanney & Ackermann, 1982). Plasmids are autonomously-replicating DNAs however and do not require any recombinational process to establish themselves in recipient cells. Furthermore, even the presence of restriction systems in receptor cells (Arber & Linn, 1969) may not pose an insuperable barrier to the epidemic spread of conjugative plasmids, since many plasmids multiply so rapidly that some of their DNAs acquire the protective label before all copies are degraded.

Table 4. *Host ranges of some conjugative plasmids*

Representative	Inc group	Host range[a] (genus)[b]
R40a	C	*Erwinia*[c] *Escherichia* *Klebsiella* *Providence* *Pseudomonas* *Salmonella* *Serratia* *Vibrio*
R390	N	*Escherichia* *Klebsiella* *Providence* *Proteus* *Salmonella* *Shigella*
RP1	P	*Acinetobacter* *Agrobacterium* *Alcaligenes*[+] *Bordetella* *Caulobacter* *Chromobacterium* *Escherichia* *Erwina* *Klebsiella* *Neisseria* *Proteus* *Pseudomonas* *Rhizobium* *Rhodopseudomonas* *Rhodospirillum* *Shigella* *Vibrio*
R300B	Q	*Salmonella* *Escherichia* *Proteus* *Providence* *Pseudomonas*

[a] Host range in this context means that members of the Inc groups shown have been reported to propagate in strains of the taxa listed. The representative plasmid shown cannot necessarily colonize all the taxa listed beside it in the third column; [b] Primary data can be found in Bukhari *et al.* (1977), Reanney (1977), Holloway (1979) and in [+] Don & Pemberton (1981); [c] Kelly & Reanney, unpublished.

Plasmids are excellent examples of well-adapted selfish DNAs (Doolittle & Sapienza, 1980; Orgel & Crick, 1980). Autonomous replication and transferability uncouple the survival of a conjugative plasmid from the survival of its host cell. The survival chances of a given plasmid will be extended if the plasmid can colonize cells other than those of its original or preferred host. Accordingly, many plasmids, especially those belonging to Inc groups, Q, C, N, W and P (Table 5) can invade many different species of bacteria. The prototype IncP plasmid, RPl, has an extraordinary host range that allows this DNA to replicate in cells as taxonomically far apart as the genus *Escherichia*, the soil bacterium *Azotobacter* and the stalked bacterium *Caulobacter* (Olsen & Shipley, 1973; Alexander & Joklik, 1977 – Table 4). These wide host ranges spread the survival options of IncP plasmids across so many taxonomic baskets that their long-term survival is more secure than that of any individual species of susceptible host (Davey & Reanney, 1980).

The mechanisms that promote the self-transfer of conjugative plasmid DNA also influence other DNAs in cells. Many conjugative plasmids can integrate into the chromosomes of their hosts (Holloway, 1979). Normally this interaction is transient but IncJ plasmids are so stably integrated into chromosomal DNA that they cannot normally be recovered as closed circular DNA from host cells again (Nugent, 1981). When such conjugative plasmids leave the chromosome they may take chromosomal markers with them. For example, a derivative of the plasmid R68.45 carries about 2 to 3 minutes of the *Pseudomonas* chromosome (Haas & Holloway, 1978; Holloway, 1979). These incorporated chromosomal genes now acquire the cell-to-cell mobility and the extended host range of the carrier replicon. It has been suggested that virtually all prokaryotic genes diffuse through various hosts in a common ecosystem by mechanisms like this (Reanney, 1976, 1977).

Conjugative plasmids probably constitute a minority of plasmids. Most independently-replicating closed circular DNAs in cells are non-conjugative i.e. they are incapable *per se* of promoting the transfer of their DNA. These small plasmids normally occur as multiples of between 10 and 40 copies per cell (Bukhari *et al.*, 1977). Non-conjugative plasmids can be mobilized by conjugative plasmids in either of two ways. First, the smaller replicon may recombine with the larger to give a cointegrate structure which may pass *en bloc* into susceptible host cells (Broda, 1979). Second, the conjugative plasmid may provide a passive transfer bridge for the smaller replicon without recombining with it (Smith, 1977). Significantly, non-conjugative plasmids have been shown to colonize cells which a mobilizing conjugative plasmid is unable to enter (Stanisch, personal communication).

The mechanisms which bring about the cell-to-cell transfer of DNA must be seen in the context of mechanisms which exchange

DNA sequences between replicons in a single cell. Such mechanisms include the translocation of insertion sequences (IS modules) and transposons (Tn modules). These transpositional events can occur in the absence of the cell's normal recombination mechanism (*rec*A in *Escherichia coli*) (Cohen, 1976; Kleckner, 1981). If a common IS module, for example, inserts into two unrelated plasmids then the region of homology so created permits fusion of DNAs. Such an insertion removes the barrier that normally prevents recombination between DNAs of different evolutionary ancestry and makes possible a variety of DNA-DNA interactions. Plasmid-plasmid, plasmid-phage, plasmid-chromosome, phage-chromosome interactions are all promoted by the infective spread of selfish modular elements (Davey & Reanney, 1980) although the various DNAs can also recombine through other mechanisms (Nash, 1981). The genetic flexibility of prokaryotes is also enhanced by the ability of restriction nucleases to mediate processes of 'natural genetic engineering' (Reanney, 1976) in which non-homologous DNAs cleaved *in vivo* by a common restriction nucleases are welded into novel genetic combinations (Chang & Cohen, 1977).

How widespread is conjugation in nature? The conjugal transfer of DNA requires the formation of a mating pair or aggregate (Clark & Warren, 1979) in which the donor and recipient cells are stabilized by cross-linking pili or other factors. Thus the potential of conjugation for DNA transfer in nature can be assessed by considering the factors which effect physical contact between cells. Here one strikes again the recurring theme of this chapter; that is, the disparity between *in vitro* models and *in vivo* realities. Estimates of mating efficiency made in the usual way using broth cultures are frankly misleading when applied to their ecological equivalents. Bradley (1980) has shown that the pili which promote conjugation can be divided into three morphological categories which correlate to some degree with plasmid classifications based on incompatibility: these are, thin flexible, thick flexible and rigid. As indicated in Table 5 plasmids with rigid pili transfer on solid surfaces, such as agar plates or disc filters, with frequencies up to 3.6×10^4 times higher than the corresponding crosses in broth (Bradley, 1980). It is worth noting that the broad host range plasmids belonging to IncP, X and C all have marked preferences for surface matings. In most cases transfer frequencies were at de-repressed levels: indeed, the broth matings so often used to assess transfer frequencies can lead to the spurious impression that the plasmids under consideration are

Table 5. *Classification of plasmid mating systems based on optimum environment for conjugal transfer*

Type of mating system[a]	Pilus morphology	Inc group[b]	Representative plasmid	Transfer frequency ratio plate/ broth[c]
Universal	Thin flexible	I	R64	0.9
		K$^{\alpha}$	pTM559	0.51
	Thick flexible	FII	R100	0.73
		HI	R27	5.5
		J	R391	0.9
		V	R753	0.35
		com9	R71	1.55
Surface preferred	Thick flexible	C	RA1	45
		D	R711b	180
		T	Rtsl	265
		X	R6K	250
Surface obligatory	Rigid	M	R446b	16 150
		N	N3	10 200
		P	RPI	2 100
		U	RA3	7 900
		W	Sa	36 450

[a] 'Universal', transfer equally good in a liquid or on a solid surface; 'surface preferred', transfer significantly better on a solid surface compared with in a liquid; 'surface obligatory', transfer fairly low in a liquid and very high (derepressed) on a surface; [b] Single representatives only are included for incompatibility group complexes I, F and H. IncU is tentative and unpublished (R. W. Hedges, personal communication); [c] Transfer frequencies on plates divided by frequencies in broth.
(Printed with permission from Bradley, 1980.)

naturally repressed. De-repressed plasmids of course transfer DNA much more effectively than repressed replicons and the question of repression is important when assessing the efficiency of plasmid dissemination in nature.

Are bacteria in nature likely to occur on stabilizing surfaces? For once the answer is tolerably clear and it is a qualified affirmative. One of the best documented aspects of microbial ecology is the propensity of bacteria to concentrate at interfaces. These include liquid-air interfaces, solid-liquid interfaces or solid-solid interfaces (Marshall, 1976). Large colloidal particles surrounded by water films are reasonable analogs for the kind of surface-liquid appositions that occur in filter matings. Bacteria may also be grouped together in microcolonies under rather stable conditions on root surfaces. Direct microscopy of root samples suggested that the root surface had an irregular bacterial cover of 4 to 10% (Rovira *et al.*, 1974). Using TEM microscopy Foster & Rovira (1978) estimated cell frequency as 1.2×10^{11} per cm^3 in the rhizoplane 0 to 1 μm from the root. The population consists of a mixed community which includes *Rhizobium, Pseudomonas* and *Achromobacter* species (Gray & Williams, 1971).

While the high concentrations of bacteria in nutritionally favoured niches like the root surface may encourage DNA transfer, other factors undoubtedly restrain conjugation. Plant roots excrete mucigels which can entrap bacteria (Greaves & Darbyshire, 1972) and fibrillar structures have been observed to immobilize a saprophytic bacterium in plant leaves (Sing & Schroth, 1977). The production of capsular material of microbial origin in aging roots (Old & Nicholson, 1978) may make the formation of mating aggregates more difficult. Pili may be sensitive to the shear effects of ground water in the same way that phage tails seem to be. Transferred DNA also faces genetic barriers which include general mechanisms such as restriction systems, and specific barriers, such as plasmid incompatibility and phage immunity, which tend to protect cells against infection with homologous DNAs. Incompatibility is a post-entry phenomenon and is to be distinguished from surface phenomena, such as entry exclusion (Falkow, 1975), which operate to prevent DNA passage into the cell. The dissemination of plasmids is restrained by the ability of certain plasmids to inhibit the transfer functions of other plasmids, classically the F plasmid (fertility inhibition – Broda, 1979). Some lysogenic strains may be poor recipients for particular plasmids: for example, the presence of prophage B3 interferes with the inheritance of IncP plasmids (Krishnapillai, 1977; Stanisich, personal communication). Conversely some plasmids inhibit phage multiplication (Taylor & Grant, 1976). There are even barriers to the transfer of DNA sequences between replicons in one cell: a given transposon may hinder a plasmid's ability to accept similar transposons (Robinson *et al.*, 1977). Thus there exist powerful genetic mechanisms which both promote and inhibit the transfer of DNA sequences and the outcome of a particular potential genetic cross in nature is influenced by a huge spectrum of interlocking variables.

Temperature seems likely to inhibit many natural transfers. Most plasmids studied seem to transfer best at temperatures between 25 to 38 °C (Shendarov, 1971; Yoshika & Nakatani, 1976). However, mating assays seldom assess the optimal temperatures for transfer and there are rather few cases in which the 'true' temperature range of a conjugative plasmid has been carefully determined. The R plasmid Rl*drd-19* did not transfer detectably at 17 °C in a 3 h mating but when the mating period was extended to 24 h, transconjugants were observed, although no transconjugants were obtained from a 48 h mating at 15 °C (Singleton & Anson, 1981). Kelly & Reanney (1982) found a moderate number of transconjugants (approximately 3.0×10^{-5}) when the soil-derived plasmid pWKl was incubated with *Erwinia herbicolor* 1414 at 12.5 °C for 20 h but none when the corresponding cross was performed at 10 °C. These data must be set in the context of meteorological data which suggest that, in temperate climates, surface soil temperatures are usually rather low. A survey of soil at 0.05 m depth conducted at the New Zealand Soil Bureau's Taita Experimental Station showed a mean temperature range extending from 5.8 °C (July) to 19.3 °C (January) (Aldridge, 1978). Similar data also show, of course, that there can be a wide daily and seasonal spread of soil temperatures and the fluctuations are undoubtedly exaggerated in soil microniches exposed to a heavy solar load or able to absorb sunlight strongly due to presence of phenolic materials like tannins and other heat-absorbing substances.

One of the chief barriers to the conjugal transfer of DNA in nature is the relative quiescence of bacteria in many natural habitats. While little is known of the dynamics of bacterial growth in soil, available data suggest that only about 15 to 30% of soil bacteria are active even under favourable conditions (Clarholm & Rosswall, 1980). Gray & Williams (1971) estimated that bacteria in woodland soil had a maximal rate of reproduction of about one division per day whilst Bowen & Rovira (1976) reported that the generation time of bacteria in untreated soil was 14.5 h. Babuik & Paul (1970) suggested that the energy input of a grassland prairie soil was sufficient to allow bacteria to divide only a few times per year. While conjugation is not coupled to cell division, the transfer of plasmid DNA by a rolling circle mechanism is an energy-requiring process and hence must, to some degree, be pegged to the metabolic fitness of the donor and recipient cells. The same is true of transfer by transformation and transduction. Koch (1971) has suggested a 'feast or famine' regime for bacteria in which bursts of rapid growth are interspersed with long periods of dormancy. DNA transfers in soil are thus severely restricted in time. They are also severely

Table 6. *The selection of strains with reduced plasmid size by continuous culture growth of* Escherichia coli *containing initially plasmid TP120*

Host strain (plasmid markers)	Presence of plasmid	Molecular weight (Md)
E. coli K–12 (TP120) (Ap,Sm,Su,Tc)	+	31.6
E. coli K–12 (TP120A) (Ap,Sm,Su)	+	20.0
E. coli K–12 (TP120B) (Sm,Su,Tc)	+	24.9
E. coli K–12 (TP120C) (Ap,Sm,Su)	+	17.0
E. coli K–12 (TP120D)	+	24.0
E. coli K–12 (TP120E) (no plasmid)	–	–

restricted in space. Bacteria in soil tend to occur in small colonies on organic particles, partly because of nutrient requirements and partly because of the ability of clays and other minerals to adsorb cells (Marshall, 1976). Gray *et al.* (1968) calculated that organic particles provided only 15% of available soil surfaces but carried about 60% of detectable soil bacteria. These foci of microbial growth are sparsely distributed in soil since only 0.02% of the surface areas in a sandy loam was colonized by microorganisms (Gray *et al.*, 1968).

Another question is why plasmid-containing populations possess plasmids under environmental conditions where they might be at a disadvantage by retaining them. In fact, in many cases it can easily be shown that the plasmid's presence constitutes a disadvantage, especially under conditions where the functions encoded are not required. For example, the introduction of plasmid R68–45 into a strain of *Pseudomonas putida* reduced the population's specific growth rate by 30%. Moreover the effect is greater with decreasing nutrient availability, suggesting that under the normal nutrient-poor conditions experienced in natural environments, the plasmid's presence is even more deleterious. This has been clearly demonstrated in studies with continuous-flow culture system (Melling *et al.*, 1977; Godwin & Slater, 1979; Adams *et al.*, 1979; Wouters *et al.*, 1978). In general the slower the growth rate, the greater the competitive advantage of plasmid-minus populations over plasmid-containing populations. Most studies have demonstrated that it is unusual to select for plasmid-minus populations, although various functions coded by the plasmids may be readily lost, reducing the overall plasmid size (Table 6). Normally these data are interpreted in terms of selection pressures favouring plasmid elimination (or partial elimination) in order to conserve the limiting quantities of growth resources for more essential cellular biosynthesis rather than in wasteful production of unnecessary DNA. That is, competition experiments between plasmid-minus and plasmid-containing isogenic populations demonstrate that the synthesis of redundant DNA is a disadvantage.

CONCLUSIONS

Bacterial evolution poses a paradox. On the one hand, there is the abundant evidence from microbial genetics that bacteria can exchange DNA by multiple mechanisms both within and between different cell types. This should, if unchecked, lead to an intercommunication within the microbial biosphere (Sonea & Panisset, 1978) and to the subsequent blurring of taxonomic identities. On the other hand, there is the impressive evidence of micropaleontology and ecology which show that bacteria have preserved recognizable phenotypes for billions of years.

In this chapter we have summarized those cellular and ecological factors which either promote or hinder genetic exchange. Gene transfer is promoted by:

(i) high population densities in microbially-active environments, such as the rhizosphere or on surfaces;

(ii) the enhanced efficiency of mating on surfaces and other interfaces where bacterial biomass tends to be concentrated in nature;

(iii) the physical bulk of DNA released by dead or moribund biomass;

(iv) the extended half-lives of extracellular DNA in soils with a colloidal content;

(v) the wide host range and stability of plasmids, especially in incompatibility groups P, X, C and W;

(vi) the high replicability of selfish DNA units and their inherent tendency of spread at epidemic proportions.

Gene transfer is hindered by:

(i) the spatial and temporal separation of bacteria in most natural environments;

(ii) the immobilization of viruses and bacteria on soil particles;

(iii) the presence of mucigels, lectins, slimes and other substances which entrap cells;

(iv) genetic barriers, such as restriction systems and surface discrimination which tend to restrict the uptake of DNA to homologous or near-homologous sequences and those such as phage immunity which discriminate against homologous sequences.

The interaction between these opposing factors is subject to an enormous spectrum of complex and variable aspects and what is true for one niche or group may be false for another. On balance however it seems to us that selection has groomed bacteria to have the best of both worlds. To a remarkable degree prokaryotes can use their potential for genetic interchange where necessary, while at the same time guarding their biological individuality.

The 'strategy' which reconciles change and stability in bacteria in our view is chiefly the division of genetic information between a *fixed* repository of genes (the chromosome) and a *floating* 'library', of genes (extrachromosomal DNA in all its forms). The issue of change and stability is, to a degree, mirrored in the division of prokaryotic DNA into its two characteristic classes (Reanney, 1977). Thus the chromosomes of *E. coli* and *Salmonella typhimurium* share a similar map order and have considerable sequence homology despite an estimated evolutionary divergence of millions of years (Sanderson, 1978). On the other hand there is little evidence of a mandatory taxonomic correlation between the DNA of many conjugative plasmids and the DNA of their host cells. This has been demonstrated for *Agrobacterium* by Currier & Nestor (1976); they found that while chromosomal DNA of *A. tumefasciens* 27 showed only 10% homology with that of *A. tumefasciens* C–58, their plasmids were 81–88% homologous and conversely, chromosomal homology between the 11BV7 and A6 strains was 81% while their plasmids were only 27% related. Also RPl can colonise cells which are taxonomically unrelated to the *Pseudomonas* line in which this plasmid was first detected (see Table 1).

Campbell (1981) has noted that the genes characteristically carried by ECEs do not look like a random sample. This suggests that most of the genes picked up during DNA/DNA interactions and transfers tend to be lost during later generations unless their incorporation into the 'floating' gene pool confers survival advantages in the long term. The chromosome encodes the essential functions needed for the everyday life of the cell while adventitious genes such as those for antibiotic resistance, tend to be located on ECEs (Davey & Reanney, 1980). This division parallels the separation of eukaryote genes into 'essential' and 'luxury' classes.

Another major factor in the stability of genotypes follows from the fact that much bacterial DNA in nature is in a nondividing, resting state, which means that bacteria in nature pass through far fewer generations per annum than they do in continuous laboratory culture (Drake, 1974). Since all successful DNA transfer processes require replication/recombination it is evident that, in the widest evolutionary perspective, DNA/DNA interactions and exchanges have been extremely rare. This is not to minimize the importance of those exchanges that do occur, merely to emphasise their sporadic nature.

While compiling this paper we have been impressed by the degree to which the transfer processes reviewed act *conservatively* rather than innovatively. We have laid some stress on the apparent ability of extracellular DNA to persist in the environment. If our inferences have any validity, then transformation, encapsidation and transduction not only disperse variation but also promote the long-term survival of genes. The multiple barriers which discriminate against foreign DNA probably ensure that extracellular DNA passes chiefly into cells with homologous or related genomic sequences. Thus while the *potential* of these transfer mechanisms to generate genetic novelty remains unchallenged, the *extent* to which this potential is realised in nature should not be over-stated. Even conjugation seems to have evolved in such a way that essential chromosomal genes are seldom disturbed while adventitious extrachromosomal genes come and go freely (Campbell, 1981). The stability paradox then is, in our view, no paradox at all. Bacterial genomes remain stable because the transfer processes which endow them with the formidable adaptive flexibility do not normally affect their identifying genotypes and appear to act at least in part to preserve pre-evolved genes from the hazards of a variable environment.

REFERENCES

ACKERMANN, H. W. & SMIRNOFF, W. A. (1975). Lysogeny in *Bacillus thuringiensis*. *Third International Congress of Virology Abstracts*, C406, p. 266.

ADAMS, J., KINNEY, T., THOMPSON, S., RUBIN, L. & HELLING, R. B. (1979). Frequency-dependent selection for plasmid-containing cells of *Escherichia coli*. *Genetics*, **91,** 627–37.

ALDRIDGE, C. (1978). Climatological data for Taita experimental station 1973 to 1976 and 20 year means. *New Zealand Soil Bureau Scientific Report*, 32.

ALEXANDER, J. L. & JOKLIK, J. D. (1977). Transfer and expression of the *Pseudomonas plasmid* RPI in *Caulobacter*. *Journal of General Microbiology*, **99**, 325–31.

AMBLER, R. P. (1980). The structure of β-lactamases. *Philosophical Transactions of the Royal Society of London B*, **289,** 321–31.

ANDERSON, G. (1958). Identification of derivatives of deoxyribonucleic acid in humic acid. *Soil Science*, **86,** 169–74.

ANDERSON, G. (1961). Estimation of purines and pyrimidines in soil humic acid. *Soil Science*, **91,** 156–61.

ANDERSON, E. S. (1975). Problems and implications of chloramphenicol resistance in typhoid bacillus. *Journal of Hygiene*, **74,** 289–99.

ANDERSON, E. S. & SMITH, H. R. (1972). Chloramphenicol resistance in typhoid bacillus. *British Medical Journal*, **3,** 329–31.

ARBER, W. & LINN, S. (1969). DNA modification and restriction. *Annual Reviews of Biochemistry*, **38,** 467–500.

BABUIK, L. A. & PAUL, E. A. (1970). The use of fluorescein isothiocyanate in the determination of the bacterial biomass of grassland soil. *Canadian Journal of General Microbiology*, **16,** 57–62.

BAGGI, G., BERETTA, L., GALLI, E., SCOLASTICO, C. & TRECCANI, V. (1978). Biodegradation of alkylphenol polyethoxylates. In *The Oil Industry and Microbial Ecosystem*, ed. K. W. F. Chater & H. J. Somerville, pp. 129–36. London: Heyden & Sons.

BAKER, P. B. & WOODS, D. R. (1977). Cometabolism of the ixodicide Amitraz. *Journal of Applied Bacteriology*, **42,** 187–96.

BALBA, M. J. & EVANS, W. C. (1977). The methanogenic fermentation of aromatic substrates. *Biochemical Society Transactions*, **5,** 302–4.

BARBER, D. A. & MARTIN, J. K. (1976). The release of organic substances by cereal roots in soil. *New Phytologist*, **76,** 69–80.

BAUCHOP, T. (1980). Scanning electron microscopy in the study of microbial digestion of plant fragments in the gut. In *Contemporary Microbial Ecology*, ed. D. C. Ellwood, J. N. Hedger, M. J. Latham, J. M. Lynch & J. H. Slater, pp. 305–26. London: Academic Press.

BEAM, H. W. & PERRY, J. J. (1973). Co-metabolism as a factor in microbial degradation of cycloparaffinic hydrocarbons. *Archive für Mikrobiologie*, **91,** 87–90.

BEAM, H. W. & PERRY, J. J. (1974). Microbial degradation of cycloparaffinic hydrocarbons via cometabolism and commensalism. *Journal of General Microbiology*, **82,** 163–9.

BERLAND, B. R., BONIN, D. J., FIOLA, M. & MAESTRINI, S. Y. (1978). Importance des vitamines en mer. Consummation et production par les algues et les bacteries. In *Les Substances Organiques Naturelles Dissontes dans l'eau de Mer*, ed. H. J. Ceccaldi, pp. 121–46. Paris: Centre National de la Recherche Scientifique.

BISWAS, G. D. & RAVIN, A. W. (1976). Genetic hybridisation of the *leu-ilv* region of bacilli. *Journal of General Microbiology*, **92,** 398–404.

BITTON, G. (1974). Adsorption of viruses onto surfaces in soil and water. *Water Research*, **9,** 473–84.

BITTON, G. & MITCHELL, R. (1974). Effects of colloids on the survival of bacteriophages in seawater. *Water Research*, **8,** 227–9.

BORDELEAU, L. M & BARTHA, R. (1968). Ecology of a pesticide transformation: synergism of two soil fungi. *Soil Biology and Biochemistry*, **3,** 281–4.

BOWEN, G. D. (1980). Misconceptions concepts and approaches in rhizosphere biology. In *Contemporary Microbial Ecology*, ed. D. C. Ellwood, J. N. Hedger, M. J. Latham. J. M. Lynch & J. H. Slater, pp. 283–304. London: Academic Press.

BOWEN, G. D. & ROVIRA, A. D. (1976). Microbial colonisation of plant roots. *Annual Reviews of Phytopathology*, **14,** 121–44.

BOWEN, G. D. & THEODOROU, C. (1973). Growth of ectomycorrhizal fungi around seeds and roots. In *Ectomycorrhizae – Their Ecology and Physiology*, ed. G. C. Marks & T. T. Kozlowski, pp. 107–50. London: Academic Press.

BRADLEY, D. E. (1980). Conjugative pili of plasmids in *Escherichia coli* K–12 and *Pseudomonas* species. In *Molecular Biology, Pathogenicity and Ecology of Bacterial Plasmids*, ed. S. B. Levy, R. C. Clowes & E. L. Koenig, pp. 217–26. New York: Plenum.

BRODA, P. (1979). *Plasmids*. Oxford: Freeman.

BUKHARI, A. I., SHAPIRO, J. A. & ADHYA, S. L. (1977). *DNA Insertion Elements, Plasmids and Episomes*. New York: Cold Spring Harbor Laboratory.

BULL, A. T. & SLATER, J. H. (1982). *Microbial Interactions and Communities*, vol. I. London: Academic Press.

BYSTRICKY, V., STOTSKY, G. & SCHIFFENBAUER, M. (1975). Electron microscopy of T–1 bacteriophage adsorbed to clay minerals: application of the critical point drying method. *Canadian Journal of Microbiology*, **21,** 1278–82.

CALOS, M. P. & MILLER, J. H. (1980). Transposable elements. *Cell*, **20,** 240–3.

CAMPBELL, A. (1981). Evolutionary significance of accessory DNA elements in bacteria. *Annual Review of Microbiology*, **35,** 55–83.

CAREY, A. E. & HARVEY, G. R. (1978). Metabolism of polychlorinated biphenyls by marine bacteria. *Bulletin of Environmental Contamination Toxicology*, **20,** 527–34.

CATLIN, B. W. (1960). Transformation of *Neisseria meningitidis* by deoxyribonucleates from cells and from culture slime. *Journal of Bacteriology*, **79,** 579–90.

CHANG, S. & COHEN, S. N. (1977). *In vivo* site-specific genetic recombination promoted by Eco Rl restriction endonuclease. *Proceedings of the National Academy of Sciences, USA*, **74,** 4811–15.

CHATTERJEE, D. K. & CHAKRABARTY, A. M. (1981). Plasmids in the biodegradation of PCBs and chlorobenzoates. In *Microbial Degradation of Xenobiotics and Recalcitrant Compounds*, ed. T. Leisinger, A. M. Cook, R. Hütter & J. J. Nüesch, pp. 213–19. London: Academic Press.

CLARHOLM, M. & ROSSWALL, T. (1980). Biomass and turnover of bacteria in a forest soil and a peat. *Soil Biology and Biochemistry*, **12,** 49–57.

CLARK, A. J. & WARREN, G. J. (1979). Conjugal transmission of plasmids. *Annual Reviews of Genetics*, **13,** 99–125.

CLARK, R. R., CHIAN, E. S. K. & GRIFFIN, R. A. (1979). Degradation of polychlorinated biphenyls by mixed microbial cultures. *Applied Environmental Microbiology*, **37,** 680–5.

CLARKE, P. H. (1974). The evolution of enzymes for the utilisation of novel

substrates. In *Evolution in the Microbial World*, ed. M. J. Carlisle & J. J. Skehel, pp. 183–127. Cambridge University Press.
COHEN, S. N. (1976). Transposable genetic elements and plasmid evolution. *Nature, London*, **263,** 731–8.
COSSAR, D., BROWN, C. M. & WATKINSON, R. J. (1981). Some properties of a marine microbial community utilizing benzoate. *Society for General Microbiology Quarterly*, **8,** 47.
CURRIER, T. C. & NESTOR, E. W. (1976). Evidence for diverse types of large plasmids in tumour-inducing strains of *Agrobacterium. Journal of Bacteriology*, **126,** 157–65.
DAGLEY, S. (1975). A biochemical approach to some problems of environmental pollution. In *Essays in Biochemistry*, vol. II, ed. P. N. Campbell & W. N. Aldridge, pp. 81–138. London: Academic Press.
DALTON, H. & STIRLING, D. I. (1982). Co-metabolism. *Philosophical Transactions of the Royal Society of London B*, **297,** 481–96.
DATTA, N. (1979). Plasmid classification: incompatibility grouping. In *Plasmids of Medical, Environmental and Commercial Importance*, ed. K. N. Timmis & A. Puhler, pp. 3–12. Amsterdam: Elsevier North Holland.
DAUGHTON, C. G. & HSIEH, D. P. (1977). Parathion utilisation by bacterial symbionts in a chemostat. *Applied Environmental Microbiology*, **34,** 174–84.
DAVEY, R. B. & REANNEY, D. C. (1980). Extrachromosomal genetic elements and the adaptive evolution of bacteria. In *Evolutionary Biology*, ed. M. K. Hecht, W. C. Steere & B. Wallace, pp. 113–47. New York: Plenum.
DAVID, W. A. L. & GARDINER, B. O. C. (1967). The persistence of a granulosis virus of *Pieris brassicae* in soil and sand. *Journal of Invertebrate Pathology*, **9,** 342.
DELLAPORTA, S. L. & PESANO, R. L. (1981). Plasmid studies in crown gall tumorigenesis. In *Biology of the Rhizobiaceae*, ed. K. Giles & A. Atherly, pp. 83–104. New York: Academic Press.
DON, R. H. & PEMBERTON. J. M. (1981). Properties of six pesticide degradation plasmids isolated from *Alcaligenes paradoxus* and *Alcaligenes eutrophus. Journal of Bacteriology*, **145,** 681–5.
DOOLITTLE, W. F. & SAPIENZA, C. (1980). Selfish genes, the phenotype paradigm and genome evolution. *Nature, London*, **284,** 601–3.
DRAKE, J. W. (1974). The role of mutation in microbial evolution. In *Symposium of the Society for General Microbiology 24*, pp. 41–58. Cambridge University Press.
DUBOISE, S. M., MOORE, B. E., SORBER, C. A. & SAGIK, B. P. (1979). Viruses in soil systems. In *CRC Critical Reviews in Microbiology*, vol. 7, ed. H. D. Isenberg, pp. 245–85. CRC Press.
DURAND, G. (1966). Contribution a l'Etude de la Biologie du Sol Sur le Catabolisms des Acides Nucleiques et leurs Derives. Thesis, University of Toulouse.
ENSMINGER, L. E. & GIESEKING, J. E. (1942). Resistance of clay-adsorbed proteins to proteolytic hydrolysis. *Soil Science*, **53,** 205–9.
EPHRATI, E. (1968). Spontaneous transformation in *Bacillus subtilis. Genetical Research*, **11,** 83.
FALKOW, S. (1975). *Infectious Multiple Drug Resistance.* London: Pion Ltd.
FASMAN, G. D. (1981). In *CRC Handbook of Biochemistry and Molecular Biology*, third edition, *Nucleic Acids*, vol. 11, pp. 923. Cleveland, Ohio: CRC Press.
FERRY, J. C. & WOLFE, R. S. (1976). Anaerobic degradation of benzoate to methane by a microbial consortium. *Archive für Mikrobiologie*, **107,** 33–40.
FINBERG, E. L., RAMAGE, P. I. N. & TRUGGILL, P. W. (1980). The degradation of

n-alkylcycloalkanes by a mixed bacterial culture. *Journal of General Microbiology*, **121,** 507–11.
FOGG, G. E. (1966). The extracellular products of algae. *Oceanography and Marine Biology. Annual Review*, **4,** 195–212.
FOSTER, R. C. & ROVIRA, A. D. (1978). The ultrastructure of the rhizosphere of *Trifolium subterraneum L.* In *Microbial Ecology*, ed. M. Loutit & J. Miles pp. 278–90. Berlin: Springer-Verlag.
GAUSE, G. F. (1934). *The Struggle for Existence.* New York: Dover Publications.
GERBA, C. P. & SCHAIBERGER, G. E. (1975). The effects of particles on viruses survival in seawater. *Journal of Water Pollution and Control Federation*, **41,** 93.
GODWIN, D. & SLATER, J. H. (1979) The influence of growth environment on the stability of a drug resistance plasmid in *Escherichia coli* K12. *Journal of General Microbiology*, **111,** 201–10.
GORING, C. A. I. & BARTHOLOMEW, W. V. (1952). Adsorption of mononucleotides, nucleic acids and nucleoproteins by clays. *Soil Science*, **74,** 149.
GOWLAND, P. C. & SLATER, J. H. (1982). Transfer and stability of plasmids in *Escherichia coli* K12. *Microbial Ecology*, (in press).
GRAY, T. R. G., BAXBY, P., HILL, I. R. & GOODFELLOW, M. (1968). Direct observation of bacteria in soil. In *The Ecology of Soil Bacteria*, ed. T. R. G. Gray & D. Parkinson, pp. 171–97. Liverpool University Press.
GRAY, T. R. G. & WILLIAMS, S. T. (1971). *Soil Microorganisms.* Edinburgh: Oliver and Boyd.
GREAVES, M. P. & DARBYSHIRE, J. F. (1972). The ultrastructure of the mucilaginous layer on plant roots. *Soil Biology and Biochemistry*, **4,** 443–9.
GREAVES, M. P., DAVIS, H. A., MARSH, J. A. P. & WINGFIELD, G. I. (1976). Herbicides and soil microorganisms. *CRC Critical Reviews in Microbiology*, **5,** 1–38.
GREAVES, M. P. & WEBLEY, D. M. (1965). A study of the breakdown of organic phosphates by microorganisms from the root region of certain pasture grasses. *Journal of Applied Bacteriology*, **28,** 454–65.
GREAVES, M. P. & WILSON, M. J. (1970). The degradation of nucleic acids and montmorillonite-nucleic-acid complexes by soil microorganisms *Soil Biology and Biochemistry*, **2,** 257–68.
GRINDLEY, N. D., GRINDLEY, J. N. & ANDERSON, E. S. (1972). R-factor compatability groups. *Molecular and General Genetics*, **119,** 287–97.
GUNNER, H. B. & ZUCKERMAN, B. M. (1968). Degradation of 'Diazinon' by synergistic microbial action. *Nature, London*, **217,** 1183–4.
HAAS, D. & HOLLOWAY, B. W. (1978). Chromosome mobilisation by the R plasmid R68.45: A tool in *Pseudomonas* genetics. *Molecular and General Genetics*, **158,** 229–37.
HARTLEY, B. S. (1974). Enzyme families. In *Evolution in the Microbial World*, ed. M. J. Carlile & J. J. Skehel, pp. 151–82. Cambridge University Press.
HARTMANN, J., REINEKEL, W. & KNACKMUSS, H. J. (1979). Metabolism of 3-chloro-, 4-chloro and 3,5-dichlorobenzoate by a pseudomonad. *Applied Environmental Microbiology*, **37,** 421–8.
HEALY, J. B. & YOUNG, L. Y. (1978). Catechol and phenol degradation by a methanogenic population of bacteria. *Applied Environmental Microbiology*, **35,** 216–18.
HEALY, J. B. & YOUNG, L. Y. (1979). Methanogenic biodegradation of aromatic compounds. In *Microbial Degradation of Pollutants in Marine Environments*, ed. A. W. Bourquin and P. H. Pritchard, pp. 348–59. Gulf Breeze: USEPA.

HELLEBURST, J. A. (1974). Extracellular products. In *Algal Physiology and Biochemistry*, ed. W. D. P. Stewart, pp. 838–63. Oxford: Blackwell Scientific Publications.

HOLLOWAY, B. W. (1979). Plasmids that mobilise bacterial chromosome. *Plasmid*, **2,** 1–19.

HOLLOWAY, B. W. & KRISHNAPILLAI, V. (1975). Bacteriophages and bacteriocins. In *Genetics and Biochemistry of Pseudomonas*, ed. P. H. Clarke & M. H. Richmond, pp. 99–132. New York: John Wiley and Sons.

HOROWITZ, A., GUTWICK, D. & ROSENBERG, E. (1975). Sequential growth of bacteria on crude oil. *Applied Microbiology*, **30,** 10–19.

JACQUES, R. P. (1964). The persistence of a nuclear-polyhedrosis virus in soil. *Journal of Invertebrate Pathology*, **6,** 251.

JACQUES, R. P. (1967). The persistence of a nuclear polyhedrosis virus in the habitat of the host insect: II: polyhedra in soil. *Canadian Entomology*, **99,** 820–9.

JENSEN, H. L. (1957). Decomposition of chloro-substituted aliphatic acids by soil bacteria. *Canadian Journal of Microbiology*, **3,** 151–64.

JOHANIDES, V. & HRSAK, D. (1976). Changes in mixed bacterial culture during linear alkylbenzenesulfonate (LAS) degradation. In *Abstracts of Papers, Fifth International Fermentation Symposium, Berlin*, ed. H. Dellweg, p. 124.

JOLLEY, E. T. & JONES, A. K. (1977). The interaction between *Navicula muralis* Grunow and an associated species of *Flavobacterium. British Phycological Journal*, **12,** 315–28.

JONES, A. K. (1982). The interaction of algae and bacteria. In *Microbial Interactions and Communities*, vol. 1, ed. A. T. Bull & J. H. Slater, pp. 189–247. London: Academic Press.

JONES, D. & SNEATH, P. H. A. (1970). Genetic transfer and bacterial taxonomy. *Bacteriological Reviews*, **34,** 40–81.

JUNI, E. (1972). Interspecies transformation of Acinetobacter: genetic evidence for a ubiquitous genus. *Journal of Bacteriology*, **112,** 917–31.

KAISER, D. & DWORKIN, M. (1975). Gene transfer to a myxobacterium by *E. coli* phage Pl. *Science*, **187,** 653–4.

KELLOGG, S. T., CHATTERJEE, D. K. & CHAKRABARTY, A. M. (1981). Plasmid-assisted molecular breeding: new technique for enhanced biodegradation of persistent toxic chemicals. *Science*, **214,** 1133–5.

KELLY, W. J. & REANNEY, D. C. (1978). Virulent, temperate and defective bacteriophages from a common soil biota. In *Microbial Ecology*, ed. M. Loutit & J. Miles, pp. 113–20. Berlin: Springer-Verlag:

KELLY, W. J. & REANNEY, D. C. (1982). The ecology and transferability of genes encoding resistance to heavy metals and antibiotics in soil bacteria. *Soil Biology and Biochemistry*, (in press).

KILPI, S. (1980). Degradation of some phenoxy acid herbicides by mixed cultures of bacteria isolated from soil treated with 2-(2-methyl-4-chloro)-phenoxypropionic acid. *Microbiology Ecology*, **6,** 261–70.

KLECKNER, N. (1977). Translocatable elements in prokaryotes. *Cell*, **11,** 11–23.

KLECKNER, N. (1981). Transposable elements in prokaryotes. *Annual Review of Genetics*, **15,** 341–404.

KNACKMUSS, H. J. (1981). Degradation of halogenated and sulfonated hydrocarbons. In *Microbial Degradation of Xenobiotics and Recalcitrant Compounds*, ed. T. Leisinger, A. M. Cook, R. Hütter & J. Nüesch, pp. 189–212. London: Academic Press.

KOCH, A. L. (1971). The adaptive responses of *E. coli* to a feast and famine existence. *Advances in Microbial Physiology*, **6,** 147–217.

KRISHNAPILLAI, V. (1977). Superinfection inhibition by prophage B3 of some R plasmids in *P. aeruginosa. Genetical Research*, **29**, 47–54.
LACEY, R. W. (1975). Antibiotic resistance plasmids of *Staphylococcus aureus* and their clinical importance. *Bacteriological Reviews*, **39**, 1–25.
LINTON, J. D. & BUCKEE, J. C. (1977). Interactions in a methane-utilizing mixed bacterial culture in a chemostat. *Journal of General Microbiology*, **101**, 219–25.
LOVATT, D., SLATER, J. H. & BULL, A. T. (1978). The growth of a stable mixed culture on picolinic acid in continuous-flow culture. *Society of General Microbiology Quarterly*, **6**, 27–8.
LOWBURY, E. J. L., KIDSON, A., LILLY, H. A., AYLIFFE, G. A. J. & JONES, R. J. (1969). Sensitivity of *Pseudomonas aeruginosa* to antibiotics: Emergence of strains highly resistant to gentamycin. *Lancet*, **2**, 448–52.
LYNCH, D. L. & COTNOIR, L. J. (1956). The influence of clay minerals on the breakdown of certain organic substrates. *Proceedings of the Soil Science Society of America*, **20**, 367–70.
MAHLER, H. R. & CORDES, E. H. (1966). *Biological Chemistry*. New York: Harper and Row.
MARSHALL, K. C. (1976). *Interfaces in Microbial Ecology*. Cambridge, Massachusetts: Harvard University Press.
MATTHEW, M. & HEDGES, R. W. (1976). Analytical isoelectric focussing of R factor-determined β-lactamases: correlation with plasmid compatibility. *Journal of Bacteriology*, **125**, 713–18.
McLAREN, A. D. & ESTERMANN, E. F. (1957). Influence of pH on the activity of chymotrypsin at a liquid solid interface. *Archives of Biochemistry and Biophysics*, **68**, 157–60.
McLAREN, A. D. & SKUJINS, J. (1968). The physical environment of microorganisms in soil. In *The Ecology of Soil Bacteria*, ed. T. R. G. Gray & D. Parkinson. The University of Toronto Press.
McCLURE, G. W. (1970). Accelerated degradation of herbicides in soil by the application of microbial nutrient broths. *Contributions of Boyce Thompson Institute*, **24**, 235–44.
MELLING, J., ELLWOOD, D. C. & ROBINSON, A. (1977). Survival of R factor carrying *Escherichia coli* in mixed cultures in the chemostat. *FEMS Microbiology Letters*, **2**, 87–9.
MILLER, T. L. & JOHNSON, M. J. (1966). Utilisation of gas oil by a yeast culture. *Biotechnology and Bioengineering*, **8**, 567–80.
MIYAMOTO, Y. (1958). The nature of soil transmission in soil-borne plant viruses. *Virology*, **7**, 250.
MIYAMOTO, Y. (1959). Further evidence for the longevity of soil-borne plant viruses adsorbed on soil particles. *Virology*, **9**, 290.
MUNNECKE, D. M. (1978). Detoxification of pesticides using soluble or immobilized enzymes. *Process Biochemistry*, **13**, 14–16, 31.
MUNNECKE, D. M. (1981). The use of microbial enzymes for pesticide detoxification. In *Microbial Degradation of Xenobiotics and Recalcitrant Compounds*, ed. T. Leisinger, A. M. Cook, R. Hütter & J. Nüesch, pp. 251–69. London: Academic Press.
MUROOKA, Y. & HARADA, T. (1979). Expansion of the host range of coliphage Pl and gene transfer from enteric bacteria to other gram-negative bacteria. *Applied and Environmental Microbiology*, **38**, 754–7.
NASH, H. A. (1981). Integration and excision of bacteriophage λ: the mechanism of conservative site-specific recombination. *Annual Review of Genetics*, **15**, 143–67.

Neu, H. J. & Ballschmiter, K. (1977). Degradation of chlorinated aromatics – microbial degradation of polychlorinated biphenyls (PCBs): two diphenylols as PCB metabolites. *Chemosphere*, **6**, 419–23.

Newman, E. I. (1978). Root microorganisms: their significance in the ecosystem. *Biological Reviews*, **53**, 511–54.

Notani, N. K. & Setlow, J. K. (1974). Mechanisms of bacterial transformation and transfection. *Progress in Nucleic Acid Research and Molecular Biology*, **14**, 39–100.

Novick, R. P., Clowes, R. C., Cohen, S. N., Cutiss, R., Datta, N. & Falkow, S. (1976). Uniform nomenclature for bacterial plasmids. *Bacteriological Reviews*, **40**, 168–89.

Nugent, M. E. (1981). A conjugative 'plasmid' lacking autonomous replication. *Journal of General Microbiology*, **126**, 305–10.

Okamoto, K., Mudd, J. A., Mangan, J., Huang, W. M., Subbaiah, T. V. & Marmur, J. (1968). Properties of the defective phage of *Bacillus subtilis*. *Journal of Molecular Biology*, **34**, 413–28.

Old, K. M. & Nicholson. T. H. (1978). The root cortex as part of a microbial continuum. In *Microbial Ecology*, ed. M. Loutit & J. Miles, pp. 291–4. Berlin: Springer-Verlag.

Olsen, R. H. & Shipley, P. (1973). Host range and properties of the *Pseudomonas aeruginosa* R factor R1822. *Journal of Bacteriology*, **118**, 772–80.

Orgel, L. E. & Crick, F. H. C. (1980). Selfish DNA: the ultimate parasite. *Nature, London*, **284**, 604–7.

Osman, A., Bull, A. T. & Slater, J. H. (1976). Growth of mixed microbial populations on orcinol in continuous culture. In *Abstracts of Papers, Fifth International Fermentation Symposium, Berlin*, ed. H. Dellweg, p. 124.

Ostle, A. G. & Holt, J. G. (1979). Elution and inactivation of bacteriophages on soil and cation-exchange resin. *Applied and Environmental Microbiology*, **38**, 59–65.

Patterson, A. C. (1965). Bacteriocinogeny and lysogeny in the genus *Pseudomonas*. *Journal of General Microbiology*, **39**, 295–303.

Pawlowsky, U. & Howell, J. A. (1973a). Mixed culture biooxidation of phenol. I. Determination of kinetic parameters. *Biotechnology and Bioengineering*, **15**, 889–96.

Pawlowsky, U. & Howell, J. A. (1973b). Mixed culture biooxidation of phenol. II. Steady state experiments in continuous culture. *Biotechnology and Bioengineering*, **15**, 897–903.

Pawlowsky, U. & Howell, J. A. (1973c). Mixed culture biooxidation of phenol. III. Existence of multiple steady states in continuous culture with wall growth. *Biotechnology and Bioengineering*, **15**, 905–13.

Pickaver, A. H. (1976). The production of n-nitrosoiminodiacetate from nitrilotriacetate and nitrate by microorganisms growing in mixed culture. *Soil Biology and Biochemistry*, **8**, 13–17.

Porter, R. D., Shoemaker, N. B., Rampe, G. & Guild, W. R. (1979). Bacteriophage-associated gene transfer in Pneumococcus: transduction or pseudotransduction? *Journal of Bacteriology*, **137**, 556–67.

Puck, T. T. & Sagik, B. P. (1953). Virus and cell interaction with ion exchangers. *Journal of Experimental Medicine*, **93**, 65–88.

Reanney, D. C. (1976). Extrachromosomal elements as possible agents of adaptation and development. *Bacteriological Reviews*, **40**, 552–90.

Reanney, D. C. (1977). Genetic engineering as an adaptive strategy. In *Genetic Interaction and Gene Transfer, Brookhaven Symposia in Biology No. 29*, pp. 248–71.

Reanney, D. C. (1982). Evolution of RNA viruses. *Annual Reviews of Microbiology*, **37**, (in press).
Reanney, D. C. & Ackermann, H. W. (1982). The comparative biology and evolution of bacteriophages. *Advances in Virus Research*, **27**, (in press).
Reanney, D. C., Kelly, W. J. & Roberts, W. P. (1982). Genetic interactions among microbial communities. In *Microbial Interactions and Communities*, vol. 1, ed. A. T. Bull & J. H. Slater, pp. 287–322. London: Academic Press.
Reanney, D. C. & Marsh, S. C. N. (1973). The ecology of viruses attacking *Bacillus stearothermophilus* in soil. *Soil Biology and Biochemistry*, **5**, 399–408.
Reanney, D. C. & Tan, J. S. H. (1976). Interactions between bacteriophages and bacteria in soil. *Soil Biology and Biochemistry*, **8**, 145–50.
Reineke, W. & Knackmuss, H. J. (1978a). Chemical structure and biodegradability of halogenated aromatic compounds. Substituent effects on 1,2-dioxygenation of benzoic acid. *Biochimica et Biophysica Acta*, **542**, 412–23.
Reineke, W. & Knackmuss, H. J. (1978b). Chemical structure and biodegradability of halogenated aromatic compounds. Substituent effects on dehydrogenation of 3,5-cyclohexadiene-1,2-diol-1-carboxylic acid. *Biochimica et Biophysica Acta*, **542**, 424–9.
Reineke, W. & Knackmuss, H. J. (1979) Construction of haloaromatic utilizing bacteria. *Nature, London*, **277**, 385–6.
Roberts, W. P. (1982). Molecular basis of Crown Gall. *International Reviews in Cytology*, (in press).
Robinson, M. K. Bennett, P. M. & Richmond , M. H. (1977). Inhibition of TnA translocation by TnA. *Journal of Bacteriology*, **129**, 407–14.
Robinson, J. B. & Corke, C. T. (1959). Preliminary studies on the distribution of actinophages in soil. *Canadian Journal of Microbiology*, **5**, 479–84.
Robinson, J. B., Corke, C. T. & Jones, D. A. (1970). Survival of bacteriophage in soil 'sterilized' with gamma irradiation. *Soil Scientific Society of America Proceedings*, **34**, 703–4.
Rovira, A. D., Foster, R. C. & Martin, J. K. (1979). Note on terminology: origin, nature and nomenclature of the organic materials in the rhizosphere. In *The Soil-Root Interface*, ed. J. C. Harley & R. S. Russell, pp. 1–4. London: Academic Press.
Rovira, A. D., Newman, E. I., Bowen, H. J. & Campbell, R. (1974). Quantitative assessment of the rhizoplane microflora by direct microscopy. *Soil Biology and Biochemistry*, **6**, 211–16.
Saunders, J. R. (1979). Specificity of DNA uptake in bacterial transformation. *Nature, London*, **278**, 601–2.
Savage, D. C. (1977). Microbial ecology of the gastrointestinal tract. *Annual Reviews of Microbiology*, **31**, 107–33.
Schwarz, J. R., Walker, J. D. & Colwell, R. R. (1975). Deep-sea bacteria; growth at *in situ* temperature and pressure. *Canadian Journal of Microbiology*, **21**, 682–7.
Seaman, E., Tarmy, E. & Marmur, J. (1964). Inducible phages of *Bacillus subtilis*. *Biochemistry*, **3**, 607–13.
Senior, E., Bull, A. T. & Slater, J. H. (1976). Enzyme evolution in a microbial community growing on the herbicide Dalapon. *Nature, London*, **263**, 476–9.
Shendarov, B. A. (1971). The effect of temperature, pH and composition of the medium on transmission of drug resistance by the method of conjugation. *Zh. Mikrobiol. Epidemiol. Immunobiol.*, **48**, 94–8.
Sielicki, M., Focht, D. D. & Martin, J. P. (1978). Microbial transformation of styrene and (^{14}C)-system in soil and enrichment culture. *Applied and Environmental Microbiology*, **25**, 124–8.

SING, V. O. & SCHROTH, M. N. (1977). Bacteria-plant cell surface interactions: active immobilisation of saprophytic bacteria in plant leaves. *Science*, **197,** 759–61.

SINGER, B. & FRAENKEL-CONRAT, H. (1961). Effects of bentonite on infectivity and stability of TMV nucleic acid. *Virology*, **14,** 59.

SINGLETON, P. & ANSON, A. E. (1981). Conjugal transfer of R-plasmid Rl*drd-19* in *Escherichia coli* below 22°C. *Applied and Environmental Microbiology*, **42,** 789–91.

SISCO, K. L. & SMITH, H. O. (1979). Sequence-specific DNA uptake in *Haemophilus* transformation. *Proceedings of the National Academy of Sciences, USA*, **76,** 972–6.

SLATER, J. H. (1978). The role of microbial communities in the natural environment. In *The Oil Industry and Microbial Ecosystems*, ed. K. W. A. Chater & H. J. Somerville pp. 137–54. London: Heyden and Sons.

SLATER, J. H. (1981). Mixed cultures and microbial communities. In *Mixed Culture Fermentations*, ed. M. E. Bushell & J. H. Slater, pp. 1–24. London: Academic Press.

SLATER, J. H. & BULL, A. T. (1982). Environmental microbiology: biodegradation. *Philosophical Transactions of the Royal Society of London B*, **297,** 575–97.

SLATER, J. H. & LOVATT, D. (1982). Biodegradation and the significance of microbial communities. In *Biochemistry of Microbial Degradation*, ed. D. T. Gibson. New York: Marcel Dekker and Sons. (in press.)

SLATER, J. H. & SOMERVILLE, H. J. (1979). Microbial aspects of waste treatment with particular attention to the degradation of organic carbon compounds. In *Microbial Technology: Current Status and Future Prospects*, ed. A. T. Bull, C. R. Ratledge & D. C. Ellwood, pp. 221–61. Cambridge University Press.

SMITH, H. J. (1977). Mobilisation of non-conjugative tetracycline streptomycin spectinomycin and sulphonamide resistance determinants of *E. coli. Journal of General Microbiology*, **100,** 189–96.

SMITH, H. O., DANNER, D. B. & DEICH, R. A. (1981). Genetic transformation. *Annual Reviews of Biochemistry*, **50,** 41–68.

SONEA, S. & PANISSET, M. (1976). Pour une nouvelle bacteriologie. *Reviews of Canadian Biology*, **35,** 103–67.

STIRLING, L. A., WATKINSON, R. J. & HIGGINS, I. J. (1976). The microbial utilisation of cyclohexane. *Proceedings of the Society for General Microbiology*, **4,** 28.

STORCK, W. J. (1980). Pesticide profits belie mature market status. *Chemical and Engineering News*, **58(17),** 10–13.

SWIFT, M. J. (1982). Microbial succession during the decomposition of organic matter. In *Experimental Microbial Ecology*, ed. R. G. Burns & J. H. Slater, pp. 164–177. Oxford: Blackwell Scientific Publications.

SYKES, R. B. & RICHMOND, M. H. (1970). Intergeneric transfer of a β-lactamase gene between *Pseudomonas aeruginosa* and *Escherichia coli. Nature, London*, **226,** 952–4.

TAYLOR, D. E. & GRANT, R. B. (1976). Inhibition of bacteriophage lambda. T1 and T7 development by R plasmids of the H incompatibility group. *Antimicrobial Agents and Chemotherapy*, **10,** 762–4.

THUNG, T. H. & DIJKSTRA, J. (1958). Bindung van virusremstoffen aen leinmineralen tijdschr. *Plantezichten*, **64,** 411–15.

TORSVIK, V. L. & GØKSOYR, L. (1978). Determination of bacterial DNA in soil. *Soil Biology and Biochemistry*, **10,** 7–12.

TULP, M. T. M., SCHMITZ, R. & HUTZINGER, O. (1978). The bacterial metabolism of 4,4′-dichlorobiphenyl and its suppression by alternative carbon sources. *Chemosphere*, **7,** 103–8.

WATSON, G. K. & JONES, N. (1979). The microbial degradation of nonionic surfactants. *Society for General Microbiology Quarterly*, **6,** 78.

WEISS, W. & RHEINHEIMER, G. (1978). Scanning electron microscopy and epifluorescence investigation of bacterial colonisation of marine sand sediments. *Microbial Ecology*, **4,** 175–88.

WILLIAMS, P. A. (1981). Genetics of biodegradation. In *Microbial Degradation of Xenobiotics and Recalcitrant Compounds*, ed. T. Leisinger, A. M. Cook, R. Hütter & J. Nüesch, pp. 97–107. London: Academic Press.

WILLIAMS, P. A. (1982). Genetic interactions between mixed microbial populations. *Philosophical Transactions of the Royal Society of London, B*, **297,** 631–9.

WILLIAMS, P. A. & MURRAY, K. (1974). Metabolism of benzoate and the methylbenzoates by *Pseudomonas putida (arvilla)* mt-2: evidence for the existence of a TOL plasmid. *Journal of Bacteriology*, **120,** 416–23.

WILSON, G. A. & YOUNG, F. E. (1972). Intergenotic transformation of the *Bacillus subtilis* genospecies. *Journal of Bacteriology*, **111,** 705–16.

WODZINSKI, R. S. & JOHNSON, M. J. (1968). Yields of bacterial cells from hydrocarbons. *Applied Microbiology*, **16,** 1880–91.

WOUTERS, J. T. M., ROPS, C. & VAN ANDEL, J. G. (1978). R-plasmid persistence in *Escherichia coli* under various environmental conditions. *Proceedings of the Society for General Microbiology*, **5,** 61.

YOSHIKA, H. & NAKATANI, R. (1976). The role of R plasmids in temperature sensitive conjugation. *Japanese Journal of Bacteriology*, **31,** 192.

METABOLIC COMMUNICATION BETWEEN BIODEGRADATIVE POPULATIONS IN NATURE

J. GREG ZEIKUS

Department of Bacteriology, University of Wisconsin, Madison, Wisconsin 53706, USA

INTRODUCTION

By and large, the challenge for microorganisms to decompose or transform various chemical elements in the environment is in itself a result of the evolution of biological light-driven carbon synthesis mechanisms on earth. The majority of biochemical transformation reactions that occur in the biosphere involve the natural cycling of carbon by life processes. Hence, the biodegradation of organic molecules into carbon dioxide is coupled to the reductive biosynthesis of organic carbon from carbon dioxide. Because organisms have very diverse mineral requirements for nutritional needs, other major and minor chemical elements greatly influence carbon cycling in nature. The microbial ecology of the carbon cycle is very complex because the component biochemical transformation reactions are integrative in nature but they are performed by a metabolically diverse population.

The dynamics of microbial carbon decomposition in aerobic and anaerobic niches of the biosphere are dramatically different. In aerobic environments, carbon which is naturally synthesized is generally completely decomposed. For example, the major chemical form of organic carbon in the biosphere, lignocellulose, is completely decomposed in aerobic soils. In anaerobic ecosystems, natural organic carbon (e.g., lignin) which enters the environment may be recalcitrant to biodegradation (Zeikus *et al.*, 1982*b*) or transformed into recalcitrant carbon (e.g., methane) which accumulates as a result of anoxic microbial processes (Zeikus, 1977).

Complete mineralization of polymeric organic carbon by microorganisms in aerobic ecosystems can be completely achieved by the activity of populations of single species. For example, both the carbohydrate and lignin polymers of wood are completely decomposed by species of white-rot fungi (Kirk, 1971). However, complete biodegradation of major photosynthetic polymers into water, methane and/or carbon dioxide requires mixed populations of anaerobic bacteria. For example, a hydrolytic species and a methanogenic species is required to mineralize pectin to carbon dioxide, methane and water in anoxic sediments (Schink & Zeikus, 1982*a*). The metabolic diversity of anaerobic bacteria far exceeds that of aerobic organisms largely because of their ability to couple dehydrogenative carbon transformation reactions to hydrogenation of multiple electron acceptors (i.e., carbon dioxide, diverse organic compounds, inorganic sulphur compounds and inorganic nitrogen compounds) in lieu of oxygen. Thus, the same carbohydrate polymers can be fermented by acidogenic or solventogenic species and the same one- and two-carbon compounds fermented by methanogenic, acidogenic or sulphidogenic species (Zeikus, 1980*a*, 1982*a*). Consequently, the anaerobic environment offers the unique opportunity to study the ecophysiology of microbial carbon mineralization and to define the metabolic features that account for both the composition and activity of the biodegradative food chain, and the behaviour and dominance of individual species in the environment.

In the absence of excess exogenous inorganic electron acceptors (i.e., oxygen, NO_3^{2-}, SO_4^{2-}, etc.) organic matter is decomposed by a complex microbial food chain into methane, carbon dioxide and water as final end products (Zeikus, 1977; Zehnder, 1978). At least four different trophic groups, which can be identified by the specific pathways utilized for controlling catabolic carbon and electron flow, are important to active decomposition of organic matter in anaerobic ecosystems (Zeikus, 1980*a*, 1982*b*). These four groups include: the hydrolytic bacteria which ferment complex multi-carbon compounds (e.g., polysaccharides, lipids, proteins) into a variety of end products including acids (e.g., lactate, acetate, propionate), neutral compounds (e.g., ethanol, methanol) and hydrogen and carbon dioxide; the hydrogen-producing acetogens which ferment alcohols larger than methanol and organic acids larger than acetate into hydrogen and acetate; the homoacetogens which ferment multi-carbon compounds, hydrogen and carbon dioxide or one-carbon compounds into organic acids via acetyl-CoA as an intermediate; and the methanogens which ferment hydrogen and carbon dioxide, one-carbon compounds and acetate into methane and carbon dioxide (Fig. 1). As a whole, the function of this anaerobic food chain can be likened to a metabolic symphony of carbon and electron flow reactions. The direction of carbon and electron flow through the

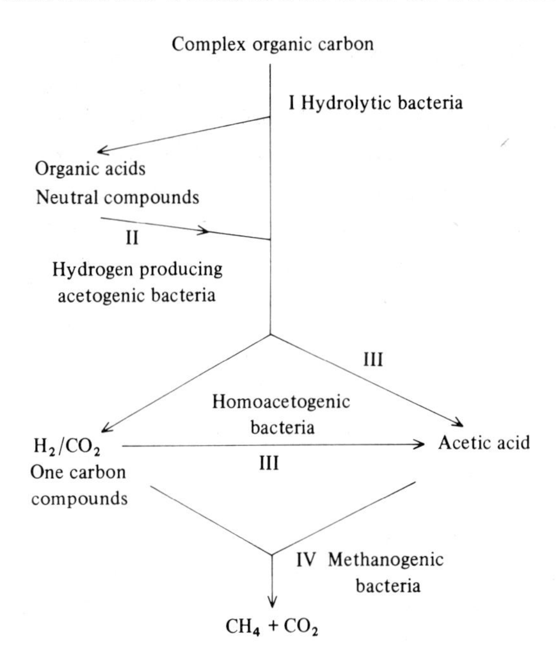

Fig. 1. Distinction of the biodegradative microbial food chain into four metabolic groups (after Zeikus, 1980*b*).

food chain depends on the chemical composition of the initial organic electron donors and the exogenous electron acceptors present in the environment. All four trophic groups display active metabolic communication during anaerobic digestion of organic matter into methane and carbon dioxide. Notably, the microbial species that act as conductors of the metabolic symphony are bacteria which catabolize the terminal intermediates (i.e., hydrogen and carbon dioxide, acetate and one-carbon compounds).

The purpose of this chapter is to explore metabolic communication between biodegradative populations in anaerobic ecosystems; that is, to discover how the biochemical transformation reactions of individual anaerobic species share in the complete mineralization of organic matter. With this goal is mind, the same scientific approach is employed to compare the ecophysiology of anaerobic bacteria in several physicochemically distinct aquatic environments in order to understand how environmental parameters dynamically alter microbial populations, metabolic communication and biochemical cycling of organic matter.

Emphasis is placed on defining the metabolic basis for species interactions and community structure during anoxic biodegradation of environmental primary production (i.e., photosynthetate). The influence of environmental physics, and sulphur and metal elements on *in vivo* carbon and electron flow during the mineralization of organic matter is described. The *in vitro* metabolism of prevalent anaerobes is analysed in mono- and co-culture fermentations to explain both synergistic and antagonistic microbial behaviour in the environment. The ability of individual species to survive and proliferate in their niche is illustrated in terms of inherent biochemical transformation reactions and macromolecular properties. Environmental parameters that limit metabolic communication between populations are hypothesized to generate recalcitrant matter and to account for the diagenesis of fossil fuels.

CARBON TRANSFORMATIONS IN FRESHWATER SEDIMENTS OF LAKE MENDOTA, WISCONSIN

Environmental parameters

Lake Mendota is a moderately eutrophic hard-water lake of glacial origin in southern Wisconsin, USA. It covers about 3 900 ha, has an average depth of about 12 m and a maximum depth of 24 m, and is anaerobic below 10.5 m during summer stratification. It is perhaps the best-studied lake in the world and methanogenic activity in its sediments was first reported by Allgeier *et al.* (1932).

The *in situ* rate and/or amount of methanogenesis in the deep water, surface sediments is limited by a variety of environmental parameters that dynamically change during the annual variation in climatic seasons. These limiting parameters include: the amount of available organic carbon electron donors (Schink & Zeikus, 1982*a*); temperature (Zeikus & Winfrey, 1976); exogenous electron acceptor availability (i.e., SO_4^{2-}, NO_3^{2-}, oxygen) (Winfrey & Zeikus, 1977, 1979); hydrogen partial pressure (Winfrey *et al.*, 1977; Winfrey & Zeikus, 1977, 1979) and acetate concentration (Winfrey & Zeikus, 1977, 1979). In the deep portions of Lake Mendota the non-limnic contribution to the total photosynthate decomposed is negligible and cyanobacteria and green algae make up the greater part of sedimented primary production (Fallon & Brock, 1979).

The major climatic influence on the organic carbon decomposition process as a whole occurs when the lake stratifies and becomes anoxic below the chemocline at about 15 m. This establishes the

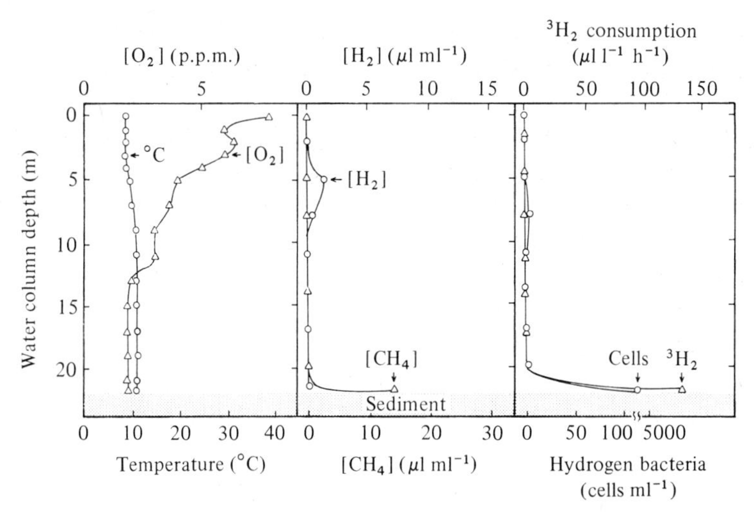

Fig. 2. Relation of Lake Mendota water column depth to *in situ* temperature, oxygen, hydrogen, methane and aerobic bacterial hydrogen consuming activity during summer lake stratification (after Schink & Zeikus, 1982*b*).

challenge for microbial decomposition of sedimented matter by anaerobic processes alone. Fig. 2 illustrates the relation of water depth to dissolved oxygen, methane and hydrogen and aerobic hydrogen consuming bacterial activity to water depth at summer stratification. Dissolved hydrogen, an energy source for both anaerobic or aerobic bacteria, is not detected in the anoxic waters or sediment but is found in the surface water layers in association with cyanobacteria that generate hydrogen in association with nitrogen fixation (Burris & Peterson, 1978). Notably, knallgas bacteria which consume hydrogen, carbon dioxide and oxygen are present in aerobic waters or the sediment surface (Schink & Zeikus, 1982*b*) but not at the chemocline where methane-oxidizing bacteria are localized (Harrits & Hanson, 1980). Hence, aerobic hydrogen-consuming bacteria appear associated with viable cyanobacteria in the aerobic surface waters and sedimented algae in the anoxic mud. Free hydrogen does not escape the anoxic sediment and is not detectable because fermentogenic hydrogen is rapidly consumed by the anaerobic bacterial population (Schink & Zeikus, 1982*b*; Winfrey *et al.*, 1977; Winfrey & Zeikus, 1977, 1979; Ingvorsen *et al.*, 1981).

Carbon and electron flow

The cell wall and cytoplasmic components of green algae and cyanobacteria constitute the majority of organic matter which serves

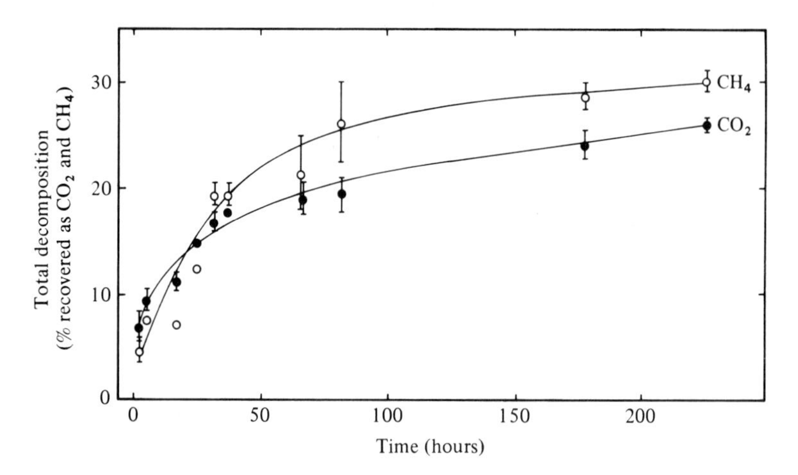

Fig. 3. Biodegradation of [U-^{14}C]pectin to $^{14}CH_4$ and $^{14}CO_2$ in anoxic Lake Mendota sediment at 15°C temperature (after Schink & Zeikus, 1982*a*).

as the initial electron and carbon sources for anoxic biodegradation in the sediments. The *in situ* turnover time for biodegradation of photosynthetic biopolymers is much slower than soluble, transportable substrates or immediate carbon precursors for methanogenic bacteria. For example, the turnover time for biodegradation of ^{14}C-pectin to $^{14}CH_4$ and $^{14}CO_2$ was 100 h (Fig. 3), while values for ^{14}C-glucose and ^{14}C-acetate were 12 h and 0.22 h (Schink & Zeikus, 1982*a*; Winfrey & Zeikus, 1979). The decomposition of cellulose was slower than pectin in Lake Mendota sediment (Nelson & Zeikus, 1974). Biodegradation of the significant, heteropolymeric glycopeptide fraction of the cyanobacterial wall may be a rate-limiting step in methanogenesis but this hypothesis remains to be proven.

Not all the reduced carbon that enters or is produced in the anoxic sediment is biodegraded. Methane which is rapidly generated from anaerobic digestion is not significantly transformed to carbon dioxide in the presence or absence of excess sulphate (Winfrey & Zeikus, 1977, 1979). However, methane is transformed to carbon dioxide when it reaches the oxygenated water and is no longer detectable (e.g., above the chemocline). Recently, the anoxic transformation of $^{14}CH_4$ to $^{14}CO_2$ reported by both methane-producing (Zehnder & Brock, 1979) and sulphate-reducing bacteria (Panganiban *et al.*, 1979) has been explained as phenomena linked to active exchange reactions associated with C_1 metabolism of anaerobes and not as a result of net depletion of methane (Zeikus,

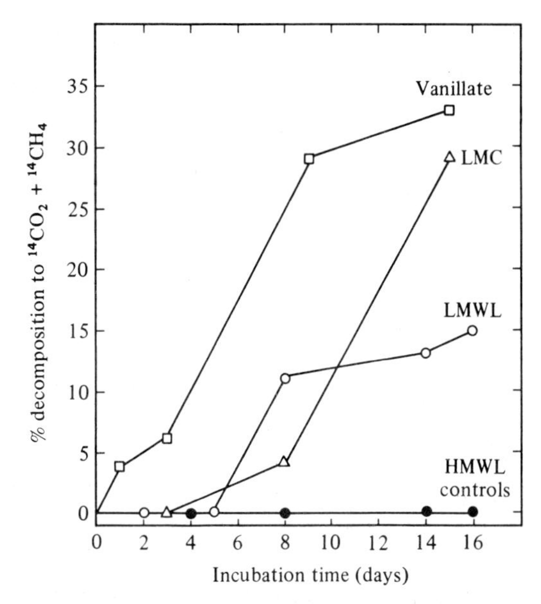

Fig. 4. Biodegradation of lignin derived aromatic substrates in anoxic sediments of Lake Mendota. Experimental conditions: anaerobic pressure tubes contained 10 ml sediment, an initial nitrogen head space, and 1 to 10×10^5 DPM ^{14}C-aromatic substrate. Controls contained 1.0% (v/v) formalin to inhibit biological activity. Abbreviations: LMC, guiacyl-glycerol-B-(O-methoxyphenyl) ether; LMWL, low molecular weight alkaline-heat treated lignin; and HMWL, high molecular weight alkaline-heat treated lignin (after Zeikus *et al.*, 1982*b*).

1982*a*). It is worth noting that both methanogens and sulphate-reducing bacteria produce natural gas and the ability of either trophic group to produce and oxidize methane does not appear significant to metabolic communication within the anaerobic ecosystem.

Lignin and related high molecular weight aromatic polymers, formed by free radical chemical condensation reactions associated with plant photosynthesis, are not significantly decomposed in anaerobic environments (Zeikus, 1981). High molecular weight ^{14}C-synthetic lignins were not degraded in anoxic Lake Mendota sediment to $^{14}CO_2$ or $^{14}CH_4$ in long term (i.e., months) incubations (Hackett *et al.*, 1977). The molecular basis for the biodegradative recalcitrance of lignin in anaerobic environments appears related to the requirement for active oxygen species during the microbial mediated depolymerization of lignin (Zeikus *et al.*, 1982*b*). Fig. 4 illustrates that high molecular weight lignin was not decomposed in anoxic lake sediment, but during the same incubation period, lignin derived soluble products including vanillate, the lignin model guiacyl glycerol-β-(O-methoxyphenyl) ether, and low molecular weight (~300) alkaline heat treated lignin were significantly biodegraded to methane and carbon dioxide. Hence, the inability of anaerobic microbes to communicate metabolically with polymeric lignins as the carbon and energy source appears inherent to the low redox potential and reducing conditions of the environment.

Sulphate as an exogenous electron acceptor in anoxic sedimentary environments plays a pivitol role in channelling electrons generated from biomass decomposition away from methanogenesis. MacGregor & Keeney (1973) demonstrated that 150 μg sulphate ml^{-1} dramatically inhibited methanogenesis in Lake Mendota sediments. This information was used by Winfrey & Zeikus (1977) to test the hypothesis that the addition of excess sulphate to freshwater lake sediment altered the flow of carbon and electrons from methanogens to sulphidogens. We showed that hydrogen consumption and acetate consumption continued in the presence of excess sulphate but that carbon flow was altered and the ^{14}C-methane produced from $^{14}CO_3^-$ or 2-^{14}C-acetate was decreased, while $^{14}CO_2$ production from 2-^{14}C-acetate increased. The major conclusions of these studies were novel at the time in that they indicated methanogens and sulphate-reducers competed for electron donors (i.e., acetate and hydrogen). These findings appear to have been misinterpreted by other investigators to indicate that sulphate reduction and methanogenesis are mutually exclusive in aquatic sediments (Smith & Klug, 1981; Kosuir & Warford, 1979; Oremland *et al.*, 1982; K. I. Ingvorsen, personal communication). However, Winfrey & Zeikus (1977) demonstrated that in the presence of excess hydrogen or acetate methanogenesis continued even with excess sulphate added to lake sediment (Fig. 5). Our results established the understanding that the amount and rate of *in situ* methanogenesis depends upon the amount of electron acceptors and donors available to the anaerobic digestion process. Ingvorsen *et al.* (1982) established that under *in situ* conditions of sulphate concentration both methanogenesis and sulphate reduction were very active in Lake Mendota sediment, and both processes accounted for nearly the same amount of terminal carbon and electron removal.

The physiological explanation which explains why sulphate-reducers and methanogens compete for hydrogen and hence display antagonistic behaviour remains to be established in laboratory cultures. However, this competitive metabolic communication is consistent with a kinetic model that predicts enhanced hydrogen

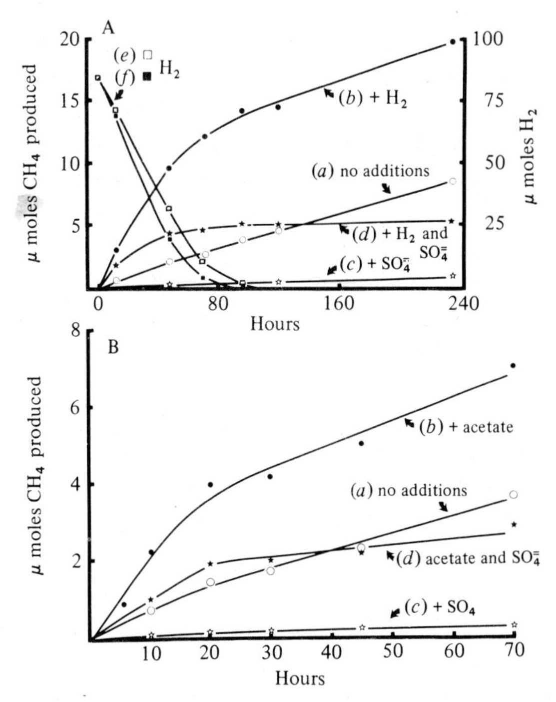

Fig. 5. Effect of hydrogen and acetate additions on sulfate inhibition of methanogenesis in Lake Mendota sediments. (A) Addition of hydrogen. Key: CH_4 produced with no additions (*a*); addition of 15% hydrogen (*b*); addition of 10 mM SO_4^{2-} (*c*); addition of 10 mM SO_4^{2-} and 15% hydrogen (*d*); hydrogen depletion from sediment without additions (*e*); hydrogen depletion from sediment containing 10 mM SO_4^{2-} (*f*). (B) Addition of 1.0 mM acetate. Key: methane from sediment with no additions (*a*): with 1.0 mM acetate added (*b*); 10 mM SO_4^{2-} added (*c*); 10 mM SO_4^{2-} and 1.0 mM acetate added (*d*). (After Winfrey & Zeikus, 1977.)

consumption rates by sulphate-reducers at excess sulphate concentrations; whereas, at low sulphate concentrations in fresh-water sediments both methanogens and sulphate reducers co-exist on hydrogen as electron donor. Methanogens were also shown to interact synergistically with sulphate-reducers in aquatic sediments (Cappenberg, 1974*a*, *b*; Cappenberg & Prins, 1975). The primary reason for non-competitive interactions between sulphate-reducers and methanogens is that sulphate-reducing bacteria function as hydrogen-producing acetogens at low sulphate concentrations when their electron flow is directed by methanogens (Bryant *et al.*, 1977). It is evident in nature that both competitive and synergistic metabolic behaviour is displayed by methanogens and sulphate-reducers such that the *in situ* population of these bacteria and the amount of *in situ* acetate and hydrogen/carbon dioxide (i.e., major terminal electron and carbon donors) transformed to methane ultimately depends on the environmental chemistry (i.e., available electron donors and acceptors, pH, salinity, etc.).

Biomass in the lake sediment is decomposed via the hydrolytic and acetogenic populations into immediate carbon precursors utilized as substrates for the methanogens. Radiotracer studies established that less than 41% of the methane produced in the surface sediment of the deep portions of the lake was formed via hydrogen reduction of ^{14}C-carbon dioxide while greater than 42% came from 2-^{14}C-acetate (Winfrey & Zeikus, 1979). Acetate and hydrogen/carbon dioxide are simultaneously transformed to methane in the sediment. The amount of methane derived from other methane precursors was not established. However, all known immediate precursors are actively transformed to methane in sediments including: formate and methanol (Winfrey *et al.*, 1977), methylmercaptan (Zinder & Brock, 1978) and methylamine (T. Phelps, unpublished observations).

The surface sediments of Lake Mendota are actively sulphidogenic. The daily $^{35}SO_4$ reduction rate varied from 50 to 600 mmol sulphate reduced (ml sediment)$^{-1}$ depending on the *in situ* temperature and sampling date. Rates of bacterial sulphate reduction were not sulphate limited at sulphate concentrations greater than 0.1 mM. Notably, the *in vitro* sulphate reduction rates in sediments were stimulated by the addition of hydrogen or alcohols but not acetate or lactate as electron donors. The seasonally most active rates of sulphate reduction in Lake Mendota surface sediment are in the same order of magnitude as those reported in sedimentary marine environments, a productive salt marsh and bacterial mats of a saline tropical lake (Ingvorsen *et al.*, 1981). A mass balance model based on detailed chemical analysis of sulphur and carbon cycle data for Lake Mendota (Ingvorsen & Brock, 1982) indicated that equivalent amounts of terminal carbon and electron donors were transformed by methanogens and sulphate reduction, and that during summer stratification electron flow via sulphate reduction in the hypolimnion constituted at least 25% of the electron flow via sediment methanogenesis.

Microbial dynamics

Temperature limits the rate of *in situ* anaerobic digestion in Lake Mendota sediment as *in situ* temperatures never approximates the optimal, *in vitro* rates for hydrolytic bacteria (Schink & Zeikus,

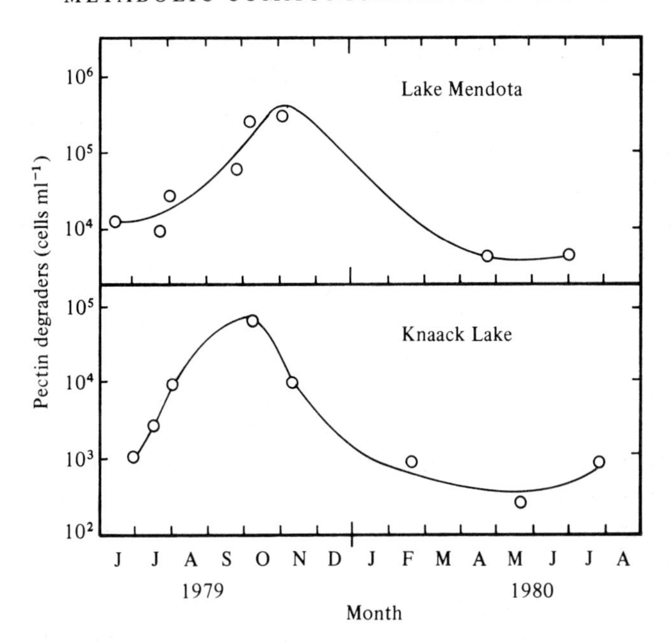

Fig. 6. Seasonal variation in total number of pectinolytic anaerobes in lake sediments (after Schink & Zeikus, 1982a).

1982*a*), sulphate-reducing bacteria (Ingvorsen *et al.*, 1981) or methanogenic bacteria (Zeikus & Winfrey, 1976). The optimal growth temperature of the prevalent methanogens (Zeikus & Winfrey, 1976; Weimer & Zeikus, 1978) or pectinolytic bacteria (Schink & Zeikus, 1982*a*) is at ⩾30 °C. The number of methanogens generally increased from late winter months to late summer months in response to increased temperature. However, the number of sediment pectinolytic anaerobes increased from minimal values in spring to maximal in fall (Fig. 6). Thus, the population dynamics of hydrolytic bacteria corresponds to the maximum input of initial electron and carbon donors (i.e., pectinous substrates) sedimented as dead algal blooms (Schink & Zeikus, 1982*a*).

As a natural model compound, pectin is a useful substrate for microbial decomposition analysis of primary biomass (i.e., photosynthate) in anoxic aquatic ecosystems. Pectin is a polymer of galacturonic acid linked by 1,4-α bonds and methoxylated to a varying extent. It is found naturally in both plant leaves and algal cell walls that are commonly deposited into freshwater sediments. Pectin can be abundant in cyanobacterial sheaths (Desikachary, 1959; Wolk, 1973) and green algal walls (Gooday, 1971; Green &

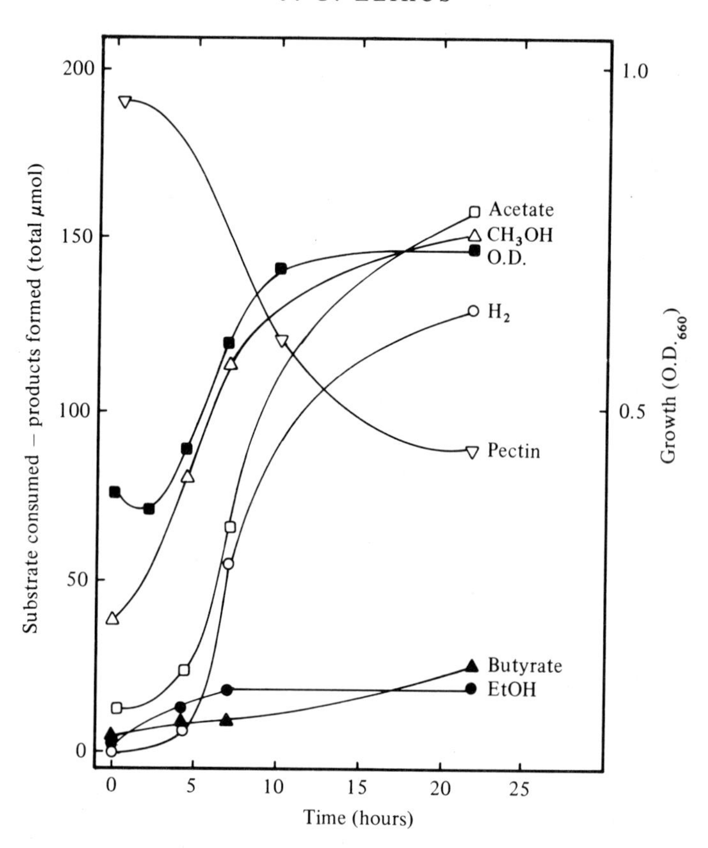

Fig. 7. Pectin fermentation time course for pure-cultures of *C. butyricum* (after Schink & Zeikus, 1982*a*).

Jennings, 1967; Prescott, 1968; Sikes, 1978). Notably, many obligately anaerobic, aerobic or facultative pectinolytic bacterial species produce methanol as a major end product of pectin fermentation but do not consume methanol as a carbon and electron source (Schink & Zeikus, 1980). The biochemical transformation that accounts for this biodegradative event in nature is pectin methylesterase activity. Hence, pectin metabolism, in part, establishes the niche for methylotrophs in nature. In anaerobic ecosystems, pectin biodegradation enables synergistic metabolic communication between hydrolytic and methylotrophic bacterial species.

Clostridium butyricum and *Methanosarcina barkeri* were isolated from Lake Mendota sediment and identified as the prevalent

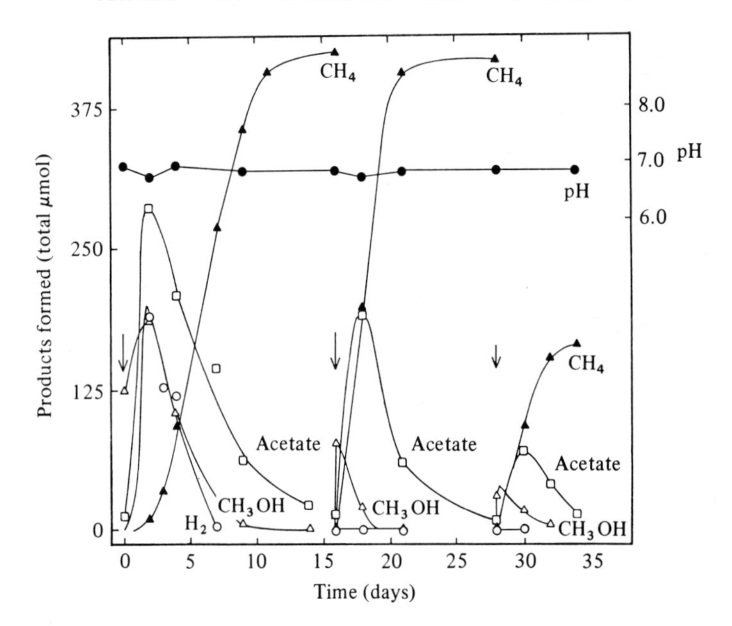

Fig. 8. Pectin fermentation time course for mixed-cultures comprised of *C. butyricum* and *M. barkeri*. The amount of pectin indicated by arrows was fed to cultures at 3 intervals. (After Schink & Zeikus, 1982*a*.)

pectinolytic bacterium and methylotrophic methanogen, respectively (Schink & Zeikus, 1982*a*). A typical batch fermentation course for pectin degradation by *C. butyricum* in pure culture is shown in Fig. 7. Notably, pure culture growth ceases in the presence of excess pectin because of the toxic low pH and not by inhibition from hydrogen or carbon end products. In mixed culture with *M. barkeri,* pectin is completely degraded by *C. butyricum* with only hydrogen, methanol and acetate being formed as intermediary metabolites. *M. barkeri* displayed a mixotrophic metabolism and simultaneously transformed acetate, methanol and hydrogen/carbon dioxide into methane. During the three experimental feedings employed the pH remained constant and neutral (Fig. 8). At the termination of the experiment the following pectin decomposition stoichiometry was achieved: 160 μmol pectin $\rightarrow$ 475 μmol CH_4 + 480 μmol CO_2 + 7.5 mg cells. This was to be expected as the balanced equation for complete anaerobic biodegradation of pectin is: $C_{6.7}H_{12}O_6 + 0.67\ H_2O \rightarrow 3.32\ CO_2 + 3.32\ CH_4$ (Schink & Zeikus, 1982*a*).

This methanogenic fermentation of pectin established that an anaerobic food chain for complete biopolymer mineralization can

Function performed	Metabolic reaction	Process significance
I. Proton regulation	$CH_3COO^- + H^+ \rightarrow CH_4 + CO_2$	Removes a toxic metaboli[te]
II. Electron regulation	$4H_2 + CO_2 \rightarrow CH_4 + 2H_2O$	Increases thermodynamic efficiency and range of substrates metabolized
III. Nutrient regulation	CH_3COOH, H_2/CO_2, CH_3OH $\rightarrow$ Vitamins amino acids	Stimulates growth of heterotrophs

Fig. 9. Role of methanogens as bioregulators of anaerobic digestion (after Zeikus, 1980*b*).

be comprised of just two species provided that they display synergistic metabolic communication. Also, this mixed culture serves to illustrate the metabolic behaviour of methanogens in anoxic ecosystems. Namely, the performance of three important bioregulatory roles distinguishes the methanogens as the conductors of the metabolic symphony of carbon and electron flow in anaerobic digestion (Fig. 9). As a consequence of their carbon transformation reactions, methanogens influence the decomposition of organic matter by other anaerobes by removing toxic metabolites, directing electron flow to limited reduced end products, enhancing growth rates and yields, and supplying essential growth factors (Zeikus, 1977, 1980*b*). The most important bioregulatory function of the methanogen during pectin fermentation was hydrogen ion consumption, for the cessation of growth of *C. butyricum* in pure culture was due to accumulation of hydrogen ions. In mixed culture, the intraspecies electron flow of *C. butyricum* was directed by the interspecies control elicited by *M. barkeri* to hydrogen in lieu of ethanol or butyrate. The alteration of electron flow when coupled to methanogenesis enabled *C. butyricum* to generate more energy in mixed culture than pure culture from an equivalent amount of substrate metabolized. Nutrient excretion by the methanogens is suggested because the mixed culture grows readily on pectin in defined medium but *C. butyricum* grows poorly in pure culture unless yeast extract is added.

CARBON TRANSFORMATIONS IN THERMAL BACTERIAL MATS OF OCTOPUS SPRING, WYOMING

Environmental parameters

Octopus Spring is an alkaline (pH 8.3), boiling volcanic pool located in Yellowstone National Park, USA. Notably, a bacterial mat ecosystem develops in Octopus Spring where volcanic source water provides constant thermal temperatures and mineral nutrients that enable the establishment of a dense prokaryotic population at temperatures below 70 °C. This thermophilic ecosystem appears to be very stable and habitats like this may have existed since the Cambrian Age. The Octopus Spring mat is an interesting environment for microbial ecologists to model a thermophilic carbon cycle because organic carbon is synthesized and decomposed by obligately thermophilic microorganisms. The chemical, thermal and microbial characteristics of the primary production portion of the bacterial mat ecosystem have been well characterized (Brock, 1978; Doemel & Brock, 1977).

The organisms responsible for primary production in this hot spring mat are *Synechococcus*, a cyanobacterium, and *Chloroflexus*, a green photosynthetic bacterium. The thickness of the mat is constant and is maintained by both limited light penetration and microbial decomposition. Biomass is apparently completely decomposed by microorganisms alone because the extreme temperature limits higher life forms (Brock, 1978). Biodegradation of primary production in this ecosystem has been examined in detail at the 60 to 65 °C portions of the mat (Zeikus, 1979; Zeikus *et al.*, 1980).

Carbon and electron flow

The relation of incubation temperature to gas formation during anaerobic decomposition of the algal-bacterial mat is shown in Fig. 10. Both methane and hydrogen were formed via microbial decomposition and these gases were not detected in controls that contained formaldehyde to inhibit biological activity. The temperature optimum observed for methanogenesis correlated with *in situ* temperature (i.e., about 65 °C). The observed temperature optimum for hydrogen formation (i.e., 80 °C) was probably a result of microbial hydrogen utilization at temperatures below 80 °C and inhibition of hydrogen consumption at 80 °C. The association of high partial pressures of hydrogen during active organic decomposi-

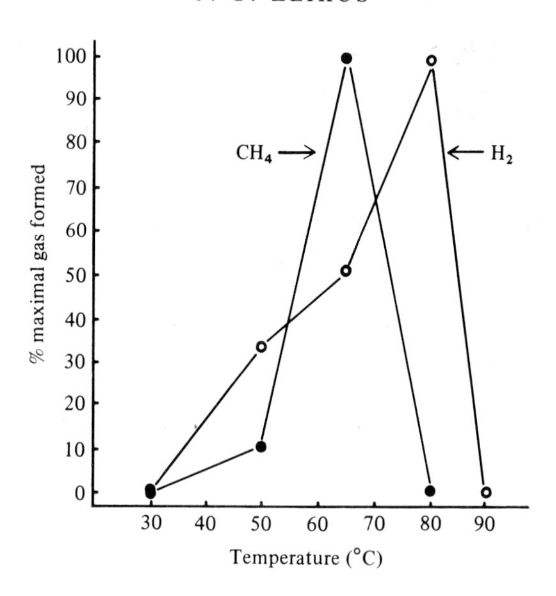

Fig. 10. Relation of *in vitro* incubation temperature to hydrogen and methane formation during anaerobic digestion of the Octopus Spring algal-bacteria mat (after Zeikus *et al.*, 1980).

tion is not characteristic of anaerobic digestion processes in neutral, mesophilic environments. It appears that methanogen metabolism is nutrient limited, perhaps by an essential trace mineral(s), in this organic rich environment because free hydrogen accumulates (Zeikus *et al.*, 1980).

Radiotracer studies established that organic matter was actively decomposed to carbon dioxide and methane when mat slurries were incubated at *in situ* temperature. Both ^{14}C-glucose and ^{14}C-carbon dioxide were transformed into ^{14}C-methane (Zeikus *et al.*, 1980), and $^{35}SO_4^{2-}$ was readily reduced to ^{35}S-sulphide (Zeikus *et al.*, 1982a). The rate of ^{14}C-pectin decomposition was significantly higher in this bacterial mat ecosystem than in surface sediments of Lake Mendota (B. Schink, unpublished observations). Notably, ^{14}C-acetate was not significantly transformed to methane in this environment (Zeikus *et al.*, 1980). This feature is also unique because the majority of carbon and electrons which flow to methane in neutral, mesophilic, freshwater environments is derived from this organic precursor. The phototrophs in the algal bacterial mat actively incorporate acetate (Sandbeck & Ward, 1981) which makes any methanogenesis from this precursor insignificant to total carbon mineralization. Furthermore, Sandbeck & Ward (1981) reported that approximately 80% of the methane was derived from carbon dioxide as precursor in the Octopus Spring ecosystem. Thus, the close spacial position of microbial phototrophs with hydrolytic bacteria enables direct metabolic communication between primary producers and decomposers in this unique thermal ecosystem.

Microbial dynamics

In thermal environments the carbon cycle is usually characterized as having short food chains and a limited species composition (Brock, 1978). These features greatly facilitate ecological studies on productivity, trophodynamics, population fluctuations and species interactions (Brock, 1970). Little is known about the diversity and ecology of obligately anaerobic thermophiles. Hence the Octopus Spring ecosystem was an ideal environment to initiate studies on the ecophysiology of thermophilic anaerobes.

High numbers of anaerobic bacteria were present in the mat (Zeikus *et al.*, 1980). Enumeration of hydrolytic bacteria in a complex medium with both proteins/amino acids and carbohydrates demonstrated $\geqslant 1.0 \times 10^9$ bacteria (g dry weight)$^{-1}$ of mat, whereas enumeration of methanogenic bacteria demonstrate $\geqslant 1.0 \times 10^6$ methanogens (g dry weight)$^{-1}$ in either basal or enriched medium that contained hydrogen/carbon dioxide. Microscopic analysis of 10^{-4} dilution tubes with complex medium revealed a diverse population of rod-shaped bacteria including sporeformers and long cells that fluoresced yellow-green under ultraviolet light.

The prevalent hydrolytic bacterium isolated from an end dilution tube with a complex chemoorganotrophic culture medium was identified as a unique species, *Thermobacteroides acetoethylicus* strain HTB2 (Ben Bassat & Zeikus, 1981). The prevalent methanogen was identified as strain YTB which differed from *Methanobacterium thermoautotrophicum* type strain ΔH in DNA guanosine plus cytosine content and immunological properties (Zeikus & Wolfe, 1972; Zeikus *et al.*, 1980). Three different hydrolytic bacteria were isolated from enrichment cultures and included: a heterolactic-producing species, *Thermoanaerobium brockii* strain HTD4 (Zeikus *et al.*, 1979); a xylose-fermenting species, *Clostridium thermohydrosulfuricum* strain 39E (Zeikus *et al.*, 1980); and a pectin-degrading species, *Clostridium thermosulfurogenes* (Schink & Zeikus, 1982*d*). A unique sulphate-reducing species isolated from a lactate enrichment culture was identified as *Thermodesulfotobacterium commune* (Zeikus *et al.*, 1982*a*).

The general physiological properties of the Octopus Spring anaerobes are compared in Table 1. All species isolated were obligate thermophiles and displayed growth temperature from ⩾60 °C to ⩽70 °C which correlates well with the temperature of the mat site studied. The temperature range for all species listed was ⩾35 °C and ⩽85 °C. The four hydrolytic bacteria isolated fermented a variety of saccharides, had a DNA G + C composition in the low 30%, were rod-shaped and contained glycerol ester type lipids. Species were distinguished by cell wall architecture, motility, endospore formation and fermentation products. Most species used thiosulphate as an electron acceptor and reduced it to hydrogen sulphide. It is worth noting that *Clostridium thermosulfurogenes* transformed thiosulphate into elemental sulphur which accumulated as a dense white precipitate during glucose fermentation. Notably, the spores of *C. thermohydrosulfuricum* are extremely heat resistant. Spores in test tube cultures withstand autoclaving for 20 min at 120 °C, and have a D_{10} (i.e., 10% survivors) value of 11 min at 121 °C (H. H. Hyun, unpublished observations). It is very common to have media and stocks of thermophilic anaerobes contaminated unless special precautions are taken to sterilize culture media. The saccharide fermentations of the individual species varied considerably and the end product yields are used in part to readily distinguish species. The major end product(s) of the species used for differentiation include: lactic acid, *T. brockii;* ethanol, *C. thermohydrosulfuricum;* ethanol and methanol from pectin, *C. thermosulfurogenes;* and ethanol and acetate but without lactate, *T. acetoethylicus.* The prevalent hydrolytic species isolated from the mat, *T. acetoethylicus,* displayed a doubling time of <25 min. Several physiological properties of this species suggest that it may have a selective advantage over the other hydrolytic bacteria isolated from the mat. This includes a faster growth rate and greater ATP yield during glucose metabolism as a consequence of higher acetate and lower ethanol-lactate yields.

Other trophic groups in the mat included methanogens, sulphate-reducers and acetogens but only limited species were isolated. The methanogen and sulphidogen displayed higher DNA G + C contents than hydrolytic species and contained ether linked lipids. *M. thermoautotrophicum* lipids are comprised of phytanyl di- and tetra-ethers of glycerol (Tornabene *et al.*, 1978) while *T. commune*

Table 1. *Physiological characteristics of Octopus Spring anaerobes*

Species – strain	Optimum temperature (°C)	Cellular morphology	DNA composition (mol % G+C)	Major lipids[a]	Energy sources	Fermentation products
Hydrolytic bacteria						
Thermoanaerobium brockii strain HTD4	65	Gram-positive,	30.0 ± 1	Glycerol esters	Hexose, starch pentoses Saccharides + $S_2O_3^{2-}$	Lactate, ethanol, acetate, H_2/CO_2 H_2S + acetate + CO_2
Clostridium thermohydrosulfuricum strain 39E	65	Gram-variable, motile, rod, spores	30.5 ± 1	Glycerol esters	starch, pentose, hexoses Saccharide + $S_2O_3^{2-}$	H_2/CO_2, ethanol, lactate, acetate H_2S, acetate + CO_2
Clostridium thermosulfurogenes strain 4B	60	Gram-negative, motile rod spores, deposits sulphur	32.6 ± 0.4	Glycerol esters	pectin, starch pentoses hexoses	ethanol, lactate, H_2/CO_2, methanol, acetate
Thermobacteroides acetoethylicus strain HTB2	65	Gram-negative, motile rod	31.0 ± 1	Glycerol esters	Pentose, hexose, starch Saccharides + $S_2O_3^{2-}$	Ethanol, H_2/CO_2, acetate H_2S + acetate + CO_2
Methanogenic bacteria						
Methanobacterium thermo-autotrophicum strain YT1	65	Gram-positive, rod, yellow green flu-orescence	48.0 ± 1	Phytanyl glycerol di- and tetra ethers	H_2/CO_2	$CH_4 + CO_2$
Sulphidogenic bacteria						
Thermodesulfotobacterium commune strain YSRA-I	70	Gram-negative, straight rod, outer wall membrane	34.4 ± 1	Non-iso-propanoid glycerol di- and mono-ethers	H_2, lactate, pyruvate + $S_2O_3^{2-}$ or SO_4^{2-} pyruvate	Acetate + H_2S + CO_2 $H_2 + CO_2$ + acetate

[a] Data of T. Langworthy.

contains non-isopropanoid mono- and di-ethers of glycerol (Langworthy *et al.*, 1982). The ether linked lipids of *T. commune* appear unique and are not reported in other prokaryotes or eukaryotes. *M. thermoautotrophicum* has a very restricted substrate range for energy metabolism and proliferates as an autotroph on hydrogen/carbon dioxide (Zeikus & Wolfe, 1972). *T. commune* uses hydrogen/sulphate as energy source but requires acetate as a carbon source under these conditions. This species also consumes pyruvate or lactate as electron donor and thiosulphate or sulphate as electron acceptor for sulphidogenesis. In addition, *T. commune* ferments pyruvate alone to hydrogen/carbon dioxide and acetate as end products. Notably, the bisulphate reductase of *T. commune* is structurally distinct from desulphoviridin or desulphorubidin or P852 (C. Hatchikian, personal communication). Homoacetogenic clostridia that ferment hydrogen/carbon dioxide or saccharides were isolated from the mat but not characterized (Zeikus, 1979). *T. brockii* can also ferment ethanol, its saccharide fermentation product, and functions as a hydrogen-producing acetogen mixed culture with *M. thermoautotrophicum* (Ben-Bassat *et al.*, 1981).

Metabolic communication during biodegradation of organic matter by mixed populations of thermophilic anaerobes is expressed at several different trophic levels. One of the most dynamic behavioural responses observed in mixed culture experiments is the influence of hydrogen consumption by *M. thermoautotrophicum* on the hydrolysis of complex organic carbon by *C. thermocellum* or *T. brockii* (Weimer & Zeikus, 1977; Ben-Bassat *et al.*, 1981). In physiological terms, utilization of hydrogen by the methanogen alters carbon and electron flow in the hydrolytic bacterium such that thermodynamic efficiency of metabolism is improved (i.e., more total growth and ATP synthesis per mol organic substrate metabolized), rates of metabolism are increased (i.e., faster organic hydrolysis rates) and the major reduced fermentation products are drastically altered (i.e., methane in lieu of ethanol, lactate and hydrogen).

Notably, hydrogen metabolism differs significantly in *C. thermocellum* strain LQRI and *T. brockii* strain HTD4 when the intraspecies path of electrons is compared at the biochemical and physiological level (Lamed & Zeikus, 1980). Figure 11 compares the biochemical electron flow circuits and the reduced end-product ratios when these species fermented cellobiose in complex medium. Both species make the same fermentation products but totally

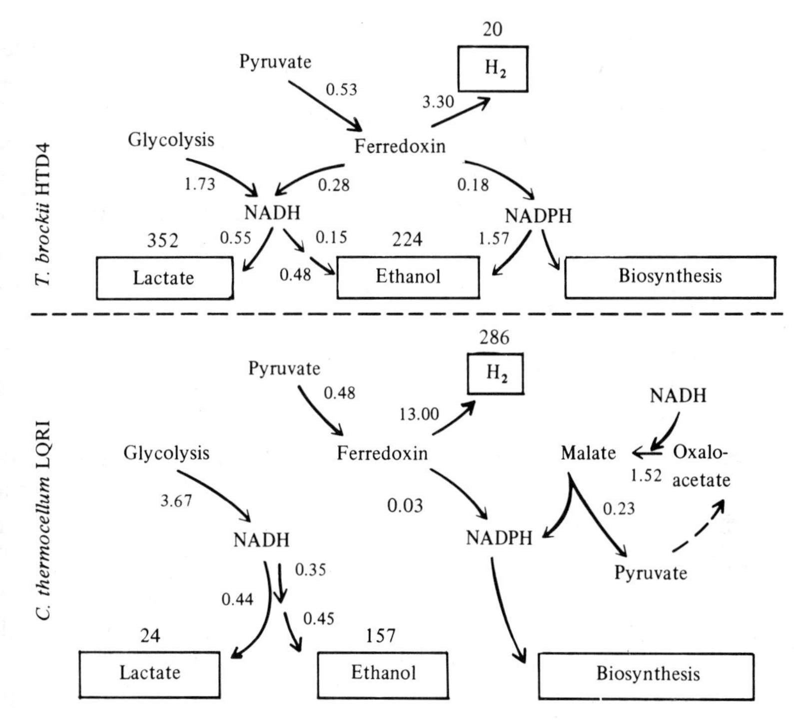

Fig. 11. Relation of the catabolic enzyme activities and proposed electron flow scheme to the reduced fermentation product yields of *T. brockii* HTD4 and *C. thermocellum* LQRI. Numbers represent final end product yields (total micromoles formed) of cellobiose fermentations and the specific activities of the enzymes indicated by the arrows in μmol min^{-1} (mg protein)$^{-1}$ at 40 °C. (After Lamed & Zeikus, 1980.)

different proportions account for carbon balance. Hydrogen is the major reduced end-product of *C. thermocellum* fermentations, but is only a minor product of heterolactic *T. brockii* fermentations. Both species contain the same glycolytic paths and transform saccharides into pyruvate and NADH; however, electron flow from these intermediary metabolites differs drastically in the two species. Most importantly, electron flow is physiologically interconnected between end-products and intracellular electron carriers in *T. brockii* but not *C. thermocellum*. The major difference is that electrons in pyruvate flow to lactate and ethanol in *T. brockii*, but only to hydrogen or cell carbon in *C. thermocellum* because ferredoxin-NAD reductase activity is not active in this strain. Consequently, hydrogenase of *C. thermocellum* is quite active and hydrogen predominates, while NADH is available from pyruvate oxidation in *T. brockii* and ethanol and lactate predominate as end-products.

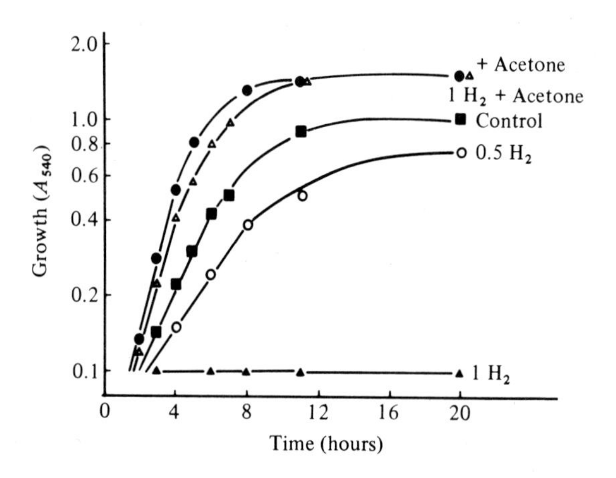

Fig. 12. Effect of hydrogen on *T. brockii* growth. Pressure tubes contained LPBB medium, 0.1% (w/v) yeast extract, 0.5% (w/v) glucose (250 μmol), and a nitrogen gas phase. Acetone (700 μmol) or hydrogen (0.5 or 1 atm) or both were added as indicated. The tubes were inoculated with 0.2 ml of culture grown on TYEG medium. (After Ben-Bassat *et al.*, 1981.)

As a consequence of biochemical differences in the electron flow circuits of these species, hydrogen at one atmosphere pressure totally inhibits growth and product formation of *T. brockii* strain HTD4 but does not significantly influence saccharide metabolism of *C. thermocellum*. Stable methanogenic mixed cultures of *T. brockii* but not *C. thermocellum* can be maintained on cellobiose without hydrogen accumulation. However, mixed cultures of *C. thermocellum* and *M. thermoautotrophicum* are stable on cellulose because cellulolysis limits the rate of hydrogen production and methane is the only significant reduced end-product (Weimer & Zeikus, 1977).

The hydrogen dependent inhibition of *T. brockii* saccharide fermentations can be overcome by the addition of exogenous electron acceptors that are reduced by enzymes of the electron flow pathway. For example, the addition of acetone in the presence or absence of hydrogen (Fig. 12) stimulates growth of *T. brockii* during saccharide fermentation because the NADP linked alcohol dehydrogenase of this species recognizes this ketone with high affinity and reduces it to isopropanol (Lamed & Zeikus, 1981). Thus, excess acetone thermodynamically alters intraspecies electron flow in *T. brockii* fermentations from hydrogen, lactate and ethanol formation to isopropanol as the major reduced end-product. Figure 13 compares a glucose fermentation time course of *T. brockii* in pure culture and in mixed culture with *M. thermoautotrophicum*. Clearly, the methanogen functions as an exogenous electron acceptor and

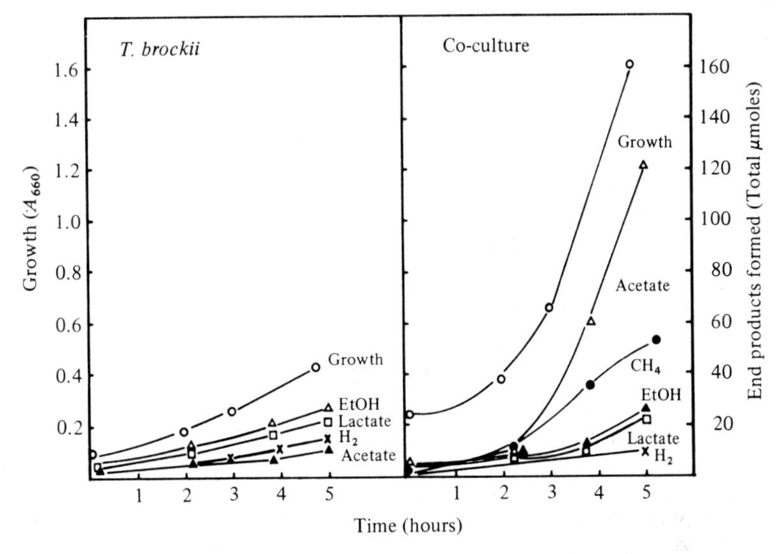

Fig. 13. Comparison of *T. brockii* glucose catabolism in pure and mixed culture with *M. thermoautotrophicum*. *T. brockii* pure cultures were grown in pressure tubes that contained 5 ml of TYEG medium and nitrogen gas phase. Mixed culture experiments were performed as follows: the methanogen was grown to an optical density of 0.18 at 540 nm in pressure tubes that contained 5 ml of TYE medium and 3 atm of H_2–CO_2 (80 : 20), the gas phase was replaced with nitrogen and 0.5 ml of inoculum of *T. brockii* and glucose (125 μmol) was added. All experimental tubes were incubated at 65 °C without shaking. EtOH, ethanol. (After Ben-Bassat *et al.*, 1981.)

it alters intraspecies electron flow of *T. brockii* from lactate and ethanol to hydrogen with interspecies hydrogen transfer resulting in methanogenesis by *M. thermoautotrophicum*. This type of metabolic communication clearly reflects a mutualistic behavioural response by both species. If one compares metabolism at five hours, *T. brockii* is just beginning to eat alone but with its methanogenic friend it has already finished dinner and has obtained more nourishment from eating the same amount of food. The methanogen also enables the hydrolytic species to use a wider range of carbon substrates as energy sources for growth by thermodynamic alteration of electron flow. Thus, *T. brockii* in methanogenic mixed culture consumes ethanol, its normal saccharide fermentation product, as the sole energy source for growth (Ben-Bassat *et al.*, 1981).

The physiological influence of exogenous hydrogen donors and hydrogen acceptors is related to the biochemical electron flow circuit of *T. brockii* in Fig. 14. In essence, ethanol, hydrogen, acetone and *M. thermoautotrophicum* influence electron flow in *T. brockii* fermentations by thermodynamic regulation of reversible

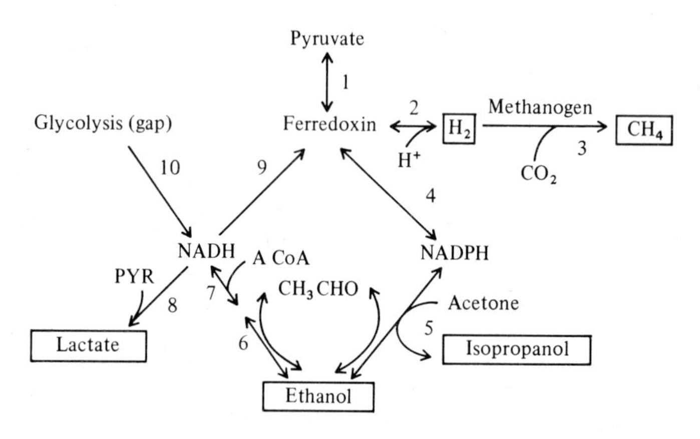

Fig. 14. Metabolic control of catabolic electron flow by reversible oxidoreductases in *T. brockii*. The numbers refer to the following enzyme activities: 1, pyruvate ferredoxin oxidoreductase; 2, hydrogenase; 3, 'methanogenases'; 4, ferredoxin-NADP oxidoreductase; 5, ethanol-NADP oxidoreductase; 6, ethanol-NAD oxidoreductases; 7, NADH-acetyl coenzyme A oxidoreductase; 8, NADH-pyruvate reductase; 9, ferredoxin-NAD oxidoreductase; and 10, glyceraldehyde-3-phosphate dehydrogenase. ACoA, Acetyl coenzyme A; GAP, glyceraldehyde 3-phosphate. Note that the specific product of enzyme 5 depends on the substrate (i.e., acetaldehyde, ethanol, or acetone). (After Ben-Bassat *et al.*, 1981.)

oxidoreductase activities. Hence, hydrogen is a metabolic inhibitor of glucose fermentation in the absence of exogenous electron acceptors (i.e., a methanogen or acetone) because hydrogen flows backward via hydrogenase and ferredoxin-NAD reductase, and this reduces intracellular electron acceptors (i.e., NAD) which are required for enzymatic glucose oxidation (i.e., glyceraldehyde 3-phosphate dehydrogenase). In the presence of acetone, hydrogen is not an inhibitor because electrons can flow from hydrogen to isopropanol via hydrogenase, ferredoxin-NADP reductase and NADP-linked alcohol dehydrogenase. Ethanol can serve as an energy source for *T. brockii* in mixed culture because electrons can flow from ethanol to hydrogen via NAD/NADP linked alcohol dehydrogenases and acetaldehyde dehydrogenases, NADH and NADPH linked ferredoxin reductases, and hydrogenase, but this requires that low hydrogen partial pressures be maintained by *M. thermoautotrophicum*.

Cooperative metabolic behaviour is also expressed among hydrolytic thermophiles (Ng *et al.*, 1981). For example, in pure culture *C. thermocellum* actively ferments cellulosic substrates but reducing sugars accumulate and growth stops because of high proton concentration. In mixed culture with *C. thermohydrosulfuricum*, reducing sugars (i.e., pentoses, glucose and cellobiose) do not accumulate from hemicellulose and cellulose hydrolysis in the fermentation broth and growth stops because of high ethanol concentration. Notably, only *C. thermocellum* ferments cellulose and only *C. thermohydrosulfuricum* can ferment pentoses; neither species can grow on hemicellulose alone and both species can ferment glucose and cellobiose. The physiological-biochemical explanations for this metabolic behaviour include: the cellulase of *C. thermocellum* cleaves β,1–4 bonds in the glycans and xylans (Ng & Zeikus, 1981); *C. thermocellum* has greater metabolic efficiency on cellulose than cellobiose than glucose since it derives more ATP/anhydroglucose consumed by conservation of the energy of hydrolysis in the β,1–4 bond via cellodextrin phosphorylase and cellobiose phosphorylase (Ng & Zeikus, 1982); and cellobiose and glucose are preferentially consumed by *C. thermohydrosulfuricum* because of the more favourable kinetics of its saccharidases (Ng & Zeikus, 1982). In fact, during batch culture on excess cellulose or cellobiose *C. thermocellum* excretes but does not consume glucose because glucose permease and hexokinase activity are inducible in this species (Ng & Zeikus, 1982).

Bacterial degradation of cellulose appears to be of little significance to carbon cycling in the Octopus Spring ecosystem because of its apparent absence as a component of primary production. However, the pectinolytic species, *C. thermosulfurogenes,* in this bacterial algal mat, actively degrades pectin via polygalacturonate hydrolase to fermentable oligomers and galacturonate which accumulates (Schink & Zeikus, 1982*d*). In mixed culture, galacturonate can be fermented as well with *C. thermohydrosulfuricum* (J. G. Zeikus, unpublished observations). Hence, metabolic communication between biopolymer-degrading species and mono- and disaccharide-fermenting species can be established as a consequence of the inherent metabolic efficiency of the former on polymers and the later on monomers, and as a consequence of the substrate-product specificity of depolymerizing enzymes.

CARBON TRANSFORMATION IN HYPERSALINE WATERS AND SEDIMENT OF GREAT SALT LAKE, UTAH

Environmental parameters

Great Salt Lake was selected as a hypersaline ecosystem suitable for a model to test the effect of high salt and sulphate on the microbial

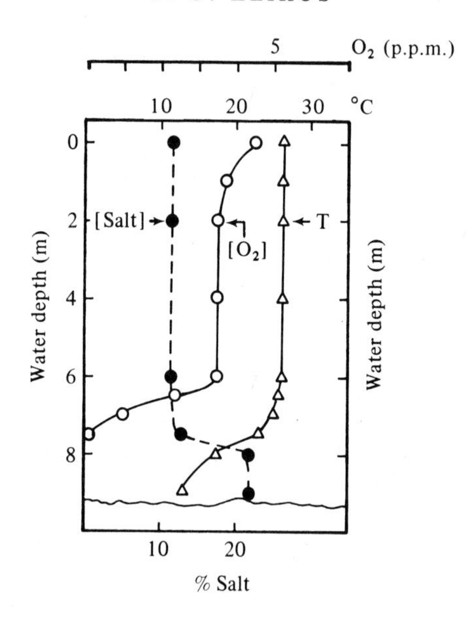

Fig. 15. Relation of water depth to the concentration of salt and oxygen and temperature in the South Arm sampling site in Great Salt Lake, Utah, taken in July 1980. (After Schink & Zeikus, 1982*c*.)

decomposition of organic matter (Phelps & Zeikus, 1980). The lake is separated by a causeway into two different surface water salinity regions, the South Arm with about 13% salt and the North Arm with about 26% salt (Post, 1977). Water and sediment was sampled from the deepest portion of the lake's South Arm. The primary producers in the lake's South Arm are algae of the genus *Dunaliella* with *D. viridis* the predominant species (Post, 1977).

Figure 15 illustrates the relation of water depth to *in situ* concentrations of salt and oxygen and temperature at the hypersaline site studied in Great Salt Lake. These chemical parameters remained constant from August 1978 to February 1982 (P. Sturm, personal communication). Notably, a pycnocline began below a water depth of 7 m and extended into the sediment (10 m) and oxygen was not detectable below 8 m. The proton activity measured in sediments and waters was 7.8 to 8.0. Notably, the sediments contained high amounts of dissolved hydrogen ($<200\,\mu$M) and low amounts of methane ($>4\,\mu$M), ethane ($0.096\,\mu$M) and propane ($0.14\,\mu$M). The surface sediments also contained 14% organic matter, 24% salinity, $162\,\mu$M sulphate and $4\,\mu$M sulphide.

Table 2. *Anoxic carbon transformation activity in Great Salt Lake sediments*[a]

Tracer	% decomposition ^{14}C-methane	^{14}C-carbon dioxide
$Na_2{}^{14}CO_3$	0	—
$U^{14}C$-glucose	0	34
$^{14}C3$-lactate	0	28
$^{14}C2$-acetate	0	37
^{14}C-methionine	34	5
^{14}C-methylmercaptan	16	26
^{14}C-methanol	50	3

[a] Experimental conditions; surface sediment was placed into anaerobic tubes that contained 1 to 3×10^6 d.p.m. of tracer and tubes were incubated for 30 days at 30 °C prior to analysis. Tubes that contained $Na_2{}^{14}CO_3$ had a H_2/CO_2 (80:20) headspace, other tubes contained nitrogen, formalin controls did not display biological transformation activity. (T. Phelps, unpublished observations.)

Carbon and electron flow

Anaerobic digestion of organic matter in Salt Lake dynamically differs from Lake Mendota because hydrogen and methane are the predominant fermentogenic gases that escape the sediments. This feature in part establishes the niche for hydrogen-consuming species in the water column. Figure 16 compares the relation of water depth with *in vivo* hydrogen consumption activities. Notably, the number of aerobic hydrogen bacteria and *in vivo* hydrogenase activity displayed a significant increase in activity between 6 and 8 m which correlated with the depletion of oxygen from these waters (Schink & Zeikus, 1982*c*). Thus, the bacteriological hydrogen consumption profile in the hypolimnion of the Great Salt Lake study site parallels that for the microbial methane oxidation profile (Harrits & Hanson, 1980) but not the hydrogen consumption profile (Schink & Zeikus, 1982b) in Lake Mendota. Microbial carbon transformation reactions in Great Salt Lake surface sediment are quite distinct from those in Lake Mendota (Table 2). Notably, ^{14}C-glucose, C2-acetate and C3-lactate were decomposed to $^{14}CO_2$ but not $^{14}CH_4$. However, $^{14}CH_4$ and $^{14}CO_2$ were formed from radioactive methanol, methylmercaptan and methionine labelled in the methyl position. Thus, the two major methanogenic precursors in Lake Mendota, acetate and hydrogen/carbon dioxide, are not significantly transformed to methane in the Great Salt Lake. However, methanogenesis does occur from one carbon methyl precursors in both sediment ecosystems. Methylmercaptan formation via the decomposition of

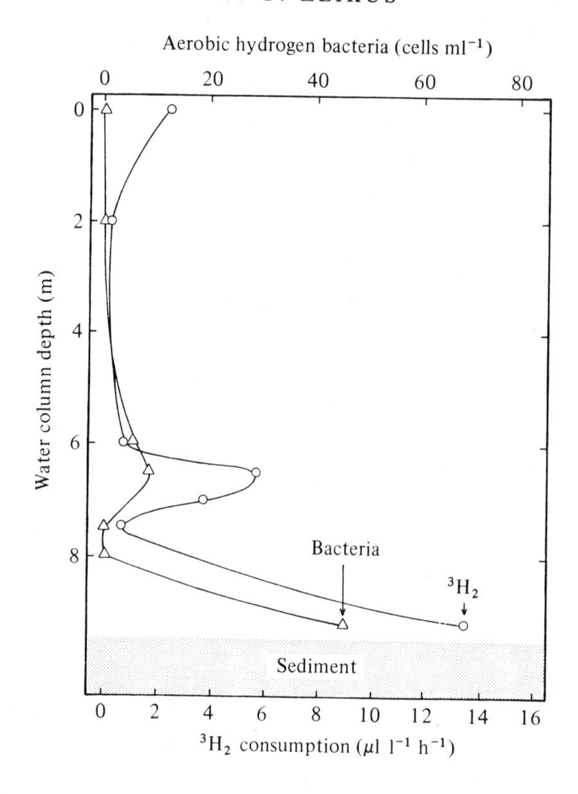

Fig. 16. Relation of aerobic bacteria hydrogen consumption activity to Great Salt Lake water depth (after Schink & Zeikus, 1982*c*).

methionine, and methanol generation via the decomposition of methoxylated substrates (i.e., proteins and carbohydrates) would appear the likely precursors for the mini-methanogenesis that occurs in the presence of excess sulphate and hydrogen in Great Salt Lake sediment. Bacterial sulphate reduction is active in surface sediments (K. Ingvorsen and J. G. Zeikus, unpublished observations. The *in vitro* rate of sediment sulphate reduction at 30 °C was 175 ± 22 nmol $^{35}SO_4^{2-}$ cm^{-3} day^{-1}. This value was enhanced approximately two fold by the addition of 7 mM ferrous sulphate or 20 mM sodium lactate which suggests that sulphidogenesis is limited in part by electron donors and end-products in the Great Salt Lake ecosystem.

Microbial dynamics

Little is known about the kinds of anaerobic bacteria in hypersaline aquatic ecosystems but Great Lake's productivity and the active biodegradation of organic matter in its sediments suggest high populations of physiologically unique anaerobes. Enumeration of hydrolytic bacteria in a complex medium with 15% salt and both proteins/amino acids and carbohydrates demonstrated approximately 1.0×10^8 bacteria (ml surface sediment)$^{-1}$. The lactate-metabolizing sulphate-reducing population was estimated about 1.0×10^6 bacteria (ml sediment)$^{-1}$ in hypersaline medium. However, methanogenic populations that consume methanol-hydrogen/carbon dioxide in hypersaline medium were at the detection limits (i.e., about 1.0×10 bacteria ml^{-1}). The population density of total anaerobic bacteria in Great Salt Lake sediments exceeds that in Lake Mendota. However, nothing is known about metabolic interactions among different trophic groups. It is clear that metabolic communication between a methanogen population in Great Salt Lake and either hydrolytic, acetogenic or sulphate-reducing populations is extremely limited by the environmental parameters (i.e., high salt and sulphate).

Enrichment cultures of sulphate-reducers in hypersaline lactate medium contained both sporeforming rods and short motile rods. Enrichment cultures of hydrolytic bacteria included Gram-negative, rod-shaped proteolytic and saccharolytic species. Only one species has been isolated and well characterized to date. The cellular growth and metabolic features of this prevalent, obligately anaerobic, hydrolytic bacterium in Great Salt Lake surface are summarized in Table 3. The unique physiological features of this organism and its abundance in hypersaline sediments suggests the species name '*Haloanaerobium prevalans*'. This species grows from 2% NaCl to saturation but optimal growth rates are observed at approximately 10% salt and optimal cell densities at about 20% salt. This growth feature correlates with the species anatomical features in that its typical Gram-negative wall maintains cell rigidity at different salt concentrations. In addition, the organism grows on amino acids and produces methylmercaptan from biodegradation of methionine.

SIGNIFICANCE OF METABOLIC COMMUNICATION IN ANAEROBIC ECOSYSTEMS

General significance

Mixed microbial populations (i.e., different trophic groups) are required to degrade complex organic matter in anaerobic niches of

Table 3. *Physiological features of haloanaerobium prevalens, gen. nov. spec. nov.*[a]

Character	Description
Cellular properties	
Morphology	Non-motile rod
Wall	Gram-negative, outer wall membrane
DNA	22 mol % G+C
Lipids[b]	Glycerol esters
Cytochromes	Absent
Growth tolerances	
Oxygen	Obligate anaerobe
Temperature	Mesophile
pH	Neutraphile
Salt	2% NaCl to saturation
Metabolic characteristics	
Nutrition	Heterotroph
Energy sources	Saccharides, amino acids
Fermentation products	Acetate, butyrate, propionate, H_2/CO_2, methylmercaptan

[a] J. G. Zeikus, unpublished observations.
[b] Data of T. Langworthy, personal communication.

the biosphere because of the chemical complexity of carbon substrates, the endogenous electron acceptor availability, and the thermodynamic and mechanistic features associated with anaerobic carbon metabolism that couples dehydrogenative transformation reactions with hydrogenative transformation reactions. In the absence of metabolic communication (i.e., different species sharing in carbon and electron transformation reactions), carbon metabolites accumulate that can alter normal carbon and electron flow such that the metabolic efficiency of biodegradative species is either impaired or totally inhibited.

The two major types of metabolic communication-behaviour displayed by biodegradative populations in nature are synergism and antagonism. Microbial synergism is exemplified by biochemical interactions between hydrolytic and methanogenic bacteria which consume different carbon and electron donors such that their joint metabolism increases individual species dominance in the environment. Microbial antagonism is exemplified by biochemical interactions between sulphate-reducing and methanogenic bacteria which consume the same carbon and electron donors such that their joint metabolism decreases individual species dominance in the environment.

Ecosystem differences including temperature, pH, salinity and the available carbon–electron donors and acceptors for biodegradation greatly alter anaerobic populations and metabolic communication between specific trophic groups. This axiom is observed by comparison of the microbial dynamics of the biodegradative populations in the Lake Mendota, Great Salt Lake and Octopus Spring anaerobic ecosystems. Metabolic communication in nature must be carefully scrutinized by the microbiologist. For example, it is easy to hypothesize, based on known metabolic interactions between methanogenic and sulphate reducing bacteria, that these trophic groups may display synergistic, antagonistic, or both behavioural modes in a given niche depending on the specific *in situ* sulphate and electron donor concentration.

Environmental significance

Metabolic communication between biodegradative populations is of importance to understanding how carbon is mineralized in the anoxic portions of the biosphere. In deep, neutral freshwater or marine sediment ecosystems, environmentally synthesized organic carbon appears to be completely decomposed to gaseous carbon and water. However, the metabolic interactions of hydrolytic bacteria with methanogenic and sulphate-reducing bacteria differ dramatically in freshwater and marine ecosystems. The importance of microbial sulphate reduction and methanogenesis in aquatic sediments is well established and varies with freshwater or marine environments (Abram & Nedwell, 1978; Cappenberg, 1974; Cappenberg & Prins, 1975; Ingvorsen *et al.*, 1981; Ingvorsen & Brock, 1982; Jones & Paynter, 1980; King & Wiebe, 1980; Martens & Berner, 1974; Mountfort *et al.*, 1980; Oremland *et al.*, 1982; Schink & Zeikus, 1982*a*; Smith & Klug, 1981; Strayer & Tiedje, 1978; Winfrey & Zeikus, 1977–1979). The significance of methanogens in controlling environmental carbon and electron flow is itself influenced greatly by sulphate concentration. Sulphate-reducing bacteria are more ubiquitous than methanogens in nature because they have a broader substrate range and derive energy via consumption of hydrogen/carbon dioxide or acetate in excess sulphate, or via production of hydrogen and acetate in limiting sulphate. Sulphate reduction is a more biogeochemically significant process in marine sediments where sulphate is abundant (Martens & Berner, 1974) and methanogenesis in freshwater sediments where sulphate is

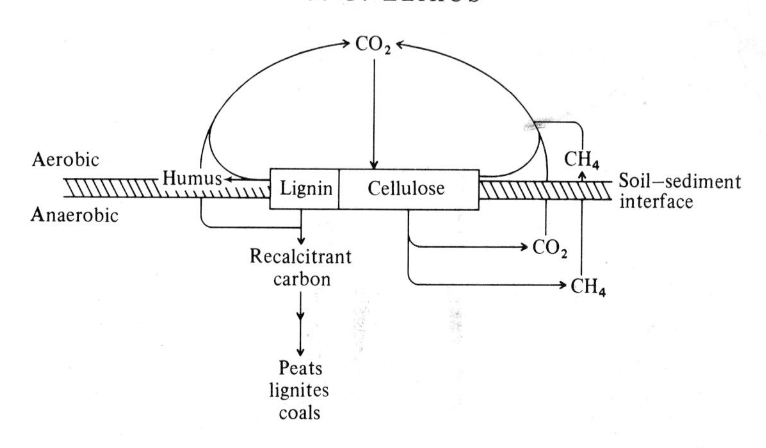

Fig. 17. Importance of lignocellulose decomposition in the biosphere (after Zeikus, 1981).

limiting (Winfrey & Zeikus, 1977). However, neither process is mutually exclusive. Nonetheless, the experimental trend still remains that in neutral freshwater sediments the amount of sulphate largely regulates the flow of carbon and electrons to methane while in marine sediments the amount of electron donors is more important.

Certain organic molecules in anaerobic environments are recalcitrant. These molecules accumulate and account in part for the diagenesis of fossil fuels. Methane is the most cosmopolitan fossil fuel and is generated in both marine and freshwater environments. Certain carbon precursors are transformed to methane regardless of the environmental sulphate concentration. For example, methanol and methylmercaptan are transformed to methane in freshwater, marine and hypersaline sediments. It is worth noting that acidogenic and sulphidogenic bacteria in addition to methanogens form methane as a trace gas (Zeikus, 1982*a*). Lignin, the second most abundant biopolymer on earth, is not significantly biograded anaerobically into the one and two carbon precursors for sulphidogenic or methanogenic bacteria (Hackett *et al.*, 1977; Zeikus *et al.*, 1982*a*). Thus, the inability of anaerobes to solubilize this polymer limits metabolic communication and lignin-containing molecules accumulate and lead to the diagenesis of coals and peats (Fig. 17). The microbial carbon transformation reactions that are associated with the diagenesis of oils remain to be clarified in the future. One trend for future research is to better understand carbon biodegradation in hypersaline carbon rich sedimentary environments that resemble the salinity of Devonian oil-bearing sedimentary rocks.

Applied significance

Metabolic communication between anaerobic microbial populations is of importance in understanding how to enhance or limit important biodegradations. Anaerobic digestion by mixed bacterial populations is a common process for treatment of concentrated organic matter in industrial, municipal and agricultural wastes (Hobson *et al.*, 1974). Effective biodegradation of organic matter into biogas requires the combined and coordinated metabolic activity of at least four different trophic groups (Zeikus, 1980*b*, 1982*b*). In single-stage methane digestors rapid and complete destruction of long chain, volatile fatty acids requires synergistic metabolic communication between hydrogen-producing acetogens and methanogens (McInerney & Bryant, 1980). Thus, low intermediary levels of hydrogen and volatile fatty acids are often diagnostic signs of digestor efficiency and high amounts indicative of digestor failure (Hobson *et al.*, 1974; Kasper & Wuhrmann, 1978). A trend for future applications research is to improve the rate and yield of methane from biodegradation of organic matter by selective modulation of metabolic communication between certain trophic groups such that carbon and electrons flow to methane in lieu of hydrogen sulphide or long chain volatile fatty acids.

The importance of metabolic interactions in the biodegradation of organic matter in the anoxic gastrointestinal tracts of animals is well recognized (Hungate, 1966; Wolin, 1974). Indeed, synergistic metabolic communication between cellulose-degrading microorganisms and the host animal is essential for biodegradation of plant biomass by mammalian secondary consumers. Improvements in ruminant nutrition are expected by selective enhancement of metabolic communication between hydrolytic and acidogenic bacteria in lieu of methanogens. In addition, a possible trend for future applied research on prevention of biomass destruction is to focus on disrupting the synergistic metabolic communication which is essential to maintenance of anoxic biodegradation in the intestinal tracts of wood-destroying termites (Breznak, 1975).

Finally, it is worth noting that metabolic communication is also important to the establishment of wetwood, an anaerobic bacterial disease syndrome in living trees (Zeikus & Ward, 1974; Schink *et*

Fig. 18. Ignited methane gas emanating from a hollow increment borer bit drilled 20 cm into a cottonwood tree located on the shore of Lake Wingra, Wisconsin. Photograph is from a time exposure taken at night. (From Zeikus & Ward, 1974.)

al., 1981*a*). Wetwood is an atypical condition of the heartwood of many living trees but is common in cottonwood trees. The high moisture content of wetwood in cottonwoods is associated with high pressures of methane gas (Fig. 18) and a fetid liquid that eminates from increment core holes bored into trees (Zeikus & Ward, 1974). Dead xylem tissue from wetwood, unlike heartwood, contains dense bacterial populations and destroyed vessel-to-ray-pit membranes (Schink *et al.*, 1981*a*). The alkaline wetwood environment is anoxic and contains significant quantities of bacterial fermentation products including acetate, butyrate, propionate, ethanol, isopropanol and methane. In addition, bacterial nitrogenase and pectinase activity were detected in wetwoods. Table 4 compares the numbers of the hydrolytic, methanogenic and nitrogen fixing anaerobes in two trees that have wetwood. Anaerobic bacteria were 10^4 times more numerous in wetwood than sapwood (i.e., living xylem tissue). Thirteen strains of heterotrophic bacteria were isolated and characterized from wetwood and included *Clostridium, Bacteroides, Erwinia, Edwardsiella, Klebsiella* and *Lactobacillus*.

Table 4. *Enumeration of anaerobic populations associated with wetwood in living trees*[a]

Tree tissue sampled	Total anaerobic bacteria		
	Hydrolytic	Methanogens	Nitrogen fixers
Cottonwood			
sapwood	3×10^2	<10	3×1^{-2}
wetwood	2×10^6	2.5×10^3	6×10^4
fetid liquid	1×10^4	1×10^2	ND
soil	6×10^5	<10	3×10^4
Elm			
sapwood	6.5×10^2	<10	6.5×10^2
wetwood	8×10^7	2.3×10^4	2.3×10^6
fetid liquid	3×10^5	2.3×10^2	1×10^5

[a] Represents cells per ml or gram as determined by 3 tube MPN analysis. ND, not determined (from Schink & Zeikus, 1981).

The significant biodegradative event associated with wetwood appears to be the bacterial destruction of the pectinous vessel-to-ray pit membranes in xylem tissue that alters water translocation and gradually spreads the disease throughout the dead xylem tissue of the tree (Schink & Zeikus, 1981*b*). A prevalent pectinolytic species in alkaline wetwoods was identified as *Clostridium butyricum* and it produced the same type of pectate lyase activity expressed in fetid liquid.

Metabolic communication appears important to the establishment of the wetwood disease syndrome because the symptoms include a diverse bacterial population, multiple fermentation products, bacterial nitrogenase and pectate lyase activity. A new nitrogen-fixing species, *Propionispira arboris,* was commonly isolated from pectin fermenting enrichment cultures (Schink & Zeikus, 1982*e*). This species grew on galacturonate, a pectinase hydrolysis product, which was not fermented by the pectinolytic *C. butyricum* strain. Synergistic metabolic interactions between hydrolytic and nitrogen-fixing, saccharide-fermenting anaerobes appear of significance in wetwood because a dense bacterial population is found in a carbohydrate rich environment that is also poor in combined nitrogen.

Acknowledgements

The research was supported by the College of Agricultural and Life Sciences, University of Wisconsin-Madison, and by grants from the National Science Foundation, Department of Energy, and Shell Oil Company.

REFERENCES

Abram, J. W. & Nedwell, D. B. (1978). Inhibition of methanogenesis by sulfate reducing bacteria competing for transferred hydrogen. *Archives of Microbiology*, **117,** 87–92.

Allgeier, R. J., Peterson, W. H., Juday, C. & Birge, E. A. (1932). The anaerobic fermentation of lake deposits. *International Revue der gesamten Hydrobiologie und Hydrographie*, **26,** 444–61.

Ben-Bassat, A. & Zeikus, J. G. (1981). *Thermobacteroides acetoethylicus* gen. nov. spec. nov., a new chemoorganotrophic, anaerobic, thermophilic bacterium. *Archives of Microbiology*, **128,** 365–70.

Ben-Bassat, A., Lamed, R. & Zeikus, J. G. (1981). Ethanol production by thermophilic bacteria: metabolic control of end product formation in *Thermoanaerobium brockii. Journal of Bacteriology*, **146,** 192–9.

Breznak, J. A. (1975). Symbiotic relationships between termites and their intestinal microbiota. In *Symposia of the Society for Experimental Biology*, vol. 24, ed. D. H. Jennings & D. L. Lee, pp. 559–80. Cambridge University Press.

Brock, T. D. (1970). High temperature systems. *Annual Review of Ecological Systems*, **1,** 191–220.

Brock, T. D. (1978). *Thermophilic Microorganisms and Life at High Temperatures.* New York: Springer-Verlag.

Bryant, M. P., Campbell, L. L., Reddy, C. A. & Crabill, M. R. (1977). Growth of *Desulfovibrio* on lactate or ethanol media low in sulfate and in association with H_2-utilizing methanogenic bacteria. *Applied and Environmental Microbiology*, **33,** 1162–9.

Burris, H. & Peterson, R. B. (1978). Nitrogen-fixing blue-green algae: their H_2 metabolism and their activity in freshwater lakes. In *Environmental Role in Nitrogen-Fixing Blue-Green Algae and Asymbiotic Bacteria*, ed. U. Granhall, *Ecological Bulletin* (Stockholm), **26,** 28–40.

Cappenberg, T. E. (1974*a*). Interrelations between sulfate reducing and methane producing bacteria in bottom deposits of a freshwater lake. II. Inhibition experiments. *Antonie van Leeuwenhoek*, **40,** 297–306.

Cappenberg, T. E. (1974*b*). Interactions between sulfate reducing and methane producing bacteria in bottom deposits of a freshwater lake. I. Field observations. *Antonie van Leeuwenhoek*, **40,** 285–95.

Cappenberg, T. E. & Prins, H. (1975). Interrelations between sulfate reducing and methane producing bacteria in bottom deposits of a freshwater lake. III. Experiments with ^{14}C-labeled substrate. *Antonie van Leeuwenhoek*, **40,** 457–69.

Desikachary, T. V. (1959). *Cyanophyta.* New Delhi: Indian Council of Agricultural Research.

Doemel, W. N. & Brock, T. D. (1977). Structure, growth and decomposition of laminated algal-bacterial mats in alkaline hot springs. *Applied and Environmental Microbiology*, **34,** 433–52.

Fallon, R. D. & Brock, T. D. (1979). Decomposition of blue-green algal (cyanobacterial) blooms in Lake Mendota, Wisconsin. *Applied and Environmental Microbiology*, **37,** 820–30.

Gooday, G. W. (1971). A biochemical and butoradiographic study of the role of the Golgi bodies in thecal formation in *Platymonas tetrathele. Journal of Experimental Botany*, **18,** 359–70.

Green, J. C. & Jennings, D. H. (1967). A physical and chemical investigation of the scales produced by the Golgi apparatus within and found on the surface of cells of *Chrysochromulina cherton. Journal of Experimental Botany*, **18,** 359–70.

Hackett, W. F., Connors, W. J., Kirk, T. K. & Zeikus, J. G. (1977). Microbial decomposition of synthetic ^{14}C-labeled lignin in nature: lignin biodegradation in a variety of natural materials. *Applied and Environmental Microbiology*, **33,** 43–51.

HARRITS, S. M. & HANSON, R. S. (1980). Stratification of aerobic methane oxidizing organisms. *Limnology and Oceanography*, **25,** 412–21.

HOBSON, P. N., BOUSEFIELD, D. & SUMMERS, R., (1974). Anaerobic digestion of organic matter. In *CRC Critical Reviews in Environmental Control*, pp. 131–91. Cleveland: Chemical Rubber Company.

HUNGATE, R. G. (1966). *The Rumen and Its Microbes.* New York: Academic Press.

INGVORSEN, K. & BROCK, T. D. (1982). Electron flow via sulfate reduction and methanogenesis in the anaerobic hypolimnion of Lake Mendota. *Limnology and Oceanography*, **27,** 559–64.

INGVORSEN, K., ZEIKUS, J. G. & BROCK, T. D. (1981). Dynamics of sulfate reduction in a eutrophic lake. *Applied and Environmental Microbiology*, **2,** 1024–39.

JONES, W. J. & PAYNTER, M. J. B. (1980). Populations of methane producing and *in vitro* methanogenesis in salt marsh estuarine sediments. *Applied and Environmental Microbiology*, **39,** 864–71.

KASPER, H. F. & WUHRMANN, K. (1978). Kinetic parameters and relative turnovers of some important catabolic reactions in digesting sludge. *Applied and Environmental Microbiology*, **36,** 1–7.

KING, G. M. & WIEBE, W. J. (1980). Tracer analysis of methanogenesis in salt marsh soils. *Applied and Environmental Microbiology*, **34,** 877–81.

KIRK, T. K. (1971). Effects of microorganisms on lignin. *Annual Review of Phytopathology*, **9,** 185–210.

KOSIUR, D. R. & WARFORD, A. L. (1979). Methane production and oxidation in Santa Barbara Basin sediments, estuarine and coastal marine. *Science*, **8,** 379–85.

LAMED, R. J. & ZEIKUS, J. G. (1980). Ethanol production by thermophilic bacteria: relationship between fermentation product yields and catabolic enzyme activities in *Clostridium thermocellum* and *Thermoanaerobium brockii. Journal of Bacteriology*, **144,** 569–78.

LAMED, R. J. & ZEIKUS, J. G. (1981). Novel NADP-linked alcohol-aldehyde/ketone oxidoreductase in thermophilic ethanologenic bacteria. *Biochemical Journal*, **195,** 183–90.

LANGWORTHY, T. A., HOLZER, G., ZEIKUS, J. G. & TORNABENE, T. G. (1982). Iso and anteiso-branched glycerol diethers of the thermophilic anaerobe *Thermodesulfotobacterium. Commune. Zbl. Bakt. Hyg. I. abt. Orig.* (in press).

MACGREGOR, A. N. & KEENEY, D. R. (1973). Methane formation by lake sediments during in vitro incubation. *Water Research Bulletin*, **9,** 1153–8.

MCINERNEY, M. J. & BRYANT, M. P. (1980). Syntrophic associations of H_2-utilizing methanogenic bacteria and H_2-producing alcohol and fatty acid degrading bacteria in anaerobic degradation of organic matter. In *Anaerobes and Anaerobic Infections*, ed. H. Gottschalk, pp. 117–26. Stuttgart: Gustav Fisher Verlag.

MARTENS, C. S. & BERNER, R. A. (1974). Methane production in the interstitial waters of sulfate-depleted marine sediments. *Science*, **185,** 1167–9.

MOUNTFORT, D. O., ASHER, R. A., MAYS, E. L. & TIEDJE, M. J. (1980). Carbon and electron flow in mud and sandflat intertidal sediments at Delaware Inlet, Nelson, New Zealand. *Applied and Environmental Microbiology*, **39,** 686–94.

NELSON, D. R. & ZEIKUS, J. G. (1974). Rapid method for the radioisotopic analysis of gaseous end products of anaerobic metabolism. *Applied Microbiology*, **28,** 258–61.

NG, T. K. & ZEIKUS, J. G. (1981). Comparison of extracellular cellulases of

Clostridium thermocvellum LQRI and *Trichoderma reesei* QM9414. *Applied and Environmental Microbiology*, **42.**
NG, T. K. & ZEIKUS, J. G. (1982). Differential metabolism of cellobiose and glucose by *Clostridium thermocellum* and *Clostridium thermohydrosulfuricum. Journal of Bacteriology*, **150,** 1391–9.
NG, T. K., BEN-BASSAT, A. & ZEIKUS, J. G. (1981). Ethanol production by thermophilic bacteria: fermentation of cellulosic substrates by co-cultures of *Clostridium thermocellum* and *Clostridium thermohydrosulfuricum. Applied and Environmental Microbiology*, **41,** 1337–43.
OREMLAND, R. S., MARSH, L. M. & BOLUN, S. (1982). Simultaneous methane formation and sulphate reduction in marine sediments. *Nature, London*, **296,** 143–5.
PANGANIBAN, A. T., PATT, T. E., HART, W. & HANSON, R. S. (1979). Anaerobic oxidation of methane in freshwater lake. *Applied and Environmental Microbiology*, **37,** 303–9.
PHELPS, T. & ZEIKUS, J. G. (1980). Microbial ecology of anaerobic decomposition in Great Salt Lake. In *Abstracts Annual Meeting American Society for Microbiology*, Miami.
POST, F. J. (1977). The microbial ecology of Great Salt Lake. *Microbial Ecology*, **3,** 143–65.
PRESCOTT, G. W. (1968). *The Algae: a Review.* Boston, Mass.: Houghton Mifflin Co.
SANDBECK, K. & WARD, D. M. (1981). Fate of immediate methane precursors in low sulfate hot spring algal-bacterial mats. *Applied and Environmental Microbiology*, **41,** 775–82.
SCHINK, B., WARD, J. C. & ZEIKUS, J. G. (1981*a*). Microbiology of wetwood: role of anaerobic bacterial populations in living trees. *Journal of General Microbiology*, **123,** 313–22.
SCHINK, B., WARD, J. C. & ZEIKUS, J. G. (1981b). Microbiology of wetwood. Importance of pectin degradation and *Clostridium* species in living trees. *Applied and Environmental Biology*, **42,** 526–32.
SCHINK, B. & ZEIKUS, J. G. (1980). Microbial methane formation: a major end product of pectin metabolism. *Current Microbiology*, **4,** 387–9.
SCHINK, B. & ZEIKUS, J. G. (1982*a*). Microbial ecology of pectin decomposition in anoxic lake sediments. *Journal of General Microbiology*, **128,** 393–404.
SCHINK, B. & ZEIKUS, J. G. (1982*b*). Ecology of aerobic hydrogen consuming bacteria in two freshwater lake ecosystems. *Applied and Environmental Microbiology* (in press).
SCHINK, B. & ZEIKUS, J. G. (1982*c*). An *in vivo* radioassay for hydrogenase activity and its application in extreme environments. *Applied and Environmental Microbiology* (in press).
SCHINK, B. & ZEIKUS, J. G. (1982*d*). *Clostridium thermosulfurogenes* sp. nov., a new thermophile that produces elemental sulfur from thiosulfate. *Journal of General Microbiology* (in press).
SCHINK, B. & ZEIKUS, J. G. (1982*e*). Characterization of *Propionispira arboris* gen. nov. spec. nov., a nitrogen fixing anaerobe common to wetwoods of living trees. *Journal of General Microbiology* (in press).
SIKES, C. S. (1978). Calcification and cation sorption of *Cladophora glomerata* (*Chlorophyta*). *Journal of Phycology*, **14,** 325–9.
SMITH, R. L. & KLUG, (1981). Electron donors utilized by sulfate reducing bacteria in eutrophic lake sediments. *Applied and Environmental Microbiology*, **42,** 116–21.
STRAYER, R. F. & TIEDJE, J. M. (1978). Kinetic parameters of the conversion of

methane precursors to methane in hypereutrophic lake sediment. *Applied and Environmental Microbiology*, **36,** 330–40.
TORNABENE, T. G. & LANGWORTHY, T. A. (1978). Diphytanyl and dibiphytanyl glycerol ether lipids of methanogenic archaebacteria. *Science*, **203,** 51–3.
WEIMER, P. J. & ZEIKUS, J. G. (1977). Fermentation of cellulose and cellobiose by *Clostridium thermocellum* and *Methanobacterium thermoautotrophicum. Applied and Environmental Microbiology,* **33,** 289–97.
WEIMER, P. J. & ZEIKUS, J. G. (1978). One carbon metabolism in methanogenic bacteria: cellular characterization and growth of *Methanosarcina barkeri. Archives of Microbiology*, **119,** 49–51.
WINFREY, M. R., NELSON, D. R., KLEVICKIS, S. C. & ZEIKUS, J. G. (1977). Association of hydrogen metabolism with methanogenesis in Lake Mendota sediments. *Applied and Environmental Microbiology*, **33,** 312–18.
WINFREY, M. R. & ZEIKUS, J. G. (1977). Effect of sulfate on carbon and electron flow during microbial methanogenesis in fresh water sediments. *Applied and Environmental Microbiology*, **33,** 275–81.
WINFREY, M. R. & ZEIKUS, J. G. (1979). Anaerobic metabolism of immediate methane precursors in Lake Mendota. *Applied and Environmental Microbiology*, **37,** 244–53.
WOLIN, M. J. (1974). Metabolic interactions among intestinal microorganisms. *American Journal of Clinical Nutrition*, **27,** 1320–8.
WOLK, P. (1973). Physiology and cytological chemistry of blue-green algae. *Bacteriological Reviews*, **37,** 32–107.
ZEHNDER, A. J. B. (1978). Ecology of methane formation. In *Water Pollution Microbiology*, vol. 2, ed. R. Mitchell, pp. 349–76. New York: John Wiley and Sons.
ZEHNDER, A. J. B. & BROCK, T. D. (1979). Methane formation and methane oxidation by methanogenic bacteria. *Journal of Bacteriology*, **37,** 420–32.
ZEIKUS, J. G. (1977). The biology of methanogenic bacteria. *Bacteriological Reviews*, **41,** 514–41.
ZEIKUS, J. G. (1979). Thermophilic bacteria: ecology, physiology and technology. *Enzyme and Microbial Technology*, **1,** 243–52.
ZEIKUS, J. G. (1980*a*). Chemical and fuel production by anaerobic bacteria. *Annual Review of Microbiology*, **34,** 423–64.
ZEIKUS, J. G. (1980*b*). Microbial populations in digestors. In *First International Symposium on Anaerobic Digestion*, ed. D. A. Stafford *et al.*, pp. 75–103. Cardiff: A. D. Scientific Press.
ZEIKUS, J. G. (1981). Lignin metabolism and the carbon cycle: Polymer biosynthesis, biodegradation and environmental recalcitrance. *Advances in Microbial Ecology*, **51,** 211–43.
ZEIKUS, J. G. (1982*a*). C_1 metabolism of chemotrophic anaerobes. *Advances in Microbial Physiology* (in press).
ZEIKUS, J. G. (1982*b*). Microbial intermediary metabolism in anaerobic digestion. I. *Anaerobic Digestion 1981*, ed. D. E. Hughes *et al.*, pp. 22–35. Elsevier Biomedical Press.
ZEIKUS, J. G., BEN-BASSAT, A. & HEGGE, P. (1980). Microbiology of methanogenesis in thermal volcanic environments. *Journal of Bacteriology*, **143,** 432–40.
ZEIKUS, J. G., DAWSON, M. A., THOMPSON, T. E., INGVORSEN, K. & HATCHIKIAN, E. C. (1982*a*). Microbial ecology of volcanic sulfidogenesis: isolation and characterization of *Thermodisulfotobacterium commune* gen. nov. and spec. nov. *Journal of General Microbiology* (in press).
ZEIKUS, J. G., HEGGE, P. W. & ANDERSON, M. A. (1979). *Thermoanaerobium*

brockii gen. nov. and spec. nov. a new chemoorganotrophic, caldocactive, anaerobic bacterium. *Archives of Microbiology*, **122,** 41–8.
ZEIKUS, J. G. & WARD, J. C. (1974). Methane formation in living trees: a microbial origin. *Science*, **184,** 1181–3.
ZEIKUS, J. G., WELLSTEIN, A. C. & KIRK, T. K. (1982*b*). Molecular basis for the biodegradative recalcitrance of lignin in anaerobic environments. *FEMS Microbiology Letters* (in press).
ZEIKUS, J. G. & WINFREY, M. R. (1976). Temperature limitation of methanogenesis in aquatic sediments. *Applied and Environmental Microbiology*, **31,** 99–107.
ZEIKUS, J. G. & WOLFE, R. S. (1972). *Methanobacterium thermoautotrophicus* sp. nov. an anaerobic extreme thermophile. *Journal of Bacteriology*, **109,** 707–13.
ZINDER, S. & BROCK, T. D. (1978). Production of methane and carbon dioxide from methane thiol and dimethyl sulfide by anaerobic lake sediments. *Nature, London*, **273,** 226–8.

THE CARBON CYCLE IN AQUATIC ECOSYSTEMS

JOHN G. ORMEROD

Botanical Institute, University of Oslo, Blindern, Oslo 3, Norway

In this article, emphasis is placed on microbiology of the carbon cycle in lakes.

Lake ecosystems are extremely complex. In temperate areas they go through a yearly cycle, involving complete mixing (circulation) of the water masses alternating with stabilization of water strata of differing density. The main cause of these alterations is seasonal variations in temperature. In many lakes, the water circulates for just a short period (days or weeks) in spring and autumn. During summer and winter, two main strata are present: the upper one is referred to as the epilimnion and may be several metres thick; the lower stratum is the hypolimnion. Between these is a transition zone known as the metalimnion, usually two or more metres thick.

Some lakes (and fjord basins) are permanently stratified owing to the presence of salts in the hypolimnion, giving a higher density irrespective of the temperature of the epilimnion. Lakes of this type are called meromictic. The salts may be of Fe or have arisen from seawater and the condition is accentuated if the lake is sheltered from the wind.

The microbial processes occurring in lakes usually show drastic seasonal variations as a result of the alternating stratification and circulation. Changes in light flux, temperature and oxygen concentration are the most important factors affecting the microbial activities. Lakes that are poor in nutrients are called oligotrophic. Those with high concentrations of nutrients are eutrophic. The key nutrient in the development of eutrophication is phosphate because the other major elements of living cells, namely carbon, nitrogen, hydrogen and oxygen, are usually more freely available in the environment. Thus the availability of phosphate in the photic zone (i.e. that part of the lake into which sufficient light penetrates to permit photosynthesis) generally determines how much of the energy of incident solar radiation will be converted into chemical energy by photosynthesis. This in turn largely determines the rate of carbon cycling in the system.

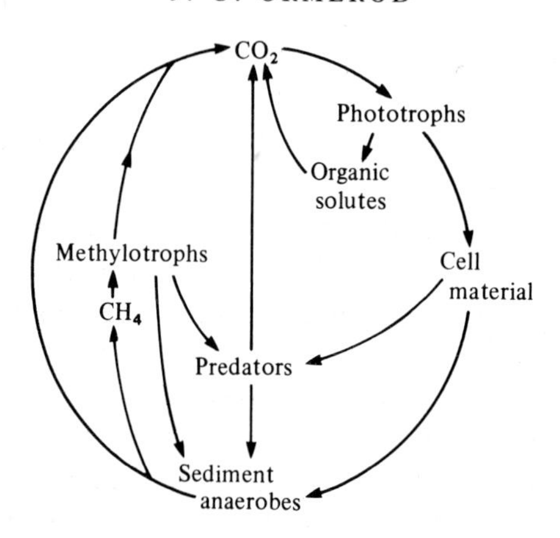

Fig. 1. The carbon cycle in lakes.

An outline of the carbon cycle in lakes is shown in Fig. 1. Complications due to interaction with the sulphur cycle are not shown. In the following, emphasis will be placed on lakes which are stratified for at least part of the year and in which the hypolimnion becomes anoxic.

PHOTOSYNTHESIS IN THE EPILIMNION

Eutrophication involves changes in the types of phototrophic organisms as well as in their total biomass. In oligotrophic lakes, various kinds of eukaryotic algae and diatoms often dominate in the epilimnion and cyanobacteria may be undetectable. As eutotrophication proceeds, cyanobacteria begin to dominate. The underlying reasons for the transition from dominance by eukaryotic phototrophs to prokaryotic are not well understood. The phenomenon is relevant to the theme of this chapter because the organisms concerned are involved in a large sector of the carbon cycle.

Cyanobacteria and eukaryotic algae both perform the same kind of oxygenic photosynthesis, involving two light reactions, and CO_2 is fixed by the Calvin cycle. There are, however, a number of features of cyanobacteria that appear to be important in their competitive relationship with the algae. Among these may be mentioned low energy of maintenance (about one tenth that of eukaryotic algae; van Liere & Mur, 1979) enabling more efficient growth at low light intensities and better overwintering abilities.

Winter survival is also facilitated by the formation of akinetes and by the ability of some cyanobacteria to grow anaerobically with H_2S as electron donor in anoxygenic photosynthesis (Padan, 1979). The importance of successful overwintering is two-fold. First, a large inoculum is available when the spring growth season begins (Preston, Stewart & Reynolds, 1980) and second, cells that survive the winter probably sequester nutrients thus depriving their competitors of them. Cyanobacteria possess phycobiliproteins which are antenna pigments that absorb light around 500–650 nm, the spectral region least absorbed by water. This permits more photosynthesis at depth. Many cyanobacteria can position themselves in the water column, by means of gas vacuoles, at the depth most favourable for their photosynthesis (Walsby, Utkilen & Johnsen, 1982). A large number of species of cyanobacteria can develop in the absence of fixed nitrogen because they can fix atmospheric N_2. Finally, the filamentous nature of many cyanobacteria makes them difficult to consume by predators (Å. Brabrand *et al.*, unpublished observations).

Which factors are the most important in the competition is not known. At all events the net result of eutrophication is an increase in photosynthetic production in the photic zone.

According to Fig. 1, the products of photosynthesis can be either cell material or solutes that are excreted from the cells into the water. The soluble excreted products lead to a short circuit in the carbon cycle because they mostly become oxidized to CO_2 in the epilimnion.

Many determinations of extracellular products of photosynthesis have been made (see Fogg, 1971 for review) and in some cases over half of the total CO_2 fixed may be excreted. The commonest low molecular weight extracellular product is glycollate, which appears to be formed mainly from phosphoglycollate produced by the reaction of ribulose bisphosphate (RuBP) with oxygen. The cost of production of one molecule of phosphoglycollate from two of CO_2 by the oxygenase reaction and Calvin cycle is extraordinarily high: 8 ATP + 5 NADPH. In comparison, the corresponding figures for the conversion of two CO_2 molecules to carbohydrate by the Calvin cycle are 6 ATP + 4 NADPH. Obviously, the excretion of glycollate from the cells represents not only a loss of carbon but also a utilization of energy for a purpose other than formation of cell material.

The main factor governing the oxygenase activity of RuBP carboxylase and therefore the production of phosphoglycollate is the concentration ratio of O_2 to CO_2. This ratio is affected by many processes. For example, both green algae and cyanobacteria can actively concentrate inorganic carbon intracellularly (Badger, Kaplan & Berry, 1978) thus decreasing the O_2:CO_2 ratio. The respiration of excreted glycollate by heterotrophic bacteria will have a similar effect, as will photorespiration of phosphoglycollate by the producing cell. Photorespiration, which occurs in cyanobacteria as well as in eukaryotic phototrophs, involves conversion, via glycine and serine, of two molecules of glycollate to one each of 3-phosphoglyceric acid (3-PGA) and CO_2. A molecule of O_2 is consumed by respiration of NADH produced in the process. (In higher eukaryotic phototrophs, an oxidase is involved.) The result of the activity of RuBP oxygenase is a short circuiting of the carbon cycle. Because the effect can be reduced by adding CO_2, it can be viewed as a form of carbon (CO_2) limitation. Furthermore, it is clear that although phosphate is usually the overall limiting factor determining the size of the standing crop of microorganisms in lakes, there must inevitably be frequent periods when CO_2 'limits' growth of the phototrophs.

Other carbon compounds than glycollate are also released into the medium during photosynthesis. There seems to be a great variety of these (Fogg, 1971) but little is known about how or why this release takes place. Substances that chelate iron or other metal ions are included here. Not all of this organic material need be amenable to biological degradation, at least in the short term. An illustration of the presence of such refractory material in lake water is given by some observations made by the author and K. Ormerod on water from Aurevatn, a lake near Oslo that is used as a public water supply. The water is slightly yellow-brown and an ozonization plant was therefore installed by the local authority to bleach it. This treatment led to massive growth of heterotrophic bacteria (mostly *Hyphomicrobium* and *Zooglea*) in the pipes of the distribution network, and a comparison of the treated and untreated water revealed that bleaching caused a significant increase in the amount of biologically degradable material, as measured by the biological oxygen demand method. This result was interpreted to mean that biologically refractory organic material in the lake water was converted by ozone into a form easily degradable by bacteria.

Although organic solutes figure as significant products of photosynthesis, the major product is usually cell material, the chemical composition of which is dependent upon the nutritional status of the cells. This in turn depends on what limits their growth. Zevenboom, Bij de Vaate & Mur (1982) have assessed the factors limiting growth of the cyanobacterium *Oscillatoria agardhii* in a highly eutrophic lake. On the basis of determinations of nutrient uptake rates in natural populations harvested at weekly intervals from the lake, they concluded that light was commonly the limiting factor for growth in the autumn and winter, phosphate in spring and late summer and nitrogen in the summer proper. In view of what has been said above it is likely that carbon would also frequently have been a limiting factor.

The effect of a growth-limiting factor on biosynthesis of cell material can be reasonably well described in biochemical terms. For example, in nitrogen-limited situations where, for one reason or another nitrogen-fixing organisms have not appeared, the phototrophs will synthesize a cell material rich in carbohydrate and poor in protein.

Most of the intermediates of the Calvin cycle are carbohydrates and the cycle provides proximate precursors for the biosynthesis of nucleic acids and polysaccharides. Acetyl-CoA, tetrapyrroles and most of the amino acids are formed by reaction sequences leading from 3-PGA. Synthesis of the amino acids of the aspartate and glutamate families involves carboxylation of phospho*enol*pyruvate (PEP) and since these comprise more than half of the total amino acids of the cell protein and are also involved in nucleotide synthesis, the PEP carboxylase reaction is quantitatively important under conditions of nitrogen sufficiency. However when nitrogen limits growth, draining of 3-PGA from the Calvin cycle will be minimal, and hexose phosphate then gives rise to storage polysaccharide (starch in green algae, glycogen in cyanobacteria).

Phosphate limitation is also characterized by exaggerated formation of storage carbohydrate although here the reason is presumably limitation of ribosome formation resulting in a low specific rate of protein synthesis. Under these conditions, the cyanobacterial cell also stores nitrogen in the form of cyanophycin, which is a polymer containing arginine and aspartic acid. Its formation is non-ribosomal (Simon, 1973). The nitrogen content of dry cell material of cyanobacteria can vary from about 4% to about 15% and the phosphorus content from 0.1 to 0.8% (Staub, 1961), depending on what has limited growth.

The photosynthetic production discussed above occurs primarily in the epilimnion, which itself may be subject to daily mixing for limited periods. The phototrophs find their optimum level either through gas vacuole buoyancy or, in the case of eukaryotes, flagellar swimming. Some cyanobacteria may confine themselves to the metalimnion (see Caldwell, 1977) where the supply of CO_2 and other nutrients is higher and conditions are more reducing.

PHOTOSYNTHESIS IN THE HYPOLIMNION

In meromictic lakes, the hypolimnion is permanently anoxic and in highly eutrophic lakes it usually becomes seasonally anoxic. Under these conditions, H_2S is produced by bacterial action and, provided that light can reach these depths, anoxygenic photosynthesis by green and purple sulphur bacteria will occur. Since cell material is produced from CO_2 in this process, it represents part of the primary production of the lake. Considering that only a small fraction of the incident light penetrates so deeply (usually >5 m) it is surprising how great a fraction of the total production in the lake this anoxygenic photosynthesis can represent. A table published by Biebl & Pfennig (1979) shows that this fraction may vary from a few to over 80% of the total production on an annual basis.

The organisms responsible, particularly the green sulphur bacteria, are well adapted to growing at very low light intensities. In some cases, the bacterial 'plate', which may be a metre or so thick, contains purple sulphur bacteria in the upper part and green sulphur bacteria in the lower. At these depths, the wavelengths of light filtering through are generally restricted to blue to blue-green (400–550 nm) and this means that the carotenoids will be the main antenna pigments absorbing light (Pfennig, 1967). Some green sulphur bacteria possess more carotenoid than others and as a result are browner in colour. These species are better adapted than the greener ones to the type of situation described (Trüper & Genovese, 1968). In spite of this, these bacteria contain very large amounts of antenna bacteriochlorophyll which absorbs practically no light in the wavelength range available.

An interesting hypothesis was put forward recently (Raven & Beardall, 1981) to explain how the green sulphur bacteria can grow at extremely low light intensities. It was pointed out that the presence of chlorosomes and absence of intracytoplasmic thylakoid membranes in these organisms means that the total membrane surface area is much less than in purple bacteria. This is believed to

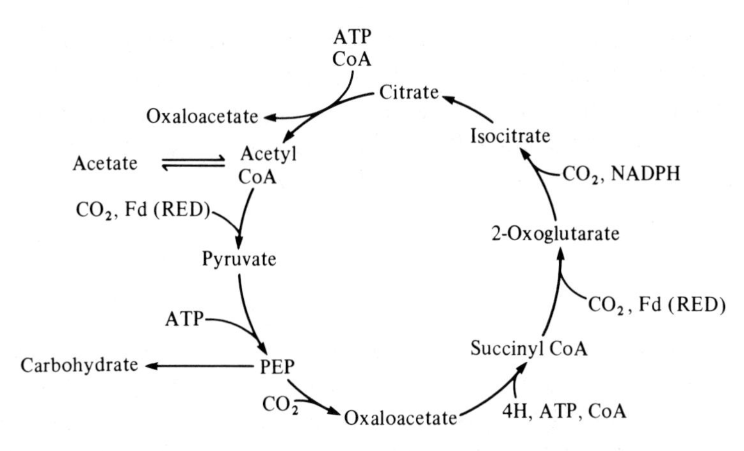

Fig. 2. The reductive tricarboxylic acid cycle of green sulphur bacteria.

have the effect of limiting re-entry of H^+, and consequent decay of the transmembrane proton gradient. This could be of great quantitive importance under the conditions of severe energy limitation prevailing where these bacteria are normally found.

Anoxygenic photosynthesis in purple and green sulphur bacteria is fairly well understood. Only one photosystem (analogous to photosystem I of oxygenic phototrophs) is present. In the green bacteria, the redox potential of the primary reductant is low enough to reduce ferredoxin (Fd) and NAD directly. The purple bacteria, on the other hand, seem unable to reduce NAD directly by the photosystem and it is believed that NAD reduction occurs by the more roundabout process of reverse electron transfer, involving a proton gradient generated by light energy. For this reason, it seems likely that conditions inside the illuminated cell of a green sulphur bacterium will be more reducing than in a purple bacterium, and this is reflected in the fundamental difference in their carbon metabolism and relations towards oxygen.

Purple bacteria fix CO_2 by the Calvin cycle during autotrophic growth. Green sulphur bacteria utilize a CO_2-fixing cycle (Fig. 2) that involves two Fd-dependent carboxylations, and is in essence a reversed Krebs cycle. One turn of the cycle produces a molecule of oxaloacetate from four of CO_2. Cycle intermediates include precursors for lipids and amino acids, while carbohydrate is formed from pyruvate by reversed glycolysis (Sirevåg, 1975).

The green sulphur bacteria are strict anaerobes, but the purple sulphur bacteria are not killed by oxygen and some can even grow chemoautotrophically in the dark, oxidizing H_2S with O_2 (Kämpf & Pfennig, 1980). Their ribulose bisphosphate carboxylase, like that of all other organisms, possesses oxygenase activity and the cells have been shown to excrete glycollate in the presence of oxygen (Takabe & Akazawa, 1977). The extent of excretion of organic products of CO_2 fixation in nature by purple bacteria has not been studied.

The depth at which the purple sulphur bacteria occur in the water column of stratified lakes has been shown to vary throughout the day (Sorokin, 1970). On sunny days, these organisms use up the H_2S in the upper layers of the hypolimnion during the morning and swim downwards, perhaps reaching a depth 3 m lower later in the day. The hypolimnion generally contains sufficient ammonia, phosphate and CO_2 such that these nutrients will not limit growth. In this situation, the purple sulphur bacteria are living in opposed gradients of light and H_2S and these two factors probably alternate as the limiting growth factor. The ability to store elemental sulphur intracellularly is of advantage under these shifting environmental conditions and it is likely that light is the most frequent growth-limiting factor (see also Sorokin, 1970).

The growth situation of the non-motile types of purple sulphur bacteria and of green sulphur bacteria, none of which possess flagella, is less well understood. Some of these organisms can form gas vacuoles which presumably permit regulation of cell buoyancy according to optimum depth. This mode of regulation is, however, too slow to react to acute changes in nutrient concentration or light intensity. It is particularly important for green sulphur bacteria to stay within the anoxic zone since they are strict anaerobes. It has been shown by field experiments that the growth of green sulphur bacteria in stratified lakes is strongly light-limited (Bergstein, Henis & Cavari, 1979).

Green sulphur bacteria, unlike their purple counterparts, excrete elemental sulphur during the metabolism of H_2S. The full ecological implication of this excretion is not known at present. In mixed cultures of these bacteria with suitable sulphur-reducing organisms, no elemental sulphur can be detected (Biebl & Pfenning, 1978) yet there must be a rapid turnover of this element in the cultures. Elemental sulphur may also be undetectable in natural populations of green sulphur bacteria (Indrebø, Pengerud & Dundas, 1979) suggesting that species interactions form an integral part of the ecology of the green sulphur bacteria in nature.

The green sulphur bacteria have also been reported to liberate organic matter during growth (Lippert, 1967) and nitrogen starved suspensions excrete 2-oxoacids during CO_2 fixation (Sirevåg & Ormerod, 1970). Such processes may also be significant for species interaction. Recently, G. O. Fjeldheim and I have found that *Chlorobium* releases acetate during growth in the light on a thiosulphate-ammonia-bicarbonate medium in which thiosulphate is the limiting factor for growth. Apart from its probable ecological importance, the excretion of acetate is of biochemical interest in an organism which readily assimilates added acetate and converts it into the same products as those formed from CO_2 (Sirevåg & Ormerod, 1970).

When acetate (2–5 mM) is added to an autotrophically growing culture, the growth rate shows an *immediate* increase of about 40%. This means that the enzymes for acetate activation and acetyl-CoA metabolism via the cycle (Fig. 2) must be present constitutively in amounts commensurate with the new growth rate. How, then, can the cells excrete acetate under autotrophic conditions? In a preliminary investigation of this problem, we have examined the acetate activating enzyme of the cells, aceto-CoA kinase (E.C.6.2.1.1.) and have shown that its K_m for acetate is high enough (0.25 mM) to explain why free acetate, once formed, would accumulate in the amounts measured experimentally (0.05–0.08 mM). However, free acetate is not an intermediate in the CO_2 fixing cycle, since the product of the citrate lyase reaction, acetyl-CoA (Ivanovsky, Sinstov & Kondratieva, 1980) is the substrate of the next reaction of the cycle, pyruvate synthase. Therefore, for acetate to accumulate, the activity of pyruvate synthase would have to be lower than that of citrate lyase. The fact that addition of acetate leads to an increased growth rate could be explained if acetate were to stimulate the pyruvate synthase reaction. This would then lead to an immediate increase in the rate of cycling. In the absence of added acetate and presence of other organisms that utilize acetate, *Chlorobium* would continue to produce acetate. The effect of acetate on pyruvate synthase has not yet been investigated.

The utilization of the excreted acetate by *Desulfuromonas* (Pfennig & Biebl, 1976) would be of no advantage to *Chlorobium* because the amount of H_2S formed per mol of acetate is no greater than that required to reduce the equivalent amount of CO_2 to acetate phototrophically. More recently, a novel type of bacterium was reported by Panganiban, Patt, Hart & Hanson (1979) which may conceivably live syntrophically with green sulphur bacteria.

The organism concerned is an anaerobic methane oxidizer which obtains energy by reducing sulphate with methane but requires acetate as carbon source. If *Chlorobium* were to 'feed' such an organism with acetate and sulphur or sulphate in the presence of methane, relatively large amounts of H_2S would be formed for the further growth of the phototroph.

A third type of anoxygenic phototroph should be mentioned here. Some cyanobacteria can be induced to perform anoxygenic photosynthesis in which H_2S is the ultimate reductant for CO_2 (Padan, 1979). Hydrogen sulphide inhibits oxygenic photosynthesis very strongly. In the anoxygenic photosynthesis, photosystem II is inactivated and the cells excrete elemental sulphur (Garlick, Oren & Padan, 1977). *Oscillatoria limnetica* was the first organism of this kind to be described (Cohen, Padan & Shilo, 1975) and it forms a highly productive 'plate' in Solar Lake (Israel) during the winter stratification, below blooms of purple and green sulphur bacteria. It is thought that anoxygenic cyanobacterial photosynthesis may be typical of aquatic situations characterized by alternating aerobic/anaerobic conditions (Padan, 1979).

Before ending this section on photosynthesis in the hypolimnion, an example of a eukaryotic phototroph that lives in an H_2S-containing environment will be given. In south east Norway, the landlocked fjord basin Hunne-bunnen is permanently stratified owing to a seawater hypolimnion with H_2S concentrations of over 0.2 mM (Braarud & Føyn, 1958). During summer large numbers of a species of Euglenaceae are present in the H_2S containing layer. Nothing is known about the physiology of this organism, except that it produces oxygen when illuminated in the absence of H_2S (Ormerod & Paasche, unpublished observation).

Finally, it should be mentioned that chemoautotrophic bacterial activity in lakes, although of great importance as far as the sulphur and nitrogen cycles are concerned, is probably of minor direct quantitative significance in the carbon cycle.

FATE OF FIXED CARBON

The breakdown of cell material produced by oxygenic phototrophs in the epilimnion may occur aerobically or anaerobically.

Rudd & Hamilton (1979) showed for a Canadian Lake (Lake 227) that three quarters of the CO_2 fixed in the epilimnion was quickly recycled, i.e. the cell material was degraded to CO_2 in the epilimnion and refixed there, while only one quarter of the cell material sedimented to the hypolimnion. Healthy cells of phototrophs in the epilimnion will continue to grow and divide throughout the growing season unless attacked or consumed by predators. The latter can be viruses (cyanophages), bacteria (primarily myxobacteria and cytophaga) or zooplankton (protozoa, crustaceans etc.). The activity of one or the other of these must vary a great deal from lake to lake. Instances of proliferation of heterotrophic bacteria and zooplankton in response to increases in primary production are well documented (Stewart, Sinada, Christofi & Daft, 1977; Ormerod, 1978; O'Brien & Noyelles, 1974). In the cases of bacterial predators, the attack will lead to lysis of the phototroph with consequent leakage of soluble organic material (Stewart *et al.*, 1977). This results in increases in other heterotrophic bacteria and most of the organic matter becomes oxidized to CO_2. Consumption of cells by zooplankton will also lead largely to oxidation to CO_2, with excretion of nutrients such as NH_3. Protozoal and crustaceal predators also consume phototrophic bacteria from the upper hypolimnion by diving. In this way, some of the organic matter in the anoxic layers of lakes may be transferred to the upper layers.

Cell material of phototrophs that are not destroyed or consumed in the epilimnion, together with the bodies of zooplankton, will eventually sediment to the bottom part of the lake and the fate of this material will now be considered.

Some decomposition will occur during the descent of the material to the bottom. The degradation of the sedimented detritus occurs mostly in the upper layers of the sediment (Robertson, 1979).

It is customary to view the anaerobic breakdown of complex organic material as occurring in stages but most of the steps and organisms involved are interdependent. The early steps in the degradation are hydrolytic and carried out by bacteria that produce extracellular enzymes or enzymes anchored to the surface of the bacterial outer membrane. Clostridia, anaerobic cytophagas and other genera are active here, breaking down polysaccharides and whatever may be left of proteins, nucleic acids and lipids. The primary products of these hydrolyses are fermented by the hydrolytic organisms themselves and other anaerobes. The products of the fermentations include fatty acids like acetate, propionate and butyrate (Jørgensen, 1982) together with, lactate, H_2 and CO_2.

Further breakdown of the products of the primary fermentations can proceed by two different mechanisms. In systems with sufficient sulphate, the final stages of conversion of the organic molecules into CO_2 are due to sulphate-reducing bacteria (Jørgensen, 1982). Until quite recently, it was thought that these bacteria were very restricted in the range of organic substrates they could use as electron donors and that acetate was the main product of substrate oxidation. Thanks to the work of Pfennig & Widdel (1981), we now know that there are sulphate-reducing bacteria that can oxidize long and short chain fatty acids and even aromatic compounds to CO_2. Furthermore, as shown by Biebl & Pfennig (1977) elementary sulphur can also act as electron acceptor in place of sulphate.

In ecosystems where the amount of sulphate present is low, as in many fresh water sediments, sulphate and sulphur reduction play a smaller part and depend on regeneration cycles such as the proposed light-dependent system described above and involving methane and green sulphur bacteria. In these low sulphur ecosystems, mineralization does not proceed to completion in the anaerobic part of the lake. Instead, between one and two thirds of the carbon reaching the sediment appears as methane and most of the rest as CO_2.

The main substrates for methane production are acetate and H_2 + CO_2 but estimates vary as to the relative importance of these (Jones, Simon & Gardener, 1982; Cappenberg & Prins, 1974).

It is quite clear that the bacterial flora of this anaerobic, sedimentary ecosystem and the reactions occurring within it are extremely complex and there are intricate energetic relationships present. Hydrogen and hydrogen donors and acceptors play a vital role in these interrelationships. The maximum energy yields of many fermentations depend on the formation of H_2. Yet, in many cases, H_2 is formed only if its partial pressure in the system is very low. If the P_{H_2} builds up, the H generated by fermentation is transferred to intracellular acceptors such as acyl CoA, with consequent sacrifice of ATP. This is what happens in pure cultures of anaerobes in the laboratory. Low P_{H_2} in nature is maintained by the presence of H_2-utilizing bacteria. There are several organisms of this type thought to be important in the degradation of organic matter, the methanogens, acetogenic bacteria of the type represented by *Acetobacterium woodii* (Balch, Schoberth, Tanner & Wolfe, 1977) and the sulphate-reducing bacteria. In the absence of sulphate, the latter organisms have been shown to act as hydrogen (and acetate) donors for other bacteria (Bryant, Campbell, Reddy & Crabill, 1977).

This part of the lake ecosystem also presents the microbiologist with bioenergetic problems. It is generally possible to understand the thermodynamic feasibility of the reactions involved in the fermentations, particularly when interspecies hydrogen transfer is taken into account, but the biochemical mechanisms of ATP synthesis are unknown in almost all of those organisms present that do not have substrate level phosphorylation. In methanogens and acetogenic bacteria it is assumed that some form of electron transport phosphorylation occurs and Gottschalk, Schoberth & Brann (1977) have suggested that a succinate-fumarate cycle may form the basis for this. In *Desulfovibrio* there is some evidence for vectorial electron transfer and it has been suggested that this is also the mechanism of energy coupling in this group of organisms (Odom & Peck, 1981).

The significance for the degradation process of the composition of the sedimented detritus will now be discussed. As already pointed out, cell material of phototrophs may contain from 4 to 15% N. Assuming a Y_{ATP} of 10 for the anaerobes and a maximum ATP yield of 4 per glucose equivalent it can be calculated that even with detritus containing 4% N, the nitrogen requirements of the anaerobes would be more than satisfied. Similar considerations apply to other nutrients such as phosphorus. It may be concluded, therefore, that anaerobic decomposition of detritus is probably always energy limited.

There are comprehensive data on methane production for a number of lakes (see Rudd & Taylor, 1980) and the rates of production vary from a few μmoles to several millimoles per m^2 per hour. A large part of the methane produced stays in the hypolimnion during stratification, although there is now good evidence for anaerobic methane oxidation with sulphate as electron acceptor (Panganiban, Patt, Hart & Hanson, 1979; see above). Depending on the nature of the sediment and the depth of the lake, a proportion of the methane produced escapes as bubbles to the atmosphere. If and when the lake circulates, the accumulated methane becomes distributed throughout the (now aerobic) water masses and is oxidized to CO_2 by aerobic methylotrophic bacteria. The immense proportions of this process, which may lead to total depletion of oxygen in the lake (Rudd & Hamilton, 1974), have only become apparent in recent years, partly as a result of generally increased awareness of these organisms in nature following the work of Whittenbury and his colleagues (Whittenbury, Phillips & Wilkinson, 1970a) and also consequent upon the development of methods for measuring the rate of methane oxidation in lakes (Patt, Cole, Bland & Hansen, 1974; Rudd, Hamilton & Campbell, 1974). During stratification, most of the methylotrophs are present in the metalimnion, oxidizing that fraction of the methane that moves upwards by diffusion, bubbling and turbulence (Rudd & Taylor, 1980).

Aerobic methane oxidizing bacteria have been classified into two types according to the distribution of membranes within the cell (Whittenbury, Davies & Davey, 1970b). This is correlated with the mechanism by which carbon is assimilated, the types I organisms, with stacked membranes, utilizing the ribulose monophosphate (RuMP) pathway and the type II organisms, with paired membranes, using the serine mechanism.

Methane is oxidized to CO_2 via formaldehyde by a series of reactions in which redox carriers are reduced and function in oxidative phosphorylation. The first step, however, is catalysed by a monooxygenase and represents a considerable loss of energy. Carbon assimilation occurs at the level of formaldehyde.

The ecological significance of the two types of aerobic methane oxidizers is not understood. Judging by the difficulties various workers have met with in obtaining pure cultures of these bacteria it would seem likely that they live in close consort with other bacteria in nature. There is evidence that methanol is liberated during methane oxidation at low oxygen concentrations (Harwood & Pirt, 1972), and this may indicate syntrophy with methanol oxidizers. The activities of methane-oxidizing bacteria could have beneficial effects on oxygenic phototrophs in the epilimnion. In particular, one could envisage the possibility that in methylotrophs possessing the serine pathways, glycollate, produced by the phototrophs, could feed in at the glyoxylate stage by oxidation and in this way act as a favourable carbon source for growth with methane as energy source. At the same time, the O_2:CO_2 ratio of the system would be reduced, thus favouring the phototrophs. At least one methane oxidizing bacterium, *Methylococcus capsulatus*, is known to be able to utilize glycollate as carbon source during methane oxidation (Taylor, Dalton & Dow, 1981). Although this bacterium is a type I methylotroph it also possesses small amounts of the enzymes of the serine pathway and in addition enzymes of the Calvin cycle.

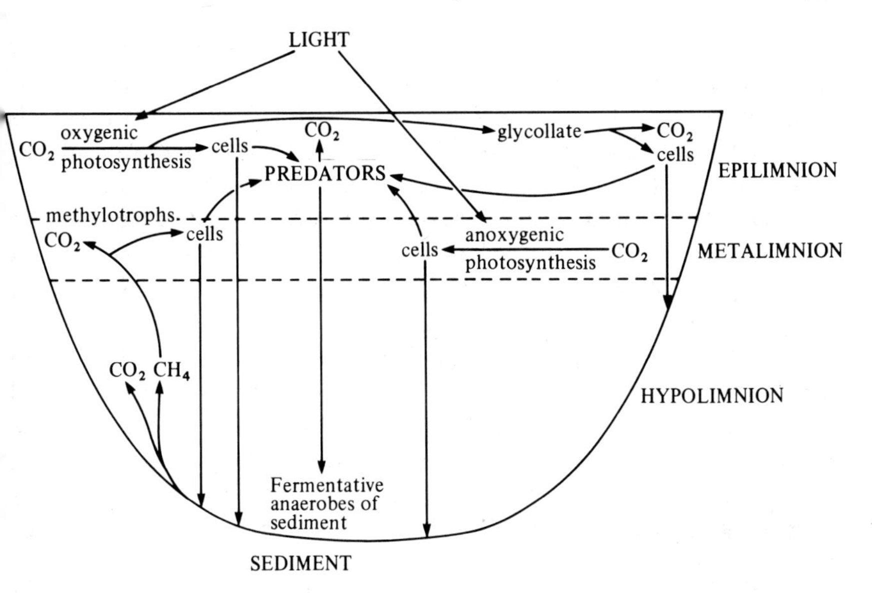

Fig. 3. Trophic structure of a stratified lake ecosystem.

Whittenbury (1981) has suggested that the latter cycle may function here not as a CO_2 fixing mechanism but as a means of lowering the intracellular oxygen tension by forming phosphoglycollate, which, after dephosphorylation is utilized via the serine pathway.

The relatively inefficient mechanism used for methane oxidation means that the yield of bacterial cell material per mol of CH_4 oxidized is low. This cell material, in addition to that of ordinary heterotrophic bacteria and zooplankton in the lake, constitutes the secondary production and is either consumed by predators in the epilimnion or sinks to the sediment where it joins the decomposition cycle.

TROPHIC STRUCTURE AND ENERGY RELATIONS IN LAKE ECOSYSTEMS

Figure 3 presents a summary of the main trophic levels in a lake ecosystem and Fig. 4 is intended to give an idea of the principal pathways of energy. It will be evident from the latter figure that of the total amount of solar radiation entering the lake, only a very small fraction (0.01–1%) is converted into chemical energy by photosynthesis. The rest is dissipated in the water, mostly as heat.

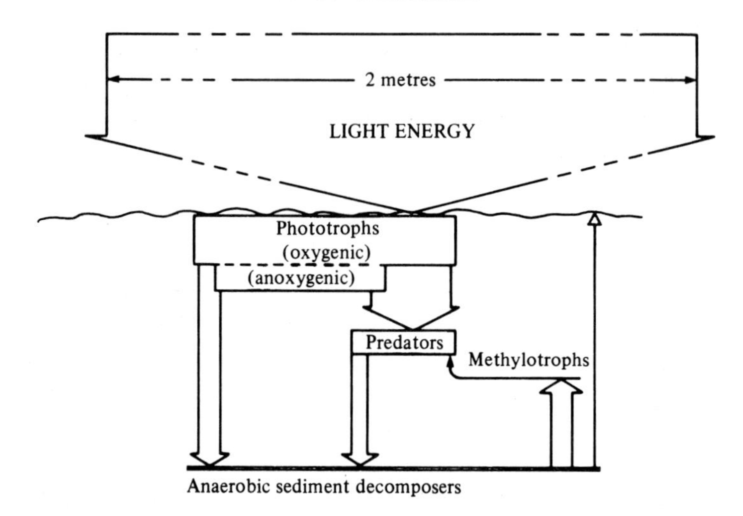

Fig. 4. Lake energy budget.

Even extreme eutrophication will only lead to a maximum conversion of a few percent of the solar energy by photosynthesis.

This does not mean that the photosynthetic process itself is so inefficient. Under light-limiting conditions, over 25% of the absorbed light energy may end up as chemical energy of cell material (Gons, 1977). But in lakes, the low efficiency at the energy input stage is the greatest single factor limiting production.

The reason that the aquatic phototrophs do not harvest a greater fraction of the incoming light energy is bound up with the potentially damaging effect of light. This effect is mediated by sensitizers such as chlorophyll and is believed to be counteracted to some degree by the carotenoids. The synthesis of chlorophyll is regulated primarily by the light intensity, in inverse fashion. When the light intensity increases, chlorophyll synthesis is inhibited, to begin again when the light intensity decreases. Temperature and nutritional factors also affect chlorophyll synthesis. This is one of the ways the cells adapt to their environment, but this method of adaptation cannot cope adequately with short-term variation in light intensity, such as are common in nature, and a phototroph in the epilimnion will usually be poorly adapted for much of the time. There will be either more or less light energy absorbed than the biosynthetic apparatus of the cells is capable of utilizing. In either case the result is a reduction in growth rate. If the amount of light energy is greatly in excess, photoinhibition of photosynthesis occurs. This is only slowly reversible. Moderate excess of light energy, in the presence of high levels of oxygen, will lead to the production of phosphoglycollate by oxygenolysis of ribulose bisphosphate. As already pointed out, this process consumes much energy in the form of NADPH and ATP and represents a loss of organic material if glycollate is excreted.

Therefore, glycollate production and photorespiration are, in a sense, a manifestation of energy regulation, analogous to the photoproduction of molecular hydrogen in anoxygenic photodiazotrophs in the absence of a nitrogen source (Gest, Ormerod & Ormerod, 1962).

Although it is widely recognized that photorespiration represents a considerable waste of energy in both aquatic and terrestrial photosynthetic systems, it is important to realize what the underlying reasons for this wastage are. Research in improving production should not be concentrating on the isolation or construction of organisms with 'improved' ribulose bisphosphate carboxylase, but on finding phototrophs that are better able to cope with variations in light intensity as a result of less sluggish adaptatory mechanisms.

REFERENCES

Badger, M. R., Kaplan, A. & Berry, J. A. (1978). A mechanism for concentrating CO_2 in *Chlamydomonas reinhardtii* and *Anabaena variabilis* and its role in photosynthetic CO_2 fixation. *Carnegie Institute of Washington Yearbook*, **77,** 259–61.

Balch, W. E., Schoberth, S., Tanner, R. S. & Wolfe, R. S. (1977). *Acetobacterium,* a new genus of hydrogen oxidizing, carbon dioxide-reducing, anaerobic bacteria. *International Journal of Systematic Bacteriology*, **27,** 355–61.

Bergstein, T., Henis, Y. & Cavari, B. Z. (1979). Investigations on the photosynthetic sulfur bacterium *Chlorobium phaeobacteroides* causing seasonal blooms in Lake Kinneret. *Canadian Journal of Microbiology,* **25,** 999–1007.

Biebl, H. & Pfennig, N. (1977). Growth of sulphate reducing bacteria with sulphur as electron acceptor. *Archives of Microbiology*, **112,** 115–17.

Biebl, H. & Pfennig, N. (1978). Growth yields of green sulfur bacteria in mixed cultures with sulfur and sulfate reducing bacteria. *Archives of Microbiology*, **117,** 9–16.

Biebl, H. & Pfennig, N. (1979). Anaerobic CO_2 uptake by phototrophic bacteria. A review. *Archiv für Hydrobiologie/Beihefte Ergebnisse der Limnologie*, **12,** 48–58.

Braarud, T. & Føyn, B. (1958). Phytoplankton in a brackish water locality of south-east Norway. *Nytt Magasin for Botanikk*, **6,** 47–73.

Bryant, M. P., Campbell, L. L., Reddy, C. A. & Crabill, M. R. (1977). Growth of *Desulfovibrio* in lactate or ethanol media low in sulfate in association with H_2-utilizing methanogenic bacteria. *Applied and Environmental Microbiology*, **33,** 1162–9.

Caldwell, D. E. (1977). The planktonic microflora of lakes. *C.R.C. Critical Reviews in Microbiology*, **5,** 305–70.

CAPPENBERG, T. E. & PRINS, R. A. (1974). Interrelations between sulphate-reducing and methane-producing bacteria in bottom desposits of freshwater lake. III. Experiments with ^{14}C-labelled substrates. *Antonie van Leeuwenhoek Journal of Microbiology and Serology*, **40,** 457–69.
COHEN, Y., PADAN, E. & SHILO, M. (1975). Facultative anoxygenic photosynthesis in the cyanobacterium *Oscillatoria limnetica. Journal of Bacteriology*, **123,** 855–61.
DAVIES, S. L. & WHITTENBURY, R. (1970). Fine structure of methane and other hydrocarbon-utilizing bacteria. *Journal of General Microbiology*, **61,** 227–32.
FOGG, G. E. (1971). Extracellular products of algae in freshwater. *Archiv für Hydrobiologie/Beihefte Ergebnisse der Limnologie*, **5,** 1–25.
GARLICK, S., OREN, A. & PADAN, E. (1977). Occurrence of facultative anoxygenic photosynthesis among filamentous and unicellular cyanobacteria. *Journal of Bacteriology*, **129,** 632–9.
GEST, H., ORMEROD, J. G. & ORMEROD, K. S. (1962). Photometabolism of *Rhodospirillum rubrum:* Light-dependent dissimilation of organic compounds to carbon dioxide and molecular hydrogen by anaerobic citric acid cycle. *Archives of Biochemistry and Biophysics*, **97,** 21–33.
GONS, H. J. (1977). On the light-limited growth of *Scenedesmus protruberans* Fritsch. Thesis, University of Amsterdam.
GOTTSCHALK, G., SCHOBERTH, S. & BRANN, K. (1977). Energy metabolism of anaerobes growing on C_1-compounds. In *Second International Symposium on Microbial Growth on C_1-compounds*, ed. G. K. Skryabin, M. V. Ivanov, E. N. Kondratieva, G. A. Zavarzin, Y. A., Trotsenko & A. I. Nestov. Pushchino: USSR Academy of Sciences.
HARWOOD, J. H. & PIRT, S. J. (1972). Quantitative aspects of growth of the methane oxidizing bacterium *Methylococcus capsulatus* on methane in shake flasks and continuous chemostat culture. *Journal of Applied Bacteriology*, **35,** 597–607.
INDREBØ, G., PENGERUD, B. & DUNDAS, I. (1979). Microbial activities in a permanently stratified estuary. I Primary production and sulfate reduction. *Marine Biology*, **51,** 295–304.
IVANOVSKY, R. N. SINSTOV, N. V. & KONDRATIEVA, E. N. (1980). ATP-linked citrate lyase activity in the green sulfur bacterium *Chlorobium limicola* forma *thiosulfatophilum. Archives of Microbiology*, **128,** 239–41.
JONES, J. G., SIMON, B. M. & GARDENER, S. (1982). Factors affecting methanogenesis and associated anaerobic processes in the sediments of a stratified eutrophic lake. *Journal of General Microbiology*, **128,** 1–11.
JØRGENSEN, B. B. (1982). Mineralisation of organic matter in the sea bed – the role of sulphate reduction. *Nature, London*, **296,** 643–5.
KÄMPF, C. & PFENNIG, N. (1980). Capacity of chromatiaceae for chemotrophic growth. Specific respiration rates of *Thiocystis violacea* and *Chromatium vinosum. Archives of Microbiology*, **127,** 125–35.
LIPPERT, K. D. (1967). Die Verwertung von molekularem Wasserstoff durch *Chlorobium thiosulfatophilum*. Thesis, University of Göttingen.
O'BRIEN, W. J. & NOYELLES, F. (1974). Relationship between nutrient concentration, phytoplankton density and zooplankton density in nutrient enriched experimental ponds. *Hydrobiologia*, **44,** 105–25.
ODOM, J. M. & PECK, H. D. JR. (1981). Hydrogen cycling as a general mechanism for energy coupling in the sulphate-reducing bacteria, *Desulfovibrio* sp. *FEMS Microbiology Letters*, **12,** 47–60.
ORMEROD, K. S. (1978). Relationship between heterotrophic bacteria and phyto-

plankton in a eutrophic lake with water blooms dominated by *Oscillatoria agardhii*. *Verhandlungen, Internationale Vereiningung für Theoretische und Angewandte Limnologie*, **20,** 788–93.
PADAN, E. (1979). Impact of facultatively anaerobic photoautotrophic metabolism on ecology of cyanobacteria (blue-green algae). *Advances in Microbial Ecology*, **3,** 1–48.
PANGANIBAN, A. T. JR., PATT, T. E., HART, W. & HANSON, R. S. (1979). Oxidation of methane in the absence of oxygen in lake water samples. *Applied and Environmental Microbiology*, **37,** 303–9.
PATT, T. E., COLE, G. C., BLAND, J. & HANSON, R. S. (1974). Isolation of organic compounds as sole sources of carbon and energy. *Journal of Bacteriology*, **120,** 955–64.
PFENNIG, N. (1967). Photosynthetic bacteria. *Annual Review of Microbiology*, **21,** 285–324.
PFENNIG, N. & BIEBL, H. (1976). *Desulfofuromonas acetoxidans* gen. nov. and sp. nov., a new anaerobic, sulfur reducing, acetate oxidizing bacterium. *Archives of Microbiology*, **110,** 3–12.
PFENNIG, N. & WIDDEL, F. (1981). Ecology and physiology of some anaerobic bacteria from the microbial sulfur cycle. In *Biology of Inorganic Nitrogen and Sulfur*, ed. H. Bothe & A. Trebst, pp. 169–77. Berlin: Springer-Verlag.
PRESTON, T., STEWART, W. D. P. & REYNOLDS, C. S. (1980). Bloom-forming cyanobacterium *Microcystis aeruginosa* overwinters on sediment surface. *Nature, London*, **288,** 365–7.
RAVEN, J. A. & BEARDALL, J. (1981). The intrinsic permeability of biological membranes to H^+: Significance for the efficiency of low rates of energy transformation. *FEMS Microbiology Letters*, **10,** 1–5.
ROBERTSON, C. K. (1979). Quantitative comparison of the significance of methane in the carbon cycles of two small lakes. *Archiv für Hydrobiologie/Beihefte Ergebnisse der Limnologie*, **12,** 123–35.
RUDD, J. W. M., HAMILTON, R. D. & CAMPBELL, A. (1974). Measurement of microbial oxidation of methane in lake water. *Limnology and Oceanography*, **19,** 519–24.
RUDD, J. W. M. & HAMILTON, R. D. (1979). Methane cycling in Lake 227 in perspective with some components of carbon and oxygen cycles. *Archiv für Hydrobiologie/Beihefte Ergebnisse der Limnologie*, **12,** 115–22.
RUDD, J. W. M. & TAYLOR, C. D. (1980). Methane cycling in aquatic environments. *Advances in Aquatic Microbiology*, **2,** 77–150.
SIMON, R. D. (1973). The effect of chloramphenicol on the production of cyanophycin granule polypeptide in the blue green alga *Anabaena cylindrica*. *Archiv für Mikrobiologie*, **92,** 115–22.
SIREVÅG, R. (1975). Photoassimilation of acetate and metabolism of carbohydrate in *Chlorobium thiosulfatophilum*. *Archives of Microbiology*, **104,** 105–11.
SIREVÅG, R. & ORMEROD, J. G. (1970). Carbon dioxide fixation in green sulphur bacteria. *Biochemical Journal*, **120,** 399–408.
SOROKIN, Y. I. (1970). Interrelations between sulphur and carbon turnover in meromictic lakes. *Archiv für Hydrobiologie*, **66,** 391–446.
STAUB, R. (1961). Ernährungsphysiologisch-autökologische Untersuchungen an der planktischen Blaualge *Oscillatoria rubescens* DC. *Schweizerische Zeitschrift für Hydrologie*, **23,** 82–199.
STEWART, W. D. P., SINADA, F., CHRISTOFI, N. & DAFT, M. J. (1977). Primary production and microbial activity in Scottish fresh water habitats. In *Aquatic Microbiology*, ed., by F. A. Skinner & J. M. Shewan, pp. 31–54. London: Academic Press.

TAKABE, T. & AKAZAWA, T. (1977). A comparative study on the effect of O_2 on photosynthetic carbon metabolism by *Chlorobium thiosulfatophilum* and *Chromatium vinosum. Plant & Cell Physiology*, **18,** 753–65.

TAYLOR, S. C., DALTON, H. & DOW, C. S. (1981). Ribulose-1,5-bisphosphate carboxylase/oxygenase and carbon assimilation in *Methylococcus capsulatus* (Bath). *Journal of General Microbiology*, **122,** 89–94.

TRÜPER, H. G. & GENOVESE, S. (1968). Characterization of photosynthetic sulfur bacteria causing red water in Lake Faro (Messina, Sicily). *Limnology and Oceanography*, **13,** 225–32.

VAN LIERE, L. & MUR, L. R. (1979). Growth kinetics of *Oscillatoria agardhii* Gomont in continuous culture, limited in its growth by the light supply. *Journal of General Microbiology*, **115,** 153–60.

WALSBY, A. E., UTKILEN, H. C. & JOHNSEN, I. J. (1982). Buoyancy changes of a red coloured *Oscillatoria agardhii* in Lake Gjersjøen, Norway. *Archiv für Hydrobiologie*, (in press).

WHITTENBURY, R. (1981). The interrelationship of autotrophy and methylotrophy as seen in *Methylococcus capsulatus* (Bath). *Microbial Growth on C_1 Compounds*, ed. H. Dalton, pp. 181–90. London: Heyden.

WHITTENBURY, R., PHILLIPS, K. C. & WILKINSON, J. F. (1970a). Enrichment, isolation and some properties of methane-utilizing bacteria. *Journal of General Microbiology*, **61,** 205–18.

WHITTENBURY, R., DAVIES, S. L. & DAVEY, J. F. (1970*b*). Exospores and cysts formed by methane-utilizing bacteria. *Journal of General Microbiology*, **61,** 219–26.

ZEVENBOOM, W., BIJ DE VAATE, A. & MUR, L. R. (1982). Assessment of factors limiting growth rate of *Oscillatoria agardhii* in hypertrophic Lake Wolderwijd, 1978, by use of physiological indicators. *Limnology and Oceanography*, **27,** 39–52.

INDEX